MEDITERRANEAN DESERTIFICATION
AND LAND USE

MEDITERRANEAN DESERTIFICATION AND LAND USE

Edited by

C. JANE BRANDT

and

JOHN B. THORNES

Department of Geography, King's College, London, UK

JOHN WILEY & SONS
Chichester · New York · Brisbane · Toronto · Singapore

Other Wiley Editorial Offices

John Wiley & Sons, Inc., 605 Third Avenue,
New York, NY 10158-0012, USA

Jacaranda Wiley Ltd, 33 Park Road, Milton,
Queensland 4064, Australia

John Wiley & Sons (Canada) Ltd, 22 Worcester Road,
Rexdale, Ontario M9W 1L1, Canada

John Wiley & Sons (Asia) Pte Ltd, 2 Clementi Loop #02-01
Jin Xing Distripark, Singapore 129809

Library of Congress Cataloging in Publication Data

Mediterranean desertification and land use / [edited by] C. Jane Brandt
and John B. Thornes.
    p. cm.
  Includes bibliographical references and index.
  ISBN 0-471-94250-2
  1. Desertification—Mediterranean Region.   I. Brandt, C. Jane.
II. Thornes, John B.
GB618.68.M43M43 1996
333.73′6′091822—dc20   95-26001
                                                                       CIP

British Library Cataloguing in Publication Data

A catalogue record for this book is available from the British Library

ISBN 0-471-94250-2

Typeset in 10/12pt Times by Dobbie Typesetting Limited, Tavistock, Devon
Printed and bound in Great Britain by Bookcraft (Bath) Ltd
This book is printed on acid-free paper responsibly manufactured from sustainable forestation, for which at least two trees are planted for each one used for paper production.

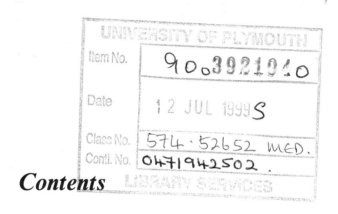
Contents

List of Contributors

C. Aguilera Technical School of Agriculture, University of Almería, Spain

J. M. Alonso Estación Experimental de Zonas Aridas, CSIC, C/. General Segura 1, 04001 Almeria, Spain

A. Aru Dipartimento Scienze Terra, Università di Cagliari, Via Trentino 51, 09127 Cagliari, Italy

A. J. Baird Department of Geography, University of Sheffield, Sheffield, S10 2TN, UK (formerly Department of Geography, University of Bristol)

J. Bastida Departamento de Termodinámica, Universitat de València, S/N Dr Moliner 46100 Burjassot (Valencia), Spain

J. C. Bathurst Water Resource Systems Research Unit, Department of Civil Engineering,University of Newcastle upon Tyne, Newcastle Upon Tyne, NE1 7RU, UK

M. M. Boer Estación Experimental de Zonas Aridas, CSIC, C/. General Segura 1, 04001 Almeria, Spain (formerly VFGB, Universiteit van Amsterdam, The Netherlands)

C. J. Brandt Department of Geography, King's College London, Strand, London, WC2R 2LS, UK (formerly Department of Geography, University of Bristol)

A. Brenner Department of Pure and Applied Biology, University of Leeds, Woodhouse Lane, Leeds, LS2 9JT, UK

K. Bunte Laboratory for Experimental Geomorphology, Katholieke Universiteit te Leuven, Redingenstraat 16 bis, B-3000 Leuven, Belgium

L. H. Cammeraat VFGB, Universiteit van Amsterdam, Nieuwe Prinsengracht 130, 1018 VZ Amsterdam, The Netherlands

J. Casimiro Mendes Instituto de Meteorologia, Rua C do Aeroporto, 1700 Lisboa, Portugal

M. Chabart Bureau de Recharches Géologiques et Minières (BRGM), 1039 Rue de Pinville, 34000 Montpellier, France

N. CLARK	Science Policy Research Unit, University of Sussex, Falmer, Brighton, BN1 GRF, UK
S. C. CLARK	Department of Pure and Applied Biology, University of Leeds, Woodhouse Lane, Leeds, LS2 9JT, UK
J. J. COLLIN	Bureau de Recherches Géologiques et Minières (BRGM), 1039 Rue de Pinville, 34000 Montpellier, France
M. CONTE	Istituto Fisica Atmosfera CNR, Piazza L. Sturzo 31, 00144 Roma, Italy
P. CORTESAO CASIMIRO	Departamento de Geografia e Planeamento Regional, Universidade Nova de Lisboa, Avenida de Berna 24, 1000 Lisboa, Portugal
M. CUETO	Estación Experimental de Zonas Aridas, CSIC, C/. General Segura 1, 04001 Almeria, Spain
A. DALAKA	Department of Ecology, Aristotelian University, 540 06 Thessaloniki, Greece
N. G. DANALOTOS	Laboratory of Soils and Agricultural Chemistry, Agricultural University of Athens, 75 Iera Odos, Botanicos 11855, Athens, Greece
L. DELGADO	Estación Experimental de Zonas Aridas, CSIC, C/. General Segura 1, 04001 Almeria, Spain
J. DIAMANTOPOULOS	Department of Ecology, Aristotelian University, 540 06 Thessaloniki, Greece
S. M. DIAMOND	Department of Animal and Plant Sciences, University of Sheffield, Sheffield, S10 2TN, UK
F. DOMINGO	Estación Experimental de Zonas Aridas, CSIC, C/. General Segura 1, 04001 Almeria, Spain
F. ESPIRITO SANTO	Instituto de Meteorologia, Rua C do Aeroporto, 1700 Lisboa, Portugal
S. GANDIA	Departamento de Termodinámica, Universitat de València, S/N Dr Moliner, 46100 Burjassot (Valencia), Spain
M. A. GILABERT	Departamento de Termodinámica, Universitat de València, S/N Dr Moliner, 46100 Burjassot (Valencia), Spain
C. M. GOODESS	Climatic Research Unit, School of Environmental Sciences, University of East Anglia, Norwich, NR4 7JT, UK
A. T. GROVE	Department of Geography, University of Cambridge, Downing Place, Cambridge, CB2 3EM, UK
L. GUTIÉRREZ	Estación Experimental de Zonas Aridas, CSIC, C/. General Segura 1, 04001 Almeria, Spain
A. R. HARRISON	Department of Geography, University of Bristol, Bristol, BS8 1SS, UK
S. J. HURCOM	Department of Geography, University of Bristol, Bristol, BS8 1SS, UK

G. IATROU	Department of Ecology, Aristotelian University, 540 06 Thessaloniki, Greece
A. C. IMESON	VFGB, Universiteit van Amsterdam, Nieuwe Prinsengracht 130, 1018 VZ Amsterdam, The Netherlands
L. INCOLL	Department of Pure and Applied Biology, University of Leeds, Woodhouse Lane, Leeds, LS2 9JT, UK
C. KILSBY	Water Resource Systems Research Unit, Department of Civil Engineering, University of Newcastle upon Tyne, Newcastle Upon Tyne, NE1 7RU, UK
M. J. KIRKBY	School of Geography, University of Leeds, Leeds, LS2 9JT, UK
C. S. KOSMAS	Laboratory of Soils and Agricultural Chemistry, Agricultural University of Athens, 75 Iera Odos, Botanicos 11855, Athens, Greece
R. LÁZARO	Estación Experimental de Zonas Aridas, CSIC, C/. General Segura 1, 04001 Almeria, Spain
J. G. LOCKWOOD	School of Geography, University of Leeds, Leeds, LS2 9JT, UK
F. LÓPEZ-BERMÚDEZ	Departamento de Geografía Física, Universidad de Murcia, Campus de la Merced, 30001 Murcia, Spain
A. LOPEZ BUENDIA	Departamento de Termodinámica, Universitat de València, S/N Dr Moliner, 46100 Burjassot (Valencia), Spain
J. P. MARCHAL	Bureau de Recherches Géologiques et Minières (BRGM), 1039 Rue de Pinville, 34000 Montpellier, France
N. S. MARGARIS	Department of Environmental Studies, University of the Aegean, Karatoni 17, 81 100 Mytilini, Lesvos, Greece
JOSÉ MARTÍNEZ-FERNÁNDEZ	Departamento de Geografía Física, Universidad de Murcia, Campus de la Merced, 30001 Murcia, Spain
JULIA MARTÍNEZ-FERNÁNDEZ	Departamento de Geografía Física, Universidad de Murcia, Campus de la Merced, 30001 Murcia, Spain
M. D. MCMAHON	School of Geography, University of Leeds, Leeds, LS2 9JT, UK
J. MELIA	Departamento de Termodinámica, Universitat de València, S/N Dr Moliner, 46100 Burjassot (Valencia), Spain
P. L. MITCHELL	Department of Animal and Plant Sciences, University of Sheffield, Sheffield, S10 2TN, UK
J. M. MOUSTAKAS	Laboratory of Soils and Agricultural Chemistry, Agricultural University of Athens, 75 Iera Odos, Botanicos 11855, Athens, Greece
J. M. NICOLAU	Estación Experimental de Zonas Aridas, CSIC, C/. General Segura 1, 04001 Almeria, Spain

J. P. PALUTIKOF	Climatic Research Unit, School of Environmental Sciences, University of East Anglia, Norwich, NR4 7JT, UK
J. PANTIS	Department of Ecology, Aristotelian University, 540 06 Thessaloniki, Greece
E. PAPATHEODOROU	Department of Ecology, Aristotelian University, 540 06 Thessaloniki, Greece
F. PÉREZ-TREJO	UNITAR, Palais de Nations CH1211, Geneva 10, Switzerland
S. PIRINTSOS	Department of Ecology, Aristotelian University, 540 06 Thessaloniki, Greece
J. POESEN	Laboratory for Experimental Geomorphology, Katholieke Universiteit te Leuven, Redingenstraat 16 bis, B-3000 Leuven, Belgium
J. PUIGDEFÁBREGAS	Estación Experimental de Zonas Aridas, CSIC, C/. General Segura 1, 04001 Almeria, Spain
A. ROMERO-DÍAZ	Departamento de Geografía Física, Universidad de Murcia, Campus de la Merced, 30001 Murcia, Spain
M. J. ROXO	Departamento de Geografia e Planeamento Regional, Universidade Nova de Lisboa, Avenida de Berna 24, 1000 Lisboa, Portugal
G. SANCHEZ	Estación Experimental de Zonas Aridas, CSIC, C/. General Segura 1, 04001 Almeria, Spain
S. SGARDELIS	Department of Ecology, Aristotelian University, 540 06 Thessaloniki, Greece
J. SHAO	Department of Geography, King's College London, Strand, London, WC2R 2LS, UK
J. E. SHEEHY	Creative Scientific Solutions, 38 Meadow View, Marlow Bottom, Buckinghamshire, SL7 3PA, UK
R. SOEIRO DE BRITO	Departamento de Geografia e Planeamento Regional, Universidade Nova de Lisboa, Avenida de Berna 24, 1000 Lisboa, Portugal)
A. SOLÉ	Estación Experimental de Zonas Aridas, CSIC, C/. General Segura 1, 04001 Almeria, Spain
G. P. STAMOU	Department of Ecology, Aristotelian University, 540 06 Thessaloniki, Greece
M. TABERNER	Department of Geography, University of Bristol, Bristol, BS8 1SS, UK
J. B. THORNES	Department of Geography, King's College London, Strand, London, WC2R 2LS, UK (formerly Department of Geography, University of Bristol)
S. VIDAL	Estación Experimental de Zonas Aridas, CSIC, C/. General Segura 1, 04001 Almeria, Spain
S. WHITE	Instituto Pirenaico de Ecologia, CSIC, Campus de Aula Dei, Avda Montaña 177, 50080, Zargoza, Spain

F. I. WOODWARD
Department of Animal and Plant Sciences, University of Sheffield, Sheffield, S10 2TN, UK

N. YASSOGLOU
Laboratory of Soils and Agricultural Chemistry, Agricultural University of Athens, 75 Iera Odos, Botanicos 11855, Athens, Greece

M. T. YOUNIS
Departamento de Termodinámica, Universitat de València, S/N Dr Moliner, 46100 Burjassot (Valencia), Spain

Preface

Both the vegetation and land use in the Mediterranean areas of Europe strongly reflect human activity since at least the Bronze Age. By Classical times, the land had already been extensively deforested and eroded. The climate, with hot, dry summers and highly variable rainfall in the winters, lent itself to the extensive production of drought-adapted tree crops like olives and cork and rain-fed cereals grown in the wetter years on terraces.

Since the middle of this century, there have been great changes to this traditional character. Firstly there were great migrations from the rural areas to cities, then there was the arrival of tourists at the coasts. There has also been the modernization and intensification of agriculture, bringing with it new demands for fertilizer, water and heavy machinery and new, highly water-demanding horticultural crops. So the environment is being transformed. Pollution, water supplies and land abandonment have become major problems and the landscape progressively more sensitive to external markets and water supply.

On top of this, European Union policies are marginalizing many of the traditional areas of pastoral and dry-land farming, leading to land abandonment. The farmer is being forced into a role of environmental caretaker or has to leave. In addition, there is a threat of climate change as a result of the suspected global warming. Land degradation and water shortages are seen as major threats to the welfare of the Mediterranean states. This has led Spain, Italy and Greece to claim special status in the United Nation's draft Convention on Desertification.

Faced with these difficulties, the European Union decided that more information on the desertification in southern European countries was essential if regional policy is to reflect these changing conditions. It therefore established under the Directorate General for Science, Research and Development, a series of programmes of research into land degradation in southern Europe. These programmes comprise several projects, the largest of which is MEDALUS, Mediterranean Desertification and Land Use. This project has as its ultimate goals the understanding, prediction and mitigation of desertification in the Mediterranean countries of the Union.

This book presents the work carried out in the first phase of MEDALUS. It describes the historical environmental context in which current concerns about desertification are set. It looks at models of climate change and future climate scenarios. It describes the development of geomorphological models which can be used to simulate future landscapes, under different climate or land use scenarios. It describes the network of field sites stretching round the Mediterranean from Portugal to Greece, each carrying out the same monitoring programme to provide the models with data and to look at fundamental geomorphological and biological processes. It looks at the use of remote sensing to determine land use, vegetation and lithology, and of geographic information systems to identify areas particularly susceptible to degradation. It looks at models of groundwater and of socio-economic changes.

One of the greatest challenges of the project, and of this book, has been the bringing together and integration of the work, not only of modellers and empirical scientists, but also of climatologists, geomorphologists, biologists, and social scientists. The editors would like to thank everyone in all seventeen research groups for their commitment and kindness.

Jane Brandt
John Thornes   August 1996

1

Introduction

J. B. THORNES

Department of Geography, King's College London, UK

1.1 DESERTIFICATION IN THE MEDITERRANEAN?

Desertification has become a major environmental issue in scientific, political and even popular circles and the term captures a sense of moving deserts, drying lakes and starving people; of impending or actual crisis. It is above all thought of as a problem of the Sahel and one that is far removed from the daily livelihoods and concerns of the states of Europe. In discussing the 'myth' of desertification, Thomas and Middleton (1994) draw attention to three potential misunderstandings about desertification: *first* that statements about its extent and activity are at best inaccurate and at worst based on nothing better than guesswork; *second*, that drylands are necessarily fragile ecosystems and highly susceptible to degradation and desertification, arising from the confusion between loss of vegetation cover on the one hand and soil degradation on the other; and *third*, that desertification is a major, if not the primary cause of human suffering and misery in drylands. Although Thomas and Middleton are mainly addressing the issue of the role of international agencies (and especially the United Nations) in resolving the problem, these misunderstandings often obscure the real physical, social and economic issues and therefore the efforts to mitigate them.

This confusion has led to a plethora of attempts to uniquely define desertification in a way that will apply to all cases. This is probably impossible, in part because some definitions deal with a condition, some with a cause and some with a set of processes. Moreover the phenomenon is related to particular geographical and physical conditions, the processes are context specific and climate sensitive and the probability and/or onset of desertification is a function of local biotic and abiotic exchanges at the regional level, and human activity at the local level. Notwithstanding these difficulties and without accepting it as unique, the definition used by UNEP (1992) as *'land degradation in arid, semi-arid and dry subhumid areas resulting from various factors, including climatic variations and human activities'* provides a workable definition for the evaluation of the phenomena in the European Mediterranean. 'Land' means the terrestrial bio-productive system that comprises soil, vegetation, other biota and the ecological and hydrological processes that operate within the system. 'Land degradation' means reduction and loss of the biological or economic productivity caused by land-use change, from a physical process,

Mediterranean Desertification and Land Use. Edited by C. Jane Brandt and John B. Thornes.
© 1996 by John Wiley & Sons, Ltd.

or a combination of the two. These include processes arising from human activities and habitation patterns, such as soil erosion, deterioration of the physical, chemical and biological or economic properties of the soil, and long-term loss of vegetation.

The history of land degradation in the Mediterranean has been the subject of speculation, debate and research for centuries (Grove, 1986). It is only in recent years however that a serious effort has been made to identify and understand the phenomenon in the context of desertification and its implications. This effort dates from the Symposium of the Commission of the European Communities on *Desertification in Europe* held in Mytilene, Greece, in 1984 (Fantechi and Margaris, 1986). In the opening remarks to this conference Roberto Fantechi noted the aptness of the title of the paper by Professor H.G. Mensching 'Desertification in Europe?' where the question mark could take on all shades of meaning from candid astonishment to perplexity and beyond. Mensching himself concluded that in Europe, desertification is only widespread in the semi-arid regions of Mediterranean countries. This includes the southern and eastern parts of the Iberian Peninsula, parts of Mediterranean France, the whole of the Mezzogiorno in Italy, Sardinia and Corsica and almost all of Greece, including the islands. Generally the most critical areas were identified as having less than 600 mm of rainfall per year, distributed over a few months, with a long dry hot summer.

In the 10 years since then, the issue of Mediterranean desertification has received much greater prominence both on the international and the European stage. Individual countries, notably Spain with its Campaign Against Desertification in Mediterranean Areas (Perez-Soba and Barrientos, 1986), had initiated efforts to deal with the problem identified in the UNEP Conference on Desertification held in 1977. Italy had a National Research Council Quinquennial Project on soil conservation (Mancini, 1986). The UNEP *World Atlas of Desertification* (UNEP, 1992) identified most of the Mediterranean as having 'very high', 'high' or 'medium' levels of soil degradation severity. The Fourth Annexe to the International Convention on Desertification (UNEP, 1992) identified Portugal, Spain and Greece as countries with a marked problem of desertification, which, while not needing international financial support, require national action programmes. At the same time, the European Parliament has called for substantial effort by the European Union to mitigate the economic and social problems arising from desertification in the southern European states.

The reasons for this prominence are not difficult to see. In the last 30 years there has been an enormous shift in economic priorities and the corresponding use of natural resources. Improved standards of living, the rise of tourism, the intensification of agriculture, the shift to irrigated agricultural production for export and the continued drift to an urban society, accompanied by rural depopulation, have all led to an emphasis on land-use conflicts ultimately linked to the problem of water (Grenon and Batisse, 1989). At the same time the growth of conservation/environment movements in Mediterranean countries has sharpened the debate about land management by individuals and the state. The ever-deepening crisis of groundwater withdrawal has led to inter-regional conflicts exacerbated by the droughts of the early and mid-1980s and to renewed droughts in Spain in the 1990s.

By the late 1980s a new bogey had appeared on the horizon in the shape of the threat of global warming generated by the greenhouse effect. Scientists were quick to claim its importance and the need for more research. Climatic variation and drought are intrinsic characteristics of semi-arid environments, and the Mediterranean environment is no exception. The problem is one of separating inherent climatic variability from climatic change, evaluating the regional effects of global warming (if any) and then their possible impact on natural and socio-economic production systems.

Notwithstanding the difficulties, forecasts of global warming should be taken seriously, especially in conditions as marginal as the semi-arid Mediterranean. The work of Palutikof et al. (1994) in the MEDALUS project clearly shows that, as a result of global warming, there will be a reduction in moisture availability. This change can and probably will have important implications for human activities.

Finally, in the early 1990s there were a series of extensive forest fires throughout the Iberian Peninsula, Italy and Greece. This loss of forest vegetation cover seemed to confirm the real fears of the public and to alert the politicians to the threat of a combination of land degradation and enhanced runoff. However, man is clearly responsible for much of the increase in forest fires, at least in Spain (Sala and Rubio, 1994), partly through land-use changes (Moreno and Oechel, 1992) and partly through deliberate firing to allow redevelopment of the land. Although the role of forest fires has been exaggerated, both in terms of soil erosion and in terms of the longevity of post-fire recovery, there is no doubt that fires have an enormous impact on public perception of hazard. López-Bermúdez (1995) reports that *El Pais*, a major Spanish newspaper, estimated in October 1994 that 'soil erosion with forest fires and torrential rain cost Spain more than 55 000 million pesetas per year'. Whatever the reality, southern Europe, its politicians and pressure groups have awoken to the problem of desertification.

Given these social and economic pressures, it is hardly surprising that the discussions of desertification in these advanced and complex societies have become ever more confused. Many aspects of this problem reach back into antiquity. Already by Plato's time, erosion was identified as a major cause of flooding, and the removal of forest and intensive grazing had been identified as the main culprits. That the picture is more complex than this has been demonstrated many times since but, as long as populations remained small and demand remained at subsistence levels, problems of land degradation were not perceived to be critical. Now that water is the mainstay of tourism and agriculture, and hence a key component of Mediterranean regional and even national economies, the issue of desertification *has* become critical. It is a remarkable testimony to the perceptiveness of Roberto Fantechi that he foresaw the seriousness of this problem some 10 years ahead of the rest of the European Union and initiated a research programme that provides the basis for addressing it.

1.2 THE EUROPEAN COMMUNITY RESEARCH INITIATIVE

In order to better understand the processes and problems of desertification in southern Europe, the European Community established a series of research programmes in the context of its 5-year research frameworks. Initially (in the EPOCH programme) these focused on the climatic hazards in southern Europe, but in the Third and Fourth Frameworks desertification was separately identified as an important research priority. Two large and a number of smaller but still important projects were established.

One of the major projects, EFEDA (European Field Experiment in Desertification Threatened Areas), had as its primary objective the understanding of the role of near-surface conditions on the atmospheric boundary layer (Bolle et al., 1993). The intention was to ask whether or not, in Mediterranean-type conditions, the land-use at the surface has a significant control on the climate above the ground. Previous experiments had shown that extensive areas of uniform land-use, such as the pine forests of the Bordeaux area, the prairie grasslands of Kansas or the Amazon rain forest, can have important effects on the atmospheric energy and water vapour fluxes. These fluxes are of sufficient magnitude to perturb the atmosphere so that changes in land-use could be important for climatic change. Mediterranean land-use is a much more varied and complex mosaic. Except where there is very extensive irrigation, it seems unlikely that land-use changes are likely to have much impact on the atmosphere. Moreover, the Mediterranean landscape is extremely varied topographically, so that the boundary layer is much more complex than in other areas. In the Sahara, for example, it has been argued that desertification processes are self-enforcing, as the drying out and dying of the vegetation cover leads to lower soil moisture storage capacity.

The second major research programme initiated by the Community is the MEDALUS programme. MEDALUS is the acronym for Mediterranean Desertification and Land Use. This programme has been undertaken in two phases, Phase I from 1991 to1992 and Phase II from 1993 to 1995. MEDALUS I involved groups from 17 universities and research institutes and its results form the contents of this book. MEDALUS II comprised 44 groups, including some from industry and commerce, and is further discussed below.

1.3 THE EUROPEAN MEDITERRANEAN BASIN

The lands surrounding the Mediterranean Sea have long been recognized as having a distinctive character that arises both from the physiographic conditions and the history of development. In particular climatologists recognize a 'Mediterranean climate' with winter rainfalls (between September and April), mild winters and high summer temperatures. The rain is characteristically intense, especially in the drier season and in the drier parts of the region. Rainfall regimes are mainly cyclonic or convectional. The cyclonic storms do not tend to penetrate from the Atlantic, rather they arise within the basin, for example in the Gulf of Genoa, and like convectional storms developed on the land masses, may produce quite intense rainfalls (Barry and Chorley, 1992). The rainfall varies from about 1000 mm in the more northerly areas and areas above about 800 m, to 250 mm in the southern drylands. Southeast Spain is one of the driest regions, with 250–300 mm rainfall coming in the winter months and five to seven dry months a year. Mediterranean drylands (those with less than 500 mm rain per year) also have a large inter-annual variability, so that for a long record period the rainfall may vary from 150 to 600 mm and coefficients of inter-annual variation of 35% may occur. Runs of wet and dry years are also a characteristic climatological feature of these regions, with the early to mid-1980s being the last drought period. Generally the rainfall has decreased overall since the end of the last century, and this can be related to changes in atmospheric pressure, the position of the Azores High and sea surface temperatures. The eastern and western parts of the Mediterranean tend to behave somewhat differently, oscillating about an axis through the Aegean, so that anomalies related to the El Niño in the west may not be reflected in the east (Conte and Guifredda, 1989).

These climatic characteristics vary greatly over short distances in the Mediterranean because of the basin and range character of the landscape and because of the impact of the sea in coastal areas. Onshore and offshore breezes play an important role in micro-climates near the coast. The intense storms in mid-October on the Spanish Levant, for example, are caused by intense lows moving in from the east and being brought to intense instability over the land.

Throughout the Mediterranean mountains play a key role in providing a great diversity of land types, and generally they reflect the underlying geological structure. The mountains of the drier Mediterranean mainly date from the Alpine orogeny and many areas are still tectonically active as shown by earthquakes and subsidence. The great basins are normally filled with the erosional products of the mountain-building and are therefore even younger and many are unconsolidated. During the Miocene large areas of the land surrounding the present-day Mediterranean Sea were also inundated and as a consequence the basins often have saline or fresh-water marls (as in Almeria, Spain) or marine clays (as in the Agri valley, Italy), both of which are highly susceptible to badland erosion. Limestones, schistose metamorphics, continental conglomerates, sands and gravel and marls and clays form the main lithologies, though deep intrusions occur, such as the Corsican granites. The result is a very wide range of lithological types, which play an important role in the diversity of response to the agents of desertification.

There is an important legacy of past geomorphological processes in the valley floors and on the lower hillslopes of the Mediterranean. Messerli (1967) showed that in the Quaternary, snow-lines were much lower than today, and glaciers were present as far south as the Sierra Nevadas in Spain. Moreover, as Vita Finzi (1969) indicated, the river valleys of the region have experienced many cut and fill sequences, though whether or not these have some degree of contemporaneity around the Mediterranean is still a matter of speculation. Certainly there is evidence that by 4000 BP extensive gullying had already occurred in some areas (Wise, et al., 1982). This indicates that the simple assumptions that severe erosion is coupled to either grazing, climatic change or desertification, prevalent in the 1960s, can no longer be regarded as tenable. Erosion has varied substantially throughout historical times and varies enormously today across space, as indicated by the accumulation of coastal deltas, the filling of lakes and the silting of reservoirs.

Under any circumstances, vegetation cover plays a key role in land degradation (Francis and Thornes, 1990) and in fact reduction in the perennial plant cover is regarded as an indicator of the onset of desertification by some. There is virtually no natural vegetation left in the Mediterranean since man has been cutting, grazing and burning for at least the last 10 000 years. Rackham (1983, cited in Grove, 1986) states that in Boetia (Greece) 'the landscape and the vegetation appear to have changed less in 2500 years than those of England in the last 1000 years or of New England in the last 180 years'. This is probably becoming less true of many other parts of the Mediterranean, where the abandonment of grazing is leading to an increase in scrubland and thence woodland (see Mairota and Papadimitriou, in press).

Despite the lack of natural vegetation, there is an important and distinctive set of vegetation types that are ubiquitous today and which play a key role in the desertification question. Essentially, apart from cultivation, three dominant vegetation types occur: woodlands, typically of trees such as oak and chestnut; high scrub, usually about 2 m in height, with bushes and occasional trees; and sparse scrub including aromatic plants such as thyme and rosemary, comprising lands that have been cultivated and abandoned. The process of use, regeneration and destruction has occurred many times throughout the Mediterranean region, so that even the most natural-looking forest may have regenerated since the Middle Ages. In Chapter 2 of this volume, Grove highlights some key elements that have shaped the land-use in the areas in which the MEDALUS field sites are situated.

In addition to the climate, lithology and vegetation, the fourth distinctive dimension of the Mediterranean environment is the development that has taken place since the late 1950s. Three important changes have transformed life and livelihoods. First there has been a major change in agriculture. Extensive agriculture based on grazing and dryland wheat has given way to intensive agriculture based on tree crops, horticulture and irrigation. This has produced a spatial concentration of agriculture and an accompanying change in the spatial distribution of the demand for rural labour. These changes have reflected the increased wealth of western Europe, changed dietary habits and the replacement of traditional sources of fruit with horticultural produce (Margaris, 1996). Second, there has been a huge social change, coupled to a dramatic shift in the standards of living. The rural lifestyle is in the final stages of transformation and the shift from countryside to city or migration overseas has paralleled that in other countries in the world. The phenomenon of second homes is now a fundamental feature of many rural areas in the Mediterranean. Third, as mentioned earlier, the growth of tourism and the littoralization of the Mediterranean economy has added to the problems of rural environments. All of these changes have led to (a) the abandonment of land for traditional agriculture and (b) the rise of demand for water for urban expansion, tourism and above all for irrigation, which are not unrelated. The distant rural areas are the gathering ground for water for irrigation, but also the areas in which flood waters generate and from which erosion produces sediment that fills the downstream reservoirs. The understanding of the impacts of changes on these areas is therefore economically critical. Even without their strategic importance for water supplies, the abandonment of traditional terrace systems may have unforeseen impacts, especially on runoff rates, erosion and downstream flooding. According to Braudel (1979) 'In the Mediterranean the land is dying when the protection of agricultural cultivation is absent'. This provides a neat twist on the argument that desertification is only man-induced. Margaris (Chapter 3 in this volume) details the implications for desertification of the recent trends in Mediterranean agriculture.

1.4 MEDALUS I OVERVIEW

In the first phase of MEDALUS, the main task was to improve understanding of the climatic, hydrological and ecological processes of desertification; under subhumid or semi-arid conditions; on cultivated land and land in a semi-natural state which had been withdrawn from agriculture. In order to do this the project adopted a field research and modelling strategy (Figure 1.1). Starting with the atmosphere, the parameters of a changing climate are passed through a model of hillslope processes which has been parametrized through field experimental data. The results are constrained by land-use changes determined by models of socio-economic change and by field studies that provide both an opportunity to validate the models and to obtain more refined statements about the processes. The programme thus involved five main components:

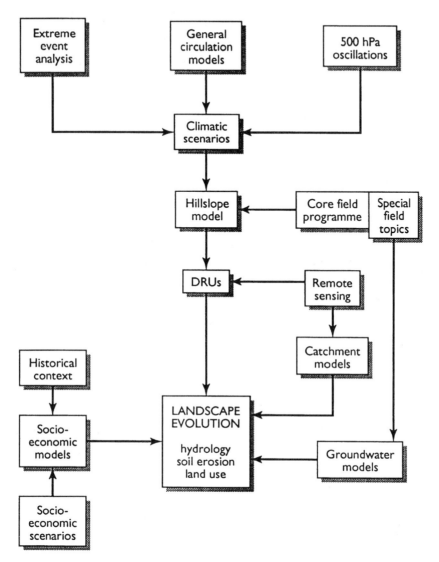

FIGURE 1.1   Components and organization of the MEDALUS I programme

1.   *Climatological studies* (Chapter 4). These involve investigations of present, past and future trends in rainfall and temperature to obtain estimates of the range of fluctuations of contemporary climates within the Mediterranean region and the likely range of future climatic conditions under various model scenarios. Investigations of series of climatological anomalies in all countries have been used together with predictive models to estimate immediate future trends. In addition scenarios from global circulation models have been downscaled to provide regional scenarios at the sub-grid scale using regression techniques. The climatological studies have also provided syntheses of extreme event data (rainfall and temperature) for the field sites

and the atmospheric conditions under which they occur. They have also endeavoured to provide a historical context for the contemporary monitoring of the field site micro-climates using conventional weather stations.

2. *Field observations* of climatic, ecological, erosional and hydrological processes at nine sites throughout the Mediterranean. These were located in Portugal (Almocreva and Vale Formoso, Alentejo), in Spain (Rambla Honda, Almeria and El Ardal), in Sardinia and Greece (Spata, Athens and Petralona/Hortiatis, Thessaloniki). Each field site was instrumented with basic weather recording facilities and collected a set of basic climatic, soil and ecological parameters (described in detail in Chapter 5). In addition each site had special topics of research relating to desertification which are described in Chapters 6 to 11. Work in the inner Alentejo (Chapter 6) is located at the Vale Formoso Soil Erosion Centre with its 30-year-long record of soil erosion plot measurements and in the wheat-producing area at the Herdade de Almocreva. The Rambla Honda field site in Almeria, Spain (Chapter 7), has been the focus for studies of the ecophysiology of Mediterranean species, particularly of *Stipa* (esparto grass), and of biomass productivity and phenology in relation to climatic and soil moisture regimes. The vegetation structure and dynamics of semi-natural regenerated land has been a focus of interest at the nearby El Ardal field site (Chapter 8) as has the definition of desertification response units (as described in Chapters 18 and 19). At the Santa Lucia site (Chapter 9) additional studies were made of the impact of fire on erosion and also of heavy metal pollution and salt-water intrusion into groundwater. The Spata site (Chapter 10) is the main agricultural site in MEDALUS I. The primary effort has been to measure the effects of rainfall exclusion on wheat in preparation for developing an agricultural component for the plant growth model. Finally the Petralona and Hortiatis field sites (Chapter 11) have specialized in investigating the impact of grazing on regenerating shrubland and the regime of decomposition of matorral species. Work on the sites in Mediterranean France which are located in vineyards is not reported here, but some details can be found in the *Atlas of Desertification* (Thornes and Mairota, 1996).

3. *Modelling investigations* have involved the development of the MEDALUS hillslope model. The hillslope model provides a mechanism for examining the role of changes of climate, vegetation and/or soils on runoff and sediment yield at the scale of an erosion plot, field or hillslope. It is based on inputs of hourly climatic variables, and is especially sensitive to rainfall. The vegetation cover (if any) determines the interception, evapotranspiration and throughfall, after which infiltration and overland flow is partitioned by soil properties. Soil moisture determines available moisture for plant growth. Plant growth rate is determined for functional types (i.e. trees, bushes, grasses and bare ground). Overland flow determines erosion rates over an irregular surface and the capacity for sediment transport of different size fractions determines deposition. This model has been tested in a variety of situations using data from the field sites (especially El Ardal and Santa Lucia).

4. Evaluation of the impact of various scenarios of climatic, economic and social change on *land-use dynamics* using complex systems models. This has focused on the island of Crete to develop the model and validate its implementation.

5. Development of *remote sensing techniques*, calibrated against the field sites, for the analysis of vegetation and soil conditions where vegetation is sparse and the

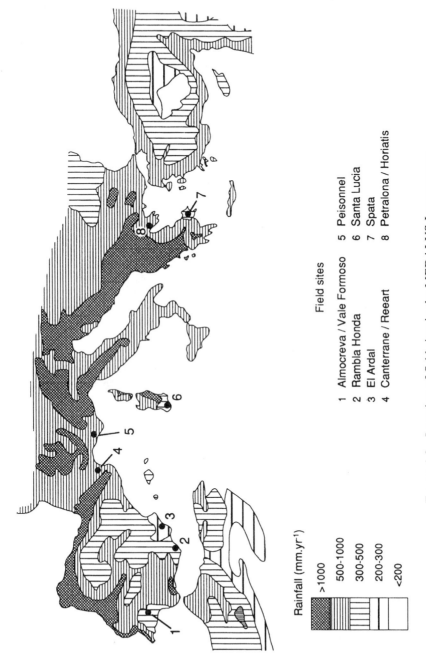

Rainfall (mm.yr^{-1})

>1000

500-1000

300-500

200-300

<200

Field sites

1 Almocreva / Vale Formoso   5 Peisonnel
2 Rambla Honda   6 Santa Lucia
3 El Ardal   7 Spata
4 Canterrane / Reeart   8 Petralona / Horiatis

FIGURE 1.2   Location of field sites in the MEDALUS I programme

underlying lithology makes up an important part of the signal emitted back from the ground to the satellites. These remote sensing studies provided a wider context into which to set the field studies and offer the possibility of extending the models beyond the field sites themselves. They have also provided inputs, together with field mapping and digital elevation models, to the determination of areas of common response to desertification processes called *desertification response units*. These mapping units are designed to be used in the organization and management of efforts to mitigate desertification.

1.5 REFERENCES

Barry, R. and Chorley, R.J. 1992. *Atmosphere, Weather and Climate*, 6th edition. Routledge.

Bolle, H-J. et al., 1993. EFEDA: An Echival experiment on desertification threatened areas. *Annales Geophysicae,* **11**, 173–189.

Braudel, F. 1975. *The Mediterranean and the Mediterranean World in the Age of Phillip II.* Fontana/ Collins, London.

Conte, M. and Giufredda, A. 1989. The long term evolution of the Italian climate outlined by using the standardised anomaly index. In Askew, A. and Lemmela, R. *Conference on Climate and Water.* Helsinki, Government Printing Office, Volume 1, 187–1195.

CORINE 1990. *Soil Erosion Risk and Important Land Resources.* Commission of the European Communities, Directorate General Environment, Consumer Protection and Nuclear Safety.

Fantechi, R. and Margaris, N.S. (eds) 1986. *Desertification in Europe.* Reidel, Dordrecht, 229 pp.

Francis, C.F. and Thornes, J.B. 1990. Runoff hydrographs from three Mediterranean vegetation cover types. In Thornes, J.B.(ed.) *Vegetation and Geomorphology.* Wiley, Chichester, 363–385.

Grenon, M. and Batisse, M. (eds) 1989. *Futures for the Mediterranean Basin.* Oxford University Press, 279 pp.

Grove, A.T. 1986. Desertification in southern Europe. *Climate Change,* **9**, 49–57.

Lopéz-Bermúdez, F. 1995. Desertification/degradation. Is this important research for the European Mediterranean countries? Unpublished paper to the *MEDALUS* Conference on Mediterranean Desertification, Myth or Reality, Almeria, November 1994.

Mairota, P. and Papadimitriou, F. 1996. Contemporary agriculture. In Thornes, J.B. and Mairota, P. (eds) *Atlas of Mediterranean Desertification: A Research Synthesis.* Wiley, Chichester.

Mancini, F. 1986. Soil conservation problems in Italy after the Council of Research finalised project. In Fantechi, R. and Margaris, N.S. (eds) *Desertification in Europe.* Reidel, Dordrecht, 147–151.

Margaris, N.S. 1996. Agricultural transformations. In Thornes, J.B. and Mairota, P. (eds) *Atlas of Mediterranean Desertification: A Research Synthesis.* Wiley, Chichester.

Messerli, B. 1967. Die eiszeitliche und gegenwärtige vergleletscherung in Mittelmeerraum. *Geographica Helvetica,* **22**, 3.

Moreno, J.C and Oechel, W.C. 1995. *The role of fire in Mediterranean-type ecosystems,* Springer-Verlag, Berlin, 205 pp.

Palutikof, J.P., Goodess, C.M. and Guo, X. 1994. Climate change, potential evapotranspiration and moisture availability in the Mediterranean Basin. *International Journal of Climatology,* **14**, 853–869.

Perez-Soba, A. and Barrientos, F. 1986. The Lucdeme programme in the southeast of Spain to combat desertification in the Mediterranean region. In Fantechi, R. and Margaris, N.S. (eds) *Desertification in Europe.* Reidel, Dordrecht, 138–147.

Rackham, O. 1983. Observations on the historical ecology of Boetia. *Annual of the British School at Athens,* **78**, 291–351.

Sala, M. and Rubio, J.L. (eds) (1994) *Soil Erosion as a Consequence of Forest Fires.* Geoforma Ediciones, Logrono.

Thomas, D.S.G. and Middleton, N.J. 1994. *Desertification: Exploding the Myth.* Wiley, Chichester.

Thornes, J.B. and Mairota, P. (eds) 1996. *Atlas of Mediterranean Desertification.* Wiley, Chichester.

United Nations 1994. *Convention on Desertification*. Paris, 43 pp.
UNEP (United Nations Environment Programme) 1992. *World Atlas of Desertification*. Edward Arnold, Sevenoaks.
Vita-Finzi, C. 1969. *The Mediterranean Valleys*. Cambridge University Press.
Wise, S., Thornes, J.B. and Gilman, A. 1982. How old are the Badlands? A case study from southeast Spain, (eds) A. Yair and R. Bryan, *Piping and Badland Erosion*, Geo Abstracts, Norwich, 29–56.

2

The Historical Context: Before 1850

A. T. GROVE

Department of Geography, University of Cambridge, UK

2.1 INTRODUCTION

Environmental change in the Mediterranean region has involved both natural events and human action, natural events being dominant until about 5000 BC, and man's influence increasing towards the present. Remote geological times are relevant in so far as they have seen the evolution of plant species and the main features of the rocks and relief which, with the climate, have always had and continue to have an important bearing on the evolution of soil conditions and water availability, and hence on the pattern of plant cover and agricultural potential. The last important changes in climate and vegetation in the lands bordering the northern shores of the Mediterranean, which took place at the end of the Pleistocene period, are part of the background. The main stages in the early development of agriculture left their marks on the landscape. The relevance of historical events to present processes in the landscape increases with proximity to our own times.

Important break-points in the history of the environment can be distinguished at the middle of the 19th century and the middle of the 20th centuries. After the 1840s, plague and famine in southern Europe were no longer important demographically, populations in most countries were steadily rising, and the old political regimes were being dismantled. More significant globally and from our point of view was the adoption of fossil fuels as the source of energy for transportation and the mechanization of agriculture. This made it possible to distribute over wide areas the surplus farm products of the main agricultural areas and the manufactures of industrial centres. Local and regional food shortages disappeared and prices of foodstuffs and raw materials diminished. The second break-point occurred after the Second World War with the industrialization of agriculture and the integration of Community economies.

2.2 PREHISTORIC TIMES

2.2.1 The physical background

The lands bordering the Mediterranean are characterized now, and have been for several thousands of years, by hot dry summers and mild wet winters, a climatic regime which is experienced by only 1% of the earth's surface (Perez, 1990). In the autumn, when the sea is

Mediterranean Desertification and Land Use. Edited by C. Jane Brandt and John B. Thornes.
© 1996 by John Wiley & Sons, Ltd.

still warm and atmospheric depressions move across the region from west to east, the stage is set for intense storms. Forest and maquis provide good protection, but slopes bared by fires, cultivation and heavy grazing are much more vulnerable to erosion.

Tectonically, most of the region is active, with local vulcanism and widespread seismic instability. Sardinia and Les Maures are exceptions, being remarkably stable. The potential for erosion has been accentuated at certain stages in recent earth history, notably during the Messinian salinity crisis 5 million years ago when, for a few hundreds of thousands of years, the western portals of the Mediterranean were closed and the resultant lake dried up to create an enormous desert basin. Rivers such as the Po excavated canyons into what is now the sea floor and lowered base levels for erosion far upstream (Cita and Corselli, 1990).

Fault displacements, for example on the north side of the Guadalentín valley in southeast Spain and alongside the Gulf of Corinth, have continued into historic times, uplifting beds of soft rock and exposing them to gullying and landslipping, catastrophic erosion for which human activity is scarcely at all responsible.

2.2.2 The Late Pleistocene environment

The successive glaciations of the Pleistocene, though they were experienced in a less severe form around the shores of the Mediterranean than in other parts of the world, were accompanied by strong cooling and repeated lowerings of sea-level. Recent dating of large debris fans in Crete, some of which extend well below present sea-level and others confined to interior basins, indicate they were accumulating during cold periods at various times between 300 000 and 100 000 years ago (Hempel, 1989). Fluvial terraces in northern Greece, of the kind called 'Older Fill' by Vita-Finzi (1969), have been shown to date back as far as the Middle Pleistocene, with the last of them formed by a major period of Late Wurmian alluviation related to glaciation of the Pindus ranges about 30 000 to 25 000 years ago (Lewin et al., 1991).

During this last cold episode, ice occupied cirques at about 2400 m on the north side of Psiloritis in Crete and glaciers descended to 2000 m on the northern flanks of the Sierra Nevada. The snow-line stood at about 2200 m on Parnassus and descended to 1300 m in the Albanian coastal ranges (Messerli, 1967). Pollen evidence shows much of southern Europe was covered by *Artemesia* steppe interspersed by patches of forest and scattered stands of trees. Temperatures were about 8°C lower than today. Precipitation is more difficult to estimate. It may have been only 60% of the present, but the levels of certain lakes such as Lake Joannina were high and it is possible that precipitation was similar to that of today but concentrated even more in the winter months (Prentice et al., 1992).

Sea-level was over 100 m lower than at present in glacial times, reducing the distance of Corsica from the mainland and linking it to Sardinia. Cyprus, Crete and Malta, which remained isolated from the continent and free of human occupation, were roamed by dwarf elephant and pygmy hippos. In the absence of large predators, they are likely to have grazed the vegetation heavily until the first mariners landed and dispatched them.

About 15 000 years ago, glaciers were retreating, sea-level was rising, and pine and juniper were spreading. After 13 500 BP, oak forests expanded over much of southern Europe and temperatures and precipitation were for a time comparable to those of the present day. Then came a final strong cooling around 11 000 BP with precipitation much lower than now. This Younger Dryas period, which was probably caused by cold, fresh meltwater from the collapsing continental ice caps spreading over the surface of the North Atlantic, lasted for about a thousand years. Wind-borne silt from the Sahara was widely deposited and is an important constituent of many soil profiles. Charcoal found in archaeological deposits in the south of France and pollen cores from Padul in southeast Spain indicate that the surrounding country was a *Juniperus–Pinus sylvestris–Amygdalus–Pistacia* forest–steppe (Pons et al., 1990).

2.2.3 The Early and Middle Holocene

At the beginning of the Holocene, 10 000 years ago, incident solar radiation in Mediterranean latitudes was about 7% higher in the summer months than it is today and correspondingly lower in winter. Pollen evidence indicates higher temperatures and more precipitation than at the present day. Atmospheric instability in the autumn would have been even greater than now and may explain why geomorphological evidence has been found in Haute-Provence of a main torrential rainfall phase during the Preboreal, between 10 000 and 8000 BP (Jorda and Vaudour, 1980).

Pollen from *Artemesia* and other plant species indicative of arid conditions declined after 12 500 BP in Boetia and, from 10 000 BP, tree pollen predominated (Allen and Katsikis, 1992). In the south of France deciduous oak, *Quercus pubescens*, pollen reached a maximum about 8000 BP; in the Padul region of southeast Spain, the cork oak, *Q. suber*, was well established. Charcoal in the Cova de l'Or, inland of Alicante, points to abundant *Q. ilex* with deciduous oaks and some wild olive, *Olea oleaster* (Vernet and Thiébault, 1987). Lime (*Tilia*), hornbeam (*Ostrya* and *Carpinus*) and hazel (*Corylus*) appear in pollen diagrams from Crete, where these trees are now absent. With a more extensive forest cover and a climate warmer and moister than it had been since the Last Interglacial, 100 000 years ago, the period from 8000 to 6000 BP was an important one for pedogenesis around the Mediterranean.

2.2.4 Prehistoric cultures

The climatic optimum between 8000 and 6000 BP was the time when Neolithic cultures were evolving, spreading round the Mediterranean and reaching previously uninhabited islands. Grazing by sheep and goats, burning, and clearing for cultivation, all modified the plant cover and in pollen diagrams conceal indications of climate change. In the south of France and eastern Spain there are indications of forest degradation, with *Pinus halepensis* spreading and then, at about 6500 BP, diminishing (Vernet and Thiébault, 1987) and the proportion of pollen from *Quercus coccifera* and evergreen shrubs such as *Rosmarinus* and *Erica* increasing.

About 5000 BP, high-level glaciers in southern Europe expanded. This glacier expansion, known in the Alps as the Rotmoos advance, has recently been identified in the Pyrenees (Gellatly et al., 1992). The climatic shift involved was quite small, on quite a different scale from those at the end of the Pleistocene, and remains unexplained. Such fluctuations, which were to be repeated at intervals until the Little Ice Age of a few centuries ago, have involved vertical movements of the snow-line and tree-line through a few hundred metres. Such oscillations are likely to have played a less important role in the modification of Mediterranean ecosystems than variations in human activity.

By the third millennium BC, cultivators and herdsmen were established throughout the Mediterranean lands. Their implements are found high in the mountains of western Crete, for instance, and megalithic monuments are widely distributed through the islands and on the mainland. Ash (*Fraxinus*) and deciduous oak were cleared or slashed to feed stock, and the relative importance of pollen from evergreen oak and box, *Buxus sempervirens*, increases in pollen diagrams.

Copper and Bronze Age remains show that most of the cultivable areas in the Mediterranean region were being exploited by the second millenium BC. Polyculture with vines, olive trees and cereals, had been established in the Aegean and in restricted areas of Spain and possibly in parts of Italy. Wine was being made in Crete (Lambert Gócs, 1990). Sheep, kept particularly for milk and wool, were the dominant stock and were probably managed in short-distance transhumance systems. Administrative and trading centres emerged, and quite extensive areas became involved in commercial trading networks by land and sea (Barker, 1985). Locally, pressure on land resources resulted in the construction of agricultural terraces, notably in Minoan Crete, on the island of Psira in the Gulf of Mirabello. Channel irrigation was probably employed in semi-arid southeast Spain (Gilman and Thornes, 1985).

The end of the second millennium and the early first millennium BC was a disturbed period when rural dwellers in the eastern Mediterranean withdrew to defensive sites in the hills. The Phoenicians and later the Carthaginians, trading in metals and grain, made demands on woodland resources for pottery kilns and smelting ores. Between 750 and 500 BC, populations grew; Greek colonies were established in North Africa and southern Italy, and there was a great increase in Mediterranean trade.

Brunet (1990) has pointed out that terraces on Delos, a small commercial island in the Aegean, must have been constructed before 400 BC. The Greek inhabitants would have been able to use slave labour to build the terraces and it may be surmised that the trading community of antiquity would have provided a ready market for produce. Van Andel and Runnels (1987) have argued that such terrace systems in the Argolid were constructed in times of peace and prosperity and fell into disrepair at times of social and economic collapse.

2.3 EARLY HISTORIC TIMES

2.3.1 Roman imperialism

By the 1st century BC, Roman imperialism was providing an increasingly powerful impetus to ecological change. The growth of cities, especially of Rome itself, placed unprecedented demands on the resources of the Mediterranean Basin. Both rural and urban populations were greater than had ever been known in the past and were not to be exceeded for more than a thousand years. In Italy, all but the most marginal land was brought under cultivation. Engineering works, involving draining marshes, damming rivers and transferring water over long distances, modified the environment in a deliberate fashion. Highland people came down to the plains; in Provence, for example, the Ligurians abandoned their hilltop *oppida* and for a thousand years lived on the centuriated lowlands.

After four centuries of Roman hegemony the Western Empire declined, and northern peoples invaded the lands bordering the western Mediterranean Basin. The 6th century AD saw a reduction in population, in part as a result of conflict, in part as a consequence of disease, notably the Great Plague of 542 AD (Hodges and Whitehouse, 1983). Trade between the eastern and western Mediterranean greatly diminished. Rural estates were left deserted and untended, towns were abandoned, hydraulic works fell into disrepair. Widespread piracy drove people away from roads and vulnerable coastal areas to seek security in nucleated settlements on defensible hill sites and in the interior (Llewellyn, 1971).

2.3.2 Medieval times

In Sardinia, which had been a granary for Rome, supplying enough grain to feed a quarter of a million people annually (Rowland, 1990), the whole structure of society seems to have changed between the 7th and the 11th centuries. The central parts of the island acquired a subsistent agrarian system that was more Germanic than Mediterranean, with cereal growing dependent on leaving the land fallow every other year or for two years out of three.

There were, exceptionally, more intensive systems of agriculture producing surpluses for sale. The Moors who invaded southern Iberia in the 8th century were accomplished agriculturalists who introduced cotton and other crops, improved irrigation, and revived commercial activity. In Sardinia, during the Pisan occupation, between 1100 and 1300, Benedictine and other religious orders were active, draining and improving the land, and introducing new crops. With the growth of commerce and artisanry, people moved into larger villages and many small settlements were abandoned. Except for Cagliari and Sassari, however, Sardinia scarcely shared in the rise of the communes that made northern and central Italy the most urbanized part of Europe. Settlements, though numerous, remained small and dispersed; there was scarcely one hearth to the square kilometre, giving a total population of little more than 100 000 spread over 24 000 km^2.

Crete, in the early years of Venetian rule, was at times self-sufficient in grain, and exported wheat to Venice: 'In 1302, the feudal lords of Chania promised to provide Venice with 30 000 *mouzouria* [*c*. 560 tonnes] of wheat at the same price as the lords of Candia had supplied it' (Theotokes, 1936). The Turks had reached the west coast of Asia Minor, and in 1318 there began centuries of intermittent warfare between Venice and Turkey. The Cretans preferred to live at peace with their powerful neighbour, from whom they habitually bought grain as well as timber, iron, oxen, and the horses which at that time were much used by the upper class in the island. The Venetians' penchant for warring with the Turks, and expecting Crete to fight those wars and pay for them, was a source of friction and one of the causes of the various unsuccessful Cretan revolts of the 14th century. After one of these, in 1343, Venice punished Crete by forbidding anyone to cultivate the Lasíthi and Anópolis plains.

The Black Death of the mid-14th century emptied the Mediterranean countryside. In Sardinia, the size of villages halved and the mean population density fell to about three persons to the square kilometre. Customs dues were greatly diminished, sales of salt fell by two-thirds, coins ceased to be minted at Iglesias, workshops fell into disuse. A renewed onset of plague in 1376 killed half the remaining people. War, brigandage, malaria and taxation by Aragon reduced Sardinia to a state of destitution. In the late 14th century the number of settlements was halved. Among those that disappeared many were quite large and old. In Sulcis, villages on both good and poor soils were abandoned because there were not enough people to farm them. Capoterra disappeared and for the next 500 years the surrounding area seems to have been sparsely settled. Sardinia was no longer a part of the Mediterranean trading network but a land of subsistent hamlets and villages where people lived a harsh life scarcely influenced by developments taking place on the Italian mainland.

When the Moors had been expelled from Portugal in the 13th century, large grants of land had been made to the Church and to those Templars, Hospitallers and Knights of the Sepulchre who had helped to expel them. After the Black Death, the monarch attempted to increase production in the lands they had acquired through the 'Lei das Sesmarias', the fundamental feature of which was the expropriation of land not being exploited and its reallocation to new owners. In Spain, as the Moors were driven from the Spanish Meseta, transhumance extended south. In the 16th century, 12 000 settlers were brought from Galicia and the Asturias to settle the southern slopes of the Sierra Nevada. It may be more than a coincidence that the Motril and Adra deltas began to form, possibly as a result of erosion of the *secano* in eastern Andalusia, at about the same time.

The Venetians considered Crete to be underpopulated and tried to persuade foreigners to settle there, even Turks and their families in 1416 (Zachariadou, 1983). They encouraged the intensive production of cereals which they were unable to grow in sufficient quantity on their own territory and needed to feed their garrisons and galley crews in Crete itself. The main duties of the Venetian landowning nobles were not only to assist in the defence of the island but to exploit the land for cereals. In good years, such as 1394, 1411 and 1414, grain was exported to Venice and the Venetian cities in Greece. Experiments were made with new crops such as sugar-cane, mastic and rice (Ploumides, 1985). Lasithi was reopened; settlers were brought in from Nauplion and drainage of the plain was improved. Cereals were cultivated on the floor of the Omalos basin.

Crete was benefiting from its position within the trading networks of both the Venetians and the Ottoman superpower. The demand for wine from the countries of northwest Europe was growing and the Venetians, realizing Cretan potentialities, introduced improved techniques for wine-making. Before long, Cretan sweet wines were taking a large share of international trade (Triantafyllidou-Baladié, 1992). The prosperity of the island, especially in the vicinity of Candia, greatly increased, bringing in its train a blossoming of Cretan culture (Hulton, 1991). The rector of Canea, Leonardo de Loredan, in a report to the Senate of 25 September 1554 (J.M. Grove, 1993, personal communication), notes that though it was able to feed itself with cereals for only 4 or at most 6 months in the year, Canea produced 14 000 botte (that is 9000 tonnes) of wine annually, sent to Flanders, Venice, Constantinople, Alexandria, Messina and Malta. Yet Canea was not the chief wine-producing area of Crete, most of the vineyards lying behind Candia (now Heraklion). Furthermore 'In the Canea plain are very large quantities of oranges and lemons so that 400 botte of lemon and orange juice are sent to Constantinople annually': (the equivalent of 260 tonnes; 1 Venetian botte being equal to 149 litres according to Pryor, 1992, p. 78).

2.4 EARLY MODERN TIMES

2.4.1 Climatic events, food shortages and plague

The Cretan population grew rapidly in the first half of the 16th century, from less than 200 000 to almost 300 000. The mean population density of the rural areas, about 50 to the square kilometre, was four or five times that of Sardinia and similar to that of Venezia or Tuscany. Much of eastern Crete was sparsely settled on account of piracy and it is likely that rural populations in many areas have never since been exceeded. This was probably the time when many of the hillside terraces were constructed, not only for vines but also to grow cereals displaced by vineyards on the plains.

Reports of shortages of grain become more frequent in the course of the 16th century and reliance on imports from Turkish lands increased. In the 1550s, the population of Crete was sharply reduced by famine, revolt and suppression, and by bubonic plague. Plague, which was endemic in Asia Minor, seems to have reached Crete and other Mediterranean lands whenever there were severe food shortages and grain was imported by ships from the east. The vectors of the disease were fleas living on black rats, for which grain was the preferred food, and soldiers on the move.

Historians such as Braudel (1990a) and Clark (1985) have pointed to the importance of the climatic events in the final decades of the 16th century to western economies and societies. Wendy Bell (1980) has shown from tithe records that Valencia and Murcia experienced an exceptional run of wet years between 1589 and 1598, followed by unusually prolonged drought between 1603 and 1614. In Sicily, 1585, 1589 and 1590 were all dry years; the meagre harvest of 1590 being followed by a long spell of drought until March 1591 when torrential rain fell through the spring months causing the seed to rot. In the following decade, the years 1606 and 1607 were excessively dry (di Blasti, 1790; Paruta, 1869). The years 1590–93 were famine years in the Papal States (Burke, 1985). From French Dauphiné, Piémond's diary of 1572–99 recorded late frosts in the 1580s and 1590s and tells of seven sterile years from 1584 to 1591, the harvest of 1585 being the poorest in living memory and frost damaging the vines in 1590 and 1592 (Greengrass, 1985). Not very far away, advancing glaciers were threatening villages in the vicinity of Chamonix.

It is only recently that the significance of similar events in Crete and the Aegean region has been recognized (Grove, in press). Weather extremes, recorded in Venetian documents, involving winter and spring droughts, excessive cold and snow or over-abundant rain, were experienced most frequently in the eastern Mediterranean, as in the west, in the years between 1570 and 1610. Riverine sediments deposited by floods about this time may well form the uppermost layers of the alluvial sediments which began to accumulate a thousand or more years earlier and which Vita-Finzi (1969) has called the Younger Fill.

Venetian governors of Crete, especially after the Ottoman conquest of Cyprus in 1571 and the food shortages in the following decades, were alarmed by the island's increasing reliance on grain imports from Turkey. Seeing the shortage of grain as being caused by vines occupying land suitable for cereals, they forbade the planting of new vineyards without licences and enacted legislation for grubbing-up existing vineyards. Provveditor Mocenigo, writing in the 1580s, recognized that people were not growing grain because the price was too low; they could make more money from wine. He pointed out that good prices could be obtained for oil and more olive trees would be planted in the future; and so it turned out. However, in some years after the Ottoman conquest, Crete exported grain to Turkish garrisons in North Africa and elsewhere, though not in very large quantities. Whether the Turks were paying the growers a higher price or whether they were extracting grain more forcefully we do not know.

Woodland in southern France in the 17th century was being destroyed at such a rate in places like the Pyrenees that Colbert issued ordinances expressly forbidding the destruction of the forests in order to preserve the flow of rivers. Cultivators were short of land for their own subsistence and for growing vines. In the eastern Alpes-Maritimes, by the latter part of the century, 70% of the cultivable slopes in communes inland of Nice and Menton were terraced up to heights of 600 m, as illustrations in the *Theatrum sabaudiae ducis* of 1682 testify (Rebours, 1990).

Languedoc in the late 17th and 18th centuries was still a poor, densely settled region where cereal growing was in the hands of bourgeois or ecclesiastical landowners. It was the poor people who grew vines, often on the least promising land, for instance on the lower slopes of the Massif Central to the north (Braudel, 1990b, p. 331). The construction of the port of Sète in 1666 and the opening of the Canal du Midi in 1681 allowed an expansion of wine exports to England and the Netherlands and encouraged investment in vineyards near the coast. Soils on the plains, which had been devoted to grain and had been regarded as too heavy for vines, were found to give better yields than the lighter soils on the slopes, and between a half and two-thirds of the Biterrois around Béziers was planted with vines. This expansion of the area under vines, as in Crete was not welcome to the authorities who tried to prevent further planting, but by the early 19th century vineyards extended west into the Corbières de Lézingnan and east towards the Étang de Thau. Roussillon, except for the valleys of the Agly and Tet, remained under polyculture.

Technical advances were being made in the design and construction of ships and firearms. Cultivation of the mulberry and the rearing of silk worms were spreading in Crete, Italy and southern France. But the 17th century was a time of religious wars and other conflicts, and in the Mediterranean countries there were recurrent famines and plague culminating in the disastrous years 1690–94.

In 1709, severe frosts ruined the grapes in central and northern France. In the Midi, the vines survived and northern markets opened up to wine from the south. 'The sudden rise in the price of wine brought casks flooding north from the Midi to Paris' (Braudel, 1990b, p. 335). Many olive trees were killed in the Midi and as a result Crete, which had been exporting olive oil for several years, shipped record quantities to Marseille in 1710. The winter of 1739/40 was the longest and coldest in modern western European history, with excessive autumn precipitation in France and flooding of the Arno and Tiber. The 'year without a summer', 1816, was responsible for the *Last Great Subsistence Crisis in the Western World* (Post, 1977). It affected northern and central Europe more than the south, but Italy was unusually cold and wet, and in Spain and Portugal late summer and autumn rains damaged the grapes; prices of olive oil in Lisbon in 1817 were the highest recorded between 1750 and 1854.

All three cold events seem to have been caused by volcanic eruptions projecting large quantities of dust and sulphur into the stratosphere thereby reducing the incident solar radiation for several months. The eruptions responsible are believed to have been those of Vesuvius, Santorini and Fujiyama in 1707, volcanoes in Kamchatka immediately prior to 1740, and Tomboro, Indonesia, in 1815 (Post, 1985).

Regulations to reduce the fire hazard, requiring the replacement of wooden buildings in towns by brick and stone, were beginning to be enforced in the 17th century. Such measures incidentally reduced rat infestation and, combined with other improvements in domestic hygiene, they diminished urban mortality. Plague remained endemic in Ottoman lands, affecting Crete from 1816 to 1821, but the last outbreak in western Europe occurred in 1730, in Marseille. In the 18th century, roads and bridges were being built, and canals and maritime trade were beginning to improve the distribution and availability of food, relieving local shortages. Potatoes and maize were being more widely adopted and were also helping to reduce the frequency of food shortages and famines. Populations began to increase in France, southern Italy and Sardinia, and most other parts of southern Europe, though not in Crete and other territories held by the Ottomans.

By the middle of the 18th century, rural areas in southern France, northern Italy and Sicily were showing signs of land shortage, with overgrazing and erosion becoming apparent. Following clearings in the common woods of St Laurent de Var for instance, 'the floods of the torrent Var became more formidable, and had already carried off much land as early as 1708' (Marsh, 1965). Charles de Ribbe, in *La Provence au point de vue des bois* is quoted by Marsh (p. 207) as stating that 'clearing continued, and more soil was swept away in 1761. In 1762, after another destructive inundation, many of the inhabitants emigrated, and in 1765, one half of the territory had been laid waste'.

Carrier (1932, p. 381) notes that 'the Parlements of Languedoc and the Sovereign Council of Roussillon obtained a declaration from the King forbidding the further making of clearings in the forest, but the damage was of too long standing to be easily remedied, and the Revolution of 1789 put the State Forests into the hands of a peasantry most anxious to exploit them'. At the end of the century, 'Fabre, a civil engineer of some repute with the government of France, declaimed very emphatically against the improvident deforestation in which the inhabitants of the Departments of Var and Basses Alpes were freely indulging'. In 1853, M. de Bonville, prefect of the Lower Alps addressed to the Government a report in which the following passages occurred:

> It is certain that the productive mould of the Alps, swept off by the increasing violence of that curse of the mountains, the torrents, is daily diminishing with fearful rapidity. All our Alps are wholly or in large proportion, bared of wood. Their soil, scorched by the sun of Provence, cut up by the hoofs of the sheep, which, not finding on the surface the grass they require for their sustenance, scratch the ground in search of roots to satisfy their hunger, is periodically washed and carried off by melting snows and summer storms. (Marsh, 1965, p. 215)

2.4.2 Alentejo around 1800

The situation in southern Portugal and Spain was very different from that in southern France. The sparsity of population in Alentejo persisted and was accentuated by the demands of maritime expansion and by the struggles of the war of independence between 1640 and 1648 which were especially destructive in Alentejo. In the next century the situation deteriorated still further, with gold from Brazil allowing the importation of wheat at prices lower at Lisbon than those for the Alentejo product.

The state of the province of Alentejo at the end of the 18th century is described in much detail in a treatise by Antonio Henriques de Silveira which was summarized by Link (1801) in his account of 4 years travelling in Portugal with the botanist Count Hoffmansegg. The cause of the barrenness of eastern Alentejo, according to Link (p. 151), was the great aridity of the soil. The best time to see the country was the beginning of spring when a great variety of heath plants were in bloom. Among these he notes the heathers, *Erica australis* and *umbellata*, and numerous cisti: *helimifolivo*, *laftanthus*, *libanotis*, *sampfirifolius*, *crispus* and *verticillatus*. In addition he mentions *Lithospermum fruticosum, Lavandula stoechus,* the junipers both *oxycedrus* and *phoenica,* and the creeping oak, *Quercus humilis.* 'Notwithstanding this variety of plants these heaths soon become irksome, even where they are beautiful, for without some cultivation no country can be pleasing, unless it be sublime and romantic'.

Quoting de Silveira, (p. 155), he states that 'Alentejo is the least populous province of Portugal; for though 36 leagues long and nearly as broad, it contains only four cities, 105 towns [villas], 358 parishes, and about 300 000 inhabitants; [339 355 according to the latest lists available in 1801]'. This compares with 531 000 in 1992.

> The towns are very populous, comparatively more so than the rest of Portugal; but there is a scarcity of villages, which generally contribute most to cultivation, many of the inhabitants of towns leading idle lives. One cause of this thin population arises from its always having been the theatre of war between Spain and Portugal. It also contains a great number of fortresses, maintains ten regiments of infantry and four of cavalry, which are constantly recruited there, and forms a fourth part of the military establishment of the whole country. Every town and village in the province, except in the fortresses, now contains fewer inhabitants than in the beginning of the last century [i.e. *c.* 1700] and in all of them are empty houses.

Most of the people were landless, and both Link and Balbi (1822) comment on the bands of vagabonds roaming Alentejo in the early 19th century. To ameliorate the situation, de Silveira had advocated the establishment of small villages of about 20 houses or the granting of permission to private persons to form such establishments 'granting them manorial rights and privileges.' (p. 156). 'The lands in Alentejo are far from being well cultivated'. 'If these lands were to be divided into small lots in parcels, the soil being nearer their habitation would be better manured and cultivated, and would not be suffered to be fallow two following years, as is now practised...'.

> In that province are three kinds of soil; fruitful black solid fat earth is found in the red clay of Elvas, Campomayor, Olivenca, Fronteira, Efremoz, Beja and Serpa; a lighter earth mixed with a little sand forms the soil around Evora and Arrayaoles, where the bad kinds of wheat, barley, and rye succeed very well, and cork trees and evergreen oaks also grow; and a sandy barren soil forms the heaths of Cantarinbo, Ponte de Sor, Monte Argil, Tancos and Vendas Novas, a tract of country about thirty leagues in circumference [about 3000 km^2]. They were once full of cork trees, but these have been sold to the charcoal-burners and thus the woods have been destroyed, excepting at a distance from the rivers. These heaths serve only as a pasture for goats, and yet at a depth beneath the surface lies a deep stratum of clay, which might be brought up by the plough, and the land rendered more fit for cultivation.

Leaving Mertola, 'a small town with high enclosed walls with neither fields nor gardens', Link and Hoffmansegg rode north to Serpa. They provide us with a description of the landscape of the region around Vale Formoso as it was nearly 200 years ago (pp. 464–466).

> On removing to a small distance from the valley of the Guadiana, we found here and there well-cultivated and even fruitful spots, which produced excellent wheat. [But] a more extensive desert does not exist in Portugal; at first we only saw a couple of houses and some fields, then another house half way, but everywhere else till within a league of Serpa only hills and mountains of sandstone and argillaceous flats covered with *Cistus ladaniferus;* nor did we meet any man in this desert.... Here and there we saw traces of former cultivation; for it is the custom in Portugal, as also in the fields of Spain that are covered with broom, to burn these plants or cut them down with a kind of sickle called *fouce rocadoura,* on good land every five,

and on bad every eight years; after which it is ploughed and sown' [a kind of shifting agriculture]. The crop indeed is very poor; for the roots of the former plants remain in the earth, and soon vegetating again cover the soil. It then serves only for pasture, which however is very poor, but the extent of ground must compensate for the badness of the herbage.

2.5 19TH-CENTURY CHANGES IN LAND TENURE

2.5.1 Alentejo

The main obstacle to improvement in the productivity of Alentejo, according to Link, was landownership.

> The nobility have too large herds of small cattle as sheep and goats, for which reason they do not have the heaths cultivated, but hire other lands besides their own which are thus likewise deprived of cultivation. Some of those who do not possess pasture for above eighty sheep, keep above a thousand, the land of their neighbours supplying the deficiency [p. 160]. Many estates belong jointly to several proprietors... Hence these lands are generally covered with cistus [*mato—Cistus ladaniferus*]. [p. 161]. [Numerous monasteries] drain the country with continuous contributions.

The main change in land tenure came in the 1870s with the partition of the communal lands, the *baldios*. These were divided among the local people, each of whom had to pay a fee for his land. Many of the holdings were too remote, too unproductive or too small to warrant the investment of labour; most people were too poor to improve their properties. By the end of the century most of what had been land available to the community had fallen into the hands of private landholders, many of whom were former estate managers or members of the same families who had acquired land earlier in the century; now their properties were even larger than they had been. The majority of the people had lost their rights in what had been community land and had little alternative but to emigrate or become seasonal labourers, while the local government had lost an important source of income with which they might have improved facilities for the those people who remained.

2.5.2 Southeast Spain

The sequence of events in southeast Spain was similar to that in Alentejo. Land ownership in the 18th century was polarized, with numerous minifundia in the hands of smallholders but with most of the land in large estates owned by the Church, communities and noble families, many of them absentees living in Madrid. As ownership of the large estates was associated with juridical rights and duties, such land could not be sold; it was entailed, and many owners, lacking capital or ability, were unable to use their land productively or dispose of it. Nor were their tenants, generally with less capital and little security of tenure, able to invest in the land.

Legislation to enable land to be disentailed and converted to private property was first introduced in the 1760s but it was not until 1836 that entailment was finally abolished. In the following year, Church land was sold off and in succeeding decades between a quarter and a third of the agricultural land in Spain came onto the market. The purchasers, comprising a new agrarian élite, consisted of members of the old nobility, some of the more prosperous tenants and managers, and wealthy town-dwellers. The pre-existing landholding structure had thus been reinforced, not reformed, and agriculture remained under-capitalized and without adequate public sources of credit (Shubert, 1990).

2.5.3 Sardinia

The system of communal agriculture that had emerged in Sardinia persisted into the 19th century and its impress can still be distinguished in the present landscape. In the centre of each of the communal territories was a nucleated settlement, the *villa*, sometimes among small hedged fields, vineyards and orchards. Surrounding this core was the *habitatione*, a block of land divided into two, three or more sectors and protected against animals by a palisade. One sector of the habitatione (the *habitatione de arari*) was cropped each year, mainly with cereals. The rest (the *habitatione de pascher bestiamente masedu*) was left for grazing by horses, donkeys and oxen from July to August. The next year another sector was cultivated and in the third another sector or a return was made to the first. In the 13th century the land had been allocated for cultivation by lot; by the 17th century, allocation was for life. At the margins of the village lay the rough grazing land, the *saltus*, which might belong to the community and in part to the state or to a feudal owner, but in any case the cattle, pigs, sheep and goats of the local community could wander there freely.

Spanish customs and especially Spanish attitudes to the rights of pastoralists had a strong influence in Sardinia. In most communities flocks and herds could graze anywhere crops were not growing; quite limited areas were enclosed. In the 17th century the *Stamenti* (*les états generaux* first convoked by King Peter in 1355) enjoined every householder to engraft 10 wild olive trees annually (500 trees would support an oil press). Otherwise farmers had received little encouragement to improve the land and plant perennial crops. Most of the arable land remained under shifting cultivation, being cropped with cereals for 1 or 2 years and then reverting to pasture for 8 to 10 years.

In the 17th century a process of colonization of the island periphery began which involved farmers taking up concessions of commune land, and pastoralists converting their winter quarters into settlements which were occupied throughout the year. Some villages were still abandoned from time to time because of banditry and vendettas, both of which were a response to the long-continued weakness of any central authority. The population of Sardinia was still less than that of Crete, in spite of its area being three times as great, and the lack of pressure on land resources probably explains the small extent of terracing.

With the arrival of the Piedmontese in the 18th century, security increased and the process of resettlement accelerated. Two thousand Corsicans settled in Gallura. Pastoralists from the central highlands of Sardinia who had long been accustomed to pasture their flocks on the hills of Sulcis in winter, established more permanent settlements, called *furriadroxius*, between Iglesias and the sea. In time they brought more and more land under cultivation and constructed more robust buildings. People who came to work in the nearby mines took up farming and added their houses to these embryonic villages and eventually the area, which had formerly depended on Iglesias, was divided into five communes each with its central place. Similar settlements were established by degrees elsewhere in Sardinia, the *cuili* of Nurra, the *stazzi* of Gallura and the *baccili* of Sarrabus. Over the years many of them grew into communities depending more on arable farming than on pastoralism (Le Lannou, 1941).

From the end of the 17th century Spanish fief-holders had begun to make land available to individuals as private freehold. After the union with Piedmont in 1730, the need for enclosure was recognized officially. An ordinance of 1771 allowed individual cultivators to enclose land with hedges and walls in order to grow crops and make hay. It was stipulated that the grass must be mown, haylofts constructed and sheds built for the beasts. Only the more prosperous could fulfil such requirements and the effects of the ordinance were consequently limited. An edict of 1820 had more far-reaching effects. Communes could divide communal land between family heads or sell them or give them away. It also allowed all proprietors to fence land not generally used for grazing or for the movement or watering of animals. However, no executive authority was stipulated. Furthermore, it was only the wealthier people who could pay lawyers to get the required permits and had the resources to build walls and sheds. The land was taken over by such people who proceeded to run their flocks on land which was no longer available to the rest of the community. Villagers were unable to gather wood from what had been common land or to graze their animals on the stubble. On the other hand, the pastoralist settlers at the periphery benefited from the legislation which enabled them to legalize the ownership of their holdings.

The sole example of an agricultural improver in Sardinia was the Marquis of Villahermosa. At the beginning of the 19th century, his Villa d'Orri estate, extending from Sarroch northwards and including the Santa Lucia valley, was bare land at the foot of the hills lying to the west of the Gulf of Cagliari, covered with lentisk and cistus and with much standing water (Valéry, 1837).

> Unfortunately, no one from outside the area can live there in the spring and summer, even country people, because of the malaria with which they have to contend. In spite of this great inconvenience, which means that the proprietor has to abandon the place for much of the year, leaving it in the hands of a bailiff who is nearly always wracked by fever or even carried off by it, the Villa d'Orri with its surroundings is remarkable as a model property on the island. Comfortable domestic buildings, surrounded by gardens and orangeries, together with fruit trees from the mainland and flower and vegetable gardens, are separated from the sea by a great mass of poplars whose size is proof of the fertility of the soil. In surveying the scene one is impressed by what determination and intelligence can accomplish, though it is true that a high social standing and other special circumstances have helped in the production, as if by magic, of a model farm out of what had been barren soil. In fact, more than 18,000 almond trees and 10,000 olive trees, now mature and growing strongly, plus thousands of mulberries and extensive vineyards have replaced the brushwood that used to grow here half a century ago on loose gravelly soils that were reputedly sterile.

'If Sardinia had fifty landowners like M. de Valhermosa, then this backward island would see its ancient prosperity restored', wrote Valéry (1837, p. 239), 'a happy land, with numerous people and abundant forests'.

2.5.4 Crete

Crete seems to have been reasonably prosperous in the 18th century, with a population of about a quarter of a million. Consuls from France and Britain were based in Khania and ships from those countries traded with the island mainly in olive oil, of which about half of local production was exported, and malvasia wine. After 1714 the Cretans developed their own merchant fleet, many of the ships being built in Khora Sphakion and captained by Sphakiots.

The population of the island declined as a result of plague in the years around 1800 and then diminished drastically to little more than 100 000 during the Greek War of Independence of 1821–28, when the island was in a continuous state of revolt against the Turks. It is possible that many of the higher terraces, especially those on the hard limestone hillsides, were abandoned at this stage and have not since been brought back into cultivation. There were other revolts later in the century, but the population began to increase after 1834, doubling every 50 years to reach about 250 000 by 1900.

2.5.5 Provence and Languedoc

Between 1730 and 1850 the population of France increased by 50%, from 21 to 33 million (Price, 1981). That of Var, excluding the town of Toulon, increased by 14% between 1821 and 1851, from 274 000 to 312 000 (Agulhon, 1982). Many of the nobles had retained their land and above all their woods. Smallholdings were subdivided, marginal land cultivated, and people depended heavily on their common rights in the forests. In spite of Arthur Young's dictum (1889, p. xxv) at the time of the revolution 'Whenever you stumble upon a grand seigneur, even one that is worth millions, you are sure to find his property desert', according to Agulhon (1982, p. 12) the larger landowners of Var were the innovators in the early 19th century. The smallholders, still dependent on the bourgeoisie from whom they held their land, 'remained faithful to the hoe and the swing-plough'.

The cultivated area around Vidauban reached its greatest extent around 1835, when the population attained a maximum of 3000, a figure to which it has returned repeatedly after intervals of emigration. Woodland with clearings still occupied those parts of the commune underlain by crystalline rocks, while land use on the deeper soils derived from Permian sedimentary rocks was diversified with cereals, olives and vines. The winters of 1820 and 1830 damaged the olive trees; those which had been planted in marginal areas of Basse-Provence during the 18th-century craze for olive cultivation were never replanted (Agulhon, 1982). Animal fats were beginning to replace olive oil (Pounds, 1990), but many people continued to cling to olive cultivation even though they sensed it was threatened.

Between 1851 and 1856 the population of Languedoc and Provence had increased by 101 000 souls. The augmentation, however, was wholly in the provinces of the plains, where all the principal cities are found. In these provinces the increase was 204 000, while in the mountain provinces there was a diminution of 103 000. The reduction of arable land is even more striking. In 1842, the department of the Lower Alps possessed 99 000 ha of cultivated soil. In 1852 it had but 74 000 ha. In other words, in 10 years 25 000 ha had been washed away or rendered useless for cultivation, by torrents and the abuses of pasturage (Clave, *Études*, pp. 65–67, quoted by Marsh, 1965, p. 214).

The populations of the mountain communes in Var certainly reached a peak in the 1840s and then declined. Many young and landless people had already gone to find work in the arsenal of Toulon and in Marseille, which was benefiting from trade with newly colonized Algeria. After 1840 they were joined by the poorer farmers. Land degradation may have played a part in the decision to leave home, but in the middle of the 19th century there were other more powerful forces at work.

2.6 REFERENCES

Agulhon, M. 1982 (translated by J. Lloyd). *The Republic in the Village: The People of Var from the French Revolution to the Second Republic.* Cambridge University Press.

Allen, H. and Katsikis, A. 1992. Environmental and climatic change in the Aegean during the late Quaternary. *Petromarula*, **1**, 9–15.

Balbi, A. 1822. *Essai statistique*, 2 vols. Paris.

Barker, G. 1985. *Prehistoric Farming in Europe*. Cambridge University Press.

Bell, W.T. 1980. The climate of south-east Spain, 1580–1630. Final Report for the Rockefeller Foundation Fellowship in Environmental Affairs (unpublished).

Braudel, F. 1986. *L'identité de la France*. Flammarion, Arthaud, Paris.

Braudel, F. 1990a (translated by Sian Reynolds). *The Mediterranean and the Mediterranean World in the Age of Philip II*, 2 vols, 2nd revised edition, Collins, London.

Braudel, F. 1990b (translated by Sian Reynolds) *The Identity of France*, 2 vols. Collins, London.

Brunet, M. 1990. Terrasses de cultures antiques; l'example de Délos, Cyclades. *Méditerranée*, **71**, 5–11.

Burke, P. 1985. Southern Italy in the 1590s: hard times or crisis. In P. Clark (ed.) *The European Crisis of the 1590s*. Allen & Unwin, London, 177–190.

Carrier, E.H. 1932. *Water and Grass*. Christophers, Glasgow.

Cita, M.B, and Corselli, C. 1990. Messinian paleogeography and erosional surfaces in Italy: an overview. *Palaeogeography, Palaeoclimatology, Palaeoecology*, **77**, 67–82.

Clark, P. (ed.) 1985. *The European Crisis of the 1590s*. Allen & Unwin, London.

di Blasti, G.E. 1790. *Storia cronologica de'Vicerè*. Palermo, vol. 2, part 1.

Gellatly, A.F., Grove, J.M. and Switsur, V.R. 1992. Mid-Holocene glacial activity in the Pyrenees. *The Holocene*, **2**, 266–270.

Gilman, A. and Thornes, J.B. 1985. *Land-Use and Prehistory in South-East Spain*. Allen & Unwin, London, 217 pp.

Greengrass, M. 1985. The later Wars of Religion in the French Midi. In P. Clark (ed.) *The European Crisis of the 1590s*. Allen & Unwin, London, 106–134.

Grove, J.M. and Conterio, A. 1995. The climate of Crete in the sixteenth and seventeenth centuries, *Climate Change*, **30**, 223–247.

Hempel, L. 1989. The denuded soils in Crete—a relic of the Ice Age? *Reports of the Deutsch Forschungsgemeinschaft* 2/89, 7–10.

Hodges, R. and Whitehouse, D. 1983. *Mohammed, Charlemagne and the Origins of Europe: Archaeology and the Pirenne Thesis*. London, Duckworth.

Holton, D. (ed.) 1991. *Literature and Society in Renaissance Crete*. Cambridge University Press.

Jorda, M. and Vaudour, J. 1980. Sols, morphogenèse et actions anthropiques à l'époque historique *sensu lato* sur les rives nord de la Méditerranée. *Naturalia Monspeliensia*, hors série, 173–184.

Lambert-Gócs, M. 1990. *The Wines of Greece*. Faber, London.

Le Lannou, M. 1941. *Pâtres et paysans de la Sardaigne*. Tours.

Lewin, J., Macklin, M.G. and Woodward, J.C. 1991. Late Quaternary fluvial sedimentation in the Voidomatis basin, Epirus, northwest Greece, *Quaternary Research*, **35**, 103–115.

Link, H.F. 1801. *Travels in Portugal*, Longman and Rees, London.

Livet, R. 1962. *Habitat Rural et Structures Agraires en Base-Provence*, Publication des annales de la Faculte des Lettres, Aix-en-Provence, 465 pp.

Llewellyn, P. 1971. *Rome in the Dark Ages*. Faber and Faber, London.

Marsh, G.P. 1965 (edited by D. Lowenthal). *Man and Nature*, Belknap Press, Harvard University Press.

Messerli, B. 1967. Die eiszeitliche und die gegenwartige Vergletscherung im Mittelmeeraum. *Geographica Helvetica*, **3**, 105–228.

Paruta, F. 1869. Diario della citta di Palermo. In G. di Marzo (ed.), *Biblioteca Storica e Letteraria di Sicilia*, 1st ser., Vol. 2, 3–8.

Perez, M.R. 1990. Development of Mediterranean agriculture: an ecological approach. *Landscape and Urban Planning*, **18**, 211–220.

Ploumides, G.S. 1985. Aitemaka kai pragmatikotetes ton Ellenon tes Venetokratias (1554–1600) [Requests and actualities of the Greeks during the Venetian occupation]. *University of Joannina, School of Philosophy, Department of History and Archaeology*, **1**, 182.

Pons, A., Couteaux, M., Beaulieu, J.L. and Reille, M. 1990. Plant invasions in southern Europe from the palaeoecological point of view. In F. di Castri, A.J. Jansen and M. Debussche (eds) *Plant Invasions in Europe and the Mediterranean Basin*. Kluwer Academic Publishers, Dordrecht, 169–177.

Post, J.D. 1977. *The Last Great Subsistence Crisis in the Western World*. Johns Hopkins University Press.

Post, J.D. 1985. *Food Shortage, Climatic Variability, and Epidemic Disease in Preindustrial Europe: The Mortality Peak in the Early 1740s*. Cornell University Press.

Pounds, N.J.G. 1990. *An Historical Geography of Europe*. Cambridge University Press.

Price, R. 1981. *An Economic History of Modern France*. Macmillan, London.

Prentice, C., Guiot, J. and Harrison, S.P. 1992. Mediterranean vegetation, lake levels and palaeoclimate at the Last Glacial Maximum, *Nature*, **360**, 658–660.

Pryor, J.H. 1992. *Geography, Technology, and War: Studies in the Maritime History of the Mediterranean 649–1571*. Cambridge University Press.

Rebours, F. 1990. Versant aménagés et déprise rurale dans l'est des Alpes-Maritimes. *Méditerranée*, **71**, 31–42.

Rowland, R. 1990. The production of Sardinian grain in the Roman period. *Mediterranean History Review*, **5**, 14–20.

Shubert, A. 1990. *A Social History of Modern Spain*. Unwin Hyman, London.

Silveira, A.H. de c. 1790 *Memorias Econimas 1*. Academy of Sciences, Lisbon.

Theotokes, S. 1938. Jakobos Foscarini e Krete to 1570. [Jakobos Foscarini in Crete in 1570]. *Epeteris Etoureias Kretikon Spondon* 1, 186–206. [*Annual of the Society of Cretan Studies*, **1**, 186–206].

Triantaphyllidou-Baladie, Y. 1992. The Cretan rural landscape and its changes in late-medieval and modern times. *Petromarula*, **I**, 47–51.

Valéry, M. 1837. *Voyages en Corse, à l'Ile d'Elba et en Sardaigne*, Vol. II. Bourgeois-Maze livre III, Paris.

Van Andel, T.H. and Runnels, C. 1987. *Beyond the Acropolis*. Stanford University Press, California.

Vernet, J-L. and Thiébault, S. 1987. An approach to northwestern Mediterranean recent prehistoric vegetation and ecological implications. *Journal of Biogeography*, **14**, 117–127.

Vita-Finzi, C. 1969. *The Mediterranean Valleys: Geological Changes in Historical Times*. Cambridge University Press.

Young, A. 1890. *Travels in France During the Years 1787, 1788, 1789*. 3rd Edition. Bohn's Standard Library, London.

Zachariadou, E.A. 1983. *Trade and Crusade: Venetian Crete and the Emirates of Mentesha and Aydin 1300–1415*. Library of the Hellenic Institute of Byzantine and Post Byzantine Studies, Publication No. 1.xxxv.

3

Changes in Traditional Mediterranean Land-Use Systems

N. S. MARGARIS, E, KOUTSIDOU and CH. GIOURGA

Department of Environmental Studies, University of the Aegean, Lesvos

3.1 INTRODUCTION

As almost everywhere, the Mediterranean landscape consists of three systems: cultivated, semi-natural (non-cultivated but managed) and land that has been built on. In this report attention is directed towards cultivated and semi-natural systems, both of which are found on sloping land, in contrast to cultivated systems which exist mainly on plainlands. All three are interconnected with socio-economic factors such as grazing pressure, tourism and demographic changes.

Within natural systems the following subdivisions can be made on the basis of the decreasing availability of water (Margaris, 1981):

1. Deciduous forests (e.g. oaks).
2. Evergreen forests (e.g. pines).
3. Evergreen shrubs (e.g. maquis).
4. Seasonal dimorphics (e.g. sub-shrubs, phrygana, tomillares).

Although Eyre (1968) states that 'two main types of plant communities recur (in Mediterranean areas) with monotonous frequency', the term monotonous refers only to maquis (which has the synonyms macchia in Italy, choresh in Israel, mattöral in Chile, chaparral in the USA, mallee in Australia and fynbos in South Africa). For the other main types, various terms have been used such as phrygana (Greece), tomillares (Spain), batha (Israel), coastal sage (USA).

Woody chamaephytes, dominating phryganic ecosystems, drastically reduce their transpiring surface during summer by the mechanism of seasonal dimorphism (Orshan, 1964). These plants develop two different leaf types: winter leaves in the wet season and summer leaves during the dry one. Winter leaves are more numerous and they undertake almost all photosynthesis needed because, during the summer, water shortage and high temperatures lead to a considerable loss of plant weight (Margaris, 1976).

Mediterranean Desertification and Land Use. Edited by C. Jane Brandt and John B. Thornes.

Generally speaking, almost all the systems mentioned above occur on mountain slopes (except for the semi-arid tomillares of Spain) and, while the deciduous forests are found at the higher elevations with a more temperate climate, the other three subsystems are typical of Mediterranean climate regions world-wide.

Agricultural systems are found both on plains and on slopes where they are usually associated with terrace cultivation. For a more systematic description the plants cultivated can be divided into annuals and perennials, grown at high (above 500 m) and low (below 500 m) elevations.

At low elevations, where a Mediterranean-type climate (sometimes known as 'the climate of olives') dominates, three basic systems of plant cultivation have developed in the Mediterranean Basin over millennia of human intervention, and these interlock with each other over the cultivation cycle:

1. Wet season cereals.
2. Vineyards.
3. Olive groves.

These types are connected by the time of the year and magnitude and nature of the agricultural effort. Wet season cereals (wheat, barley), planted after autumn rains, need the farmer's labour for harvesting, threshing and processing from June to the middle of August. At that time the grape harvest starts, and when it ends in October it is replaced by the olive harvest.

Moreover we can discern two basic agricultural types based on the annual and perennial plants used for cultivation. The annuals are mainly cereals and legumes with variations in practice according to elevation. At low elevations the period of cultivation coincides with the wet season (November–May), while at high elevations, it coincides with the warm season (April–September) reflecting practices developed in 'temperate' climates where there are summer as well as winter rains. Wheat and barley are the 'wet season' cereals cultivated at low elevations while maize and sunflower are cultivated at higher elevations.

From experience, Mediterranean farmers know that leguminous species transform atmospheric nitrogen into nitrogen available for use by plants and soils using the symbiotic microbes in their roots. For this reason cultivation of legumes in the cereal fields, every third year or so, is a traditional practice which replenishes the soil with nitrogen. At high elevations intercropping of, for example, maize and beans together, provides the same results.

The perennials at higher elevations are mainly deciduous fruit trees (apples, cherries and nuts for example), while at lower elevations both deciduous (almonds and figs) and evergreen trees (olives and carobs) are grown. Vineyards are present at both high and low elevations.

High-elevation cultivation is less important today than in earlier times, both because the crops grown are in the minority in the Mediterranean Basin and because the mechanization of agriculture is not possible on the higher, steeper slopes.

Bearing in mind also the peculiarities of the Mediterranean-type climate, where the wet and mild season is followed by a dry and hot one, the analysis concentrates on low elevations. It is important to remember, however, that although the plains provide the products, hills and mountains provide the water because of the higher rainfall and snowfall there.

3.2 FOREST FIRES

In Mediterranean-type ecosystems, the combination of high temperatures and a water deficit during the summer, leads to frequently recurring fires. The association of this climatic type with fire has been known for a long time (Griesebach, 1872). Shantz (1974), in his review of the Mediterranean-type ecosystems of California, refers to them as 'fire-type ecosystems' and believes that it is unlikely that this type of ecosystem was ever free from fires. Ecosystems subject to frequent fires over thousands of years may have developed, in the course of evolution, properties which make them extremely flammable. In fact, it may be important in these ecosystems to have deliberate burning every so often in order to prevent large accumulations of fuel and the associated high-temperature fires which cause ecological catastrophe. Biswell (1974) also suggests that 'fire in chaparral is both natural and inevitable. It has always occurred and probably always will, because vegetation becomes extremely dry near the end of a long, hot, nearly rainless summer. At that time, also, humidity may be extremely low and winds high'.

Plants dominating Mediterranean-type ecosystems are equipped with adaptive strategies to enable their post-fire recovery (Margaris, 1981; Arianoutsou and Margaris, 1982). This perturbation and the recovery from it are considered to be incorporated in the ecosystem genetic information pool. Under normal fire frequencies, fire acts as a selective force in these systems. Therefore they should be considered to be fire-induced or fire-adapted. Plant adaptations are, therefore, homoeostatic responses of ecosystems following fire-induced disturbances.

Ecosystem degradation caused by a combination of frequent fires and subsequent overgrazing has led to the consideration of fire as a catastrophic event, often attributed to deliberate criminal actions. Because of this misunderstanding, the policy inevitably followed until now has been one of fire exclusion. This will undoubtedly lead to severe damage followed by desertification for the following reasons.

As a consequence of the fire-exclusion policy, biomass is accumulated in the ecosystem. When fire eventually does occur, the high temperatures occurring result in damage to the recovery mechanisms which the system has evolved. For example, extremely high temperatures damage the soil seed bank beyond recovery. This fact was realized about 15 years ago in California and today prescribed burning techniques are a common practice (Conrad and Oechel, 1981).

From the ecological point of view, data available show that in systems from which fire has been excluded the diversity of both plant and animal species is reduced. An example to illustrate that is the 'complete protection' project of Mountain Fynbos, South Africa, where threatened species were virtually eliminated as a result of the fire exclusion that was supposed to protect them (Bands, 1977).

3.3 NATURAL ECOSYSTEMS AND THEIR MANAGEMENT

In the past when the population was strongly linked to local resources, forest products such as wood, cork, acorns and resin were of crucial importance for its well-being. Fuel wood was of great importance to people mainly for heating and cooking. Socio-economic changes that have occurred during recent decades, such as introduction of oil central-heating systems, electricity and bottled gas, strongly affected wood collection from forests.

Figure 3.1 illustrates the reduction of fuel wood production in the prefectures of the Aegean Islands in Greece. These data show that today's collection of fuel wood from forests is a fraction of what it was in the past. This indicates not only that less care is being devoted to the forests but also that fuel is accumulating in the forests (Margaris, 1987a).

In all cases the interest of the local population in forest, especially pines and oaks, is diminishing mainly because the traditionally collected products such as resin, acorns and cork are losing their economic value. Figure 3.2 shows the sharp decrease in the collection of acorns and resin on the island of Lesvos in the North Aegean of Greece.

It is well known that interaction between 'natural' and 'cultivated lands', depends on the socio-economic structures existing in time and space. Disruption of traditional agro-pastoral forms of management as well as the substitution of forestry products for chemicals, changes the system into a new state, characterized by less care, more fires, overgrazing and desertification. This can be exemplified by the case of Kella in northern Greece (Vokou et al., 1984), the Asphodel Deserts of Thessaly (Pantis and Margaris, 1988) and the Aegean Islands (Margaris 1987b).

Almost all the above trends have been confirmed during the MEDALUS I programme by investigators in Corsica, Greece, Italy and part of Spain. In the south of France the forest cover is increasing. In Portugal the abandonment of cork collection for paper production has encouraged the development of extensive eucalyptus plantations. In other places in Portugal, some of them close to the MEDALUS I Vale Formoso experimental site, replacement of cork oaks by invading pine trees can be observed at the present time. Considering the relationship between pines and fire this replacement is obviously rather dangerous especially in the light of expected future climatic warming. The techniques used in the past for protecting cork oak must not be abandoned.

Since overgrazing is believed to be a trigger mechanism for desertification in southern Europe (except in the south of France and Portugal), we considered whether or not the

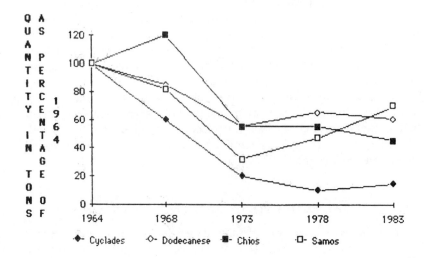

FIGURE 3.1   Trends in fuel wood collection in the Aegean Islands (Source: Greek National Agricultural Statistics, years 1964–83). The vertical axis is scaled to 1964 = 100

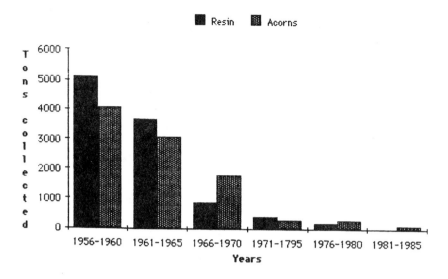

FIGURE 3.2 Collection of acorns from oaks and resins from pines in the island of Lesvos

exclusion of grazing for periods longer than a decade would induce ecosystem restoration and rehabilitation.

In the Mt Aipos area on the island of Chios (northeast Aegean, Greece) there is a desertified area of more than 25 000 ha. Today, in this area, almost all types of woody plants more than 0.5 m high are entirely absent; however, the names of some places as well as plant remnants, show that above-ground biomass was much greater in the recent past.

In 1976, and in parallel with re-afforestation, more than 50 grazing enclosures (of about 0.2 ha each) were established. Bearing in mind that, according to Margaris (1987), contemporary overgrazing is the driving force of desertification, attention was directed to assessing future trends and the likelihood of plant recovery in these grazing exclosures.

Table 3.1 contains some data dealing with the presence of woody plants in both protected and unprotected areas, from which it is obvious that recovery is rather high. The biological indices used were developed by Shannon and Weaver (1963) and Simpson (1949) and are shown in Figure 3.3. The Shannon–Weaver index measures diversity by the formula:

$$H = \sum_{i=1}^{S} p_i \log p_i \qquad (1)$$

where H is the Shannon–Weaver diversity index, S is the number of species and p_i the proportion of individuals in the total sample belonging to the ith species.

The Simpson index measures diversity with the formula:

$$A = \sum_{i=1}^{S} p_i \frac{(n_i - 1)}{(N - 1)} \qquad (2)$$

TABLE 3.1 The number of individuals for each woody plant in the experimental areas (10 samples were takes per $100\,m^2$ each according to the method developed by Parsons and Molbenke, 1975)

	Unprotected area	Protected area for 1–7 years	Protected area for 8–13 years
Anthyllis hermaniae (L.)			1
Asparagus acutifolius (L.)	3	7	11
Astragalus trojanus (L.)	1	2	2
Ballota acetabulosa (L.)	1	1	9
Clematis vitalba (L.)		113	323
Euphorbia mirsinites (L.)	70	44	68
Fumana thimyfolia (L.)			14
Helichrysum stoechas (Ten)	1	9	23
Lonicera etrusca (G. Santi)			3
Medicago arborea (L.)			11
Phillyrea latifolia (L.)	4	10	13
Prunus spinosa (L.)	23	21	8
Ptilostemon chamaepeuce (L.)			1
Pyrus amydaliformis (Vill)	303	378	297
Pyrus communis (L.)			1
Quercus coccifera (L.)	11	29	12
Rubia peregrina (L.)		2	20
Sacropoterium spinosum (L.)	2219	1974	1973
Senecio bicolor (Willd)		1	
Teucrium divaricatum (Sieber)		29	62
Teucrium polium (L.)	8		
Thymus capitatus (L.)	21		66
Total number of plants	2665	2660	2918

where A is the diversity index, S the number of species, i is the species number and p_i is the proportion of individuals in the sample belonging to the ith species. N is the total number of species and n_i the number of the individuals in the ith species.

Table 3.2 contains data relating to the herbaceous plant biomass within and outside the grazing exclosures. It is evident that a recovery is taking place because there has been a four-fold increase in the $g\,m^{-2}$ above-ground herbaceous plant biomass from the $29\,g\,m^{-2}$ observed outside the exclosures. These values were obtained from a harvest early in May at a time when herbaceous biomass is at a maximum. A comparison of this data with other areas of Greece where there are phryganic communities shows that rates of recovery are very high since herbaceous biomass contribution is generally less than $50\,g\,m^{-2}$ (e.g. Diamantopoulos, 1983; Pantis, 1987).

Biomass increase (Table 3.2) reflects not only the growth of individual species but also an increase in biological diversity. A marked augmentation is observed in the numbers of plant species as well as of individual plant numbers. Regression analysis of the number of plant species against the number of plants of all species predicts the increase of these variables in the following three years (Figures 3.4 and 3.5).

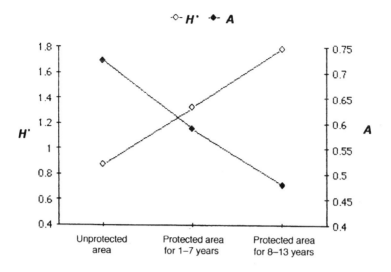

FIGURE 3.3   Changes in Shannon's (*H*) and Simpson's (*A*) indices for woody species on Chios where grazing exclosures have been established

3.4 TRENDS IN MEDITERRANEAN AGRICULTURE IN THE UPLANDS

From the mid-1950s on, the introduction of tractors and other machines radically changed the whole scene in terms of agricultural practices, efforts involved and production; from human and animal power the agricultural society turned to petrol. Generally speaking, mechanization was possible only on the lowlands and proved to be imossible in places with steep slopes which are so common round the Mediterranean. The use of machinery has led to great changes in cereal production. In Greece, for example and according to National Agricultural Statistics wheat yields in favourable areas have risen from 1 tha^{-1}y^{-1} to 4 tha^{-1}y^{-1} and maize from 3 tha^{-1}y^{-1} as much as 15 tha^{-1}. As a result, the extent of areas cultivated with annual crops (mainly cereals and legumes) and the volume of production in all areas using terrace agriculture showed a dramatic decrease. This clearly indicates the

TABLE 3.2   Herbaceous (mainly annual) plant biomass, number of plant species and numbers of individual plants observed after prohibition of grazing in an almost desertified area of Chios

Treatment	Biomass (g m^{-2})	Number of	
		plant species	individual plants
With grazing	29	22	269
Without grazing for:			
3 years	39	24	317
6 years	65	24	492
9 years	90	29	670
12 years	109	37	834

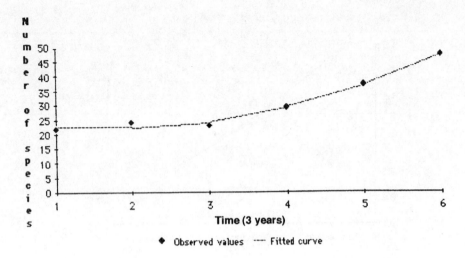

FIGURE 3.4   Trend analysis of number of plants of all species against period of protection from grazing $r = 0.99$, $P = 0.009$

abandonment of terrace cultivation which is taking place at an increasing rate all over the Mediterranean year by year. Table 3.3 contains some data showing the decline in the Aegean Islands.

   In addition to the cultivation of annual cereals and legumes, typical Mediterranean agriculture also involved the cultivation of perennial plant species, the principal examples

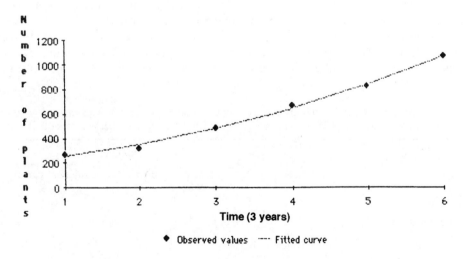

FIGURE 3.5   Trend analysis of number of plants of all species against period of protection from grazing

TABLE 3.3   Selected cases of animal crop cultivation decline in the prefectures of the Aegean, Greece

Aegean prefecture and product type	Area cultivated (ha)			Production (tonnes)		
	1961	1983	Decline %	1961	1983	Decline (%)
Cyclades (beans)	1500	300	74	800	160	80
Lesvos (lentil)	85	8	90	68	4	94
Dodecanes (beans)	275	120	56	170	45	75
Samos (lentil)	120	20	84	132	38	71
Chios (broad beans)	2500	400	84	2100	200	90
Chios (wheat)	3450	900	74	2400	500	79

Source: National Agricultural Statistics, Greece.

of which were olives, figs, almonds, plums and vines. In most cases these species are long-living. Of course, on the plains, the land was used mainly for the cultivation of cereals, potatoes and other annual plant species.

At the same time as the agricultural changes due to mechanization, socio-economic changes have been occurring over the last two decades. These have been strongly related to new technologies, and have affected almost all sectors of traditional aboriginal cultivation such as olives, figs and almonds.

Data available to date show two basic trends in rates of abandonment, one rate in Greece, Corsica, Italy and Spain, and an increased rate in southern France and Portugal. In these last two countries, abandonment of agricultural fields is followed by invasion of natural Mediterranean ecosystems. Pine forests, for example, are the end result in southern France.

3.5 MODERN MEDITERRANEAN AGRICULTURE AND WATER USE

Modern agricultural activities on Mediterranean plains consume much greater quantities of water than traditional agriculture. This water is mainly collected on the mountains and uplands. For each kg of sunflower oil for example (see Figure 3.6), 2 tonnes of irrigation water are needed. Given that olives grow without irrigation, it seems that the maintenance of olive groves is essential for the Mediterranean in relation to the potential impact of global climatic changes. Moreover, both the literature and observations indicate that the 'olive forests' (Vokou, 1988) have much higher floral and faunal diversity compared with pine forests. These results are supported by new findings from ornithological research. According to The Bird Life International, results of Phase 1 of the project Conservation of Dispersed Species in Europe have shown that second to wetland birds, birds of agricultural habitats are generally the most threatened in Europe (Tucker, 1993, personal communication).

3.6 SOCIO-ECONOMIC TRENDS OF CONSUMERS IN RELATION TO MEDITERRANEAN AGRICULTURE

Socio-economic trends related to consumer characteristics are the main driving forces for Mediterranean agriculture. Trends in some products, such as in the oil market, make it

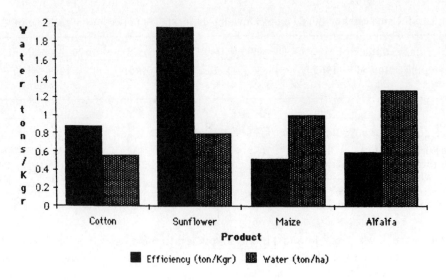

FIGURE 3.6   Tonnes of water applied to yield 1 kg of product, and tonnes of water applied to each
hectare of land for different crops

clear that the demand for and production of seed oil, for example, is constantly growing
while the use of olive oil is diminishing relatively and absolutely. For example, the
production of sunflower oil in all southern European member states is increasing
constantly. In 1983 in Greece, 20% of the oil production was from seed oil, but just 6 years
later this had risen to 36%. The ratio of sunflower to olive oil prices in 1984 was 1:1.4, in
1985 it was 1:1.7 and by 1991 it has risen to 1:4.0.

At the same time as these changes, soya bean cultivation has been introduced to the
southern European Union member states. Production figures are given in Table 3.4 and
will be another factor affecting the position of olive oil in the market.

It is important to realize that almost all the agricultural products coming from perennial
plant species in the dry parts of the Mediterranean, such as almonds, figs, carobs and
plums, with the partial exception of olive trees and the vines, provide products that reach
the consumer in their dry form. However, the consumer has reduced the consumption of
these dried products because nowadays fresh fruit is available in winter. This is a further
cause of the abandonment of traditional perennials. In the Aegean Islands, for example, in
1961 about 800 ha were cultivated with fig trees but only 327 ha remained in 1983 (Table
3.5). The same thing has happened with almond and carob groves.

TABLE 3.4   Production of soya oil (thousand tonnes) in Greece, Italy and Spain

	1980	1987	1988	1990
Greece	0	5	6	22
Italy	0	806	1589	1450
Spain	0	3	4	27

The production efficiency of some crucial Mediterranean products such as figs and almonds has been calculated. Figure 3.7 shows a decrease in yield per hectare compared with the mean values of the years 1961–65. The 5-year mean was used in order not to interfere with oscillations in production related to differences in the climatic parameters from year to year. It becomes obvious from Figure 3.7 that in all prefectures of the Northern Aegean administrative area, the productions efficiency figures show that less agricultural care is being shown.

Abandonment is considered to be a driving force to desertification because the next step is overgrazing and fire. This is shown by the photograph in Figure 3.8. It shows remains of an almond plantation in an almost desertified hill on the island of Lemnos, Greece, which 30 years ago was completely covered with almond trees.

TABLE 3.5 Cultivated area and production of figs in the Aegean Islands

Year	Cultivated area (ha)	Total as % of 1961	Production (tonnes)	Total as % of 1961
1961	806	100	6087	100
1965	726	90	4485	74
1970	637	79	3589	59
1975	431	53	2987	49
1980	424	53	2497	41
1983	327	41	2246	37

Source: National Agricultural Statistics, Greece.

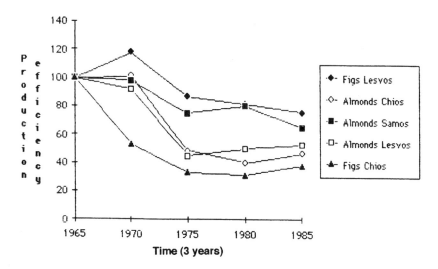

FIGURE 3.7 Production efficiency (yield per hectare) in figs and almonds in the North Aegean area, compared with the 1961–65 average (= 100%). (Source: National Agricultural Statistics, Greece)

FIGURE 3.8   The abandonment of almond trees cultivation in the island of Lemnos, Greece, causing desertification

In his book *La Méditerranéen*, Fernand Braudel (1979) states that 'in the Mediterranean the land is dying when the protection of the agricultural cultivation is absent'. It is supposed that all these trends reflect today's situation and that future changes will not be only climatic but also socio-economic.

3.7 CONCLUSIONS

During the MEDALUS I phase, working at both local and larger scales, the Aegean group realized that great changes are occurring in land-use patterns because of the new socio-economic behaviour. Those changes are causing desertification-like results and for this reason they must be taken seriously. For example, new synthetic products are replacing natural products such as cork, resin and tannins from acorns, and are causing abandonment of areas with cork oak, pines and oak forests. In these places abandonment is usually connected with minimization of the standing biomass and other members of the flora and fauna. The well-known Mediterranean biological diversity deteriorates and precious natural resources such as soil and water are lost. These phenomena, when associated with climatic change, will dramatically increase desertification.

In addition, new agricultural practices developed in recent times are substituting almost all traditional Mediterranean agricultural products. For the new cultivation, large quantities of water are needed for irrigation. For example, for each litre of sunflower oil at least 2 tonnes of water are needed while the traditional olive groves are generally not irrigated.

3.8. REFERENCES

Arianoutsou, M., and Margaris, N.S. 1981a. Producers and the fire cycle in a phryganic ecosystem. In N.S. Margaris and H.A. Mooney (eds) *Components of Productivity of the Mediterranean-Climate Regions. Basic and Applied Aspects.* Dr. W. Junk, The Hague.

Arianoutsou, M., and Margaris, N.S. 1981b. Early stages of regeneration after fire in a phryganic ecosystem (East Mediterranean). Regeneration by seed germination. *Biol. Ecol. Medit.*, **8**(3).

Arianoutsou, M., and Margaris, N.S. 1982. Phyrganic (East Mediterranean) ecosystem and fire. *Ecologia Mediterranean*, **8**, 473–480.

Bands, D.P. 1977. Prescribed burning in Cape Fynbos catchments. In H. Mooney and C.E. Conrad (Eds) *Environmental Consequences of Fire and Fuel Management in Mediterranean Ecosystems.* USDA/Forest Service, General Technical Report WO-3, Washington, DC, 245–256.

Biswell, H.H. 1974. Effects of fire on chaparral. In T.T. Kozlowski and C.E. Ahlrgen (eds) *Fire and Ecosystems.* Academic Press, New York, 321–364.

Braudel, F. 1979. *La Méditerranéen et le Monde Méditerranéen á l'epoque de Philippe II.* Tome I. Librairie Armand Colin, Paris.

Conrad, C.E. and Oechel, W.C. 1981. *Dynamics and Management of Mediterranean Type Ecosystems.* USDA/Forest Service, Pacific Southwest Forest and Range Experimental Station, Berkely.

Diamantopoulos, I. 1983. Structure and distribution of phryganic ecosystems in Greece. PhD Thesis, Thessaloniki. (In Greek with an English summary).

Eyre, S.R. 1968. *Vegetation and Soils.* Edward Arnold, London.

Griesebach, A. 1872. *Die Vegetation der Erde nach ihrer Klimatischen.* Leipzig, Arnordung, Leipzig.

Margaris, N.S. 1976. Structure and dynamics in a phryganic (East Mediterranean) ecosystem. *J. Biogeography*, **3**, 249–259.

Margaris, N.S. 1980. Structure and dynamics of Mediterranean-type vegetation. *Portug. Acta Biologica*, **16**, 45–58.

Margaris, N.S. 1981. Adaptive strategies in plants dominating Mediterranean-type Ecosystems. In F. Di Castri et al. (eds) *Mediterranean-type Shrublands.* Elsevier, Netherlands, 309–315.

Margaris, N.S. 1982. Harvesting before the fire for energy, Mediterranean ecosystems in Greece. Costs and Benefits. In G. Grassi and W. Palz (eds) *Energy from Biomass.* D. Reidel Publishing Co, The Netherlands, 95–98.

Margaris, N.S. 1987a. *Ecological Risks Affecting the European Forests.* Forecasting and Assessment in Science and Technology Occasional Commission of the European Communities DG XII. Paper No. 163.

Margaris, N.S. 1987b. Desertification in the Aegean Islands. *Ekistics*, **323**, 132-136.

Margaris, N.S., Schutt, P. 1987. *The Forest of Europe. Major Ecological Threats.* In FAST (Forecasting and Assessment in Science and Technology) No. 163.

Orshan, G. 1964. Seasonal dimorphism of desert and Mediterranean chamaephytes and its significance as a factor in their water economy. In A.J. Rutter and F.H. Whitehead (eds) *The Water Relations of Plants.* Blackwell, Oxford, 206–222.

Pantis, I. 1987. Structure, distribution and management of asphodel deserts in Thessaly. PhD Thesis, Thessaloniki Greece. (In Greek with an English summary).

Pantis, I. and Margaris, T.H. 1988. Can systems dominated by asphodels be considered as semi-deserts? *Int. J. Biometeoreology*, **32**, 87–91.

Parson, D.J. and Molclenke, A.R. 1975. Convergence in vegetation structure along analogous climatic grades in California and Chile, *Ecology*, **56**.

Shannon, C.E., and Weaver, W. 1949. *The Mathematical Theory of Communication.* University of Illinois Press, Urbana, IL.

Shantz, H.L. 1974. *The Use of Fire as a Tool in Management of the Brush Ranges of California,* Sacramento, California State Board of Forestry.

Simpson, E.H. 1949. Measurement of diversity. *Nature*, **163**, 688
Vokou, D. 1988. The olive groves as natural olive forest. In the minutes of the Conference 'Olive groves of the Aegean Islands', Mytilini 1988, 5–11.
Vokou, D., Diamantopoulos, J., Mardiris, T.H., and Margaris N.S. 1981. Desertification in Northwestern Greece: the case of Kella. In R. Fantechi and N.S. Margaris (eds) *Desertification in Europe*. Reidel, Dordrecht, 155–160.

4

Climate and Climatic Change

J. P. Palutikof[1], M. Conte[2], J. Casimiro Mendes[3],
C. M. Goodess[1] and F. Espirito Santo[3]

[1]*Climatic Research Unit, School of Environmental Sciences,
University of East Anglia, Norwich, UK*
[2]*Istituto Fisica Atmosfera CNR, Rome, Italy*
[3]*Instituto Nacional de Meteorologia e Geofisica, Lisbon, Portugal*

4.1 INTRODUCTION

The climate of the Mediterranean Basin is marginal for many of the economic activities on which the people of the region depend. The risk of land degradation and desertification is already a very real one under the present climatic regime. Desertification is not caused primarily by climate but rather by decisions at all levels in society regarding land use. However, the climatic environment provides the context for the chain of events which leads from a particular decision to the deterioration of the land. Future climate changes may lead to a climatic environment even more conducive to desertification.

The aim of this chapter is to describe the present-day climate of the Mediterranean, and possible changes which may occur over the next 50 years. Although data are analysed from the whole region, attention focuses on the area north of the Mediterranean Sea, in line with the stated aims of this book. Past climates are not considered here: they are comprehensively described in an earlier chapter (Chapter 2) of this book by A.T. Grove.

Current scientific research is focused on the enhanced greenhouse effect as the most likely cause of climate change in the short-term future. A vast literature surrounds this topic, and is best summarized by the publications of the Intergovernmental Panel on Climate Change (Houghton et al., 1990, 1992) and referred to below as IPCC90 and IPCC92 respectively. For the purposes of the discussion here, the most important questions surround our ability to predict the impacts on the climate of a relatively small region: the Mediterranean Basin.

Much of this chapter is concerned with the construction of regional scenarios of climate change due to the enhanced greenhouse effect. The word 'scenario' is used because it implicitly recognizes the uncertainty involved. Scenarios are 'internally-consistent pictures of a plausible future climate' (Wigley et al., 1986). The current level of scientific knowledge does not allow a more definite prediction at the regional level. Although scenarios can be used, for example, to assess the sensitivity of economic activities to possible climate changes, they are not forecasts. There is no doubt that the quality of regional scenarios will improve in the future, in step with advances in the scientific understanding of the enhanced greenhouse effect.

Two approaches to the problem of assessing future changes in climate are adopted here. In the first, current trends in an indicator of atmospheric pressure (the height of the 500 hPa surface) are extrapolated to around 2020 using a statistical model. The relationship identified between this indicator and precipitation in the Mediterranean Basin is used to make inferences about possible future trends in precipitation. Problems exist in justifying this approach, because it depends on present-day statistical relationships being preserved in the future.

The second approach uses the results from general circulation models (GCMs) to construct seasonal scenarios of the change in temperature, precipitation and potential evapotranspiration. GCMs are complex three-dimensional models of the atmospheric circulation, based on the fundamental laws of physics. Although this approach does not depend on statistical relationships, a new set of problems exists, related to the realism and accuracy of the models. These are fully discussed in Section 4.4.6. Despite these problems, there is no doubt that GCMs offer the best potential for the development of regional scenarios. As improved model runs become available in the future, so the quality of regional scenarios will rise.

In order to place the climate change scenarios in context, it is necessary to describe the present-day characteristics of the climate of the Mediterranean region. Thus, the first sections of this chapter are devoted to analyses of records from a network of temperature and precipitation stations, as shown in Figure 4.1. These stations have at least 20 years of data in the period 1951–90. The network is used to describe spatial patterns of temperature and precipitation over the Mediterranean Basin, at the annual and seasonal level. Time series are examined to determine the trends that have occurred in the recent instrumental period. A special case study is made of time series of temperature (1940–92) and precipitation (1931–92) from the Alentejo region of Portugal, in order to demonstrate how basin-wide trends are reflected at the local level.

Following these descriptive sections, evidence is presented to support the existence of a 'Mediterranean Oscillation' analogous to the Southern Oscillation. Analysis of data on the height of the 500 hPa surface shows a 22-year periodicity, with the oscillations in the western and eastern Mediterranean being in perfect phase opposition. Extrapolation of trends in the height of the 500 hPa surface allows us to make some general statements about future patterns of precipitation over the region. This leads naturally into an analysis of possible climate changes due to the enhanced greenhouse effect. The chapter ends with a review of the uncertainties involved in attempts to predict future climate at the regional scale.

4.2 CLIMATOLOGY OF THE MEDITERRANEAN REGION

In the Köppen classification, a Mediterranean climate is defined as one in which winter rainfall is at least three times the summer rainfall (Köppen, 1936). This is true of almost the whole of the area studied in this chapter. Indeed, over much of the Mediterranean, summer rainfall is virtually zero. This strong summer–winter rainfall contrast is echoed by a pronounced seasonal cycle in almost all climate variables.

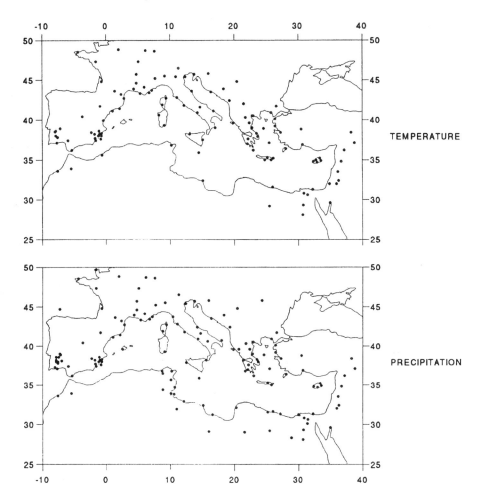

FIGURE 4.1   Location of stations with complete records of temperature and precipitation for the period 1961–85

In July, August and September the region experiences warm, dry conditions linked to the presence of a strong high-pressure ridge extending eastwards from the Azores subtropical anticyclone. Over Egypt, this ridge is displaced southward by a trough extending northwest from the Arabian Gulf towards Greece, which is associated with the Indian summer monsoon trough. The area therefore exhibits a pronounced west–east pressure gradient (Bartzokas, 1989).

The rainy season commences in mid-October. At this time, the average upper westerlies change from a three-wave to a four-wave pattern on the 5-day time scale. A trough in this wave pattern is located over Europe, although the exact position is highly variable. Winter is characterized by cyclonic disturbances and low mean pressure in the Mediterranean, with higher pressure to the east associated with the Siberian High. The main pressure contrast is between low pressure over the sea and high pressure over the land.

The rainy season continues until around the end of April. However, from the time of the equinox the major features of the upper circulation move northwards in response to the passage of the sun. By May, the polar front and the associated strong upper-air westerly flow are sufficiently far north that their influence is removed from the Mediterranean Basin. The subtropical highs and their associated ridges once more exert their influence and the rainy season ends.

In the Iberian Peninsula, it has been estimated that 54% of rain-producing depressions are of Atlantic origin (Linés Escardó, 1970). These depressions are weakened as they track across the peninsula, and seldom bring rain to the Mediterranean Basin.

In the Mediterranean Basin itself, precipitation is caused mainly by cyclonic disturbances of local origin. Orographic effects also play an important role. Wigley and Farmer (1982) describe four preferred points of origin within the basin itself, three of which spawn rain-producing depressions. Of these, the most important is the Gulf of Genoa, where depressions form in the lee of the Alps. Further east, the preferred locations for the formation of depressions are to the south of Greece and over Cyprus (termed, respectively, Central Basin and Eastern Basin depressions by Wigley and Farmer). Atlas Mountain lee depressions, which form in the spring, are seldom associated with rainfall. Rather they are accompanied by hot, dry and windy conditions, particularly when they track eastwards across North Africa into Egypt.

Furlan (1977) emphasizes the role of the zone extending from the Gulf of Genoa across the Po valley into the northern Adriatic for the formation of depressions. Cantú (1977) considers the Gulf of Genoa with, to the south, the Tyrrenhian Sea and, to the north, the Alps, as of major significance.

The formation of depressions within the Mediterranean appears to be favoured when the mid-latitude westerlies are in their blocked stage, with associated meanders in the polar front jetstream. This leads to meridional transport of cold air which, in turn, favours cyclogenesis over the warm ocean (Perry, 1981; Wigley and Farmer, 1982).

The movement of depressions is not well understood. In the western Mediterranean, they are frequently steered along the Mediterranean Front, formed when cold continental air moves over a warmer sea surface. This front is most pronounced in the spring. Depressions originating in the Gulf of Genoa seldom penetrate as far as the Eastern Basin of the Mediterranean. Around half of all Central Basin depressions are steered over the Black Sea. Fronts formed in the eastern Mediterranean tend to follow a preferred path either to the northeast or the east.

In the next section, the spatial and temporal variations of temperature and precipitation are examined. These are based on a data-set of monthly statistics, records being at least 20 years in length.

4.2.1 Spatial patterns of temperature and precipitation

The maps discussed in this section are based on the network of stations shown in Figure 4.1. The period of record used for the construction of the maps is 1961–85 for the temperature series, and September 1961 to August 1985 for precipitation. In the analyses which follow, we have used a hydrological year for precipitation, running from September (the start month of the autumn rains) to August inclusive.

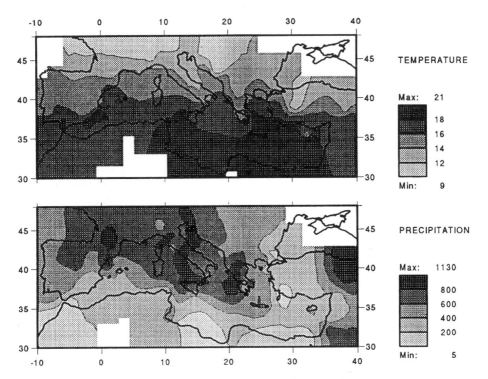

FIGURE 4.2   Upper map: annual mean temperature (°C) for 1961–85. Lower map: mean annual precipitation (mm) for the hydrological years 1961–84

Figure 4.2 shows the annual mean temperature and total precipitation for the region. Highest mean temperatures are found in the south and southeast, exceeding 18°C over Libya and Egypt. There is a gradient towards the north and northwest, to temperatures below 12°C.

Precipitation in the region does not exceed 1200 mm per annum. In the area north of the Mediterranean Sea, lowest values are found in southeastern Spain, with between 200 and 400 mm per annum, and over western Turkey and the western shore of the Black Sea, with less than 400 mm. The preferred areas for cyclogenesis are clearly seen: over the Gulf of Genoa and the Tyrrenhian Sea in the western Mediterranean (over 800 mm), to the west and south of Greece in the central Mediterranean (over 800 mm), and over Cyprus in the eastern Mediterranean (400–600 mm). The orographic effect of the Pyrenees produces enhanced precipitation amounts.

Seasonal patterns for temperature are shown in Figure 4.3. In winter (December, January and February), lowest mean temperatures in the true Mediterranean region are experienced over Italy, the north coast of the Adriatic Sea and western Greece (below 6°C). The most northerly warm conditions (over 12°C) are found in an extensive belt covering the islands of the eastern Mediterranean, and in isolated patches over Spain, Portugal and northern Italy. In the other three seasons, the spatial distribution of temperature mimics the annual pattern, with a gradient from south-east (high

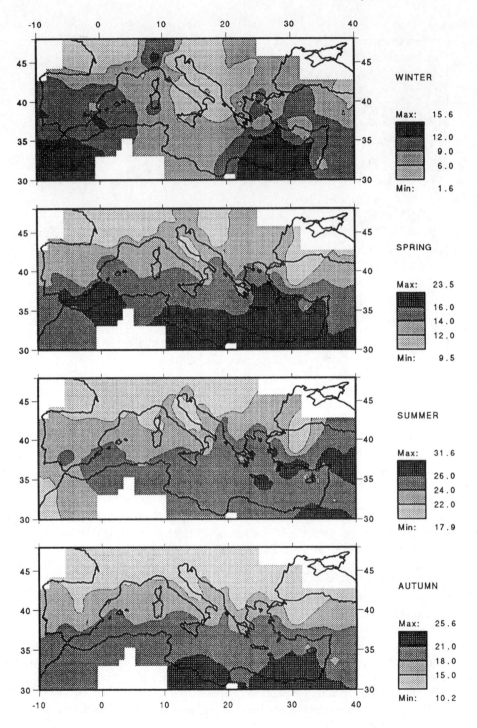

FIGURE 4.3   Seasonal mean temperature (°C) for 1961–85

temperatures) to north-west. Highest temperatures are, of course, found in summer. In this season, the most extensive areas of high temperatures for the northern Mediterranean are over Crete and southern Turkey, rising to over 26°C.

Seasonal precipitation totals are shown in Figure 4.4. In winter, precipitation is generally above 60 mm over the Mediterranean islands (except the Balearics) and the land to the north of the Mediterranean Sea. Spring precipitation is over 40 mm in most areas north of the Mediterranean Sea. The summer season demonstrates a pronounced gradient in the northern Mediterranean, from less than 20 mm in southern Spain and Turkey, to over 60 mm over northern Italy. Autumn shows a maximum over the central Mediterranean, of over 60 mm and, in isolated areas, over 90 mm.

It is clear from the above description that most of the region conforms to the Köppen definition of a Mediterranean climate. However, particularly over the northern Adriatic and Aegean, summer rainfall totals are in excess of one-third of winter precipitation. Wigley and Farmer (1982) examined the shift in the northern boundary of Mediterranean climates in the Eastern Basin over four decades. They found that the most northerly extent was in the 1940s but, in the 1970s, the boundary lay some 3–4° south of this position, across central Greece and the middle of the Anatolian Plateau.

Figure 4.5 shows the mean monthly precipitation at seven stations in the northern Mediterranean, selected to demonstrate the regional diversity which exists in the seasonal cycle. Only one, Thessaloniki, does not have a true Mediterranean-type climate (defined as the precipitation in the wettest month being three times that of the driest month). However, sites such as Ricote and Tabernas demonstrate much less seasonal variation than, for example, Mertola and Toulon. Some sites have a bimodal rainfall pattern (Ricote and, to a lesser extent, Tabernas and Thessaloniki), with a drier winter interposed between a wet autumn and spring. Elsewhere, the pattern is unimodal (Amantadis et al., 1991; Garrida and Garcia, 1992).

4.2.2 Temperature time series

In this section, we look first at temperature variations over the whole of the Mediterranean. We then focus on the Alentejo region of Portugal: an area with a relatively dense network of reliable and long meteorological records. This gives some indication of whether regional (Mediterranean-wide) variations are repeated at the local scale, and vice versa.

To examine trends in temperature over the Mediterranean Basin, annual and seasonal standardized anomaly indices (*SAI*s) were calculated, following the approach of Nicholson (1983), who developed the technique for calculating precipitation *SAI*s. At time step t, the *SAI* is given by:

$$SAI_t = (1/N) \sum_{i=1}^{N} (T_{it} - \mu_i)\sigma_i \tag{1}$$

where T is the temperature (or precipitation) at station i and N is the number of stations. The parameters μ_i and σ_i denote, respectively, the mean and standard deviation of the temperature series at the ith station. To summarize, in order to allow for differences in

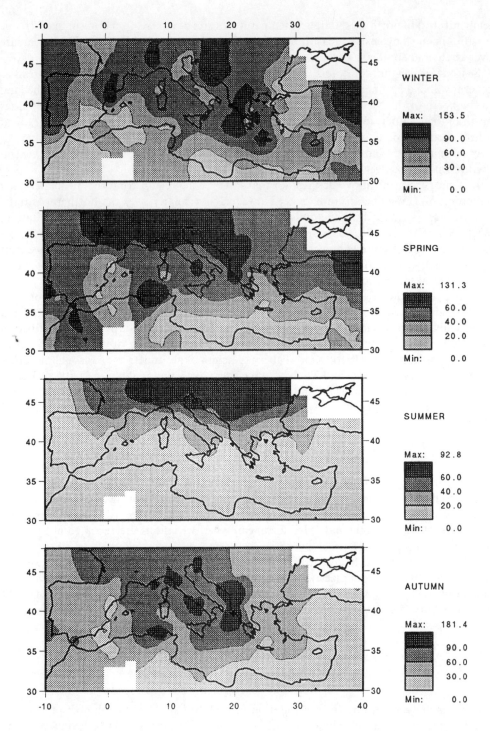

FIGURE 4.4   Mean seasonal precipitation (mm) for the hydrological years 1961–84

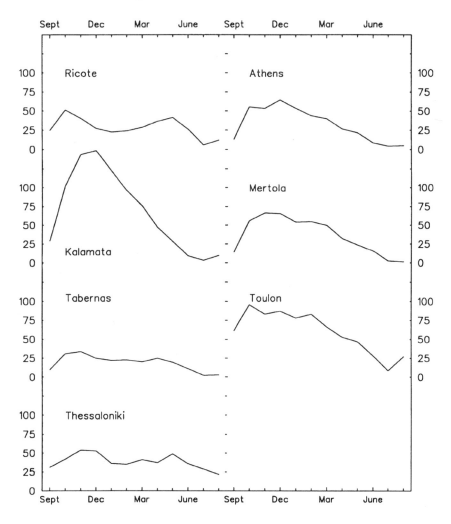

FIGURE 4.5 Mean monthly precipitation (mm) at seven stations in the northern Mediterranean, calculated from data for the hydrological years 1961–84

temperature regime between stations, each value is standardized (by dividing by the long-term standard deviation for the site) prior to incorporation in the index.

The analysis was restricted to the area 10°W to 30°E by 25°N to 45°N, which contains 80 stations. Data were available for the period 1961 to 1985. The results for the annual (calendar year) mean temperature are shown in the upper part of Figure 4.6. In the early part of the record there is a clear cooling trend, culminating in a period of persistently below-average temperatures between 1971 and 1976. Of the nine years since 1976, six have been above average, and three below. The time series for spring, summer and autumn are similar, and are not reproduced here. However, winter (shown in the lower part of Figure 4.6) fails to demonstrate the cooling trend of the early years.

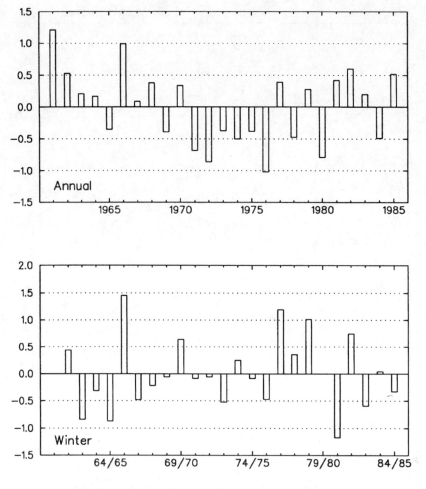

FIGURE 4.6   Standardized anomaly indices of Mediterranean annual mean temperature (upper
graph) and winter mean temperature (lower graph)

It is important to note that these time series only show the pattern of temperature
change relative to the 1961–85 mean. Longer-term studies do, however, confirm the
existence of a cooling trend terminating in the mid- to late 1970s. Metaxas et al. (1991)
used sea surface temperatures to examine trends since 1873. They found minima around
1910 and again around 1975–80, with maxima in about 1940 and 1965. Giles and
Balafoutis (1990), working with 80 years of data from three Greek stations, found a
cooling trend in the combined series between 1925 and 1975.

4.2.3 Case study of the Alentejo Region temperature

A detailed study of temperature trends has been made for the Alentejo region of Portugal,
using records from 15 stations between 1940 and 1992 to calculate an *SAI*. All 15 stations

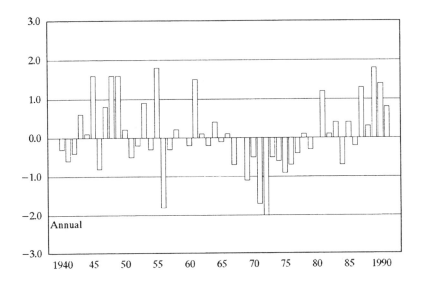

FIGURE 4.7 Standardized anomaly index of annual mean temperature for the Alentejo region of Portugal

were available from 1960 to the end of the record. Although only three stations were present in 1940, the number had increased to eight by 1941.

The annual and seasonal *SAI*s are shown in Figures 4.7 and 4.8 respectively. The annual *SAI* demonstrates a cooling trend from around 1960, culminating in 1972, and followed by a gradual warming to 1980. These trends are in agreement with those for the Mediterranean region as a whole (Figure 4.6) and are, if anything, more obvious. Although the seasonal series show less clear trends, all have a relatively cool period in the early and mid-1970s. The winter time series, unlike that for the whole Mediterranean region, does show cooling at this time.

4.2.4 Precipitation time series

Again, we first examine region-wide trends, for the whole Mediterranean. This is followed by a case study of variations over the Alentejo area of Portugal.

Annual and seasonal precipitation *SAI*s were calculated according to Equation 1 for the same area used for temperature. This region contains 100 precipitation stations. A hydrological year was used, running from September to August inclusive (summer being generally dry throughout the Mediterranean, with the autumn rains beginning in September). The precipitation *SAI*s, therefore, begin with the hydrological year September 1961 to August 1962 (termed 1961/62 in the following discussion), and end with the hydrological year September 1984 to August 1985.

The annual *SAI* is shown in Figure 4.9. Two 5-year periods of below-average precipitation occur near the beginning (1963/64 to 1967/68) and at the end of the record (1980/81 to 1984/85). The greatest rainfall deficits occur during the most recent period. Most of the intervening years have above-average rainfall (7 out of 12 years) or only just

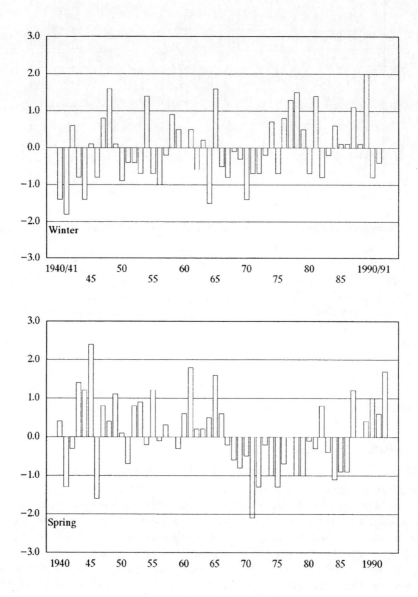

FIGURE 4.8   *(Caption opposite)*

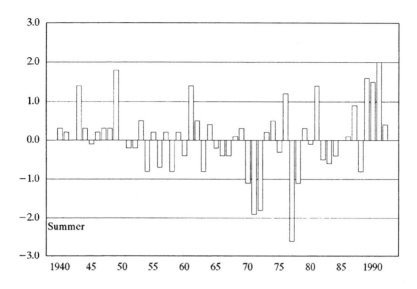

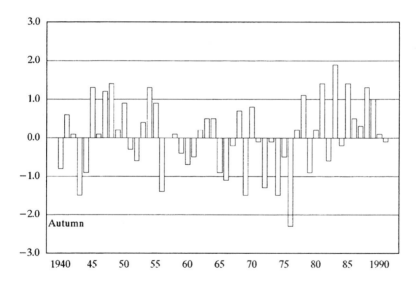

FIGURE 4.8   Standardized anomaly indices of seasonal mean temperature for the Alentejo region of Portugal

below average rainfall (3 out of 12 years). Only two of the intervening years (1970/71 and 1974/75) are relatively dry. None of the seasonal *SAI*s mirror the annual pattern exactly. A similar feature was noted by Maheras (1988) in a study of precipitation over the western Mediterranean in the last century. Thus, the annual trends must be produced by the interaction of deficits and surpluses in different seasons. During the period 1963/64 to 1967/68, which is relatively dry in the annual *SAI*, only spring rainfall is persistently below

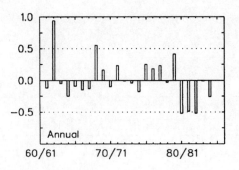

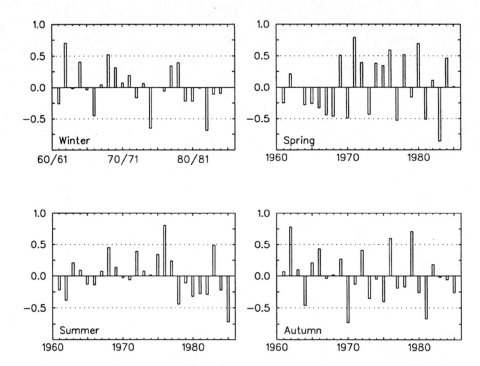

FIGURE 4.9   Standardized anomaly indices of Mediterranean annual (hydrological year) and
seasonal precipitation

average. In contrast, during the dry period 1980/81 to 1984/5, spring rainfall is below
average in only 2 years whereas winter precipitation is low throughout. In 1962/63, which
has the largest positive *SAI* of the annual series, autumn and winter were relatively wet,
while summer was relatively dry.

There are no obvious trends in the seasonal records, particularly that for autumn. The
winter series shows some evidence of a gradual decrease in precipitation over the period of
record. The variability in spring appears to increase towards the end of the record, and
there is a tendency for more dry seasons to occur in the first few years (with five
continuous below-average years from 1964 to 1968) followed by a relatively wet spell, with

seven above-average years in the decade 1969–78. Summer rainfall contributes little to the overall annual total, but there is a clear tendency for the later part of the record to be drier than the earlier years.

4.2.5 Case study of the Alentejo Region precipitation

Precipitation records from 51 stations in the Alentejo region of Portugal were used to calculate annual and seasonal *SAI*s over the period 1931 to 1992. At least 25 stations were present from 1941 onwards.

The annual and seasonal time series are shown in Figures 4.10 and 4.11 respectively. The most conspicuous feature is the clear decline in rainfall totals in the spring season, starting in the early 1960s. In the final 10 years, there were only three above-average years in this season. This decline is not reflected in the annual *SAI* (nor in the all-Mediterranean *SAI*s of Figure 4.9). The principal feature of Figure 4.10 is the dry period between 1964/65 and 1975/76, although there were three above-average years within this period: 1965/66, 1968/69 and 1969/70. This dry period overlaps with the dry period 1963/64 to 1967/68 seen in the annual Mediterranean *SAI* (Figure 4.9), but in the Alentejo the dry period starts later and lasts longer.

Table 4.1 shows the proportion of the annual rainfall occurring in each season in the periods 1931–60 and 1961–90. A decline of 7% has occurred in the spring season, while the proportions in autumn and winter have risen, by 3% in both cases. There is an increased tendency in the recent period for rainfall to concentrate into a 6-month period extending from September to February.

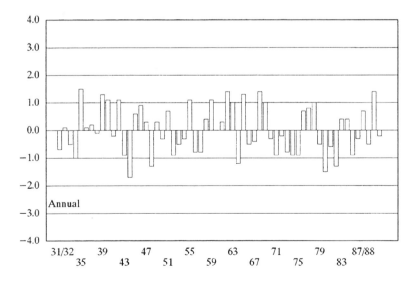

FIGURE 4.10 Standardized anomaly index of annual (hydrological year) precipitation for the Alentejo region of Portugal

TABLE 4.1 Percentage of annual precipitation occurring in each
season in the Alentejo region of portugal

	1931–60	1961–90	% Change
Autumn	26	29	+3
Winter	39	42	+3
Spring	31	24	−7
Summer	4	5	+1

(a)

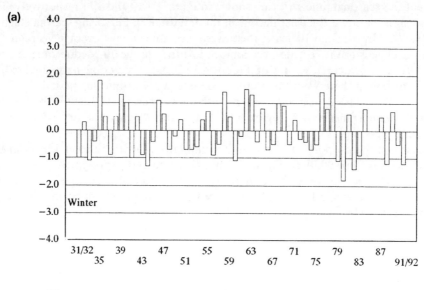

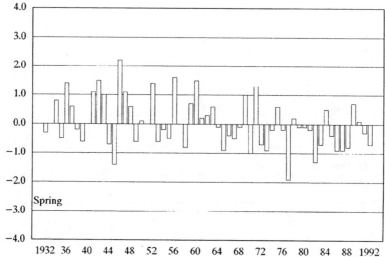

FIGURE 4.11   (*Caption opposite*)

4.3 TRENDS IN 500 HPA GEOPOTENTIAL HEIGHT
OVER THE MEDITERRANEAN

Here, we present the results of an analysis of the height of the 500 hPa geopotential surface over the Mediterranean. First, we propose the presence of an oscillation analogous to the well-known Southern Oscillation (Philander, 1990). We then suggest that, superimposed on this oscillation, there are long-term trends in 500 hPa height, increasing in the western Mediterranean and decreasing in the eastern Mediterranean. Finally, these trends are

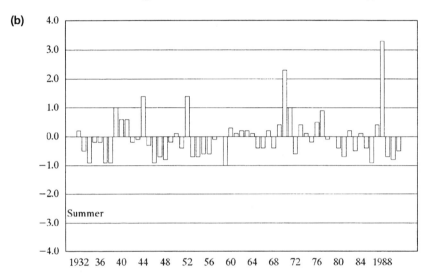

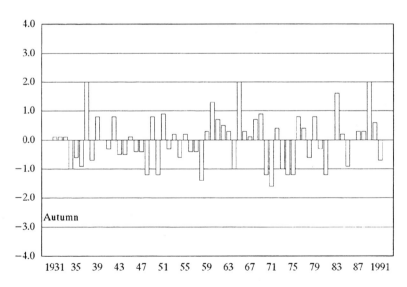

FIGURE 4.11   Standardized anomaly indices of seasonal precipitation for the Alentejo region of Portugal

related to precipitation amounts, and possible climatic effects of an extrapolation in the height trends are investigated.

4.3.1 The Mediterranean Oscillation

The existence of oscillations in the atmospheric pressure distribution over ocean basins, in particular the tropical Pacific Ocean, is well known (see, for example, Philander, 1990). The term Southern Oscillation is used to describe the sea-level pressure difference between Tahiti and Darwin. The North Atlantic Oscillation (van Loon and Rogers, 1978) is expressed by the difference in pressure between the Azores and Iceland. Characteristic weather conditions are associated with the different phases of these oscillations. For example, the 'normal' condition in the Tropical Pacific is characterized by relatively high pressure at Tahiti and relatively low pressure at Darwin, giving positive values of the Southern Oscillation Index (SOI). This is associated with an east-to-west near-surface circulation, wet conditions over the western Pacific islands and dry conditions over Peru. However, during those episodes when the SOI is negative (El Niño episodes), drought is experienced in the western Pacific and storms and floods over Peru.

An attempt was made to discover whether a similar phenomenon can exist over a smaller marine basin, in this case the Mediterranean Sea. Atmospheric pressure data over the period 1946–89 from two stations, Algiers in the extreme western Mediterranean and Cairo in the extreme eastern Mediterranean, were examined. The height of the 500 hPa

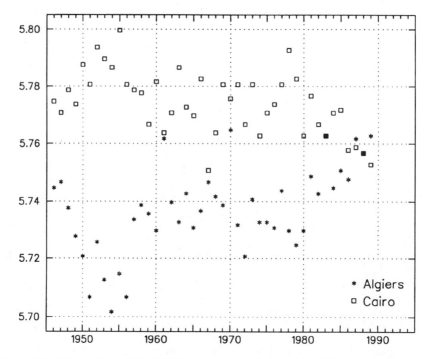

FIGURE 4.12   Time series of the 500 hPa geopotential height (m × 10³) at Algiers and Cairo

surface, taken from radiosonde observations, was used as an index of pressure. Almost half the entire atmosphere is contained below the 500 hPa height. It is therefore a good indicator of the state of the whole atmospheric column, and is not generally affected by local conditions. Generally speaking, the higher the 500 hPa surface, the higher the sea-level pressure. Missing data were patched with a 200 km grid-pass model.

Figure 4.12 shows the time series of the annual mean 500 hPa height at the two stations (44 years). Spectral analysis of these data showed that the greatest density lies at a frequency f of 0.045 cycles yr^{-1} (Figure 4.13), which implies a period of approximately 22 years. Indeed, visual inspection of Figure 4.12 suggests the existence of three peaks and two troughs in the Algiers time series, which has the greater spectral density at $f = 0.045$ cycles yr^{-1}. It is not possible to pick out a similar periodicity in the Cairo time series. The possible existence of a 22-year periodicity in the data raises the possibility of a link to sunspot variations (see Wigley, 1988, and Stuiver and Braziunas, 1992, for reviews of possible sunspot–climate links). A paper by Thomas (1993) attempts to link variations in Rome rainfall to sunspot numbers. However, it should be borne in mind that many climatologists remain sceptical about the existence of a possible link.

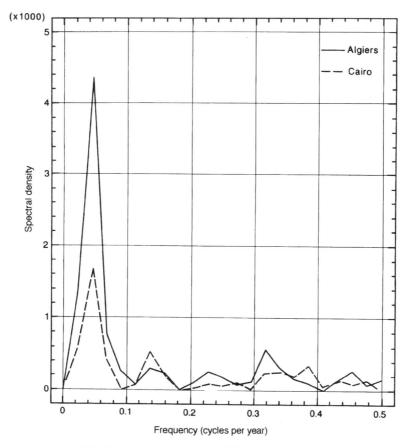

FIGURE 4.13   Periodogram of the 500 hPa height at Algiers and Cairo

Over 44 years, a periodicity of 22 years can be represented by an eighth-order polynomial. In consequence, the data were smoothed using a polynomial of this order, as shown in Figure 4.14. The interesting feature of the figure is that the oscillations at the two stations are in perfect phase opposition. This suggests the possible existence of some kind of 'flip–flop' in the dynamic field over the Mediterranean, analogous to the Southern Oscillation, and which may be termed the Mediterranean Oscillation.

The correlation coefficients between the 500 hPa height at Algiers, and at 13 other stations throughout the Mediterranean Basin, were found. The interpolated correlation surface is shown in Figure 4.15. There is a smooth decline in the correlation from Algiers eastward across the Mediterranean, culminating in an inverse correlation between the 500 hPa height at Algiers and Cairo of −0.63.

The presence of a Mediterranean pressure oscillation can be explained in part by the length of the basin. The distance from west to east is about 3000 km, approximately half

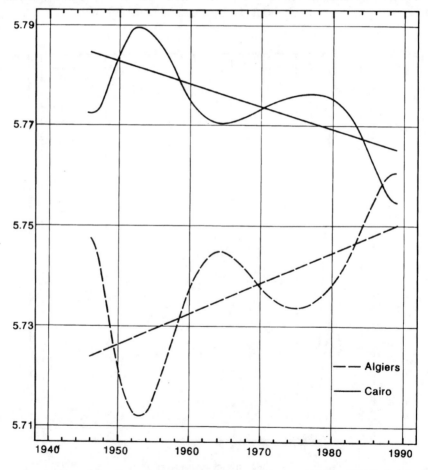

FIGURE 4.14   Eighth-order polynomial smoothing of the Algiers and Cairo 500 hPa height (m × 10³)
time series, with the fitted linear trends

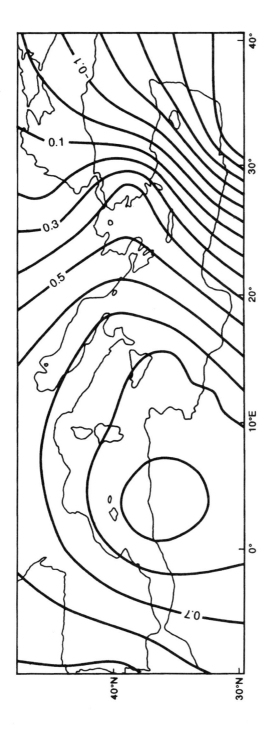

FIGURE 4.15 Correlation surface interpolated from the correlation coefficient between the 500 hPa height at Algiers and thirteen upper air stations in the Mediterranean

the long atmospheric wavelength. Long atmospheric waves are sometimes enhanced over southern Europe, principally because of the presence of the Azores anticyclone, but also because of the thermal discontinuity between continent and ocean and complex orographic effects.

4.3.2 Long-term trends in 500 hPa geopotential height

The eight-term polynomial filtering in Figure 4.14 shows clearly that the 22-year oscillations are imposed on a long-term positive trend at Algiers, and a long-term negative trend at Cairo. The linear trends were computed for both stations, and are shown in Figure 4.14. The change in the height of the 500 hPa surface is, at Algiers, about $0.6 \, \text{m yr}^{-1}$ and, at Cairo, about $-0.4 \, \text{m yr}^{-1}$. The slope of the regression equation (height against time) was calculated for 12 other stations in the Mediterranean Basin and the results contoured, as shown in Figure 4.16. The maximum positive value (indicating an increase in height) is located in the central–western part of the Mediterranean, around Sardinia. The change from positive to negative gradient occurs to the east of Crete, from where there is a gradual increase in the size of the negative gradient eastwards.

A possible interpretation of the positive trend of the 500 hPa height in the central–western basin is that high-pressure patterns, despite continuous invasions and retreats, are gradually increasing their presence. This effect determines (in agreement with the general dynamic theory of long atmospheric waves) an opposite effect in the eastern basin. Should the present trend continue in the central–western basin, a 1% increase in the height of the 500 hPa surface in the central–western Mediterranean will be reached in approximately 2030.

4.3.3 500 hPa height–precipitation relationships

It was argued at the end of the previous section that the positive trend in the 500 hPa geopotential height (2500) in the central–western basin of the Mediterranean indicates an increase in the frequency and/or intensity of anticyclonic 'blocking' episodes. In anticyclones, precipitation mechanisms are usually suppressed while, over the Mediterranean, temperatures are enhanced. Thus we would expect some relationship to exist between the 22-year cycles in 2500 on the one hand and temperature and precipitation on the other. Maxima in the 2500 record should be associated with relatively higher temperatures and lower precipitation, and vice versa.

As a preliminary investigation, we examined the situation in the Western Basin of the Mediterranean. As noted from Figure 4.16, the maximum positive trend in 2500 height is located in the region of Sardinia. We therefore used monthly mean 2500 data from Cagliari to examine pressure/temperature and pressure/precipitation relationships. The temperature and precipitation data from the WMO CLIMAT network for Italy were used to calculate an across-station average value at each monthly time step (which may be seen as a national average for Italy). Then, for each of these three variables, monthly departures from the 1951–80 mean were calculated. The correlation coefficient between the monthly 2500 and temperature anomalies was found to be $+0.60$. The correlation between 2500 and precipitation is -0.58. Both coefficients are significant at the 95% level. Although this analysis does not take into account the effects of autocorrelation on the

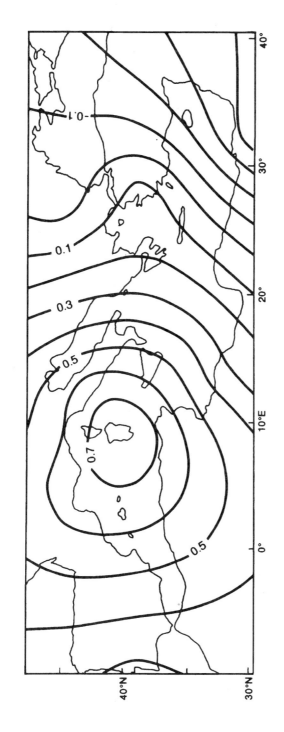

FIGURE 4.16 Slope of the regression equation (500 hPa height against time) interpolated from data at fourteen upper air stations in the Mediterranean (m yr^{-1})

significance level, the strength of the relationship demonstrates that 2500 fluctuations are directly related to temperature and inversely related to precipitation.

In order to test the relationship between 2500 and precipitation further, we extended the network of precipitation stations to cover the whole of the Western Basin. Data were obtained from 58 stations, and the annual totals were used to calculate an annual *SAI*, according to Equation 1, for the period 1951–89. The stations were divided into three geographical belts, as follows:

Southern Belt   (south of 38°N)   15 stations
Central Belt   (38N–42°N)   21 stations
Northern Belt   (42N–46°N)   23 stations

This approach was chosen on the basis that it would allow identification of trends which are not basin-wide. In the event, all three belts showed linear trends with time from initial values above the mean to final values below the mean (Figure 4.17). This suggests a decline in precipitation throughout the Western Basin. The rate of decrease appears to have been slightly higher in the Southern Belt (-0.027 yr^{-1}) than in the Central (-0.020 yr^{-1}) and Northern (-0.023 yr^{-1}) Belts.

The annual *SAI* of the 2500 was also calculated, using data from eight radiosonde stations in the Western Basin (Lisbon, Milan, Algiers, Rome, Nîmes, Brindisi, Cagliari and Tripoli). The correlation coefficients between the *SAI* of the 2500 and the three precipitation *SAI*s are: Southern Belt -0.52, Central Belt -0.56, and Northern Belt -0.52. All three coefficients are statistically significant at the 95% level although, again, this significance level does not take into account the effects of autocorrelation.

4.3.4 Future climatic trends

In Section 4.3.1 it was shown that oscillations, with an approximate 22-year periodicity, exist in the height of the 500 hPa surface over the Mediterranean region. The phase of these oscillations in the Eastern Basin is in exact opposition to the phase in the Western Basin. Superimposed on these oscillations, the time series of the 500 hPa height in the Western Basin shows a gradual increase in height with time, whereas in the Eastern Basin there is a gradual decrease. For the Western Basin it was possible to show that both temperature and precipitation are correlated (inversely in the case of precipitation) with the height of the 500 hPa surface, although the correlation coefficients are only in the range 0.5–0.6. Physically plausible reasons were presented to explain these relationships.

These results can be used to construct hypotheses about trends in precipitation in the short-term future. Winter's model (Granger and Engle, 1987) was used to extrapolate future trends in the 500 hPa surface. This model is based on three processes of exponential smoothing, and estimates the mean level, the trend and the 'seasonality' (i.e. the periodicity) of the data-set. In summary, it is a seasonally adjusted linear extrapolation based on the most recent values of the level, trend and seasonal indices. The equations are:

$$I[t] = by[t]/S[t] + (1-b)I[t-L] \tag{2}$$

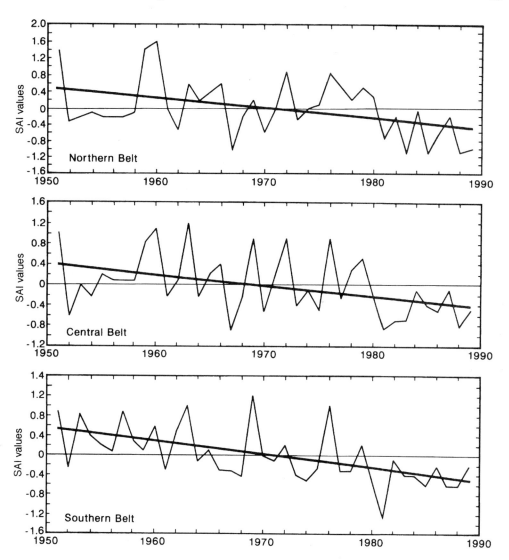

FIGURE 4.17   Standardized anomaly indices of annual precipitation for the southern, central and
northern belts of the western Mediterranean, 1951–89

$$S[t] = ay[t]/I[t - L] + (1 - a)(S[t - 1] + T[t - 1])$$ (3)

$$T[t] = c(S[t] - S[t - 1]) + (1 - c)T[t - 1]$$ (4)

where $y[t]$ is the observation at time t, L is the period of seasonality, $I[t]$ is the seasonal
index, $S[t]$ is the smoothed level, $T[t]$ is the smoothed trend, and a, b and c are smoothing
constants. The constants a, b and c are set by minimizing mean squared error over the
fitting set. The forecast for time $t + m$, from data available at time t, is given by:

$$S[t] + mT[t]I[t + m - L] \tag{5}$$

if m is less than or equal to L, and:

$$S[t] + mT[t]I[t + m - L(\text{int}(m + L)/L)] \tag{6}$$

if m is greater than L. In the case presented here, L is set equal to 11.

In the first instance, Winter's model was run on the 500 hPa height data for Algiers and Cairo up to the year 2025. The resulting time series are shown in Figure 4.18. The solid lines in this figure show the smoothed series. It can be seen that the good phase opposition between the behaviour of the 500 hPa surface over Algiers and Cairo, observed in the period 1946–89, is maintained in the prediction. When a quadratic trend is fitted to the complete data-set (1946–89 observed, 1990–2025 predicted), interesting departures from a simple linear extrapolation of the observations are apparent in the later years. In 2025, the height of the 500 hPa surface is shown by the quadratic to be higher at Algiers and lower at Cairo than indicated by a simple linear extrapolation, suggesting an acceleration of current trends.

Winter's model was applied to the data from the other 12 stations in the Mediterranean Basin for which upper air observations were available. In Figure 4.19 the prediction of the height of the 500 hPa surface for the year 2020, interpolated between the 14 data-points, is shown. A progressive increase in height is indicated for the central and western part of the Mediterranean Basin, and a decrease in the eastern part. In the central and eastern areas, this change is around $0.8\% \pm 1.0\%$. If a change of this magnitude were to occur, it would constitute an important climatic change, with large influences on the near-surface climatology of the region.

In Section 4.3.3, the height of the 500 hPa surface was shown to be inversely correlated with precipitation in the central–western Mediterranean. We cannot be certain, of course, that this relationship will be maintained in its present form up to 2025, when the 500 hPa surface in the central–western Mediterranean is predicted by Winter's model to be higher than anything recorded in the period 1946 to 1989. However, if the relationship were maintained, then we would expect precipitation in the central–western basin to decrease.

It must be stressed that this analysis is based on a statistical model, Winter's, and on the statistical relationship between precipitation and the height of the 500 hPa surface. Although we can rationalize the results in terms of the physics of the atmosphere, the basis of the analysis remains purely statistical. Furthermore, it relies on the statistical relationships observed at present being preserved in the future. This must reduce our confidence in the results. In the next section, we examine ways of developing scenarios of future climate change, based on physical models of the global climate system.

4.4 FUTURE CLIMATE SCENARIOS

In Section 4.3, an underlying assumption of the analysis, although never explicitly stated, was that the observed trend in the 500 hPa height might be due to the enhanced greenhouse effect. We have no way to determine whether or not there is a causative relationship. A statistical model was used to examine possible future trends in climate over

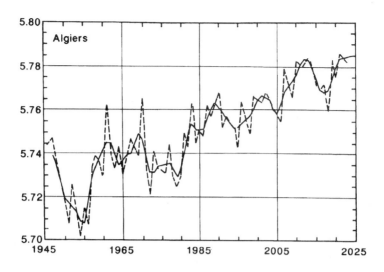

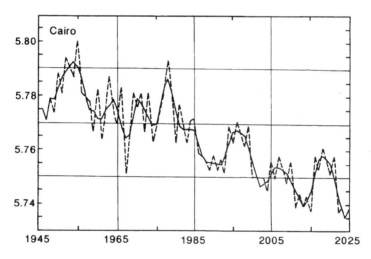

FIGURE 4.18 Extrapolation of the 500 hPa height (m×10³) at Algiers and Cairo to 2025 using Winter's model. The solid line shows the smoothed prediction

the Mediterranean Basin. It was stressed that, because Winter's model is statistical, it does not incorporate the physical attributes of the atmospheric circulation system. It assumes that trends and periodicities which exist in the historical and present-day data-set will continue in the future. Yet we have no reason to suppose that the atmosphere will continue to operate in this essentially linear manner. And the further we extrapolate the model into the future, the more unlikely this assumption is to be correct.

In order to examine the effects on climate of a particular perturbation, such as the addition of greenhouse gases to the atmosphere, we need a physical model of atmospheric

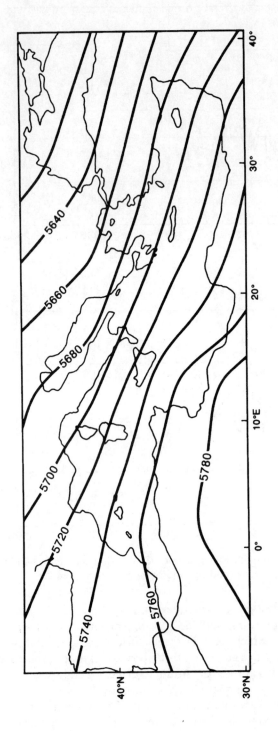

FIGURE 4.19   The height of the 500 hPa surface over the Mediterranean (m), predicted using Winter's model for the year 2020

behaviour. The most sophisticated, three-dimensional, climate models are the general circulation models (GCMs). In this section, we use the results from GCMs to create scenarios of a possible future climate over the Mediterranean Basin. The variables considered are temperature, precipitation and potential evapotranspiration.

4.4.1 Modelling future climate change with GCMs

GCMs are complex, three-dimensional, computer-based models of the atmospheric circulation which have been developed by climatologists from numerical meteorological forecasting models. Until recently, the standard approach has been to run the model with a nominal 'pre-industrial' atmospheric carbon dioxide (CO_2) concentration (the control run) and then to rerun the model with doubled (or sometimes quadrupled) CO_2 (the perturbed run). In both, the models are allowed to reach equilibrium before the results are recorded. This type of model application is therefore known as an equilibrium response prediction (see Cubasch and Cess, 1990, for a review of equilibrium GCM experiments). The great advantage of this approach is that it is parsimonious of computer time.

Many of the groups which have developed GCMs are currently engaged in time-dependent (or transient response) experiments, where the CO_2 concentration increases gradually through the perturbed run and where the oceans are modelled using ocean GCMs. The results are beginning to appear in the literature (see, for example, Cubasch et al., 1992). However, the output is, as yet, not widely available for the development of regional scenarios and impact analyses, and the scenarios presented here are based on equilibrium experiments. The consequences of modelling the general circulation in this way are discussed further in Section 4.4.6.

The four GCM experiments used in this chapter are from the following research institutions: the UK Meteorological Office model (abbreviated here to UKMO; the model version used here is as described by Wilson and Mitchell (1987a)); the Goddard Institute of Space Studies model (GISS; Hansen et al., 1984); the Geophysical Fluid Dynamics Laboratory model (GFDL; Wetherald and Manabe, 1986); and the Oregon State University model (OSU; Schlesinger and Zhao, 1989). The models vary in the way in which they handle the physical equations describing atmospheric behaviour. The UKMO, GISS and OSU GCMs solve these in grid-point form whereas the GFDL model uses a spectral method. All models have a realistic land/ocean distribution and orography (within the constraints of model resolution), all have predicted sea ice and snow, and clouds are calculated in each atmospheric layer in all models.

4.4.2 The application of GCMs to regional scenario construction

A number of authors have used GCMs to examine climate changes over the Mediterranean Basin due to the enhanced greenhouse effect. In a relatively early study, Wilson and Mitchell (1987b) used the UK Meteorological Office five-layer model, with a horizontal resolution of about 330 km, to look at climate changes over western Europe associated with $4 \times CO_2$. Temperature increases over the Mediterranean were in the range 3–6°C in winter, and in the range 4–9°C in summer. Precipitation is shown to decrease over the Mediterranean Basin in both winter and summer, except over Turkey, where a slight increase is indicated in the winter season.

One problem with GCMs lies in the coarse spatial resolution of the grid-point output: several hundreds of kilometres for this generation of GCMs at these latitudes. This limits their usefulness where regional climate change is of concern. The three studies described below, which are relevant to the Mediterranean region, have overcome this problem in one of three ways: by increasing the spatial resolution of the GCM, by nesting a limited area model inside a parent GCM, or by using a statistical 'downscaling' approach.

IPCC90 made a special study of southern Europe and Turkey (35–50°N by 10W–45°E). This was based on equilibrium-mode experiments, using three GCMs with a high spatial resolution (Mitchell et al., 1990). The results were used to predict climate changes over the region as a whole, due to the global warming expected by 2030. They indicate a warming of about 2°C in winter and 2–3°C in summer. Whereas only a small increase in precipitation is suggested for the winter season, the decrease in summer is indicated to be between 5 and 15%. Summer soil moisture was shown by the models to decrease by 15–25%.

Giorgi et al. (1992) used the technique of nesting a limited area model (LAM) inside a GCM to improve the spatial resolution, and examined the climatic changes induced by $2 \times CO_2$ over Europe. At the local and regional level, they found statistically significant differences in the temperature and precipitation changes indicated by the LAM predictions of climate change and those of the driving GCM. Despite this, the overall change in precipitation, when averaged over all land grid-points, is remarkably similar in the LAM and in the parent GCM: $+11.5$ and $+11.9\%$ of control run precipitation respectively. The warming predicted by the LAM is between 1.5 and 7°C. The change in regional precipitation amounts varied between -20% of present-day values over the Alpine region in October and $+177\%$ over the western Mediterranean in July. The widespread reductions in precipitation found in October are related to a substantial weakening of the jetstream.

Von Storch et al. (1993) use a statistical model to move from the relatively coarse resolution of GCM output to the much finer resolution required for a satisfactory regional scenario of climate change, in this case for the Iberian Peninsula (a technique known as downscaling). They use the canonical correlation technique to predict winter rainfall over the Iberian Peninsula from large-scale North Atlantic sea-level pressure. When the method is applied to GCM output from an equilibrium experiment, the statistical model indicates a slight increase in Iberian rainfall in a $2 \times CO_2$ world, whereas the four GCM grid-points representing the Iberian Peninsula show a strong decrease in rainfall. However, with a GCM transient experiment based on IPCC90 Scenario A (Houghton et al., 1990) the results from the statistical model are much closer to the GCM grid-point output for rainfall, both showing a decrease in rainfall: 7 mm per month for the statistical model, and 9 mm per month from the GCM output, after 100 years.

4.4.3 Scenario construction for temperature and precipitation

In the scenarios presented here, we have used a statistical approach to overcome the problem of coarse spatial resolution in GCMs. The method assumes that the grid-point output from GCMs can be seen as measures of regionally averaged climate, rather than values specific to that point in space. Using the records from a network of observing stations, multiple regression equations are calculated in which the independent variables are regionally averaged values of a range of climate parameters, expressed as anomalies from the long-term mean. The dependent variable is station temperature, for the construction of temperature scenarios, and station precipitation for the construction of precipitation scenarios. Then, appropriate values taken from GCM output, and expressed as the difference, $2 \times CO_2$ minus $1 \times CO_2$, are substituted in the regression equations to obtain the station-specific perturbation due to $2 \times CO_2$. This procedure is repeated for the whole network of stations, and the results plotted and contoured to produce the final scenario for the Mediterranean region. The method is described in detail by Palutikof and Wigley (1996) and is only summarized here. However, a number of points require discussion in more detail.

First, the GCM data are a composite, arrived at by consolidating and standardizing the results from the four individual GCMs. This follows the approach of Santer et al. (1990), and is based on the finding that no single model can be identified as being consistently better than the others at simulating current climate. However, the four models have different sensitivities to climate change (i.e. for the same perturbation, the global mean temperature change will be different). In order to avoid biases introduced by these differing sensitivities, it is necessary to standardize the model output.

For the temperature scenarios, at each grid-point, the temperature change T_i between the perturbed ($2 \times CO_2$) and control ($1 \times CO_2$) run for each model was first divided by the equilibrium (global annual) temperature change for that model i, prior to the calculation of the four-model average. This produces a 'standardized' model average temperature change per °C global change. If $\Delta \bar{T}^*$ is the standardized model average and $\Delta T_{eq(i)}$ is the equilibrium temperature change, this may be expressed as:

$$\Delta \bar{T}^* = \frac{1}{n} \sum_{i=1}^{n} (\Delta T_i / \Delta T_{eq(i)}) \tag{7}$$

For precipitation, it is not meaningful to express the model change between the perturbed and control run as a simple difference, because of the large differences in seasonal amounts. For each grid-point and each model, the $2 \times CO_2$ minus $1 \times CO_2$ precipitation was expressed as a percentage of the control run precipitation. Then, to allow for the different model sensitivities, the percentage change was divided by the equilibrium temperature increase for that model. Finally, the results from the four models were used to calculate a mean standardized precipitation change (p in units of % per °C global-mean warming) for that grid-point. Mathematically, this has the form:

$$p = (1/n) \sum_{i=1}^{n} 100[(P_i(2 \times CO_2) - P_i(1 \times CO_2))/P_i(1 \times CO_2)]/\Delta T_{eq(i)} \tag{8}$$

where P_i denotes the absolute precipitation.

Second, the scenarios were made site-specific by the use of methods described by Kim et al. (1984) and Wigley et al. (1990). Records for 248 temperature stations, and 328 precipitation stations, were collected, each having at least 20 years of record for 6 months out of the 12 over the period 1951–88. Regression equations were developed to predict station temperature anomalies (for the temperature scenarios) and station precipitation anomalies (for the precipitation scenarios) from regionally-averaged climate anomalies.

Of the five predictor variables tested which initially included point and gradient measures of sea level pressures, only temperature and precipitation were found to contribute usefully to the prediction of station values. After verification, the regression equations (one for each season at each station) were used to derive the detailed scenarios. This was done by inserting the appropriate interpolated GCM-derived grid-point changes into the regression equation for every station in the data-set. These procedures are fully described by Palutikof et al. (1992). It should be noted that not all the stations in the original data-sets were used in the final scenario construction. Some were discarded because they have too few near neighbours for the calculation of the regionally-averaged predictor anomalies. Others were rejected because the correlation coefficients between the predicted and observed station anomalies fell below the selected cut-off of 0.8 for temperature and 0.7 for precipitation. For these reasons, some areas remain blank in the scenario maps which follow.

We present in Figure 4.20 the seasonal scenarios of the change in temperature predicted for a 1°C global warming due to the enhanced greenhouse effect, taken from Palutikof et al. (1992). In winter, most of the region north of the Mediterranean Sea is predicted to experience warming greater than the global mean (i.e. greater than 1°C per degree global warming). Over land, the changes are smallest in the areas immediately adjacent to the Mediterranean Sea, and over the Mediterranean islands (generally less than 1°C per degree global warming). Extensive areas of warming greater than 1.2°C per degree global warming are found, particularly in the northeast of the region. Spring season changes are smaller, with most of the west and south of the region having values less than the global change. However, in much of the north and east, in a belt stretching from France around the northern Mediterranean to Jordan, changes are indicated to be greater than the global level of warming, particularly away from the coastal margin. Summer warming is greatest in the east and northeast, with a more limited area of warming over northwestern Africa and southern Spain and Portugal. In autumn the greatest warming (over 1.4°C per degree global warming) is seen in the west over northwest Africa, Spain and southwest France, and in the east over eastern Turkey.

Figure 4.21 shows the seasonal scenarios of the change in precipitation (as a percentage of the control run precipitation) for a 1°C change in global mean temperature (Palutikof et al., 1992). In winter, the areas of increased precipitation are mainly in the north of the study area, together with an area extending from Italy and Sardinia south into Tunisia and parts of Algeria. Lower precipitation over land areas is indicated mainly over the Middle East and North Africa. In spring the dividing line between higher (to the north) and lower rainfall runs along the north coast of the Mediterranean Sea, with only three spatially restricted areas of higher precipitation over Africa. In summer the patterns are not spatially coherent. Because of low correlations arising from very low (sometimes zero) rainfall in this season at the present day, no prediction could be attempted for large parts of Spain, Portugal, Turkey and North Africa. The scenario for autumn precipitation suggests a decline over the western Mediterranean and an increase over much of the central and eastern Mediterranean.

The patterns of precipitation change suggested by these scenarios are extremely complex. This is to be expected since present-day spatial patterns of precipitation in the

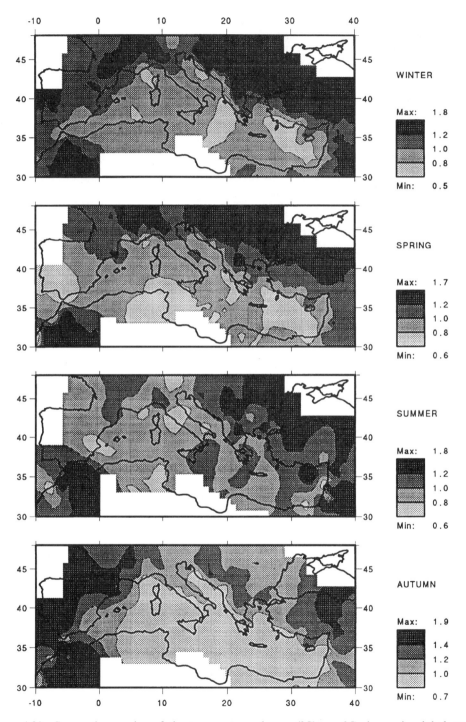

FIGURE 4.20 Seasonal scenarios of the temperature change (°C) per °C change in global mean temperature

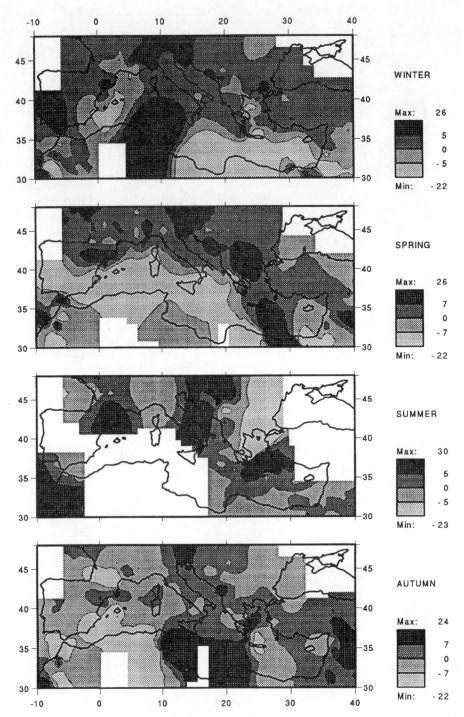

FIGURE 4.21   Seasonal scenarios of the precipitation change (%) per 1°C change in global mean temperature. Reproduced with permission of John Wiley & Sons, Ltd from Palutikof et al. (1994)

Mediterranean are highly variable, as a reflection of the varied topography and coastline. However, and in part as a result of this complexity, the reliability which we can place in these scenarios of precipitation is low. This uncertainty is discussed further in Section 4.4.6.

4.4.4 Scenario construction for potential evapotranspiration

Potential evapotranspiration is not properly assessed by GCMs, owing to their crude treatment of ground hydrology (Rind et al., 1990). However, we can place rather more confidence in GCM estimates of regional temperatures (Mitchell et al., 1990). Therefore, as a basis for the construction of scenarios of the change in potential evapotranspiration due to the enhanced greenhouse effect, we have used the temperature change scenarios described above.

This approach implies the need to establish a link between temperature and potential evapotranspiration. The principal meteorological controls of potential evapotranspiration are radiation, wind speed, vapour pressure and temperature. Penman (1948, 1953) developed sophisticated procedures to calculate potential evapotranspiration from these meteorological parameters. However, lack of data often precludes the application of these procedures, and a number of authors have devised simple methods for calculating potential evapotranspiration, based on temperature, day length and, in some cases, other readily available measures of climate (see, for example, Thornthwaite, 1948; Blaney and Criddle, 1950; Doorenbos and Pruitt, 1977).

In a recent paper, Palutikof et al. (1994) investigated the application of these simple methods in the Mediterranean Basin. Their performance was judged against potential evapotranspiration calculated using the Penman (1948) equation, modified according to Doorenbos and Pruitt (1977). It was found that, whereas the Thornthwaite method consistently underestimated potential evapotranspiration by a substantial margin, the Blaney and Criddle method, as modified by Doorenbos and Pruitt (1977), performed reasonably well. In this modified formula, the reference crop evapotranspiration in millimetres per day for the month under consideration, Et, is given by:

$$Et = c[p(0.46t + 8)] \tag{9}$$

where c is an adjustment factor which depends on minimum relative humidity, sunshine hours and daytime wind speed, p is the mean percentage of total annual daylight hours in that month, and t is the mean temperature.

At many Mediterranean sites, the information on relative humidity, sunshine and wind speed required for the calculation of c is not available. An alternative method to derive c was sought. It was found that regression equations, one for each standard season, in which t is the independent variable and c is the dependent variable, could be used to predict c from temperature data alone (Palutikof et al., 1994). In Figure 4.22 we show the relative performance of the method at two sites in Portugal. Here, the regression equations are being tested on a set of independent data (i.e. data not used to construct the equations).

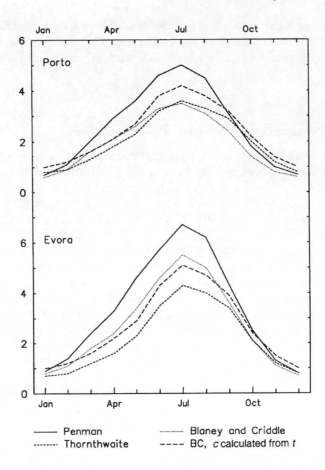

FIGURE 4.22   Comparison of monthly potential evapotranspiration estimates (mm day^{-1}) at sites in Portugal. 'BC, c calculated from t' is the method used here to produce scenarios of potential evapotranspiration (after Palutikof et al., 1994)

The regression-based Blaney and Criddle method was used to construct present-day seasonal maps of potential evapotranspiration. For each temperature record from the 248-station network, each having a record length of at least 20 years, the mean temperature in each month was calculated. These values were used to calculate the potential evapotranspiration (in mm day^{-1}), and an average seasonal figure was found. These station values were used to produce the interpolated maps shown in Figure 4.23.

All seasons show a north–south gradient in potential evapotranspiration across the region. The spatial variation in winter, the season of lowest evapotranspiration, is small, ranging between less than 1 mm day^{-1} and almost 3 mm day^{-1}. The gradient across the region is higher in the other three seasons. The most spatially complex pattern (and the greatest deviation from the straightforward north–south gradient) is found in summer, when the range is between less than 5 mm day^{-1} and over 7 mm day^{-1}. In spring and autumn, values range between under 2 mm day^{-1} and over 4 mm day^{-1}.

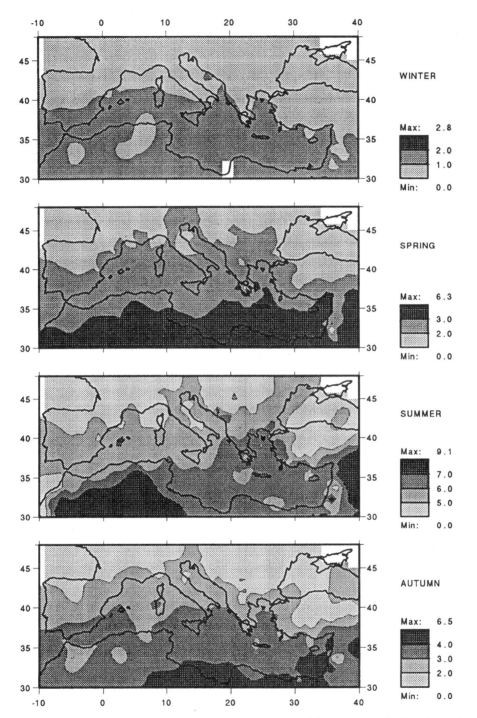

FIGURE 4.23 Seasonal maps of present-day estimated potential evapotranspiration (mm day^{-1}). Reproduced with permission of John Wiley & Sons, Ltd from Palutikof et al. (1994)

The task of constructing scenarios of the change in potential evapotranspiration from the temperature scenarios is made easier by the fact that a perturbed temperature value is available for each site in the station network. It is a relatively simple matter to substitute for temperature in the regression equation to obtain c, and then proceed to the calculation of the Blaney and Criddle potential evapotranspiration.

It is possible that, due to the warming in the perturbed case, the regression equations are being extensively applied outside the range of temperatures used in their construction. To investigate this, the number of cases where the independent variable (temperature) exceeds the maximum value used in the construction of the regression equations was determined. We found the number of occurrences to be relatively few. The figures are: winter, 7 cases; spring, 10 cases; summer, 27 cases; autumn, none.

Figure 4.24 shows the seasonal scenarios of the change in potential evapotranspiration (mm day^{-1}) indicated for a 1°C change in global mean temperature. The least change is seen in winter. This is also the season with the most straightforward spatial patterns: a north–south trend, with the areas of highest present-day potential evapotranspiration having the greatest indicated changes and vice versa. The largest change is found in summer. In this season the north–south trend is absent: lowest values are found in the central region of the Mediterranean Basin where, over Corsica, Sardinia, Sicily and southern Greece, the change is predicted to be less than 1.5 mm day^{-1}. The change in potential evapotranspiration increases towards the edges of the map, particularly towards North Africa and southern Spain in the west, and towards southern Turkey in the east, where the predicted change is greater than 2.4 mm day^{-1}. In spring and autumn, the indicated change lies between 0.8 and 2.0 mm day^{-1} (with the maximum value in autumn slightly lower than that in spring).

4.4.5 Timing of the changes

The scenarios of Sections 4.4.3 and 4.4.4 are presented as the regional change in a particular climate variable to be expected in response to a 1°C change in mean global temperature. As such, they do not provide any information on *when* such changes might be expected to occur. The results from four transient response GCMs presented in IPCC92 (Gates et al., 1992) show a constant rate of warming in the later decades of around 0.3°C per decade. This is in line with the findings of IPCC90, based on the 'business-as-usual' CO_2 forcing scenario and an energy balance atmospheric model coupled to an upwelling–diffusion ocean model (Bretherton et al., 1990). Although the impossibility of placing calendar dates on this figure must be emphasized, it suggests that a 1°C temperature change may be achieved in a period of around 30 years.

It should be noted that the figure of 0.3°C per decade does not take into account possible opposing anthropogenic influences, in particular the forcing from sulphate aerosols and stratospheric ozone depletion. Wigley and Raper (1992) made temperature projections based on IPCC92 emissions scenario IS92a (Leggett et al., 1992), taking into account the ozone-depletion feedback and best-guess sulphate aerosol effects. They used their upwelling–diffusion energy-balance climate model (as used in IPCC90, see above) and found the warming between 1990 and 2100 to be in the range 1.7–3.8°C.

The results from these time-dependent experiments can be combined with the scenarios of the magnitude of change presented in this chapter, and superimposed on a baseline

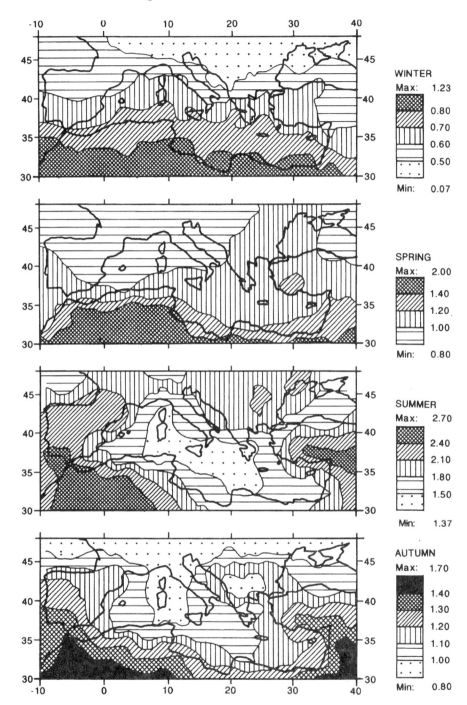

FIGURE 4.24   Seasonal scenarios of the absolute change in potential evapotranspiration (mm day^{-1}) per 1°C change in global mean temperature. Reproduced with permission of John Wiley & Sons, Ltd from Palutikof et al. (1994)

(present-day) climatology in order to arrive at a scenario of climate for a particular future time. An example of the application of this approach to the development of 'snapshot' scenarios for Europe is the ESCAPE project (Rotmans et al., 1994). This approach requires that the spatial pattern of the enhanced greenhouse signal remains constant with time, but the available model evidence suggests that this is a reasonable assumption to make (Mitchell et al., 1990; Gates et al., 1992).

4.4.6 The validity of climate scenarios

In Sections 4.4.3 and 4.4.4 we developed scenarios of the change in temperature, precipitation and potential evapotranspiration due to a 1°C global warming caused by the enhanced greenhouse effect. These are self-consistent estimates of possible future climate conditions in the Mediterranean Basin, since they are based on results from the same GCMs. The generation of GCMs used here, although powerful tools in the study of climate change, are not able to reproduce the present-day climate of the Mediterranean Basin exactly (Palutikof et al., 1992) which must therefore reduce our confidence in their predictions of future regional climates. This is particularly the case with scenarios of precipitation (Mitchell et al., 1990).

Furthermore, the model output used here is taken from equilibrium response predictions. For any given CO_2 level, the actual change will lag behind the corresponding equilibrium change for that CO_2 level. The large-scale patterns of change from transient response GCM experiments are similar to those obtained from comparable equilibrium experiments, scaled down by an appropriate factor (Gates et al., 1992). Differences do exist, largely because equilibrium model runs ignore important oceanic processes such as ocean current changes, differential thermal inertia effects between different parts of the oceans and between land and ocean, and changes in the oceanic thermohaline circulation. These differences are greatest in areas where the ocean thermal inertia is large, such as the North Atlantic and high southern latitudes (Mitchell et al., 1990). They are relatively small in most regions (and in the Mediterranean Basin in particular).

In the development of the scenarios presented here, we have introduced a further level of uncertainty by using regression models to link regionally averaged (or model grid-point) climate variables to station temperature and precipitation. For temperature, the relationship between the point anomalies and the regionally averaged predictor variables is generally strong: over 80% of the variance in point temperature is explained by the predictors over most of the region and most of the year. The temperature scenarios can therefore be regarded as a considerable improvement on those constructed directly from GCM grid-point output, because of the enhanced spatial resolution. However, for point precipitation the variance explained is more typically in the region of 60% or less (Palutikof et al., 1992). Therefore, confidence in the sub-grid-scale scenarios of precipitation must be much lower. The potential evapotranspiration scenarios are derived from the temperature scenarios and a modified form of the Blaney and Criddle equation, introducing a further source of error.

Despite all these reservations, the fact remains that these scenarios are a set of self-consistent estimates of a possible, and plausible, climate future. The construction techniques used are of general applicability, and the quality of regional scenarios will improve as more accurate GCM predictions become available in the future.

4.5 CONCLUSIONS

We have attempted in this chapter to present a comprehensive survey of present, and possible future, climates in the Mediterranean Basin. A large data-bank of monthly temperature and precipitation records, from sites throughout the region, was used to analyse the spatial patterns at the annual and seasonal level.

Time series analyses were performed on the data, with a particular aim of discovering whether trends exist. For annual temperatures over the period 1961–85, we found evidence of a clear cooling trend, culminating in the early to mid-1970s. This trend has been noted by other authors. For precipitation, two 5-year periods of below-average precipitation occur near the beginning and at the end of the record. The greatest rainfall deficits occur during the most recent period (1980/81 to 1984/85).

In a case study of the Alentejo region of Portugal over the period 1931–92, a clear tendency was noted for rainfall to become concentrated into a shorter period of the year: the proportion of annual precipitation in the autumn and winter seasons has been increasing at the expense of spring. No significant change in the annual total has occurred.

An analysis was made of the behaviour of an indicator of atmospheric pressure: the height of the 500 hPa surface. A comparison of upper air data from Algiers and Cairo suggested the existence of a Mediterranean Oscillation, analogous to the Southern Oscillation, with a periodicity of around 22 years. This is superimposed on a longer-term linear trend of $+0.6$ m yr^{-1} at Algiers, and -0.4 m yr^{-1} at Cairo. Winter's (statistical) model was used to extrapolate this trend up to 2025. A statistical relationship was shown to exist between precipitation and the height of the 500 hPa surface in the central–western Mediterranean. Because of this, and bearing in mind the uncertainties, it can be hypothesized that the increasing height of the 500 hPa surface indicated by Winter's model will lead to a reduction in precipitation in the central–western Mediterranean.

Seasonal scenarios were presented of the change in temperature, precipitation and potential evapotranspiration in response to a 1°C change in global mean temperature due to the enhanced greenhouse effect. The largest area of warming is seen in autumn, and the smallest in summer. Over the year as a whole, the warming is indicated to be greatest in the northern, particularly northeastern, parts of the study region. The calculation of potential evapotranspiration is based on a modified form of the Blaney and Criddle equation, and therefore the scenarios for this variable reflect the temperature change scenarios. Precipitation changes are complex, reflecting the complexity of present-day rainfall patterns. In summer, the changes lack spatial coherence. This is understandable given that, for much of the Mediterranean region, present-day rainfall is at or close to zero in the summer months, and the scenarios express the change as a percentage of the control-run precipitation. In winter and spring the scenarios indicate higher precipitation in the north of the study region, and lower precipitation in the south. In autumn the contrast is between the western Mediterranean (a decline in precipitation) and the central and western Mediterranean (an increase). In interpreting these scenarios, it is necessary to bear in mind the uncertainties involved.

By examining present patterns and possible future trends of climate over the Mediterranean region, it is possible to assess the changing risk of desertification. However, this risk will only be realized if the land-use practices of a region are inappropriate for the natural environment. Awareness of the possibility that global warming may cause regional climate changes which enhance the potential for desertification is the first step towards prevention.

4.6 ACKNOWLEDGEMENTS

In addition to the support provided by the European Community under the MEDALUS I project, which all contributors to this book enjoyed, we wish to acknowledge the support of the United Nations Environment Programme (UNEP) under their Mediterranean Action Plan, Contract No. FP/5101-88-01 (2862). Our thanks also go to the various National Meteorological Departments in the Mediterranean region who supplied instrumental data for the analyses.

4.7 REFERENCES

Amantadis, G.T., Housiadas, C. and Bartzis, J.G. 1991. Spatial distribution of rainfall in the Greater Athens area. *Meteorological Magazine*, **120**, 41–50.

Bartzokas, A. 1989. Annual variation of pressure over the Mediterranean area. *Theoretical and Applied Climatology*, **40**, 135–146.

Blaney, H.F. and Criddle, W.D. 1950. Determining water requirements in irrigated areas from climatological and irrigation data. *USDA (ARS) Technical Bulletin*, **1275**, 59 pp.

Bretherton, F.P., Bryan, K. and Woods, J.D. 1990. Time–dependent greenhouse-gas-induced climate change. In *Climate Change: The IPCC Scientific Assessment* (ed. J.T. Houghton, G.J. Jenkins and J.J. Ephraums), Cambridge University Press, Cambridge, 173–193.

Cantú, V. 1977. The climate of Italy. In *Climates of Central and Southern Europe, World Survey of Climatology 6* (ed. C.C. Wallén), Elsevier, Amsterdam, 127–173.

Cubasch, U. and Cess, R.D. 1990. Processes and modelling. In *Climate Change: the IPCC Scientific Assessment* (ed. J.T. Houghton, G.J. Jenkins and J.J. Ephraums), Cambridge University Press, Cambridge, 69–91.

Cubasch, U., Hasselmann, K., Höck, H., Maier-Reimer, E., Mikolajewicz, U., Santer, B.D. and Sausen, R. 1992. Time-dependent greenhouse warming computations with a coupled ocean–atmosphere model. *Climate Dynamics*, **8**, 55–69.

Doorenbos, J. and Pruitt, W.O. 1977. Guidelines for predicting crop water requirements. *FAO Irrigation and Drainage Paper*, **24**, 144 pp.

Furlan, D. 1977. The climate of southeast Europe. In *Climates of Central and Southern Europe, World Survey of Climatology Vol. 6* (ed. C.C. Wallén), Elsevier, Amsterdam, 185–223.

Garrida, J. and Garcia, J.A. 1992. Periodic signals in Spanish monthly precipitation data. *Theoretical and Applied Climatology*, **45**, 97–106.

Gates, W.L., Mitchell, J.F.B., Boer, G.J., Cubasch, U. and Meleshko, V.P. 1992. Climate modelling, climate prediction and model validation. In *Climate Change 1992. The Supplementary Report to the IPCC Scientific Assessment* (eds. J.T. Houghton, B.A. Callander and S.K. Varney), Cambridge University Press, 97–134.

Giles, B.D. and Balafoutis, C.J. 1990. The Greek heatwaves of 1987 and 1988. *International Journal of Climatology*, **10**, 505–517.

Giorgi, F., Marinucci, M.R. and Visconti, G. 1992. A $2 \times CO_2$ climate change scenario over Europe generated using a Limited Area Model nested in a General Circulation Model. 2. Climate change scenario. *Journal of Geophysical Research*, **97**, 10011–10028.

Granger, C.W.J. and Engle, R.F. 1987. Econometric forecasting: a brief survey of current and future techniques. In *Forecasting in the Social and Natural Sciences* (ed. K.C. Land and S.H. Schneider), Reidel, Dordrecht, 117–139.

Hansen, J.E., Lacis, A., Rind, D., Russell, L., Stone, P., Fung, I., Ruedy, R. and Lerner, J. 1984. Climate sensitivity: analysis of feedback mechanisms. In *Climate Processes and Climate Sensitivity* (eds J.E. Hansen and T. Takahashi), Geophysical Monograph 29, Maurice Ewing Vol. 5. American Geophysical Union, Washington, DC, 130–163.

Houghton, J.T., Jenkins, G.J. and Ephraums, J.J. (eds) 1990. *Climate Change: the IPCC Scientific Assessment*. Cambridge University Press, Cambridge, 364 pp.

Houghton, J.T., Callander, B.A. and Varney, S.K. (eds) 1992. *Climate Change 1992. The Supplementary Report to the IPCC Scientific Assessment.* Cambridge University Press, Cambridge, 200 pp.

Kim, J.W., Chang, J.T., Baker, N.L. and Gates, W.L. 1984. The statistical problem of climate inversion: determination of the relationship between local and large-scale climate. *Monthly Weather Review,* **112**, 2069–2077.

Köppen, W. 1936. Das geographische System der Klimate. In *Handbuch der Klimatologie 3* (ed. W. Köppen and R. Geiger), Gebrüder Borntraeger, Berlin, 46 pp.

Leggett, J., Pepper, W.J. and Swart, R.J. 1992. Emissions scenarios for IPCC: an update. In *Climate Change 1992. The Supplementary Report to the IPCC Scientific Assessment* (eds. J.T. Houghton, B.A. Callander and S.K. Varney), Cambridge University Press, Cambridge, 69–95.

Linés Escardó, A. 1970. The climate of the Iberian peninsula. In *Climates of Northern and Western Europe, World Survey of Climatology Vol. 5* (ed. C.C. Wallén), Elsevier, Amsterdam, 195–226.

Maheras, P. 1988. Changes in precipitation conditions in the western Mediterranean over the last century. *Journal of Climatology,* **8**, 179–189.

Metaxas, D.A., Bartzokas, A. and Vitsas, A. 1991. Temperature fluctuations in the Mediterranean area during the last 120 years. *International Journal of Climatology,* **11**, 897–908.

Mitchell, J.F.B., Manabe, S., Tokioka, T. and Meleshko, V. 1990. Equilibrium climate change—and its implications for the future. In *Climate Change: The IPCC Scientific Assessment.* (ed. J.T. Houghton, G.J. Jenkins and J.J. Ephraums), Cambridge University Press, Cambridge, 131–172.

Nicholson, S.E. 1983. Subsaharan rainfall and the years 1976–80: evidence of continued drought. *Monthly Weather Review,* **111**, 1646–1654.

Palutikof, J.P. and Wigley, T.M.L. 1996. Developing climate change scenarios for the Mediterranean region. In: *Climate Change and the Mediterranean* Volume 2 (ed. L. Jeftic, S. Kečkeš and J.C. Pernetta), Arnold, London, 27–56.

Palutikof, J.P., Guo, X., Wigley, T.M.L. and Gregory, J.M. 1992. *Regional Changes in Climate in the Mediterranean Basin due to Global Greenhouse Warming.* MAP Technical Reports Series No. 66, UNEP, Athens, 172 pp.

Palutikof, J.P., Goodess, C.M. and Guo, X. 1994. Seasonal scenarios of the change in potential evapotranspiration due to the enhanced greenhouse effect in the Mediterranean Basin. *International Journal of Climatology,* **14**, 853–869.

Penman, H.L. 1948: Natural evapotranspiration from open water, bare soil and grass. *Proceedings of the Royal Society of London,* **A193**, 120–145.

Penman, H.L. 1953. The physical bases of irrigation control. *Report of the 13th International Horticultural Congress,* **2**, 913–923.

Perry, A. 1981: Mediterranean climate— a synoptic reappraisal. *Progress in Physical Geography,* **5**, 107–113.

Philander, S.G. 1990. *El Niño, La Niña, and the Southern Oscillation.* International Geophysics Series Vol. 46, Academic Press, San Diego, 293 pp.

Rind, D., Goldberg, R., Hansen, J., Rosensweig, C. and Ruedy, R. 1990. Potential evapotranspiration and the likelihood of future drought. *Journal of Geophysical Research,* **95**(D7), 9983–10004.

Rotmans, J., Hulme, M. and Downing, T.E. 1994. Climate change implications for Europe: an application of the ESCAPE model. *Global Environmental Change,* **4**, 97–124.

Santer, B.D., Wigley, T.M.L., Schlesinger, M.E. and Mitchell, J.F.B. 1990. *Developing Climate Scenarios from Equilibrium GCM Results.* MPI Report No. 47, Max Planck Institut für Meteorologie, Hamburg, 23 pp.

Schlesinger, M.E. and Zhao, Z-C. 1989. Seasonal climate changes induced by doubled CO_2 as simulated by the OSU atmospheric GCM/mixed-layer ocean model. *Journal of Climate,* **2**, 459–495.

Stuiver, M. and Braziunas, T.F. 1992. Evidence of solar activity variations. In *Climate Since A.D. 1500* (ed. R.S. Bradley and P.D. Jones), Routledge, London, 593–605.

Thomas, R.G. 1993: Rome rainfall and sunspot numbers. *Journal of Atmospheric and Terrestrial Physics,* **55**, 155–164.

Thornthwaite, C.W. 1948. An approach towards a rational classification of climate. *Geographical Review*, **38**, 55–94.

van Loon, H. and Rogers, J. 1978. The seesaw in winter temperature between Greenland and northern Europe. Part 1: General description. *Monthly Weather Review*, **106**, 296–310.

Von Storch, H., Zorita, E. and Cubasch, U. 1993: Downscaling of global climate change estimates to regional scales: an application to Iberian rainfall in wintertime. *Journal of Climate*, **6**, 1161–1171.

Wetherald, R.T. and Manabe, S. 1986: An investigation of cloud cover change in response to thermal forcing. *Climatic Change*, **8**, 5–23.

Wigley, T.M.L. 1988. The climate of the past 10,000 years and the role of the sun. In *Secular Solar and Geomagnetic Variations in the Last 10,000 Years* (ed. F.R. Stephenson and A.W. Wolfendale), Kluwer Academic, Dordrecht, 209–224.

Wigley, T.M.L. and Farmer, G. 1982: Climate of the eastern Mediterranean and Near East. In *Palaeoclimates, Palaeoenvironments and Human Communities in the Eastern Mediterranean Region in Later Prehistory* (ed. J.L. Bintliff and W. van Zeist), BAR International Series 133, British Archeological Reports, Oxford, 3–37.

Wigley, T.M.L. and Raper, S.C.B. 1992. Implications for climate and sea level of revised IPCC emissions scenarios. *Nature*, **357**, 293–300.

Wigley, T.M.L., Jones, P.D. and Kelly, P.M. 1986. Empirical climate studies: warm world scenarios and the detection of climatic change induced by radiatively active gases. In *The Greenhouse Effect, Climatic Change and Ecosystems* (ed. B. Bolin, B.R. Doos, J. Jäger and R.A. Warrick), SCOPE 26, Wiley, New York, 271–322.

Wigley, T.M.L., Jones, P.D., Briffa, K.R. and Smith, G. 1990. Obtaining sub-grid-scale information from coarse-resolution General Circulation Model output. *Journal of Geophysical Research*, **95**(D2), 1943–1953.

Wilson, C.A. and Mitchell, J.F.B. 1987a. A doubled CO_2 climate sensitivity experiment with a global climate model including a simple ocean. *Journal of Geophysical Research*, **92**(D11), 13315–13343.

Wilson, C.A. and Mitchell, J.F.B. 1987b: Simulated climate and CO_2-induced climate change over western Europe. *Climatic Change*, **10**, 11–42.

5

The MEDALUS Core Field Programme: An Overview of Sites and Methodology

L. H. CAMMERAAT

Universiteit van Amsterdam, The Netherlands

5.1 INTRODUCTION

The general objectives of MEDALUS required the collection of data to increase the understanding of basic processes and causes of soil degradation and desertification in the semi-arid Mediterranean. To provide these data, a harmonized or core programme of standardized measurements was carried out at eight field sites along the north rim of the Mediterranean from the Alentejo, Portugal, in the west to Petralona, Greece, in the east.

The core programme covered the measurement of very many different biotic and abiotic parameters under contrasting site conditions. From the outset it was therefore necessary to develop a common programme of consistent measurements to ensure that the data could be used as a basis for modelling processes at the different locations and for making detailed inter-site comparisons. Very slight differences in methodology and procedure can cause great uncertainty when inter-site comparisons of data are made.

The field sites selected were representative of regions within the European Mediterranean for which very little harmonized information was available in earlier literature. A geo-referenced data-base has been developed containing not only the field data but also the MEDALUS hillslope model and remotely sensed images. This will provide a valuable source of information for those involved with desertification research. At the same time it is hoped that the data can serve as a benchmark reference point for monitoring future changes at the sites.

This chapter contains short comparative descriptions of the field sites and outlines the contents of the core field programme. It also gives an example of the detailed inter-site comparisons that can be made using the data.

5.2 LOCATION AND CLIMATE OF FIELD SITES

The locations of the field sites are shown in Figure 5.1. All sites are situated within 100 km of the Mediterranean or Atlantic coasts. The climatic conditions at the sites are different, because of the regional effects of land masses and mountains on atmospheric circulation patterns.

Mediterranean Desertification and Land Use. Edited by C. Jane Brandt and John B. Thornes.
© 1996 by John Wiley & Sons, Ltd.

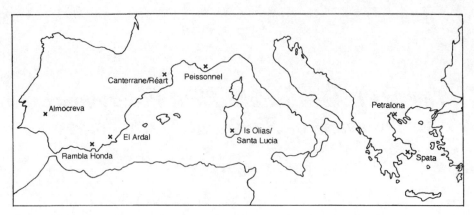

FIGURE 5.1   Map of the European Mediterranean showing the location of the core field sites

Annual precipitation across the Mediterranean is shown Figure 5.2 (from Palutikof et al., 1996) and data from the sites are given in Table 5.1. Precipitation ranges from approximately 220 mm yr^{-1} (SE Spain) to approximately 800 mm yr^{-1} for the Var region (SE France). Almost all areas are characterized by a dry period in summer, and they all are situated within the Mediterranean climatic zone (Köppen Csa climate), except for the

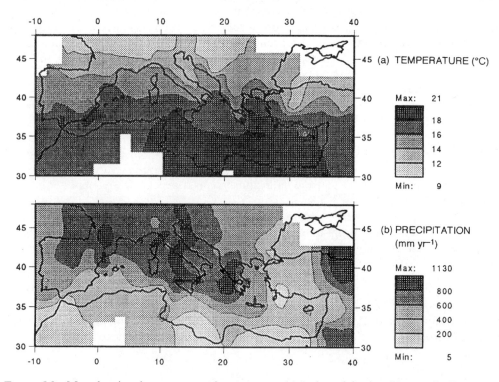

FIGURE 5.2   Map showing the mean annual temperature (a) and precipitation (b) over the European Mediterranean, 1961–85 (from Palutikof et al., this volume, chapter 4)

TABLE 5.1 An overview of some important climatic data of the field sites

Location	Nearest station	Precipitation (mm yr^{-1})	Mean annual temp. (°C)
Almocreva (P)	Beja	583	16.1
Rambla Honda (S)	Tabernas	218	15.4
El Ardal (S)	Emb. Cierva	276	17.1
Canterrane/Réart(F)	Perpignan (St. Colombe)[a]	582.6 (662)[a]	15.3
Peissonnel (F)	Le Luc–Le Cannet	840	14.5
Is Olias (I)	Cagliari	448	16.4
Spata (Gr)	Athens	390	17.8
Petralona (Gr)	Thessaloniki (Agios Mamas)[b]	464.2 (387)[b]	15.9 (16.9)[b]

[a]Nearest pluviometric station.
[b]Nearest station 1980–90.

Petralona site (N Greece) which has a wetter summer (Köppen Cfa climate). Furthermore the inter- and intra-annual variability in precipitation is very high. As a result of high summer temperatures and low summer precipitation, the potential evapotranspiration is generally high. Most of the core field sites have potential evapotranspiration figures over 1250 mm yr^{-1}, except for both the French sites which have a potential evapotranspiration between 1000 and 1250 mm yr^{-1}. Hence all field sites have a considerable water shortage not only in summer but also on a yearly basis and have 'non-leaching' soil water regimes. More detailed information on climate can be found in Palutikof et al. 1996 (this volume, Chapter 4).

5.3 CRITERIA FOR THE SELECTION OF CORE SITES

The Mediterranean offers a great variety of landscapes. In order to cover as many as possible it was decided that each field area should contain at least two types of land use that are dominant in that area: semi-natural forests and scrubland, known as matorral (Spain) maquis and garrigue (France) or macchia (Italy), abandoned agricultural land or agricultural land, with various types of culture, like cereals or (non-irrigated) fruits, for example. Each field site was located within a landscape type covering at least 5 ha. In addition the selection of the field sites was also influenced by the local circumstances such as land ownership and accessibility.

TABLE 5.2 An overview of some general characteristics of the MEDALUS field sites

Location	Region	Characteristic	Land use
Almocreva	(Alentejo, P)	Micro-catchment	Wheat/forestry
Rambla Honda	(Almeria, E)	Hillslopes	Semi-natural/abandoned
El Ardal	(Murcia, E)	Micro-catchment	Semi-natural/wheat/abandoned
Canterrane/Réart	(Roussillon, F)	Hillslopes	Vineyards/abandoned olive grove
Peissonnel	(Var, F)	Hillslopes/catchment	Vineyards/semi-natural
Is Olias	(Sardinia, I)	Hillslopes	Semi-natural/pasture/forestry
Spata	(Athens, Gr)	Hillslope	Olive grove/vineyard
Petralona	(Thessaloniki, Gr)	Hillslope	Semi-natural/wheat

TABLE 5.3   An overview of the main land-use types of the core field sites

Location	Semi-natural	Abandoned*	Agriculture/Forestry
Almocreva (P)	—	—	Wheat/eucalyptus plantations
Rambla Honda (S)	*Stipa t.* shrubland	*Anthyllis/Retama* shrubland	—
El Ardal (S)	Shrubland	Shrubland < wheat >	Wheat (almonds, vineyards)
Canterrane/Réart (F)	Shrubland	Shrubland	Vineyards < grapes/olives >
Peissonnel (F)	Pines/shrubland	—	Vineyards
Is Olias (I)	Shrubland	Shrubland < pastures >	Eucalyptus plantations
Spata	—	—	Olive grove/vineyard/(orchards)
Petralona (Gr)	Shrubland	—	Wheat/anisum

*Indication of type of shrub at present stage of succession.
< > Former type of land use (when known).
( ) Other typical land use in direct surroundings.

On each type of land use, three field measurement areas were installed, each 1 ha in size, at different topographic altitudes along the hillslope or the catchment slope (Table 5.2). The sites were selected in such a way that the substratum was uniform within each field measurement area. The choice of whether to work on a hillslope or in a micro-catchment was governed by the topographic properties and possibilities of the sites. The advantage of working in a micro-catchment was that it could also be instrumented for measuring runoff and sediment yields, helpful for validating the simulation models being used within MEDALUS.

In addition a certain degree of homogeneity was required in the vegetation cover and type, because replicate destructive and non-destructive vegetation measurements had to be made at each position along the catena. In general this resulted in a subdivision with semi-natural vegetation on the upper part of the hillslope and (abandoned) agricultural land on the (middle and) lower part of the slope.

The actual land use at the different sites varied widely as can be seen from Table 5.3.

5.4 CROSS-SITE COMPARISON OF BASIC CHARACTERISTICS

In the following paragraphs a basic description is given of the MEDALUS core programme and some cross-site comparisons are made. The descriptions and data presented are taken from the MEDALUS data-base or from reports submitted by the individual field groups to the *MEDALUS I Final Report* (pp. 312–606). For more detailed descriptions and results of the core field sites reference is made to the corresponding chapters in this book, or to previous MEDALUS reports. The photographs (Figures 5.3 to 5.7) give a first impression of some of the fields sites.

5.4.1 Lithology and soils

The lithology and soils of the individual sites are very different. Within the sites themselves, and even down the slopes, heterogeneity in soil depth and development is very common (Table 5.4). In general, it can be said that most soils have a considerable rock fragment content (slightly gravely to gravely; FAO, 1977) and that most soils belong to the Entisol or Inceptisol orders of the USDA classification (Soil Survey Staff, 1992).

TABLE 5.4 An overview of some important abiotic characteristics

Location	Lithology	Type of soils (USDA class.)	Soil depth(s) (cm)	Slope (°)
Almocreva (P)	Red schist/sediment	Aquept/Ochrept	>85/15–40	(7–10)
Rambla Honda (E)	Black schist/sediment	Fluvent/Orthent	75/15–70	(5–20)
El Ardal (E)	Limestone/sediment	Orthent/Orthid	0–45/15–>100	(5–15)
Canterrane/Réart (F)	Slates/sediment	Ochrept	30–>100	(5–15)
Peissonnel (F)	Sandstone/shale	Ochrepts	15–>100	(3–10)
Is Olias (I)	Sediments/para-metamor.	Orthent/Ochrept	>100	(7–25)
Spata (Gr)	Sandstone	Ochrept	46–>100	(4–12)
Petralona (Gr)	Limestone/marl	Orthent/Ochrept	0–40/40–>100	(5–20)

The sites at Almocreva (P) (Figure 5.3) are situated on hillslopes developed on reddish medium to highly metamorphic rocks. On the upper part of these slopes, relatively thin soils occur, sometimes developed in Tertiary Ranja deposits. They were classified as being members of the Ochrepts sub-order (Soil Survey Staff, 1992). On the lower sections of the slopes and in the valley bottom, locally derived sediments with thicker soils are present where waterlogging occasionally occurs (Aquepts).

At the Rambla Honda site (S) shallow soils (Typic Torriorthents) are developed on dark micaschists on the upper and middle parts of the slope. The lower part of the slope consists of an alluvial fan formation with shallow Typic Torrifluvents. The soils are generally acid.

The El Ardal site (S) (Figure 5.4) has well-banked limestones on the upper slopes. Downslope there is a clear break which is related to a change in lithology. In the lower section marls are present, which also have different soils and land use. The El Ardal sites show poorly developed soils belonging either to the Xerorthent sub-order or Xerollic paleorthid group. Petrocalcic horizons at limited depths are typical in this area.

FIGURE 5.3 Overview of the Almocreva site (P): Ploughed fields with isolated *Quercus suber* trees (February 1992)

FIGURE 5.4   The El Ardal field site (S): Upper part of the catena covered with *Rosmarinus* matorral and isolated *Pinus halepensis* trees. In the centre of the photograph a runoff plot is visible. In the valley bottom vineyards and almond orchards are visible (June 1991)

FIGURE 5.5   Vineyards in the Cantarane area (F) on Plio–Pleistocene fluvial terraces. In the background are hillslopes covered with maquis (footslopes of the Pyrenées) and behind them the mountains of the Canigou massif (April 1991)

The Canterrane and Réart sites (F) (Figure 5.5) are both situated on the footslopes of Plio–Pleistocene sedimentary deposits ranging from fine- and coarse-grained conglomerates to clays. In these deposits fluvial terraces are developed as well as deep active gullies. The upper slopes on slates and limestone are not instrumented. Soils are classified as *sols bruns ferrsialitiques* (French classification).

The Is Olias (I) site has a very hilly character and is developed on Palaeozoic rocks (alternations of low-grade para-metamorphics of sedimentary origin) and Plio–Quaternary detritic deposits on the footslopes. The soils belong to the Xerorthent and

FIGURE 5.6   Runoff plot in the olive grove at the Spata field site (Gr) (spring 1991)

Xerochrept groups, of which the latter is related to less steep slopes with semi-natural vegetation.

The Spata field site near Athens (Gr) (Figure 5.6) is situated on a small hillslope developed in Tertiary sedimentary parent materials (sandstone/marl and conglomerate/sandstones). The moderately deep to deep soils are classified as Typic or Calcic Xerochrepts and generally have a Calcic or Petrocalcic horizon.

The Petralona site near Thessaloniki (Gr) (Figure 5.7) is situated on hard massive limestones on the upper half of the catena and marls on the lower part, on which cereals

FIGURE 5.7   Overview of the Petralona field site (G): To the right the *Quercus coccifera* shrubland covered limestone hill slope. At the left margin of the photo the agricultural fields can be seen (October 1992)

TABLE 5.5　An overview of living biomass (production) at the MEDALUS core project sites

Location	Vegetation type	Spring (g m^{-2})	Autumn (g m^{-2})
Almocreva (P)	Wheat	1004 ± 96.8	—
Rambla Honda (S)	*Retama sphaerocarpa*	188.3 ± 14.5[b]	117.2 ± 9.8[b]
	Anthyllis cytisoides	238.7 ± 9.4[a]	133.3 ± 15.6[b]
	Stipa tenacissima	326.3 ± 18.1[a]	356.3 ± 44.2[b]
El Ardal (S)	Abandoned fields	200–250	—
	Shrubland	2469	—
Canterrane/Réart (F)	Shrubland	219.8 ± 56.9	—
Peissonnel (F)	No data	No data	
Is Olias (I)	No data	No data	
Spata (Gr)	Herbs/grasses under olive	378.8 ± 127.6	184.1 ± 26.4
	Vineyard	48.0 ± 13.4	73.8 ± 14.0
Petralona (Gr)	*Quercus cocc.* (leaves ± wood)	5270 ± 1232	8137 ± 1277
	Wheat	1173 ± 206	—

[a]Above-ground net primary production.
[b]Above-ground biomass at the annual minimum (autumn).

and other crops are grown. The soils on these slopes have considerable carbonate contents and belong either to the Xerorthent or Xerochrept group. They are very shallow to shallow in the limestone area and more deeply developed on the marly footslopes.

5.4.2 Land use and vegetation

All the sites have a different land-use history, giving rise to different present-day conditions. General characteristics of land use have already been shown in Table 5.3 and more specific data on standing biomass are given in Table 5.5.

Because of the large land-reclamation programme in Portugal in the 1930s, the Almocreva site is now completely covered with wheat. Previously it had been covered by *Cistus* spp. shrubs. At present, large areas are being planted with eucalyptus forests.

The Rambla Honda site (S) is covered entirely by semi-natural vegetation, with three successive zones where *Retama* spp. (lowest zone), *Anthyllis* spp. (middle zone) and *Stipa tenacissima* (upper zone) dominate. The two lower zones of vegetation (*Anthyllis* and *Retama*) grow on slopes which were abandoned over 40 years ago. Each of the three vegetation zones is covered by separate measurement areas.

The El Ardal field site (S) has an instrumented micro-catchment which covers three land-use zones: an upper zone with grazed *Rosmarinus officinalis*/*Thymus vulgaris* matorral (shrubland) and some patches of *Pinus halepensis*; a middle zone with abandoned fields and cereals and a lower zone with cereals; almonds, olives and viticulture.

The Canterrane site (F) is located on a long slope with vineyards on the terraced footslopes. At the top of this hill, partly burned *Quercus coccifera* shrubland (on limestone) is present. On the slates are shrub lands dominated by *Cistus* spp. The Réart sites (F) are situated on abandoned vineyards and olive groves (abandoned for 2–6 years), which are being colonized with *Cistus* spp. shrubs. The Peissonnel site (F), is an elongated catchment with three sites on different hillslopes. All these sites are at present under viticulture. The watersheds are under semi-natural vegetation, but might have been abandoned from cropping long ago.

The Is Olias site on Sardinia (I) has three types of land use: a shrubland slope, an abandoned meadow with regenerating shrub land and forest, and also areas with eucalyptus plantations.

The Spata site (Gr) covers only agricultural land, with an olive grove on the upper and a vineyard on the lower slopes. In the neighbourhood other orchards are also present.

The Petralona site (Gr) has two types of land use: the upper slope covered with grazed *Quercus coccifera* dominated shrubland and the lower part of the slope where cereals (wheat) are grown as well as some cultivation of anisum.

In Table 5.5 an overview is given of some of the biomass measurements carried out in the field sites. The methodologies applied (see Section 5.7.1) are very laborious and could only be undertaken for a limited number of species per site. The sites where wheat is produced (Almocreva and Petralona) show more or less similar biomasses of about 1000–1100 g m^{-2}.

As some of the sites are on privately owned vineyards, no standing biomass values could be obtained. For the semi-natural vegetation, data are only available for some of the dominant species, different for the different sites.

5.4.3 Runoff coefficients

On all except the French sites, runoff measurements were carried out on standardized plots of 10×2 m. On the French sites, runoff coefficients were so large that runoff concentration caused extensive rilling and, therefore, natural erosion phenomena would have been interrupted by the standard 10×2 m plot design. Consequently, plots of 2.5×40 m along the lines of the vines were used.

A summary of some of the results is shown in Table 5.6. These should be interpreted with care as the data are averages covering several months or years. Individual rainfall events bring varying runoff rates and sediment yields, related to storm intensity and duration, and antecedent soil conditions. The precipitation totals given in Table 5.6 only cover the measured events. By presenting the values in this way, extreme, but geomorphologically interesting events are averaged out.

The cover type of the runoff plots is related to the local circumstances, and in Table 5.6 the dominant vegetation species and the percentage of plant cover is given. Runoff percentages vary from less than 1 to over 30% depending on the soil surface characteristics, vegetation and the soil material. As could be expected, runoff figures on the limestone areas are very low (less than 1%). The Rambla Honda and Is Olias plots give intermediate values for runoff generation (1–6%). The French sites have, in general, the highest runoff percentages and the relation with soil tillage is clearly demonstrated by the results. Similar results were demonstrated by rainfall experiments carried out in these areas.

Sediment yields also vary widely, from as low as 0.01 t.ha^{-1} yr^{-1} to more than 30 t.ha^{-1} yr^{-1} in some of the Peissonnel sites (F) (implying an annual lowering of the soil surface of 1.2 mm). However, the results are very difficult to compare as the lengths of the measurement periods are not equal for the different sites.

TABLE 5.6   Sediment production and runoff of the bounded runoff plots (20 m²)

Location	Cover (type)	Precip. (mm)	Runoff[a] (%)	Sediment yield[a] (t.ha^{-1}yr^{-1})	Cover (%)
Vale Formoso (P)[b]	Wheat/fallow	1440	6.8 ± 2.9 [13.6]	0.70 ± 0.28 [0.84]	Variable
	Cistus		0.1 [3.4]	0.01 [0.44]	Approx. 60
	Abandoned		4.4 [14.1]	0.04 [0.07]	?
Rambla Honda (S)[c]	*Retama*	533	3.5 ± 3.2	0.05 ± 0.02	26–61
	Anthyllus	533	7.7 ± 2.8	0.16 ± 0.02	58–73
	Stipa	533	10.7 ± 2.1	0.26 ± 0.13	61–75
El Ardal (S)[d]	Wheat	319	0.6 [1.0]	0.01 [0.16]	Variable
	Fallow	319	0.6 ± 0.3 [1.0]	0.01 ± 0.01 [0.01]	10–60
	Matorral	319	0.6 ± 0.4 [1.1]	0.01 ± 0.00 [0.12]	10–75
Canterrane (F)[e]	Vineyard	1213	23.0 ± 17.4	?	5–25
Réart (F)[e]	Vineyard	?	12.4 ± 10.4	?	5–25
Blais (Peissonnel, F)[f]	Vin/her	528	4.66	0.299	>50
	Vin/che	553	18.4	3.00	<25
	Vin/til	573	11.4	3.89	<25
Rimbaud (Peissonnel, F)	Vin/che	547	4.38	33.0	<25
	Vin/til	635	9.98	12.6	<25
Ferauds (Peissonnel, F)	Vin/che	569	30.0	7.06	<25
	Vin/til	528	10.2	3.76	<25
Is Olias (I)[g]	Macchia	894	1.13 ± 0.66	0.21 ± 0.22	60 ± 28
	Machia burned	891	1.13 ± 0.24	0.18 ± 0.07	71 ± 15
	Eucalyptus	892	3.03 ± 0.91	0.40 ± 0.26	44 ± 4
Spata (Gr)[h]	Olives	724	1.63 ± 1.61	0.03 ± 0.01	>70
	Vineyard	724	11.0 ± 2.7	3.44 ± 0.98	Variable
Petralona (Gr)[i]	Wheat	169	0.29 ± 0.20	0.48 ± 0.32	Variable
	Maquis	169	0.32 ± 0.10	1.08 ± 0.47	27–73

[a]Where there is more than one plot with the same cover type, the mean and [standard deviation] are given.
[b]Oct. 1990 to May 1994: plots 1, 2, 10, 11 (166.6 m²); plot 7X (83.1 m²); plot 13A (166.6 m²). Between brackets: average Oct. 1991 to May 1994 plots 1, 2, 10, 11; Oct. 1988 to May 1994 plot 7X and 13A.
[c]Sept. 1991 to May 1993.
[d]Average 1991 and 1992: plot 12; plots 2, 13, 14; plots 6, 7, 9, 15, 16, 17. Between brackets average 1989 to 1992: plot 2; plots 6, 7, 9: average 1990 to 1992: plot 12.
[e]5/92–11/93 (120 m² plots); 5/92–9/93 (200 m² plots).
[f]1991: 100 m² plots. Three sites within catchment: Les Blais, Le Rimbaud and Le Puits des Ferauds. vin/her = vineyard with herbaceous cover, vin/che = vineyard treated with herbicides, vin/til = vineyard with soil tillage.
[g]2/92–2/94; (plots 1–6, 7–12, 13–18).
[h]3/91–3/94; (plots 1–3, 4–6).
[i]Runoff: 4/92–7/92, sed. yield 4/92–4/94.

5.5 A SHORT CHARACTERIZATION OF THE CORE PROGRAMME FIELD SITES IN RELATION TO DESERTIFICATION

The characteristics of the individual field sites are very different. At each site, sets of processes such as rill and sheet erosion, overland flow concentration, land management practices, desiccation, (over) grazing and forest fires, result in varying types of land degradation. As shown in the sections above, large differences also exist in geology, land use, semi-natural vegetation and annual rainfall.

The collection of field data only started in the second half of 1991. Apart from high inter-annual precipitation variability, many important processes have a dynamic nature and therefore a measurement period of 1.5 years is too short to really indicate all important processes and their implications. Only short descriptions are given below, to highlight some important local features.

The sites at El Ardal (S) and Petralona (Gr) are similar in character; both have a matorral-covered limestone slope and a colluvial/marl footslope mainly covered by wheat. Annual precipitation is higher at the Petralona site. Although the semi-natural vegetation itself is different (the dominant species at El Ardal is *Rosmarinus officinalis* and at Petralona it is *Quercus coccifera*), these two sites give a good opportunity to compare the western and eastern part of the Mediterranean. In both areas grazing is important. With the exception of the soils developed under *Pinus halepensis* at El Ardal, the soil is better structured at Petralona and there is a higher aggregate stability. The difference in clay type and content and in (secondary) $CaCO_3$ partly explain the differences in erodibility.

Both French sites are different from the other sites, in their land use, lithology, runoff and precipitation. Although Roussillon is in the driest part of France, it is more humid than the other MEDALUS sites. Here degradation due to soil erosion plays a very important role. The erosion is caused by land management practices with alternating years of ploughing between the lines of vines and the use of herbicides. This practice increases surface crusting, and initiates overland flow concentration, rilling and gullying. In the semi-natural areas surrounding the sites, forest fires are a major source of disturbance. The Roussillon area is also well known for its high-intensity rainfall events.

The field site at Spata (Gr) has very different characteristics. There is high diversity in the types of land use, and the agriculture is intensive. This is due to the pressure of expanding Athens.

The three remaining areas, Almocreva (P), Rambla Honda (S) and Is Olias (I) all have their own specific characteristics from the point of view of land use, vegetation and lithology.

The Almocreva site (P) is characterized by its continuous cover of wheat and isolated oak trees. In this area there has been a rapid increase in the planting of eucalyptus trees. The soil tilled for wheat is subject to strong surface sealing and has a low aggregate stability and high dispersivity. In the valley bottoms, dispersed material accumulates, with signs of waterlogging of the soil. Here desiccation and rain-wash, combined with land-use practices, are the major sources of degradation. There are some indications of high salt concentrations as well.

The Rambla Honda site (S) is very mountainous with three major semi-natural vegetation zones related to hillslope morphology. The upper slopes are covered by *Stipa tenacissima*, the middle slopes are formed by a dissected alluvial fan and are covered with *Anthyllis cytisoides*, whereas the lower part of the fan is covered with *Retama sphaerocarpa*. The soils are very shallow and poor. This is agriculturally the most marginal area being investigated by the project and has only very limited possibilities for agricultural land use.

The Is Olias site on Sardinia also is hilly. In this area vegetation cover has changed from a heterogeneous pattern of forests, shrublands, meadows and agriculture to being dominated by forests and shrublands. At several places there are eucalyptus plantations. Forest fires forms a major threat to the hillslopes, whereas in the lowland plains of the catchment (where there is no core programme field site area), salinization due to salt-water intrusion is an important cause of degradation.

5.6 SITE LAYOUT

5.6.1 Replicates

Each of the eight field sites were located so that, within a short distance, two of the most typical land uses, combined with typical lithology, soil and vegetation, in the region could be found. Each land use had to cover an area of at least several hectares and to cover a complete catena. The catenas were divided into three sections and on each section three replicate measure areas were installed, each 100×100 m. Automatic weather stations were installed near each site to gather meteorological data on an hourly basis. The standard layout of the field sites is shown in Figure 5.8. In some cases the topography of the field area enabled the field groups to install their measurement hill slopes within a micro-catchment. All the field sites were built, maintained and monitored by the local research groups.

5.6.2 Subdivision of measurement areas

Figure 5.9 shows the layout of each measurement area. They were divided into nine 33×33 m blocks or sub-areas for the following purposes:

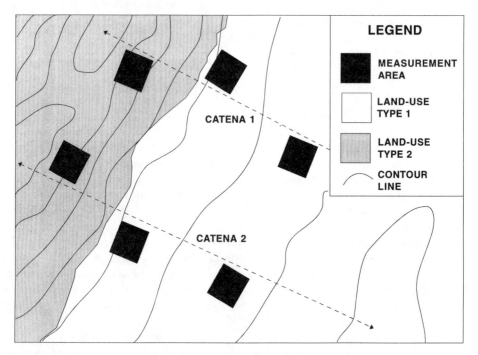

FIGURE 5.8   Schematic representation of hillslope with field site layout along two catenas on two types of geology and corresponding land use (e.g. land-use type 1: limestone with shrubland; land-use type 2: marls with agriculture)

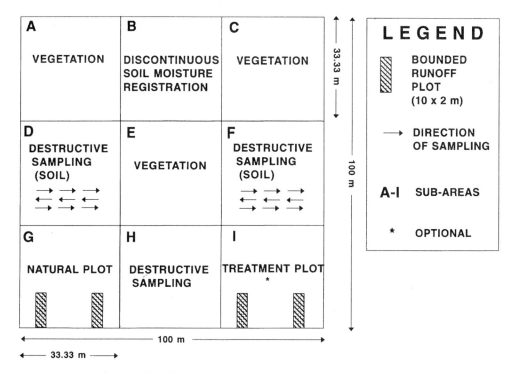

FIGURE 5.9   The standard layout of the 1ha measurement areas

—   vegetation observation and sampling (three sub-areas)
—   destructive soil (moisture) sampling (two sub-areas)
—   the measurement of runoff and sediments (one block)
—   additional plots for extensions to the core programme (three sub-areas)

In some cases the scale of the topography did not allow the use of a single, contiguous measurement areas of 1 ha and the sub-areas had to be separated.

Within some of the sub-areas, other divisions were made, for the measurement of different types of vegetation parameters (Figure 5.10), but all vegetation data were collected in sub-areas where there was no destructive soil sampling.

Runoff (overland flow) was measured in bounded plots of 10 × 2 m (or larger, depending on the local topography and source area, and vegetation cover). The runoff and sediment was collected either on an event or continuous basis depending on the level of instrumentation at the site. At some places standard Gerlach troughs were used.

5.7 FIELD AND LABORATORY METHODOLOGIES

The list of 44 parameters monitored at the field sites in the core programme was developed to provide inputs for calibrating and validating the MEDALUS hillslope model. In order to standardize the collection and analysis of the data, a field manual was produced and

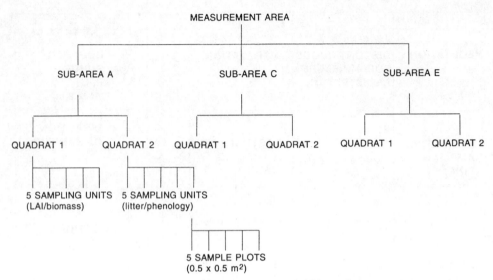

FIGURE 5.10   Vegetation sampling areas layout (within each measurement area)

kept under continuous revision (Cammeraat, 1992). In addition, several workshops were organized to improve the understanding and practical procedures of the methods used and to stimulate the exchange of experience and knowledge between groups.

In the following sections a synopsis is given of the parameters measured in the core programme. Table 5.7 shows the complete list. The parameters are separated into biotic and abiotic measurements and on the basis of sampling frequency. Basic field site characteristics were described in a set of maps reflecting topography including slope profiles at a scale 1:10 000. Detailed maps of geology, soils, land use and vegetation type and finally the layout of the (instrumented) measurement areas were made.

5.7.1 Vegetation and land use

All measurement were carried out seasonally except when stated.

Species composition was determined for all non-destructive quadrats.
Plant cover, phenology and height were measured once or twice a year, depending on the seasonal variability. Two methods were used, the first based on line or point transects, the second on vertical photographs. Also height, and non-biotic cover were recorded.
Leaf area index (LAI) was determined by surface/weight or surface/geometrical relations for the dominant species at each location. The non-destructive DEMON instrument (CSIRO, 1990) was also used but was not found to be successful for sparse vegetation.
Above-ground biomass and production was measured using various methods depending on the vegetation type (Milner and Elfyn Hughes, 1968; Anderson, 1970; Newbould, 1970; Etienne, 1989; Bencat, 1990). The biomass of perennial plants was determined using allometric methods, by combining destructive sampling and non-destructive measurements. Annual plants were sampled once, when at their peak biomass.

TABLE 5.7   Parameters being measured in the core field programme at the eight field locations

Atmospheric parameters
Continuous
1.01 autographic rainfall
1.02 dry bulb temperature
1.03 wet bulb temperature
1.04 wind speed
1.05 wind direction
1.06 net radiation
1.07 photosynthetically active radiation

1.08 soil radiation temperature
1.09 vegetation temperature*
1.10 atmospheric CO_2*
Event-based
1.11 rainfall
One-off
1.12 radiation variability with respect to aspect

Vegetation and land-use parameters
Seasonal
2.01 percentage cover
2.02 leaf area index
2.03 shoot elongation
2.04 grazing patterns*
2.05 grazing density*
2.06 shoot and leaf arrangement
2.07 litter production
2.08 litter decomposition
2.09 primary production

2.10 spectral signature*
2.11 above ground biomass
One-off
2.12 below-ground biomass
2.13 root arrangement and depth
2.14 species composition
2.15 land use
2.16 vegetation cover type
2.17 spatial pattern of vegetation

Soil parameters
Event based
3.01 soil moisture
3.02 surface runoff
3.03 sediment yield
3.04 subsurface flow*
Periodic
3.05 soil moisture (weekly)
3.06 soil organic matter*
3.07 percentage soil covered by stones
One-off
3.08 bulk density

3.09 soil water retention curves
3.10 soil profile description
3.11 soil depth
3.12 soil map
3.13 infiltration characteristics
3.14 soil texture
3.15 soil chemistry
3.16 aggregation
3.17 soil fauna*
3.18 soil temperature (at -50 cm)

Surface parameters
Continuous
4.01 nutrients*
Periodic
4.02 crusting
4.03 cracking
4.04 compaction
4.05 permeability of crust

4.06 photography of plot
4.07 site maps
4.08 plot maps
4.09 stone content/arrangement
4.10 roughness

*Not measured at all sites.

Below-ground biomass of roots was measured using an allometric method for roots larger than 5 mm. The finer roots were measured by destructive sampling as described by Milner and Elfyn Hughes (1968) and Newbould (1970).

Root arrangement and depth were determined qualitatively in soil pits and trenches.

Litter production was measured monthly in open (unbounded) quadrats, or whole plant traps, for the dominant species. A distinction was made between wood and litter. The amounts were expressed in litterfall per unit area per month. Adapted methods, as described in Newbould (1970), French et al. (1979) and Morris (1982) and in the *Field Manual*, were used.

Litter decomposition. As the measurement period was too short to work with litter bags (Mason, 1977; Mitchel et al., 1986) it was proposed to do an inter-site study of decomposition rates using the cotton strip assay method (Harrison et al., 1988). This does not give a direct measure of rates of litter decomposition, but makes a comparisons between sites possible.

Spectral signature was measured only at some core field sites (Harrison et al., 1993).

5.7.2 Abiotic measurements

One-off measurements

Soil profile description and classification was carried out using the FAO guidelines (1977) for the mineral part of the soil profile. The classification was done according the FAO–ISRIC system, FAO (1988), the USDA soil taxonomy (1992) and the local system. The ectorganic part of the soil profile, when present, was described and classified according to Klinka et al. (1981). The soil profiles were located within the measurement areas.

Soil texture was determined using the classical sieve and pipette methods using standardized size classes (FAO, 1977).

Dry bulk density was measured using soil cores sampled with a ring, when the soil did not contains many rock fragments (Blake and Hartge, 1986). When the soil was too stony, a clod was sampled, coated with paraffin or resin in order to estimate its volume. The amount of rock fragments was also determined. Where rock fragments larger than 5 cm were present, the soil was excavated step by step and the weight of the soil and the volume of the pit was determined (Hanson and Blevins, 1979; Goudie, 1981; Huntington et al., 1989).

Soil water retention curves for non-stony soils were determined in the laboratory using standard sand boxes for the lower suction trajectory. The higher suction values were determined using a high-pressure membrane (Stakman, 1974). For stony soils, clods had to be sampled. Water retention of porous rock fragments was also determined. The curves were fitted using the relationships of Van Genuchten (1980). In stony soils the measurement of *in-situ* soil moisture retention curves was carried out in some places by using a combination of Time Domain Reflectometry (TDR) and a tensiometer system (Bergkamp et al., in press).

Saturated hydraulic conductivity of the first 50 cm of soil was determined by the *in-situ* inverse auger hole method, and from plotting draw down versus time (Kessler and Oosterbaan, 1974).

Infiltration characteristics were determined by field experiments with a portable rainfall simulator (constant head drip plate with randomizer at 1.5 m height, Bowyer-Bower and Burt, 1989). The following parameters were determined: rainfall intensity and duration, time to ponding, and runoff. By using several rainfall intensities an infiltration envelope was constructed.

Soil chemistry. The following chemical characteristics were measured, using a 1:1 soil water solution: pH, pH(KCl), EC25, organic carbon content (Allison, 1935), Cat-ion

Exchange Capacity (CEC), total nitrogen and the water-soluble salts K^+, Na^+, Ca^{2+}, Mg^{2+}, NH_4^+, Fe^{2+}, Al^{3+}, HCO_3^-, Cl^-, NO_3^-, SO_4^{2-}, PO_4^{3-} according to international standards (Page, 1982).

Aggregation was determined by the degree of stability (Imeson and Vis, 1984) and by the size distribution (corrected for stone content). Both macro- and micro-aggregation were determined by dry sieving and using a Microscan particle sizer (Brey and Cammeraat, 1993). Several other methods were also used as not all methods were suitable for every situation.

Rock fragment content and arrangement were measured and noted for each horizon or per 10 cm layer.

Soil roughness was measured as an index determined by using a chain (Quinton, 1990), or by exact determination at cross sections.

Soil temperature was measured at −50 cm depth.

Seasonal abiotic measurements

Rock fragment content and position at the soil surface was measured with a point counting technique (Daniels et al., 1968).

Soil crusting was determined by a visual estimation of the soil surface using the scale of Boekel crusting/slaking index (Boekel, 1973).

Cracking, when present was quantified by measuring the depth and width of the cracks. When cracking occurred, the swelling/shrinkage characteristics of the soil were also determined (Brasher et al., 1966).

Compaction of the soil surface was measured with a pocket penetrometer.

Permeability of undisturbed soil crusts was measured using the method of Boiffin and Monnier (1985).

Photography of the runoff plot surface showed active soil surface changes and processes. All plots were completely covered with one or more series of photographs.

Continuous and event-based measurements

The following continuous measurements were carried out at different time intervals.

Meteorological measurements using automatic weather stations, with the sampling of solar radiation, net radiation, dry bulb temperature, wet bulb temperature, wind speed, hourly rainfall totals and 5-minute rainfall intensities, enabling a detailed estimation of evapotranspiration and rainfall input.

Stream discharge was measured continuously in the micro-catchments.

Gravimetric soil moisture content was sampled on a weekly basis. Metal cores were used when possible or, when the soil was too stony, clods of soil were used. The soil moisture content needed to be corrected for stones. Furthermore it was important to distinguish between vegetated and bare areas (Reinhart, 1961; Hanson and Blevins, 1979; Goudie, 1981; Huntington et al., 1989).

Plot runoff and sediment yield were collected after rainfall events, and the results were expressed as percentage runoff. The sediment concentration was corrected for dissolved solids (Brown et al., 1970; Hem, 1970) and was expressed in sediment yield per surface area (Wischmeier and Smith, 1978).

5.8 THE CHARACTERISTICS AND PURPOSE OF THE MEDALUS DATA-BASE

All data gathered in the MEDALUS core programme were stored in the MEDALUS data-base. Actually three data-bases exist, one dealing with remote sensing images, one containing meteorological and weather station data and one containing the results of field measurements as described above. The data-base was published on CD-ROM in 1995 using a Geo-Management System (Cammeraat and Prinsen, 1995). The data-base was essential for the teams who wanted to calibrate and validate their models (both the MEDALUS hillslope model and the SHE model) and also to make inter-site comparisons possible. Furthermore the data-base will be extended and remain an important data resource in the future. With the introduction of a geo-management system at the end of 1993, data based on maps, remote sensing images, (air)-photos and Geographic Information System (GIS) applications will also be incorporated and accessed.

5.9 AN EXAMPLE OF DETAILED INTER-SITE COMPARISON AGGREGATION

In this section an example of the cross-comparison of data between core field sites will be given, some results shown and briefly discussed.

Soil aggregation was studied extensively as part of the core programme at all the field sites. Aggregate stability and size distribution were measured. Aggregation can be studied in different ways: firstly reflecting soil formation and development; secondly reflecting sensitivity to soil erosion; and thirdly as a reflection of the interaction between biotic processes in and on the soil and the properties of the soil material itself. This latter approach is applied within the desertification response unit (DRU) concept (Imeson et al., this volume, Chapter 18) where aggregation is regarded as an essential parameter at the lowest hierarchical level. This requires at least a seasonal sampling strategy, or a sampling strategy related to different stages of vegetation succession.

Figure 5.11a depicts the aggregate distribution of the Ap horizons of some soils from Almocreva (P). This area was completely covered with wheat. Differences in aggregation must be the result of differences in parent materials and/or soil formation, as the area has been cultivated for at least 30 years.

Some of the samples taken from topsoils in Petralona (Gr) are shown in Figure 5.11b. Here two main soil groups could be recognized, one developed on marls the other on limestones. The land-use type on these two soils is quite different: cultivation (mainly wheat) on marl and grazing the shrubby vegetation on limestone. The largest numbers of coarse aggregates were found under the wheat, reflecting the land management and the different parent material.

In Figure 5.12a differences in aggregate distributions are shown for the Peissonnel (F) site. The samples taken at Rimbaud reflect the difference in aggregation between different types of vegetation cover, on a homogeneous soil. The sampling took place at two depths (as there is a well-developed surface crust, most readily visible for the bare site) and showed a high number of fine aggregates between 0 and 2 cm depth. Under *Pinus halepensis* there are relatively low amounts of coarse aggregates. The aggregates sampled under *Quercus* spp. showed higher numbers of larger aggregates than the areas covered

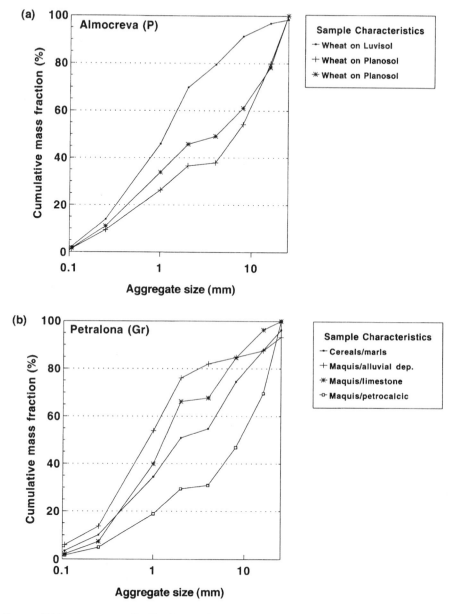

FIGURE 5.11 Aggregate distribution at the (a) Almocreva site (P), (b) Petralona site (Gr)

with *Cistus* spp. Differences between the *Pinus* and *Quercus/Cistus* cover might be related to different types of litter and related biological activity.

In Figure 5.12b a first result is shown for the aggregate distributions in a study area located in the Guadalentín area (MEDALUS II target area) on slates, very low grade metamorphic rocks. Here a similar sampling procedure to that used at the Var site was

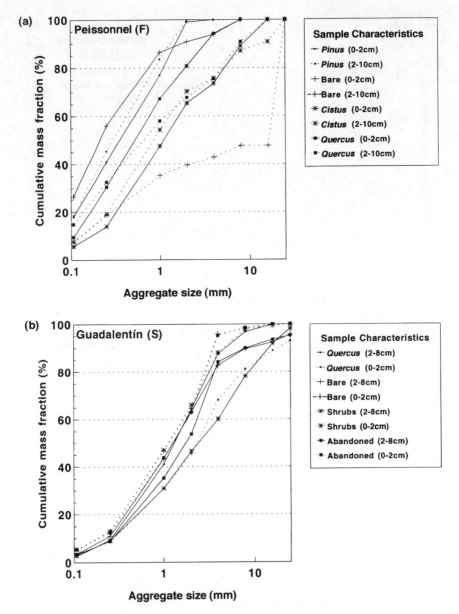

FIGURE 5.12   Aggregate distributions related to vegetation species in the (a) Peissonnel site (F), (b) Cerro de Jarita

applied, but with seasonal monitoring. In general, the plant-covered areas showed a higher number of larger aggregates compared with the bare areas.

In Figure 5.13a an example is given from the Réart site (Roussillon, F), taken from a study by Hin (1993). Samples taken under litter (*Cistus monspeliensis*) showed the highest

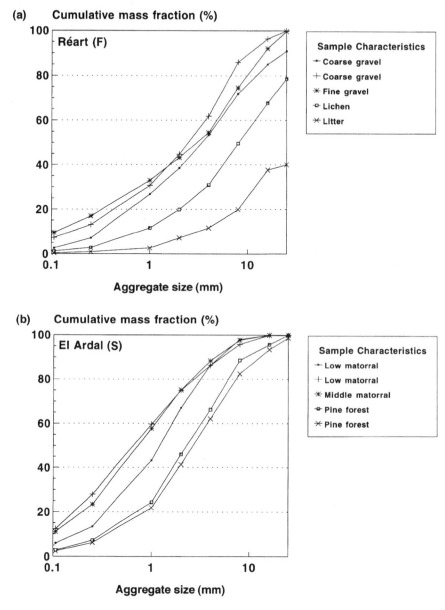

FIGURE 5.13 Aggregate distributions related to plant cover in the (a) Réart site (F), (b) El Ardal site (S) (from Hin, 1993)

number of larger aggregates. Here again it was clear that the aggregate size for the topsoil (0–10 cm) is related to characteristics or presence of the vegetation cover.

In the last figure (Figure 5.13b), taken from the same study, aggregate size distributions are shown for different types of matorral cover near the El Ardal site (S). Here it was found that a cover of *Pinus halepensis* with dense undergrowth of *Brachiopodum* spp. has different aggregation characteristics from the *Rosmarinus officinalis*-dominated matorral.

In general, direct comparisons between the different field sites still have to be made with care because local circumstances like parent material, land use and soil moisture regime determine the degree and type of aggregation of the soil. However, some general trends can be observed, confirming the importance of vegetation type and succession stage on the aggregation of the soil and showing its highly dynamic aspects. Simultaneously, important relationships were found for aggregation stability and for micro-aggregation (Hin, 1993; Imeson et al., 1996, this volume), which showed to be essential for the DRU concept.

5.10 CONCLUSIONS

A harmonized and standardized measurement programme was successfully established within the MEDALUS programme to provide:

1. A reliable and inter-site comparable data-set for process-oriented research and modelling.
2. A benchmark data-set for future research at the core programme sites.
3. An improvement of field measurement methodology especially related to areas with sparse vegetation and stony soils.

The core measurement programme was carried out at eight field sites located on areas of typical land use across the European Mediterranean. The measurement layout and methodology was harmonized to make it exactly the same at each field site. A detailed but balanced list of various biotic and abiotic parameters and variables was measured and sampled at different temporal and spatial scales at each of these locations. This was only possible because of the good level of cooperation between the different field groups on the one hand and with the modelling groups on the other.

All data were stored in a data-base which was accessible to all project members and this same data-set was released on CD-ROM in the first half of 1995.

Because the data collection only started during the second half of 1991, the length of the data series available to date is too short to formulate final conclusions on desertification, degradation and land-use problems within the European Mediterranean. However, the data-set has proved to be extremely useful and indispensable input into the physically based models developed within MEDALUS and as an information source for physical processes and relationships in the Mediterranean ecosystems.

5.11 REFERENCES

Allison, L.H. (1935) Organic soil carbon reduction of chromic acid, *Soil Science*, **40**, 311–320.

Anderson, F. (1970) Ecological studies in a Scanian woodland and meadow area South Sweden: II. Plant biomass primary production and turnover of organic matter. *Botaniska Notiser*, **123**, 8–51.

Aru, A. and Baroccu, M. (1993) The Rio Santa Lucia Catchment area. In *MEDALUS Final Report*, Bristol, 534–459.

Bencat, T. (1990) Leaf biomass and Leaf Area Index (LAI) of Black Locust (*Robinia pseudacacia* L.) in southwest Slovakia. *Ekologia (CSFR)*, **9**, 259–268.

Bergkamp, G., Cammeraat, L.H., Martinez-Fernandez, J., in press. Water movement and vegetation patterns on shrubland and an abandoned field in desertification threatened areas. *Earth Surface Processes and Landforms*.

Blake, G.R. and Hartge, K.H. (1986) Bulk density In: A Klute (ed.) *Methods of Soil Analysis* Part I, 2nd edition, ASA and SSSA, Madison, 363–375 .

Boekel, P. (1973) *De betekenis van de ontwatering voor de bodemstructuur op de -zavel- en lichte kleigronden.* Rapport 5-1973 Instituut voor bodemvruchtbaarheid Haren, 42 pp (in Dutch).

Boiffin, J. and Monnier, G. (1985) Infiltration rate as affected by soil surface crusting caused by rainfall. In F. Callebaut, D. Gabriels and M. De Boodt (eds) *Proceedings of Symposium on Assessment of Soil Surface Sealing and Crusting*, Ghent 210–217.

Bowyer-Bower, T.A.S. and Burt, T.P. (1989) Rainfall simulators for investigating soil response to rainfall. *Soil Technology*, **2**, 1–16.

Brasher, B.R., Franzmeier, D.P., Valassis, V. and Davidson, S.E. (1966) Use of SARAN-resin to coat natural soil clods for bulk-density and water-retention measurements. *Soil Science*, **101**, 108.

Brey, V. and Cammeraat, L.H. (1993) The determination of micro-aggregation with the use of the Quantachrome Microscan II Particle Size Analyzer. Internal report University of Amsterdam.

Brown, E., Skougstad, M.W. and Fishman, J. (1970) Methods for collection and analysis of water samples for dissolved minerals and gases. Tech. Water-Resources Inv. Bk 5 Chap A1.

Callebaut, F., Gabriels, D, and De Boodt, M. (eds) (1985) *Proceedings of Symposium on Assessment of Soil Surface Sealing and Crusting.* Flanders Research Centre for soil erosion and soil conservation, Ghent, 374 pp.

Cammeraat, L.H. (compiler) (1992) *MEDALUS Field Manual Version 3.1.* With contributions from S.C. Clark, A.C. Imeson, L.H. Cammeraat, J, Meliá and J.J. Collin. Printed at University of Bristol, 115 pp.

Cammeraat, L.H. and Prinsen, H. (eds) (1995) *The MEDALUS Application on CD-ROM*, Chaumont-Gistoux/Amsterdam:Da Vinci/FGBL-UvA, pp. 1–154, 110 mb.

CSIRO (1990) *DEMON. An Instrument for Estimating Leaf Area from Transmittance of Sunbeams. User's Manual.* CSIRO Centre for Environmental Mechanics Canberra, Australia, 26 pp.

Daniels, R.B., Gamble, E.E., Bartelli, L.J. and Nelson, L.A. (1968) Application of the point-count method to problems of soil morphology. *Soil Science*, **106**, 159–152.

Diamantopoulos, J. (1993) Petralona, Thessaloniki, Greece. In *MEDALUS Final Report*. Bristol, pp. 560–580.

Etienne, M. (1989) Non destructive methods for evaluating shrub biomass: a review. *Acta Oecologia*, **10**, 115–128.

FAO (1977) *Guidelines for Soil Profile Description.* Soil Survey and Fertility Branch, Land and Water Development Division, FAO, Rome.

FAO (1988) FAO/UNESCO Soil Map of the World, Revised Legend, Wageningen: World Resources Report 60, FAO, Rome, Reprinted as Technical Paper 20, ISRIC pp. 1–38.

French, N.R., Steinhorse, R.K. and Swift, D.N. (1979) Grassland Biomass Trophic Pyramids: Perspectives in Grassland Ecology, *Ecological Studies*, **32**, Springer Verlag, New York, 204 pp.

Goudie, A. (ed.) (1981) *Geomorphological Techniques.* Allen & Unwin, London.

Grieg-Smith, P. (1983) *Quantitative Plant Ecology*, (3rd edition). Cambridge University Press, Cambridge.

Hanson, C.T. and Blevins, R.L. (1979) Soil water in coarse fragments. *Soil Science Society of America Journal*, **43**, 819–820.

Harrison, A., Latter, P. and Walton, D.W.H. (eds) (1988) *Cotton Strip Assay: An Indication of Decomposition in Soils.* ITE Symposium No 24, Institute of Terrestrial Ecology, Grange-over-Sands, UK.

Harrison, A., Taberner, M. and Hurcom, S. (1993) Site-based remote sensing of vegetation and land cover. *MEDALUS I Final Report* Bristol, 225–263..

Hem, J.D. (1970) *Study and Interpretation of the Chemical Characteristics of Natural Water.* Geological Survey Water-Supply Paper 1473, US Government Printing Office, Washington.

Hin, J. (1993) Relationships between soil aggregation parameters and visually differentiated Mediterranean landscape units. MSc Thesis, University of Amsterdam, 65 pp (in Dutch with English summary and figure captions).

Huntington, T.G., Johnson, C.E., Johnson, A.H., Siccama, T.G. and Ryan, D.F. (1989) Carbon organic matter and bulk density relationships in a forested spodosol. *Soil Science*, **148**, 380–386.

Imeson, A. C. and Vis, M. (1984) Assessing soil aggregate stability by water-drop impact and ultrasonic dispersion. *Geoderma*, **34**, 185–200.

Imeson, A.C. and Vis, M. (1984) Seasonal variation in soil erodibility under different land-use types in Luxembourg. *Journal of Soil Science*, **35**, 323–331.

Kessler, J. and Oosterbaan, R. (1974) Determining hydraulic conductivity of soils. In *Drainage Principles and Applications*. Publication 16 III. International Institute for Land Reclamation and Improvement, 253–256.

Klinka, K., Green, R.N,, Trowbridge, R.L, and Lowe, L.E. (1981) *Taxonomic Classification of Humus Forms in Ecosystems of British Columbia: First Approximation*. Ministry of Forest Province of British Columbia.

Klute, A. (ed.) (1986) *Methods of Soil Analysis*. American Society of Agronomy and Soil Science Society of America Madison 1188 pp.

López-Bermúdez, F. (1993) El Ardal, Murcia, Spain. In *MEDALUS Final Report*. Bristol, 433–460.

Mason, C.F. (1977) *Decomposition*. Edward Arnold, London.

Milner, C. and Elfyn Hughes, R. (1968) *Methods for the Measurement of the Primary Production of Grassland*. Blackwell, Oxford.

Mitchel, D.T., Coley, P.G., Webb, S. and Allsopp, N. (1986) Litterfall and decomposition processes in the coastal fynbos vegetation South-Western Cape, South Africa. *Journal of Ecology*, **74**, 977–993.

Morris, J.W., Bezuidenhout, J.J. and Furniss, P.R. (1982) Litter Decomposition: Ecology of Tropical Savannas, *Ecological Studies*, **42**, Springer Verlag, New York, 689 pp.

Newbould, P.J. (1970) *Methods for Estimating the Primary Production of Forests*. Blackwell, Oxford.

Oliveros, C., Keime, M.P., Rodriguez Lahuerta, C., Chabart, M. and Eulry, M. (1993) Roussillon, France. In *MEDALUS Final Report*. Bristol, 504–533.

Page, A.L. (ed.) (1982) *Methods of Soil Analysis. Part 2: Chemical and Micro-biological Properties*, 2nd Edition. American Society of Agronomy and SSSA, Madison.

Pettijohn, E.J. (1975) *Sedimentary Rocks*, (3th edition). Harper and Row Publ. Inc, New York, 626 pp.

Poesen, J., Ingelmo-Sanchez, F. and Mücher, H. (1990) The hydrological response of soil surfaces to rainfall as affected by cover and position of rock fragments in the top layer. *Earth Surface Processes and Landforms*, **15**, 653–671.

Puigdefabregas, J. (1993) Tabernas, Almeria, Spain. In *MEDALUS Final Report*. Bristol, 433–503.

Quinton, J. (1990) European soil erosion model, validation data-base, data sheets and notes for guidance. Internal Report, EUROSEM Project.

Reinhart, K.G. (1961) The problem of stones on soil-moisture measurement. *Soil Science Society of America, Proceedings*, **25**, 268–270.

Roxo, J. and Cortesao Casimiro, P. (1993). Lower Alemtejo, Beja and Mertola, Portugal. In *MEDALUS Final Report*. Bristol, 406-432. .

Soil Survey Staff (1992) *Keys to Soil Taxonomy*, 5th edition. SMSS Technical Monograph No. 19. Pocahontas Press Inc, Blacksburg, Virginia, 556 pp.

Stakman, W. P., Valk, G. A. and Harse, G.G.u.d. (1974) *Determination of soil moisture retention curves; I. Sandbox apparatus, II. Pressure membrane apparatus*, Wageningen: I.C.W.

Van Genuchten, M. Th. (1980) A closed-form equation for prediction of the hydraulic conductivity of unsaturated soils. *Soil Science Society of America, Journal*, **44** 892–898.

Wischmeier, W.H. and Smith, D.D. (1978) Predicting rainfall erosion losses. *Agriculture Handbook 57*. US Department of Agriculture, Washington DC, 58 pp.

Yassaglou, N., Kallianou, Ch., Danalatos, N., Kosmas, C., Moustakas, N. and Tsatiris, B. (1993) Spata, Athens, Greece. In *MEDALUS Final Report*. Bristol, 581–607.

6

Inner Lower Alentejo Field Site: Cereal Cropping, Soil Degradation and Desertification

M. J. Roxo, P. Cortesão Casimiro and R. Soeiro de Brito

Departamento de Geografia e Planeamento Regional. Universidade, Nova de Lisboa, Portugal

6.1 INTRODUCTION

In the early 1960s, it became clear that in the Inner Alentejo the Portuguese national agricultural policy of widespread wheat cultivation was causing land degradation. Productivity was falling rapidly, and the most visible consequence was a steady migration of people towards major cities, escaping from agriculture that in some areas was no longer viable. The study of erosion processes and the possible mitigation actions has a strong relevance to the economy of this area, not only in terms of production and added revenue, but also because a very significant number of people, and their future livelihoods, are involved.

Within the framework of the MEDALUS field programme, the Universidade Nova de Lisboa worked at two field sites, the Vale Formoso Erosion Centre which was established in 1961 and the Herdade de Almocreva, farmland currently under wheat cultivation. As with all the other MEDALUS field sites, these provided the complete suite of data for the core field programme, but they also put special emphasis on the role of wheat cultivation and land use in soil degradation. The main objectives were:

1. To look at the effects of different land uses on soil erosion.
2. To measure soil moisture regimes under different land uses, cultivation and abandonment, and different rehabilitation schemes.
3. To describe the effects of rainfall on cereal production.

The principal theme is the relationship of land use to processes like erosion and variables like soil moisture. Land use is seen as the most important control of degradation and within the region it is changing, mainly due to abandonment of cultivation. Vale Formoso is an area which is clearly degraded and already marginal to agricultural production. The area round Almocreva is less degraded, the soils are better and agriculture will continue for some time to come. The land-use history and socio-economic

Mediterranean Desertification and Land Use. Edited by C. Jane Brandt and John B. Thornes.
© 1996 by John Wiley & Sons, Ltd.

background have been explored in some detail because they give a temporal and spatial significance to the degradation processes and provide a means for moving from the experimental plot scale to the regional scale.

The core field programme data-set was supplied to the other project groups for use in model development and validation despite the fact that most of the plots were under cereal cultivation and not under natural vegetation. However, the results show that erosion from plots with natural vegetation is much smaller than that from fields ploughed for wheat. The Cobres river basin was used for the work on the SHE model (see Chapter 15).

6.2 GEOGRAPHICAL SETTING

6.2.1 The field sites

The locations of the Vale Formoso and Herdade de Almocreva field sites are shown in Figure 6.1.

Vale Formoso Erosion Centre

Vale Formoso is located on the River Guadiana catchment area, 1.5 km from Vale do Poço (Montalvo), at km 25 on the Serpa–Mértola road. The area has a rolling topography, consisting of gentle slopes, and flattened summits with altitudes ranging from 180 to 220 m. Simple slope profiles never exceed 25%. However, in some places, the landscape monotony is broken by sectors with steep slope angles, due to outcrops of bedrock and the streams become deeply incized. There are steep slopes on the Guadiana, Limas and Cobres among other rivers.

This general morphology is related to large metamorphic schist outcrops from the Lower Devonian. Among these, luminous schists dominate, cut through by large and small quartz veins. The rock fragments to be found on the soil surface are mainly quartz, and less weathered pieces of schist.

Herdade de Almocreva

The field site is located near Beja, 37°57′N–7°55′W, in Herdade de Almocreva. The altitudes in the area range from 220 to 180 m. The hills have smooth flat tops and broad, little-incized valleys. The average slope angles are low, 6.9 to 9.4 %. The site is in the headwaters of the Sado river, which has an annual full regime. Locally the river network near Almocreva is dendritic with a seasonal regime.

The geology consists of Carbonic yellow schist, reasonably weathered, covered in some places with continental deposits (Rañas) from the Pliocene. The soils are Vx Mediterranean, red/yellow schist soils and Sr Raña, which come from similar deposits. The former cover 70% of the area and the latter 30%. The soils are shallow and dominantly influenced by schist. The A horizon is only 15 to 25 cm deep, with a gradual transition to a B–C horizon that may be very shallow or could reach, in some cases, as much as 50 cm.

The schists have low productivity and low specific retention. Permeability is usually higher on the upper weathered area, near the fracture zone where the schist is bedded with other rocks. In areas with higher hydrologic capabilities, water is not usually found deeper than 50–60 m.

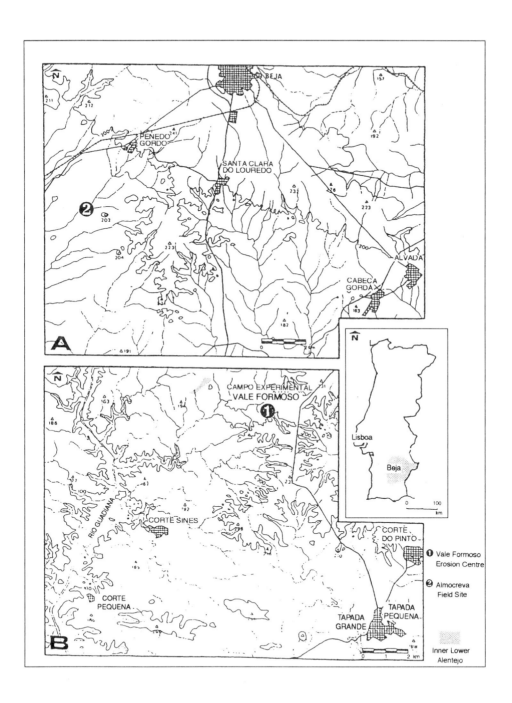

FIGURE 6.1   Location of the field sites

6.2.2 Climate

At Vale Formoso there is a meteorological station sited about 500 m from the erosion plots and this has been recording information since 1931. This 60-year record provides a very detailed description of the local weather.

The hydrological year is taken from September to August. September is usually the beginning of the agricultural year when the first autumn rain occurs following several months of drought. Figure 6.2 shows that although the annual rainfall is highly variable there are distinctly wetter periods such as the 1940s (average of 763.5 mm) and distinctly drier periods such as the 1970s and part of the 1980s (average of 461.8 mm).

If the following exceptional years are discounted (1949/50—1079.2 mm, 1946/47—1070.0 mm, 1989/90—1044.7 mm or 1980/81—236.4 mm, 1991/92—278.6 mm), there is a trend for the extremely wet years to become less wet and for the extremely dry years to become more dry although the amplitude between the extremes is increasing. The number of consecutive years with similar annual totals is variable, but is never more than 3 years. For example there was a wet period from 1945 to 1948 (842.4 mm/1070.0 mm /819.8 mm), and a dry period from 1971 to 1974 (371.4 mm/ 357.5 mm/317.3 mm).

The inter-annual irregularity is clearly illustrated when a wet year is followed by a dry one and vice versa. Some examples of dry–wet sequences are: 1944/45 (352.8 mm) followed by 1945/46 (842.4 mm), 1949/50 (1079.2 mm) and 1950/51 (404.9 mm) or more recently, 1989/90 (1044.7 mm) followed by 1990/91 (402.8 mm) and an even dryer 1991/1992 (278.6 mm).

Seasonal variability in rainfall has also been considered with autumn being the months of September, October and November; winter the months of December, January and February; spring the months of March, April and May; and summer the month of June, July and

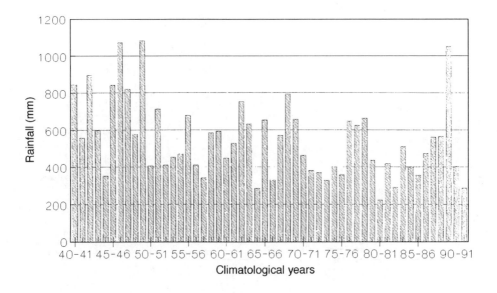

FIGURE 6.2   Annual rainfall, Vale Formoso 1940–92

August. Table 6.1 shows that most of the annual rainfall (36.7%) tends to fall during the winter months with roughly equal proportions in autumn and spring (about 29%). The summer months are much drier and contribute only 5% to the annual total. Table 6.1 also shows the maximum and minimum rainfalls for each season and that the timing of the periods of maximum rainfall really is very variable. For example while 42.0% of the annual rainfall fell in the winter of 1949/1950, only 4.6% fell in the winter of 1980/81 and, although there is frequently no rain at all during the summer, in 1938-39, 15.3% of the annual rain fell in the summer.

The timing of the rainfall is especially important in this region of Portugal because the land use is predominantly agricultural and the crop yields depend on rain coming at the right time.

Autumn is the most crucial season, as the timing and amount of the first rain events determine the start of the agricultural year. Too much rainfall will make ploughing and/or seeding technically difficult because the soil becomes waterlogged but too little means that the soil is extremely compacted and crusted. Because the field site at Herdade de Almocreva was located on agricultural land all instrumentation had to be removed and reinstalled after ploughing. The dependence of the time of ploughing on the autumn rain meant that the periods of recording at the site did not always coincide with those of other MEDALUS sites.

During the winter period the wheat plants start to develop. Too little rain restricts the growth and too much may saturate the soil and rot the roots. If the roots develop superficially on saturated soil, when the soil dries the plants will die because they cannot reach the moisture deeper in the soil.

Spring is an important season for determining the crop yield because the seeds need rain in the spring to develop. If it is too much diseases can develop and wind and the weight of water during heavy storms can break the stalks. Too little rain does not allow the seeds to develop. In northern Europe countries, the harvest takes place in August making use of more solar radiation, higher moisture availability and warmer temperatures. In Mediterranean countries the dry period starts by the middle of May, when the high temperatures have already dried the soil and the spikes so that no further growth is possible, and the harvest takes place before the middle of June. In dry years, maturation occurs in too short a period because high temperatures sometimes dry the soil and spikes as early as late April.

Clearly it is important that when future climatic scenarios are developed, they include some account of seasonal rainfall patterns. This region is heavily dependent on rain-fed wheat production and could suffer from regional-scale unemployment, land abandonment and migration of the population to the cities.

TABLE 6.1   Seasonal rainfall, Vale Formoso, 1931—92

Season	Average (mm) 1931–92	Max. total (mm) annual%	Min. total (mm) annual%
Autumn	163.5 (29.4%)	473.8 (45.3%) 1989/90	20.7 (5.9%) 1975/76
Winter	210.8 (36.7%)	453.1 (42.0%) 1949/50	10.9 (4.6%) 1980/81
Spring	156.1 (28.5%)	415.8 (49.4%) 1945/46	14.2 (2.2%) 1976/77
Summer	25.72 (4.9%)	104.4 (15.3%) 1938/39	0.0 Frequently

It is worth looking at the rainfall of 5 recent years in detail because they display not only extreme annual values but also large seasonal oscillations (Table 6.2). In fact, the first agricultural year of the MEDALUS project (1991/92) turned out to be the second driest year ever recorded in Vale Formoso, with an extremely poor wheat yield. The drought occurred all over the region, so the Almocreva field site was also subjected to a severe drought whose consequences were aggravated by a previously dry year. Groundwater and reservoirs were nearly exhausted and only survived because 1989/90 had been an extremely wet year.

The climatic characteristics of the site at Herdade de Almocreva are similar to Vale Formoso and variability is mainly due to local phenomena, such as thunderstorms, which can increase or decrease the annual totals by 10 to 20%.

6.2.3 Geomorphology and soils

Geomorphologically, the Inner Lower Alentejo is a peneplain with Quaternary river incisions, related to the proximity of the Guadiana river acting as a secondary base-level. It consists of a rolling topography, with flattened summits, at an average altitude of 200–300 m, and slope angles around 15–25% with simple profiles. The lithology is highly metamorphosed and weathered Hercinian schist, rock outcrops are common, and residual landforms are due to quartzite crests associated with overthrusts. Slope angle, valley depth and rock outcrops increase closer to the Guadiana river.

The soils are, consequently, shallow Lithosols of schists and greywackes, classified as 'red Mediterranean soils'. Rock fragments exist in high quantities, usually quartz and schist fragments brought up by the plough on the surface and subsurface. At Vale Formoso the soil texture is roughly the following (30–40% sand, 18–24% silt and 10–28% clay). Soil depth ranges from 20 to 40 cm, depending on the degree of degradation.

6.2.4 Land-use history and socio-economic background

The Vale Formoso Erosion Centre is located in the Mértola Municipality. The area's land-use changes between *c.* 1900 and 1985, which can be seen as a potential cause of environmental degradation, are summarized in Table 6.3.

By the time of the first historical record, the unpublished 1897 Land-Use Map, only 30% of the total municipal area was under natural vegetation, implying that cropping and grazing were already very significant. However, the degradation process started much earlier, in the 12th Century, as the country was being reconquered southwards from the

TABLE 6.2   Seasonal rainfall (mm), Vale Formoso, 1987–92

Year	Total	Autumn	Winter	Spring	Summer
1987/88	595.2	225.1 (38%)	230.1 (38%)	81.9 (14%)	58.1 (10%)
1988/89	597.3	237.0 (40%)	86.5 (14%)	262.9 (44%)	10.9 (2%)
1989/90	1044.7	473.8 (45%)	353.3 (33%)	224.2 (21%)	6.4 (1%)
1990/91	402.8	114.3 (28%)	141.3 (35%)	127.5 (32%)	19.7 (5%)
1991/92	298.4	86.1 (29%)	71.2 (24%)	84.6 (28%)	56.5 (19%)

TABLE 6.3   Evolution of land-use area: percentages in 1897, 1950 and 1985

Land use	1897	1950	1985
(Fallow + cereals + ploughed)	62.95	89.60	53.3
Quercus	7.13	7.47	13.2
Shrubs (natural vegetation)	29.17	1.23	31.3
Unproductive (rock + bare)	0.57	1.07	1.1
Total	99.82*[1]	99.37*[2]	98.9*[3]

*[1]other land uses
*[2]other land uses
*[3]1.1% unclassified.

Arabs, and immense feudal properties were donated to religious orders for management and protection.

By the end of the 19th century, as a result of the Liberal Revolution that confiscated almost all the Church properties and privileges (including Middle Age relics, like taxes levied when people moved within counties), the land was sold to the people (a recently emerged middle-class bourgeoisie). Following this change in the ownership of land, in 1899 specific subsidies to promote wheat production were created and wheat import was almost completely banned. This initiated a wave of occupation and colonization of large areas of land that were not productive before.

As a result, by the beginning of the 20th century, 63% of the Mértola Municipality had already been cultivated although productivity was extremely low. The land was not ideally suitable for cereal cropping, because of the general steepness of slopes, the soil type and the land-use capacity. There were areas of *Quercus* trees which were cleared of underlying shrubs and which provided foraging for pigs. Almost 30% of the area was still covered by natural vegetation, but it was intensively exploited for honey, wood and hunting, as it had been for centuries.

The First and Second World Wars concentrated the minds of many nations on self-sufficiency for food. In Portugal, the dominantly low-income, working classes depended heavily on bread for their survival. Thus, further actions were undertaken to ensure self-sufficiency in cereals, although this was never attained. Following to some extent the principles of Mussolini's Bataglia del Grano, almost all the remaining natural vegetation areas (hill tops and steepest slopes) were ploughed for cereal cultivation.

Most of the Inner Lower Alentejo turned into an area of cereal monoculture; it was generally tree-less, with some patches of *Quercus* and shrubland. This was further assisted by a Governmental Policy called the Wheat Campaign, that also supported farmers by supplying seed, fertilizer and subsidies for machinery and crop insurance, under Salazar's slogan 'wheat is the frontier that best defends us'. This policy, induced by a mixture of wheat deficiency and the need to encourage human settlements in low-density rural areas, was the first of several measures leading to a very closed economy.

The first years of this campaign, 1929–35, produced fairly good yields. Climatically some of the years were almost ideal and the soil had all the properties derived from a balanced, fairly undisturbed, natural vegetation cover. The campaign encouraged further clearance of natural vegetation, although the soils were clearly not suitable. Productivity started to fall dramatically. The population grew to a maximum never achieved before, but started to decline soon afterwards. A copper mine, São Domingos, which was owned and operated by a British firm, Mason & Berry, absorbed some of the unemployed agriculture workers. Most of the people rented rather than owned their land. Nevertheless, by 1950, 90% of the Municipality area was agricultural land, and only 12% was covered by natural vegetation.

By the 1950s it was clear that soil degradation was becoming a serious problem, and in many areas the land became unproductive, a wasteland where a stony regolith covered huge areas, and natural vegetation was given no chance of recovering. Along with the departure of many people to the main cities (Lisbon's industrial suburbs) and to foreign countries (France and Germany), agriculture started to decline spatially. Despite this, because of factors such as the colonial war, the isolation of Portugal in trade terms, and the late industrialization which began in the 1950s and early 1960s, agriculture remained the main sector of the national economy.

The intensity of agriculture depleted the soils, leading to massive use of fertilizers. This intensification, coupled with the late mechanization that rapidly became common in this region, contributed to further soil degradation mainly by structural destruction. This decline in soil productivity was a consequence of several factors:

— soils were not adapted to these kinds of crops and cropping techniques, they barely had any agricultural potential;
— the slopes were generally too steep and together with the lithology promoted intensive overland flow, which quickly eroded the shallow topsoil;
— the bare soil exposed to the local climatic conditions lost both volume and fertility.

By the beginning of the 1960s, the consequences were the following:

— massive land-abandonment and emigration which continued throughout the whole decade;
— extensive exposure of rock outcrops due to the total loss of soil in some places;
— valleys infilled with sediment carried from the slopes above;
— extended fallow periods lasting as long as 10 years in some areas ;
— low biomass production and minimum biodiversity.

After the 1974 revolution a collective type of landownership was tried, but the agrarian reform failed. In 1981 the emergence of the tertiary sector in the economy resulted in a final blow against a decaying cereal agriculture on highly degraded areas. After joining the EEC in 1986 this became even clearer, because production costs were three to four times higher than elsewhere in northern Europe, and the open-market model implied free trade. The population continued to decrease and reached such a level that in 1991 fewer people lived in the Municipality than in 1864. In 1985, natural vegetation covered 31% of the whole area, against agriculture which occupied only 53.3% of the area, most of which is fallow land, used for grazing.

At the time of writing, in 1993, the tendency is clearly towards more land abandonment. More and more fields are becoming fallow and/or natural vegetation areas. This process clearly represents a clear, though slow, environmental recovery. In socio-economic terms, hunting and related enterprises are becoming dominant.

6.3 EXPERIMENTAL LAYOUT

6.3.1 Vale Formoso Erosion Centre

The Experimental Erosion Centre of Vale Formoso was established in 1961. As can be seen in Figure 6.3, it consists of 15 Wischmeier erosion plots, 20.00×8.33 m, and a half-sized plot of 20.00×4.15 m. These are located side by side around a small hill facing east-southeast. The slope angles vary from 10 to 20%.

The soils are red schist soils, predominantly thin, with little organic matter and a high percentage of coarse schist and quartz fragments (rocks and gravel). Soil profiles dug near the plots show that the Ap horizon normally lies on a transition CB horizon, made of reasonably weathered bedrock.

Near the plots, divided into two groups, one facing south-southeast and the other east-southeast, there is meteorological equipment: several rain gauges inclined at different angles (vertical to 30), three continuously recording rain gauges and evaporation pans. This equipment continued to be used during the project.

At the end of the 1989/90 agricultural year the crop rotation on the 16 plots was as follows:

— Plots 1, 2, 7, 7x, 10 and 11: fallow/wheat
— Plots 3, 4, 5, and 6: fallow/wheat/sideration/wheat[1]
— Plots 8, 9, 12 and 13: fallow/wheat/legumes, alternated with sideration/wheat
— Plots 14 and 15: wheat/natural pasture with intercropping (3 years), clover (4 years)

In 1989/90 the land cover scheme for the centre plots was changed, although the rotation wheat/fallow on plots number 1, 2, 10 and 11 was kept as a control. The objective was to estimate soil erodibility under extreme conditions. Four plots (6, 7, 14 and 15) were left unplanted, but were ploughed up and down as soon as any vegetation started to grow on them. In reality ploughed fields are common all year round in the region, and represent a theoretical situation of maximum erodibility depending on the erosivity of possible rainfall events.

On the seven remaining plots, following the Alentejo Agricultural Regional Services directive, a new crop rotation scheme was adopted, on a basis of rotation in time and not on space. The introduction of the same culture on several plots, with different slope angles and exposure, resulted in a reduction in the number of variables and enabled a better understanding of the influence slope angle and exposure on soil erosion.

In September 1988 a new plot was built under the auspices of a previous project (EEC—EPOCH—'Climatic variability in semi-arid environments in Spain and Portugal'), to the standard dimensions of 20.0×8.33 m. It was covered with natural pasture and was protected from grazing by a fence. This plot (number 13A) is located between plots 13 and 14 and has an average slope angle of 18%. The proximity of fallow plots under continuous up and downslope ploughing (6, 7, 14 and 15) enables comparison between plots of equal exposure and slope angle but of extremely different susceptibilities to erosion.

---

[1]Sideration is a system whereby pasture is grown but is not cut, rather it is ploughed back into the soil to increase the organic matter content.

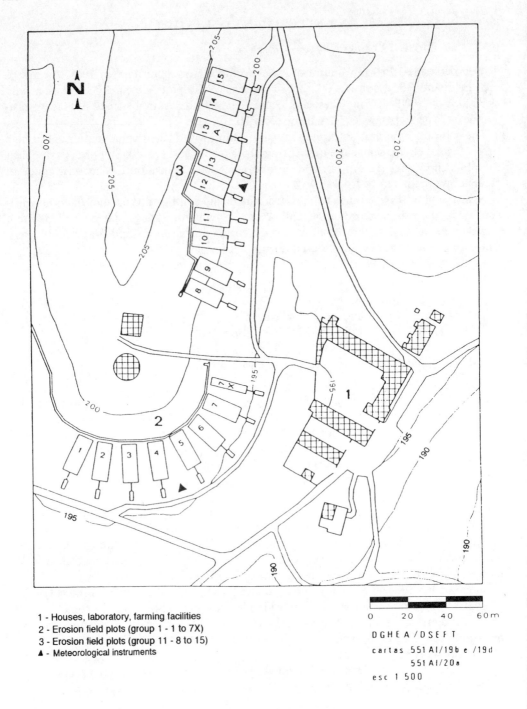

1 - Houses, laboratory, farming facilities
2 - Erosion field plots (group 1 - 1 to 7X)
3 - Erosion field plots (group 11 - 8 to 15)
▲ - Meteorological instruments

0   20   40   60 m

DGHE A /DSEFT
cartas 551 AI/19b e /19d
             551 AI/20a
esc 1 500

FIGURE 6.3   Diagram of Vale Formoso site

The remaining plot (7x) had several local species introduced onto it, *Cistus ladaniferous*, *C. rotundifolius* and *Thymus* among others. The objective was to evaluate how a forced recovery, with local species, could increase soil moisture and reduce erosion.

The following data are available for a continuous period of 30 years:

— daily rainfall and evaporation
— continuous rain-gauge graphs
— wind speed and direction (only for some periods)
— runoff and sediment loss by rainfall event
— crop records
— type and date of soil treatments
— notes on soil surface conditions, crop development, etc., per sediment collection.

Under the framework of this project the plots under study were those with the wheat/ fallow rotation. However, comparisons were also made with plots 7X and 13A, those with natural vegetation, to enable conclusions about non-agricultural land to be drawn.

6.3.2 Herdade de Almocreva

The site at Herdade de Almocreva covers several small basins:

— basins 1, 2 and 3 are agricultural land with a fallow/wheat/oat rotation, basin 4 is a mixed forest;
— basin 3, where the core programme was implemented, is 32 ha in area with 750 m of main stream line, a perimeter of 2125 m and an average slope angle of 9.4%.

These basins were instrumented under a previous United Nations Development Program to test the efficiency of different drainage methods and soil conservation techniques under local conditions. In basins 1 and 2, contour ditches dug and contour-line ploughing was practised and two measurement areas were established.

The data collected for the MEDALUS project were as follows:

— weekly soil moisture measurements, three samples taken along the catena for the two sub-fields
— monthly biomass for the two sub-fields
— runoff and sediment yield per event

6.4 RAINFALL AND SOIL EROSION

Sediment yield from the runoff plots at Vale Formoso has been recorded continuously since 1961. However, the products of several storms are frequently bulked together because it is not always possible to empty all the sediment traps between storms. While this is not an ideal situation, especially for process-orientated research, the data-set is of good quality and is extremely long. It is possible to draw some interesting conclusions from it especially concerning simple relationships between land use and soil erosion.

Erosion data, recorded since 1961 on plots 1, 2, 10 and 11, should be analysed with extreme care (Table 6.4). Over this period the plots 1, 2, 10 and 11 followed the same wheat/fallow rotation sequence with, at any time, plots 1 and 10 having the same cover and plots 2 and 11 having the other cover. Instead of grouping the data by plot, they should be grouped by cover type. Figure 6.4 shows erosion under wheat and stubble from individual storms in 1990/91; it also shows the amount of rainfall from each storm. The relationship between rainfall and sediment yield is not clear in this figure, so rainfall and sediment have been accumulated in Figure 6.5. In fact, cumulative values for the different covers do become the only means of drawing real conclusions from the data. The period of 1990–91 is the only for which all plots were subjected to the same rainfall events. If one major trend is visible, there is a positive correlation between cumulated sediment and rainfall, although there are a series of fluctuations. Generally the abandoned field plot has less sediment yield, and the wheat plots are very similar to the bare soil ones. Stubble assumes an intermediate position, growing closer to the abandoned field line over time.

The year 1990/91 was an interesting year in Vale Formoso and the values shown in Figure 6.4 can be placed within the context of a much longer series of data (Table 6.4). Attention should be drawn to several aspects of the data.

— Plot 2 lost 928 kg of sediment in 30 years, which represents a loss of 55 720 kg of soil per hectare (each plot is 1/60 of a hectare). Plot 11 lost 657 kg, and, although it is steeper than plot 2, the southerly orientation of plot 2 means that it is exposed to the prevailing winds. Plots 2 and 11 both lost more soil than plots 1 and 10, because over the 30 years they were more frequently under wheat during the extremely wet periods.

— Geomorphologically these values represent an enormous amount of sediment being removed from the slopes into the valleys and streams. It quickly causes the silting up of small dams and valley floors and increases the impact of flooding.

The aspects of plots 1 and 10 account for differences in their sediment yields. Again, plot 1 faces the direction of the prevailing rain. If we take a look at the monthly values for the two plots under each part of the rotation, things become clearer (Table 6.5).

— The highest monthly sediment yields are recorded under wheat, particularly in the month of December. By March the growing plants afford some protection to the soil.

TABLE 6.4   Vale Formoso 1961–91, sediment yield (kg) per plot, from   plots with cereal/fallow rotation

	Plot			
	1	2	10	11
Average 1961–91	1.7856	3.4018	1.0006	2.4090
Maximum per event 1961–91	61.8910	89.0840	64.6390	76.6260
SD	5.8609	11.1949	4.7973	9.3797
Total 1961–91	487.473	928.700	273.171	657.664

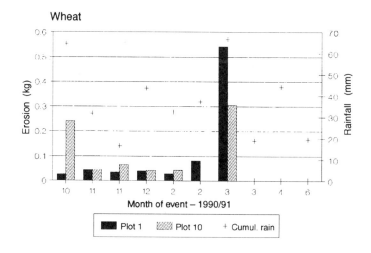

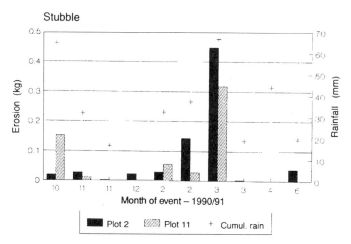

FIGURE 6.4   Erosion and rainfall events on wheat and stubble, Vale Formoso 1990/91

— The bare soil plots are also ploughed in September and March, thus accounting for the peaks of sediment yield in the autumn and again in March. On the whole there is more sediment from the bare plots than the ones under wheat.
— When the plots are not ploughed and are left under stubble they have the best protection and the erosion values are lower but still dependent on the rainfall. In March and April, when grass grows through the stubble, soil loss is less than from the other plots.

The first year of MEDALUS, 1991/92, was a year of extreme drought in the Inner Lower Alentejo and the region was declared a disaster area. The lack of rainfall occurred in autumn and winter which normally account for 60% of the annual rain. November had

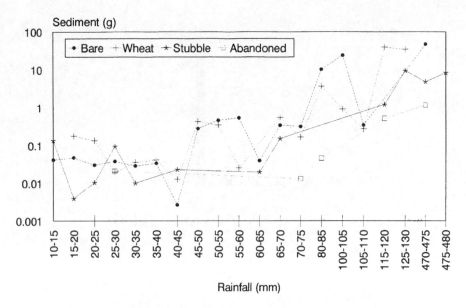

FIGURE 6.5   Erosion and rainfall, Vale Formoso 1989–92

only 8.1 mm compared with a long-term average of 70.2 mm. All the remaining months (except October, June and August) had below-average rainfall (Figure 6.6). As well as lower amounts of rain, there were fewer rain days, 55 this year compared with an average of 76 (Table 6.6).

Throughout 1992/92, the amount of sediment collected at Vale Formoso did not exceed the threshold at which it could be actually measured (the collection devices split the samples and need a certain amount to make an accurate recording). However, at the Herdade de Almocreva site all the sediment, rather than a sample, is collected and a consistent data-set is available and is shown in Figure 6.7.

TABLE 6.5   Monthly average sediment yield per event, 1961–1991, for several covers

Months	Erosion under wheat	Erosion under stubble	Erosion on bare soil
September	0.024	0.390	4.506
October	0.067	1.294	6.865
November	1.397	1.491	6.703
December	10.462	1.887	0.054
January	2.456	0.651	0.007
February	0.744	0.209	0.168
March	0.371	0.165	3.585
April	0.169	0.022	0.845
May	0.062	0.015	0.691
June	0.006	0.100	0.176
July	0.000	0.045	0.418
August	0.000	0.020	0.083

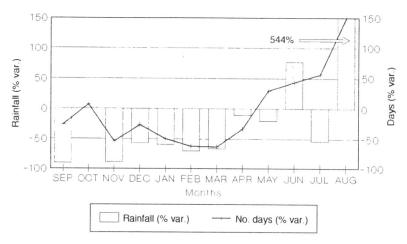

FIGURE 6.6   Percentage monthly difference in the total amount of rainfall and number of rain days in 1991/92 compared with the 1966–91 average.

The runoff data show an increase in runoff towards the bottom of the slope and the increase is more significant in Field A, which had a steeper slope angle. Future work, when the data-set is larger, will consider the influence of the exposure of the two fields to prevailing winds during storms. It is expected that this will explain the higher amounts of runoff in Field A than Field B.

Field A also had higher values of sediment yield. In Field B there is a visible increase in yield towards the bottom of the slope, but in Field A, in four of the events the top plot had higher values.

There are two anomalies in the data. The first event of the year produced a extremely high sediment yield in Field A. However, this is due to the fact that at the time of the storm, the plot was unbounded and a rill developed carrying a lot of runoff into the tank. Another anomaly shown in Figure 6.7 is that sediment but no runoff was recorded for the last event in 1992. In this case temperatures of around 38°C evaporated the water from the collection tanks.

6.5 SOIL MOISTURE

Time series of soil moisture changes during 1991/92 at both Almocreva and Vale Formoso are shown in Figures 6.8 and 6.9 respectively. Soil moisture never exceeds 16% at Almocreva or 20% at Vale Formoso. The decrease in moisture after an event is rapid in both sites due to high evapotranspiration. As already discussed above, there were few rain

TABLE 6.6   Annual rainfall events per daily amount of rainfall, Vale Formoso, 1991/92

Rainfall classes (in mm)	≤ 1	1–5	5–10	10–15	15–20	20–25
1991/92, no. of events	12	22	10	8	1	2

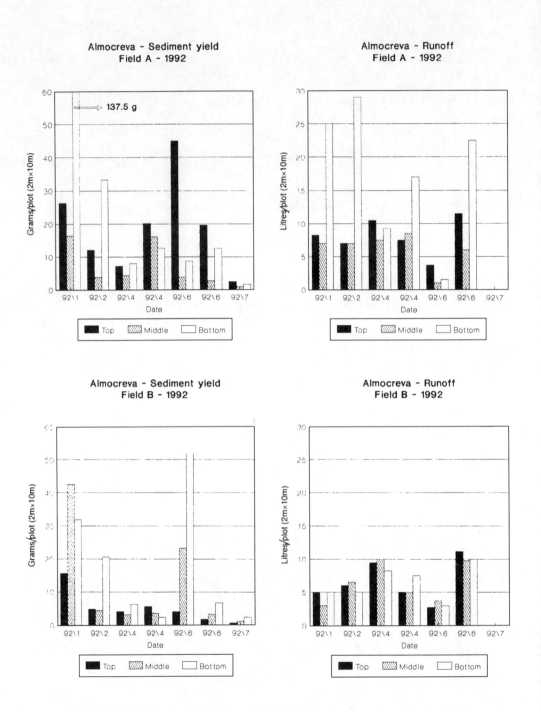

FIGURE 6.7   Sediment yield and runoff at Herdade de Almocreva sites A and B

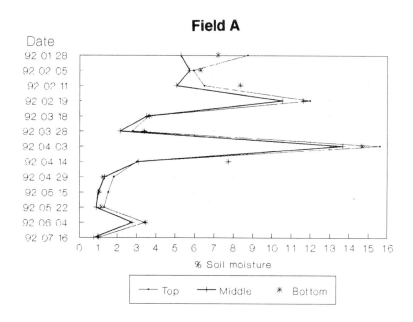

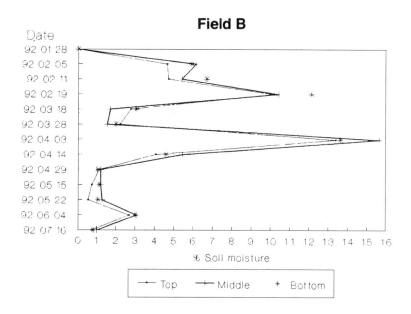

FIGURE 6.8 Soil moisture at Herdade de Almocreva sites A and B

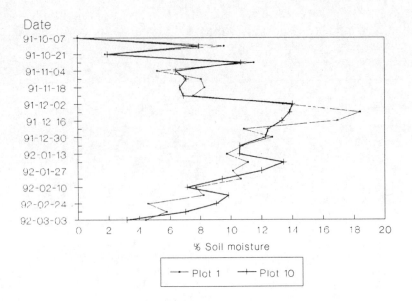

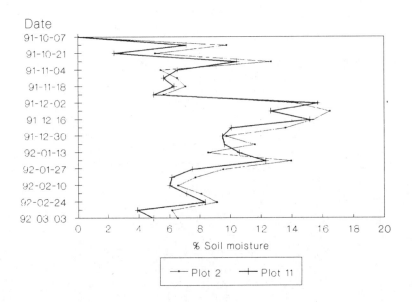

FIGURE 6.9   Soil moisture at Vale Formoso

events during the period and soil moisture recharge was predominantly achieved by heavy morning dew and sometimes frosts.

At Almocreva, the pattern of soil moisture down the two catenas is as follows. In Field B the bottom of the slope is usually the sector with higher moisture contents, the top has lower values. On Field A, because the slope profile is more regular and steep, the middle sector has the lowest values. Water moves from a reasonably flattened summit, to the lower sector on the flat valley bottom. This downslope distribution of soil moisture distribution is slightly reflected in wheat biomass production (Figure 6.10). In Field B there is an increase in biomass from the top to the bottom of the slope, in Field A the biomass is more evenly spread downslope.

The soil moisture patterns at Vale Formoso are different and Figure 6.9 shows the difference between a wheat cover (plots 2 and 11) and a stubble cover (plots 1 and 10). The plots with stubble tend to lose soil moisture slightly more slowly. Plots 1 and 2 have higher values, their slope angle is lower and soil clay content is higher. In the opposite group, plots 10 and 11, the steeper slope section and the sheltered position from prevailing winds during rainfall events, result in lower soil moisture content. Local thunderstorms can create a patchwork of different biomasses throughout the region and radically different biomasses in adjacent areas. Soil moisture values close to zero are often reached in the early summer and can persist until October.

6.6 BIOMASS

The biomass data recorded reflect the dry conditions of 1991/92. Although it started better than 1990/91, as the year went by it became clear that the wheat yield was going to be very small. The spikes appeared too soon, and the relative surplus of rain into the spring did not have any effect. However, it can be seen from Figures 6.10 and 6.11 that the April 1992 biomasses at Almocreva are higher than those at Vale Formoso. Although the humus content is very low for both field sites, 2.36% in Almocreva and 1.75% in Vale Formoso, the small improvement at Almocreva may account for the higher yields.

Figure 6.10 shows that, in Almocreva, Field A clearly has a higher biomass production, but Field B has more spike-weight, maybe due to a better solar exposition. Higher values appear at the bottom of the slope, which is to be expected, though lower soil moisture contents are found, caused by transpiration. However, the relationship between soil moisture and biomass is far from being a linear relation because, when the area reaches saturation, wheat biomass decreases. The absence of automatic weather station meteorological data does not allow further analysis.

Figure 6.12 shows both the very low soil moisture contents and the relative inverse relation between biomass increase and water availability. The larger biomass increases correspond to an increase in water consumption, leading to steep decreases in percentage moisture content. Over winter the moisture increases, but by May–June the soil moisture drops rapidly to zero. However, the importance of water to wheat development depends on its stage of growth, for example water is needed for spike development in April but too much rain physically damages the plants.

The fact that 1991/92 was so dry is useful for the development of extreme dryness scenarios. However, it should be remarked that the region is in reality in a state of catastrophe. The government took exceptional measures to diminish the effects of drought

130

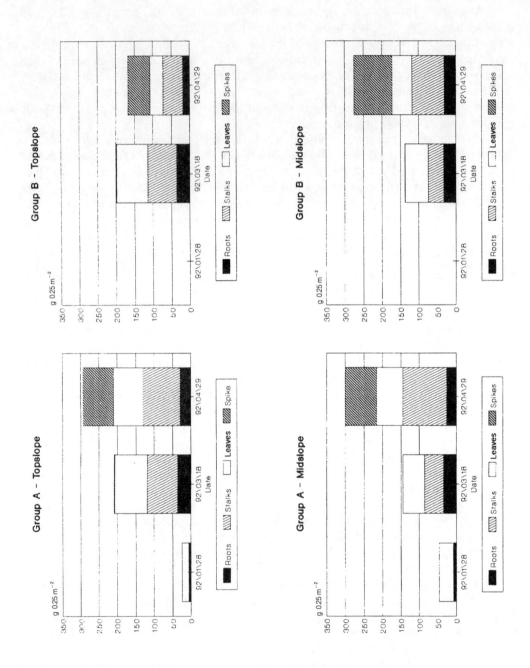

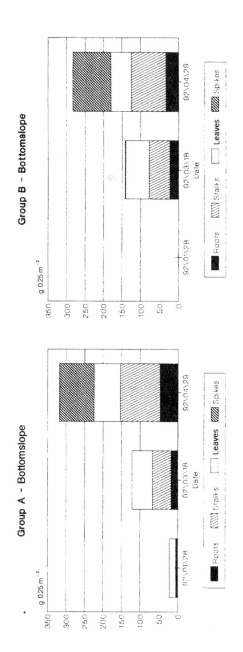

Figure 6.10 Biomass production at Herdade de Almocreva 1991/92, sites A and B

Group I

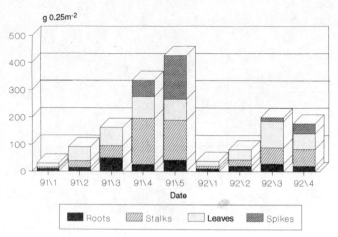

Group II

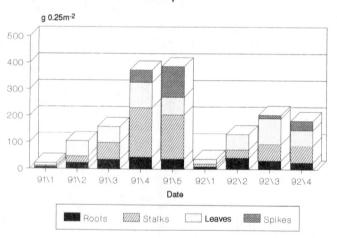

FIGURE 6.11 Biomass production at Vale Formoso, 1990/91 and 1991/92

on agriculture: an 18 million ECU direct grant (2700 ECU per hectare of lost wheat), low interest rate credit lines for up to 3 years (10%, 6% and 4% respectively, with a 27 million ECU limit). The fragility of some Mediterranean agriculture which is based on cereal cropping becomes very clear when these climatic anomalies occur, although throughout history variability has always been the only constant. Higher productivity in northern Europe, and set-aside on the horizon for low-productivity areas, put a heavy socio-economic burden on the region.

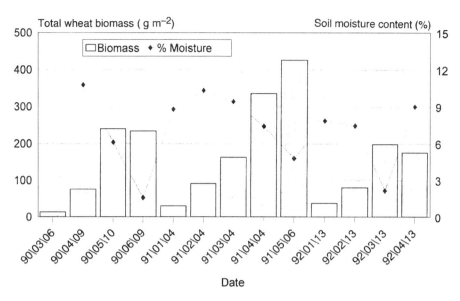

FIGURE 6.12   Total wheat biomass and soil moisture content, Vale Formoso, plots 1 and 2, 1990–92

6.7 CONCLUSIONS

Agriculture is the major factor responsible for environmental degradation in the Inner Lower Alentejo region.

Climatic variability is extreme, and has been a reality in this area as far back as meteorological data exist. Natural vegetation is regulated by water availability, but it is well adapted to severe water stress not only during the summer, but also for significant periods of low soil moisture and high evapotranspiration. Even with low growth rates, the plants themselves plus the litter produced by natural vegetation, protect the soil to a greater extent than cereal crops. The results in Table 6.7, concerning natural vegetation, and Table 6.8, concerning the cultivated plots, show this difference very clearly.

Although the period of study is significantly different, 30 years of records for the cultivated plots and only 3 years for natural vegetation, the year 1989/90 was an extreme year in the amount of rainfall (annual amount and event intensities) and erosion. Despite this, maximum sediment yields from the natural vegetation plots are well below the cultivated ones: plot 13A had 1/25th the yield and plot 7X (which is half the size of the other plots) at 1/5th.

TABLE 6.7   Vale Formoso sediment yield (kg) per event, from the natural vegetation plots, 1988–91

	Cistus plot (7X)	Abandoned plot (13A)
Average 1988–91	0.719	0.311
Maximum per event 1988–91	7.443	3.088
SD	1.782	0.795
Total 1988–91	20.145	8.732

TABLE 6.8 Vale Formoso sediment yield (kg) per event, from the wheat/fallow plots, 1961–91

	Plot			
	1	2	10	11
Average 1961–91	1.7856	3.4018	1.0006	2.4090
Maximum per event 1961–91	61.891	89.084	64.639	76.626

The cumulated sediment for the extremely wet year of 1989/90 is given in Table 6.9. It is clear that erosion is enormously greater under wheat, and even on plot 1 under stubble, than under natural vegetation on abandoned land. (Plot 10 is the exception here, but this plot has produced anomalous behaviour several times over the last 30 years.) *Cistus ladaniferous* also affords more protection than wheat but its dominance prevents the establishment of other shrubs, leaving large patches of bare soil except when the annuals are present.

One solution for the rehabilitation of extreme soil degradation on arable land would be to abandon the fields, as was done in plot 13A, and then use *Cistus* to control the erosion and to establish a natural vegetation community.

The results from Almocreva, along the slope catenas, are extremely interesting, including the differences between Fields A and B with the different orientations of their slopes, although clear conclusions are difficult to reach.

Desertification can be interpreted as a progressive decrease in the auto-regeneration capacity of natural system, where the input–output balance will result in a gradual loss of productive capacity. Because the soil supports all primary activity, the systems which depend on it will therefore become poorer. The degree of desertification will, consequently, be a direct reflection of how degraded the soil is, its thickness, and its nutrient and organic matter content.

It is considered that desertification was started by human activities, and that in this part of Portugal the soil-use capacity was originally extremely low. This fact, together with the climatic characteristics, made cereal-cropping agriculture the worst possible choice and it initiated a process that is now resulting in land abandonment, unemployment and severe social attrition.

As far as the natural ecosystem is concerned, land abandonment would certainly slow down desertification, eventually enabling some less degraded areas to recover. The future EC Common Agriculture Policy will lead, in this region, to a process of this kind, as wheat production will become almost completely unprofitable. Country to city migration will start again, creating problems in metropolitan areas. The introduction of exotic species

TABLE 6.9 Vale Formoso, cumulative sediment yield for several covers, 1989–90

	Stubble (1)	Wheat (2)	Stubble (10)	Wheat (11)	*Cistus* (7X)	Abandoned (13a)
1989–90 Total Sediment	63.704	169.210	5.611	155.580	18.103	4.687

creates some labour and local revenue, but is not a good choice in environmental terms, because it is usually done in steep areas associated with extremely high erosion rates.

6.8 ACKNOWLEDGEMENTS

We would like to acknowledge the help of the Direcção Geral de Hidráulica (Enga. Fátima Amaral), the Direcção Regional de Agricultura do Alentejo (Engo. Serra Mira), the Escola Superior Agrária de Beja (Prof. Engo. António Parreira), and Professor Doutor Mariano Feio, Augusta Jacob, and José Lourenço

The Vale Formoso Erosion Centre is the property of the Direcção Geral de Hidráulica e Engenharia Agrícola. The use of the laboratory, the implementation of new experiments and scientific instruments, the access to existing data and the use of the local facilities for accommodation purposes was negotiated with the owning institution. The Almocreva farm is the property of the Escola Superior Agrária de Beja who also came to an agreement for the use of the site.

6.9 BIBLIOGRAPHY

Boardman, J, Foster, I.D.L. and Dearing, J.A (1990) *Soil Erosion on Agricultural Land.* Wiley, Chichester, 687 pp.

Cardoso, J. and Carvalho, J (1965) *Os solos de Portugal, sua classificaçao, caracterizaçao e génese— a sul do Rio Tejo.* Direcçao Geral dos Serviços Agrícolas, Secretaria de Estado da Agricultura, Lisbon.

Feio, M. (1951) *A evoluçao do relevo do Baixo Alentejo e Algarve.* Estudo de Geomorfologia, Comunicaçaes dos Serviços Geológicos de Portugal, 32, Lisbon, 186 pp.

Feio, M. (1991) *Clima e agricultura,* Direcçao Geral de Planeamento Agrícola, Lisbon, 266 pp.

Feio, M. and Henriques, V. (1986) *As secas de 1980–81 e de 1982–83 e as principais secas anteriores. Intensidade e distribuçao regional.* Memórias do Centro de Estudos Geográficos, Lisbon.

Ferreira, I.M..M, Ferreira, A.J.R. and Sims, I.A. (1985) *Análise preliminar dos dados dos talhaes de escoamento do Posto Experimental de Vale Formoso para o periodo de 1962–63 (1979–80), em termos da equaçao universal de perda do solo.* Direcçao Geral de Hidráulica e Engenharia Agrícola, Évora.

Goudie, A . et al. (1990) *Geomorphological techniques.* British Geomorphological Research Group, Unwin Hyman, London, 570 pp.

PDCSA (1979) *Projecto de drenagem e conservaçao do solo no Alentejo—Trabalhos executados.* Relatório Técnico—Janeiro 1979 a Agosto de 1979, Beja.

Ribeiro, O., Lautensach, H. and Daveau, S. (1988) *O ritmo climático e a paisagem,* Vol. II— Geografia de Portugal. Sá da Costa, Lisbon, 296 pp.

Serviços, Geológicos de Portugal (1989) *Carta Geológica do Sul de Portugal—Escala 1:200.000— Notícia explicativa,* Lisbon.

Thornes, J.B. (1990) *Vegetation and Erosion—Processes and Environments,* Wiley, Chichester, 518 pp.

Vinhas A (1986) *Bacias experimentais de Almocreva, análise dos registos de precipitaçao e escoamento dos três anos hidrológicos (1980–81, 1981—82, 1982–83).* Ministério da Agricultura Comércio e Pescas, Direcçao Geral de Hidráulica e Engenharia Agrícola, PDCSA, Lisbon.

7

The Rambla Honda Field Site: Interactions of Soil and Vegetation Along a Catena in Semi-arid Southeast Spain

J. Puigdefábregas, J. M. Alonso, L. Delgado, F. Domingo,
M. Cueto, L. Gutiérrez, R. Lázaro, J. M. Nicolau,
G. Sánchez, A. Solé and S. Vidal

Estación Experimental de Zonas Aridas, CSIC, Almeria, Spain

C. Aguilera

Technical School of Agriculture, University of Almeria, Spain

and

A. Brenner, S. Clark and L. Incoll

Department of Pure and Applied Biology, University of Leeds, Leeds, UK

7.1 INTRODUCTION

Desertification and land degradation in the Mediterranean Basin show two contrasting trends (Le Houérou, 1991). In the southern part, continuous population growth increases the susceptibility of the landscape to degradation by enlarging the cultivated area and by increasing the grazing pressure on rangelands. In the northern part, with the social and technological changes that took place in the 1950s, traditional rural life disintegrated. Some areas support increasing intensification and immigration, while others suffer from loss of population and abandonment of land.

The interruption of the activity of man in the landscape raises some intriguing questions. It is known that man has been a very significant factor in shaping the Mediterranean landscape (Puigdefábregas, 1992). Therefore the cessation of human activity does not lead to any ideal or primeval condition, because such a condition has never existed without man. Vegetation is recovering in many places, but very little is known about the long-term response of the landscape. There are local increases in erosion

Mediterranean Desertification and Land Use. Edited by C. Jane Brandt and John B. Thornes.

and wild fires, and the destruction of man-made structures which controlled runoff and soil loss is common feature. These facts suggest that the interruption of man's activities is not a guarantee for long-term recovery of the landscape and that greater risks of instability of the landscape may be faced, especially if climate fluctuates.

The main goal of the Almeria group in the MEDALUS project is to provide information about the evolution of the semi-arid Mediterranean landscape and its sensitivity to climate once the activity of man has been interrupted. A typical catena with different land uses and elapsed times since abandonment was selected and a research programme was set up with the following objectives.

1.   To find relationships between land use, soil, vegetation and organic matter along a catena representative of a semi-arid environment.
2.   To provide information about the response of the elements of the catena, in terms of runoff, sediment outputs, leaf area index (LAI) and net primary productivity, to rainfall events and to seasonal and annual variations in climate.
3.   To describe the physiological strategies of the main building species and the micro-environmental variations created by their patchy distribution. To use this information to improve existing water-balance models.

The Almeria research programme is designed to provide both comparable data for the core MEDALUS project and information specific to the objectives of the Almeria group. In the following, the geographical setting and the experimental design are described, as are preliminary results from the first year's research.

7.2 GEOGRAPHICAL SETTING

7.2.1 The field site

The Almeria field site (Figure 7.1) is located in the contact zone between the south versant of the Filabres range, a core of Pre-Cambrian to Triassic metamorphic rocks (from the Nevado–Filabride complex), and the Neogene depression of Sorbas–Tabernas, in the eastern part of the Betic Cordillera. The area is covered by extensive alluvial fan systems which have developed since the Late Pliocene. It may be considered as a key area to study slope–fan relations in total sediment outputs and their significance at a regional scale (Harvey, 1987).

In the lower sector of the Rambla Honda (30S-WG-5509 UTM coordinates), an ephemeral river (rambla) draining a basin of $30.6 \, km^2$ in the southern versant of the Filabres range, a study area was selected. It is 300 m wide by 600 m long, from 630 m altitude at the rambla bed, to 800 m on the divide (Figure 7.1).

This sloping area consists of a catena of soils and vegetation (Figure 7.2), going from the upper sector with soils on micaschist bedrock and *Stipa tenacissima* L. tussocks, to soils on sedimentary deposits with *Anthyllis cytisoides* L. shrubs in the upper part of the alluvial fans and *Retama sphaerocarpa* (L.) Boiss. on the lower part of the fans and on drainage channels.

Although the agricultural land has been abandoned, the area is still grazed by sheep flocks and, in rainy winters, plots on alluvial fans and riverbeds may be cultivated.

The study area comprises two separate sampling sites, coded by the initials SP and MA:

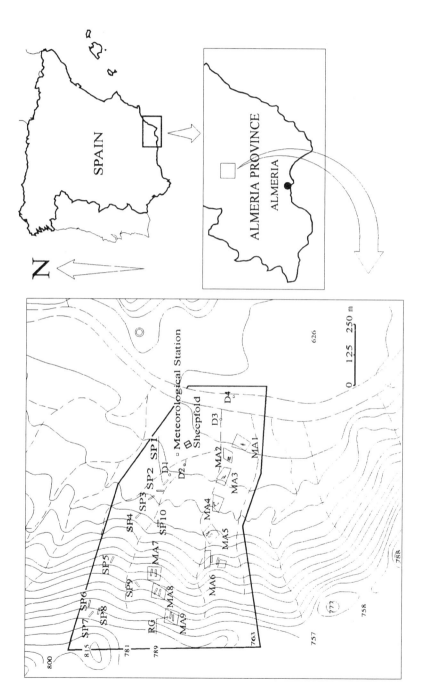

FIGURE 7.1   Location of Rambla Honda field site. SP = runoff plot in sampling site 1; MA = measurement area in sampling site 2; RG = rainfall gauge; D = bore hole; — — — = drainage line.

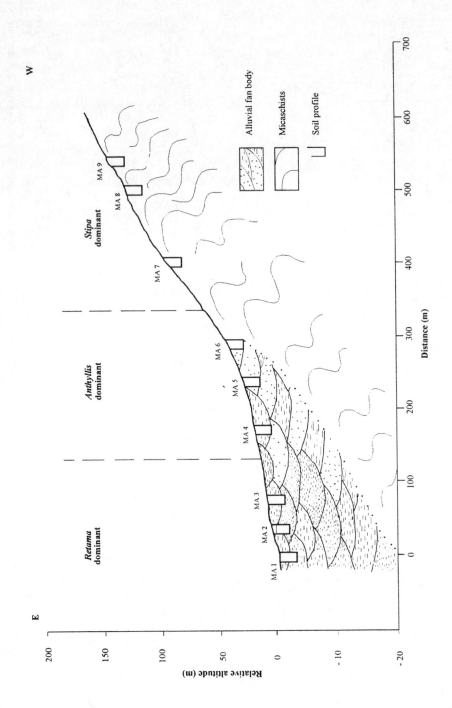

FIGURE 7.2 Diagram of the catena in the Rambla Honda showing the bedrock, the distribution of the vegetation and the positions of measurement areas (MA) and soil profiles

SP   This includes 10 runoff plots (SP1 to 10), 8×2 m, on abandoned fields on alluvial fans and slope colluvia. They were set up in November 1990 with the objective of collecting information about the role of old fields with low perennial plant cover and different gradients as potential sources of runoff and sediments. The characteristics of the plots are presented in Table 7.1.

MA   This site was set up in September 1991, and includes nine measurement areas (MA1 to 9) arranged in a line along the catena: three in *Retama sphaerocarpa*, three in *Anthyllis cytisoides* and three in *Stipa tenacissima* dominated areas (Figure 7.2). In each MA, two closed runoff plots, 10×2 m, were built, one with relatively dense and one with sparse perennial plant cover. In addition, each MA has been provided with three types of plots for sampling soil moisture, litterfall and non-destructive vegetation sampling. The main characteristics of the MAs are shown in Table 7.2.

Grazing was excluded from the whole study area from 1 September 1991 onwards through arrangements with the owners.

7.2.2 Climate

The climate is semi-arid with long hot summers. Geographical interpolations among nearby weather stations with long series (Lazaro and Rey, 1991) provide estimates of 300 to 350 mm annual rainfall, mainly in the cold season, and mean annual temperatures of 15.5 to 16.5°C. The nearby Tabernas meteorological station provides the most complete climatic record in the area (Figure 7.3). The 3-year data-set from the automatic weather station in Rambla Honda (installed in 1990), confirms the high annual and seasonal variability of rainfall (Figure 7.4).

Prevailing winds mostly come from N, NW and SE (Figure 7.5), following the alignment of the Rambla Honda valley. The wind speed is greater than 5 m s^{-1} for only 1% of the year (Figure 7.6).

Global solar radiation (Figure 7.7) reaches a maximum of 1000 W m^{-2} in summer and around 500 W m^{-2} in winter, indicating that the minimum atmospheric extinction is

TABLE 7.1   Characteristics of the runoff plots (SP) in sampling site 1

Runoff plot	Relief	Grass cover (%)	Shrub cover (%)	Altitude (m)	Slope gradient (°)
1	AF	40	5	650	12
2	AF	65	5	648	21
3	AF	20	15	655	27
4	COL	40	15	690	29
5	COL	70	5	725	33
6	COL	70	5	770	28
7	COL	90	10	773	41
8	COL	75	10	769	23
9	COL	80	10	740	35
10	COL	80	25	695	45

AF = alluvial fan; COL = colluvium.

TABLE 7.2 Characteristics of the measurement areas (MA) in sampling site 2

| | | | | | | Runoff plots | | | |
| | | | | | | Dense perennial plant cover | | Sparse perennial plant cover | |
Relief	Plant cover (%)	Surface (m²)	Altitude (m)	Slope gradient (°)	Vegetation	Slope gradient (°)	Plant cover (%)	Slope gradient (°)	Plant cover (%)
MA1 FD (ST)	8	1682	628	3	*Retama*	2	26	2	0
MA2 AF (CV)	19	1009	632	4	*Retama*	4	55	5	6
MA3 AF (CV)	15	1253	637	5	*Retama*	5	43	5	4
MA4 AF (CV)	29	778	648	7	*Anthyllis*	7	48	9	24
MA5 AF (CV)	32	644	662	12	*Anthyllis*	12	36	12	22
MA6 COL (ST)	25	324	681	20	*Anthyllis*	20	48	22	15
MA7 MO (CV)	34	611	720	25	*Stipa*	24	40	23	23
MA8 MO (CC)	39	466	730	17	*Stipa*	21	44	18	30
MA9 MO (CV)	31	437	752	27	*Stipa*	21	32	23	29

FD = fluvial deposit; AF = alluvial fan; COL = colluvium; MO = micaschists.
ST = straight; CV = convex; CC = concave.

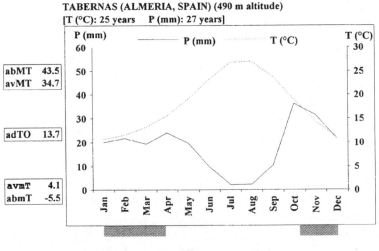

TABERNAS (ALMERIA, SPAIN) (490 m altitude)
[T (°C): 25 years   P (mm): 27 years]

| abMT | 43.5 |
| avMT | 34.7 |

| adTO | 13.7 |

| avmT | 4.1 |
| abmT | -5.5 |

abMT	Absolute maximum temperature (°C)
avMT	Average maximum temperatures from hottest month (°C)
adTO	Average of daily temperature oscillations (°C)
avmT	Average minimum temperatures from coldest month (°C)
abmT	Absolute minimum temperature (°C)
▨	Months with potential absolute minimum temperatures < 0 (°C)

17.9°C - average yearly temperature
218.3mm - average annual rainfall

FIGURE 7.3 Climatic diagram of the Tabernas weather station, according to Walter (1973)

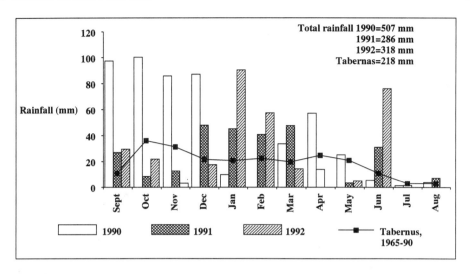

FIGURE 7.4 Monthly distribution of rainfall in Rambla Honda (630 m altitude) since the installation of the weather station. Annual periods start 1 September 1990 and finish 31 August 1992

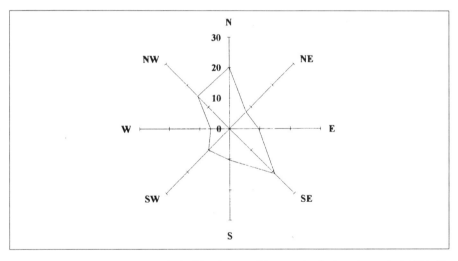

FIGURE 7.5 Main directions of wind in the Rambla Honda during the period 1991–92

around 21 and 27% respectively (for extraterrestrial radiation of 1273 W m^{-2} in summer and 689 W m^{-2} in winter, at this latitude).

7.2.3 Geology and hydrogeology

The bedrock of the area is highly convoluted and fractured, dark grey, fine-grained, Devonian–Carboniferous slaty micaschists with graphite and garnets, crossed by abundant

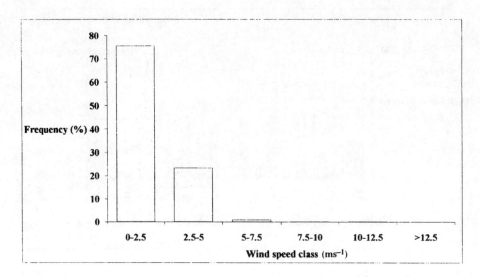

FIGURE 7.6   Frequency distribution of wind speed (m s⁻¹) in the Rambla Honda during the period
1991–92

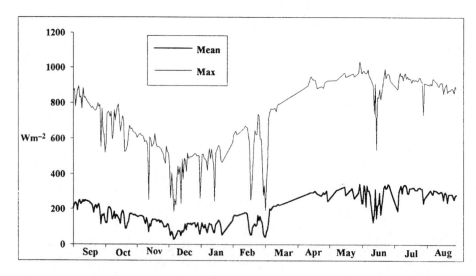

FIGURE 7.7   Solar radiation (335–2200 nm) at the Rambla Honda weather station during the study
period showing daily maxima and daily average (including night period) values

quartz veins alternating with thin phyllite layers. Its degree of weathering is related to the
layering pattern and to the proportions of garnets and quartz. When the latter is high, spurs
or shoulders are formed by differential erosion and colluvial debris is accumulated upslope
behind them. In the middle to low part of the catena, the slope colluvia gradate to an
alluvial fan formation which connects with the large Rambla Honda fan system.

Four bore holes were drilled along an altitudinal transect, from the root of the alluvial fan to the rambla floor in order to investigate the sedimentary characteristics of the alluvia and colluvia and to monitor the phreatic level. The bore holes reached the underlying micaschist basal rock. The thickness of the sediment (18, 8, 16, 28 m) and hence the altitude above sea-level of the basal rock (627, 632, 609, 597 m) show the existence of a rather pronounced palaeo-relief, antecedent to the deposition of sediment. The sedimentary columns contain alternating beds of coarse and fine material. Gravels and sands, with a planar geometry, prevail in the upper section, whereas red loams with small sandy intercalations predominate in the lower section. The intermediate section shows transitional characteristics.

In bore hole 4, a permeability test (constant head, Lefranc type) was performed in dry conditions. The estimates of permeability range from 8.61×10^{-3} cm s^{-1} for the total column to 1.9×10^{-2} cm s^{-1} for the gravel beds. Owing to the inaccuracies of this test, these figures should be treated with caution, but they suggest a moderate total permeability which almost doubles in the coarser layers.

7.2.4 Soils

The soils from Rambla Honda are essentially colluvial and alluvial in origin. Those on the slopes have developed directly from micashists and quite shallow slope deposits. Steepness of slope and variability in hardness of bedrock influence soil thickness: where slates with abundant quartz veins dominate, soils are usually shallow (up to 15 cm); where phyllite strata dominate, soils are thicker (up to 60 cm). Those on alluvial fans have developed from bedded colluvia which have originated from the erosion and sedimentation of material from the slopes above them.

All soils are mostly channery loamy sands and channery fine sandy loams (with low proportions of silt + clay). Coarse fragments (larger than 2 mm in diameter) within the soil mass range from 12.75% in the lower part of the alluvial fan to 67% in the higher part of the catena, with 2–8 mm fragments being the most abundant size (Table 7.3). Surface stoniness also increases up the catena (Table 7.4, Figure 7.8). The dominant stone type, either quartz or micaschist, is quite variable and depends on proximity to large quartz outcrops (veins of variable thickness, from a few millimetres to more than 1 m).

The soils of the whole area show little development of pedogenic horizons and may be arranged in two main groups (Tables 7.5 and 7.6): (a) the soils on the lower part of alluvial fans, classified as Eutric Fluvisols and Eutric Regosols, which exhibit a clear stratification and an irregular distribution of organic matter down the profile, and (b) the soils on the upper part of alluvial fans and hillslopes, classified as Eutric Leptosols, which are texturally somewhat finer. Both groups of soils have some common attributes: (1) pH ranges from slightly acid (6.5) to moderately alkaline (8); (2) electrical conductivity is very low, from 0.003 to 0.022 S m^{-1}; (3) cation exchange capacity is also very low, less than 10 cmol kg^{-1}, as a consequence of the low relative content of both clay and organic matter. But also they show some differences: the soils on the alluvial fans contain relatively less clay, less organic matter and less coarse fragments than the slope soils.

The organic matter content of surface horizons in all soils is quite variable (Table 7.6) being largely dependent on the vegetation type and cover: under bushes the proportion can

TABLE 7.3   Particle size distribution and bulk density of soils from the Rambla Honda catena. Bulk density was measured on samples from two depths: 0–7 cm and 20–27 cm

			Particle size distribution						
				Fine earth fraction (%)					Bulk
MA	Horizon (cm)	>2 mm (%)	c.sand	f.sand	c. silt	f. silt	clay	Horizon (cm)	density (kg l⁻¹)
	0–4	12.75	24.4	51.1	9.8	10.9	3.8	0–7	1.51
	4–11	32.21	41.5	42.6	4.9	8.3	2.9		
1	11–42	31.46	50.9	36.2	3.7	6.9	2.4	20–27	1.67
	42–73	0.74	n.a.	n.a.	n.a.	n.a.	n.a.		
	73–96	30.45	47.4	38.4	4.3	7.6	2.4		
	0–3	22.71	42.0	40.0	5.0	9.6	3.5	0–7	1.59
2	3–34	38.64	48.5	36.6	4.0	8.3	2.7	20–27	1.65
	34–60	54.71	n.a.	n.a.	n.a.	n.a.	n.a.		
	0–6	40.68	53.9	29.7	4.6	8.8	3.1	0–7	1.52
3	6–21	36.40	51.1	33.8	4.2	8.4	2.5		
	21–38	29.39	50.7	37.0	3.0	7.3	2.0	20–27	1.57
	38–70	25.50	42.1	43.3	4.1	8.2	2.3		
	0–4	40.06	40.6	36.3	5.9	12.0	5.1	0–7	1.53
4	4–20	35.96	34.6	38.5	6.0	16.9	4.0		
	20–50	45.31	31.7	38.5	5.9	17.0	6.9	20–27	1.55
	0–3	40.63	31.7	47.5	6.1	10.2	4.5	0–7	1.45
5	3–20	34.30	35.8	41.9	5.8	11.7	4.7		
	20–60	44.70	33.7	39.9	6.3	15.9	4.1	20–27	1.59
	0–5	60.12	38.4	35.2	7.7	14.5	4.3	0–7	1.45
6	5–16	40.90	32.8	39.2	7.1	16.7	4.2		
	>16	62.01	21.4	26.6	12.0	31.1	8.9	20–27	1.68
	0–5	55.83	30.0	37.8	8.9	15.9	7.5	0–7	1.34
7	5–30	58.47	29.8	36.2	6.9	18.7	8.5	20–27	1.34
	30–60	30.49	25.9	30.7	8.6	24.2	10.6		
	0–3	47.05	26.6	41.4	8.4	15.9	7.7	0–7	1.34
8	3–15	67.02	27.0	33.2	7.5	22.1	10.2		
	>15	54.06	26.1	24.0	9.6	29.9	10.4	20–27	1.21
9	0–6	55.19	37.0	38.2	6.7	15.3	2.8	0–7	1.27
	6–60	28.56	21.6	33.9	11.3	26.6	6.6	20–27	1.5

n.a. = not available. c. = coarse; f. = fine.

almost be twice that between bushes; areas with *Retama* have the lowest organic matter content (1.38 to 1.57%) while areas with *Stipa* have the highest (from 3.8 to 5.2%).

Soils from the lower part of alluvial fans have a 20 to 40 mm thick, surface layer, exclusively formed by very fine gravels (Figure 7.9). Under this gravely layer there is an horizon with a much finer texture, 2 to 10 mm thick, fine to very fine sandy, quite compacted and only crossed by vertical or sub-vertical cracks. Under these two surface layers, an undifferentiated, coarse-textured mass, containing randomly

TABLE 7.4 Surface stoniness and roughness from runoff plots in the sampling site 2. Plots A have a denser plant cover than plots B. Roughness coefficients are the average standard deviation of relative altitudes of 41 horizontal transects per plot

Plot	Surface stoniness (average ± SE) (%)	Surface roughness (cm)
1A	14.44 ± 3.92	
1B	16.11 ± 3.72	
2A	53.33 ± 5.42	1.15
2B	32.22 ± 6.74	1.13
3A	46.67 ± 7.24	
3B	63.61 ± 9.46	
4A	75.55 ± 7.77	
4B	76.67 ± 7.72	
5A	67.78 ± 6.88	3.11
5B	56.11 ± 7.42	2.64
6A	67.78 ± 7.73	
6B	75.89 ± 7.74	
7A	75.83 ± 7.66	
7B	73.89 ± 7.74	
8A	55.55 ± 6.87	3.45
8B	57.41 ± 6.78	3.07
9A	69.17 ± 7.52	
9B	73.15 ± 7.47	

FIGURE 7.8 Surface stoniness in the lower part of the alluvial fan (MA 1). Observe how rock fragments are on the soil surface and not embedded within the surface. A 50×50 cm frame with a 10×10 cm grid was used to evaluate the surface stoniness

TABLE 7.5 Soil morphology and classification (FAO, 1985) of the soils in the Rambla Honda catena

Soils	Horizons (cm)	Horizon type	Soil classification (FAO, 1990)	Solum thickness	Structure type	Structure grade	Consistency	Pores
1	0–4	A	Eutric Fluvisol	>3 m	platy very thick	low	loose	many
	4–11	2C			medium crumb	low	loose	common
	11–42	3C			medium crumb	low	loose	common
2	0–3	A11	Eutric Regosol	>3 m	fine subang. blocky	low	loose	common
	3–34	A12			medium crumb	low	loose	common
	21–38	2C			medium crumb	low	loose	common
3	0–6	A11	Eutric Regosol	>3 m	med. subang. blocky	low	loose	few
	6–21	A12			fine subang. blocky	low	loose	few
	21–38	AC			fine subang. blocky	low	loose	few
4	0–4	A11	Eutric Regosol	>1 m	fine subang. blocky	low	soft	common
	4–20	A12			med. subang. blocky	low	soft	common
	20–50	2C			med. subang. blocky	moderate	slight. hard	common
5	0–3	A11	Eutric Regosol	>1 m	medium platy	low	soft	common
	3–20	A12			med. subang. blocky	low	soft	common
	20–60	2Bca			fine prismatic	moderate	soft	common
6	0–5	A11	Eutric Leptosol	0.25 m	med. subang. blocky	low	soft	many
	5–16	A12			no structure	—	loose	many
	>16	AC			fine prismatic	moderate	slight. hard	few
7	0–5	A11	Eutric Leptosol	0.6 m	med. subang. blocky	low	soft	common
	5–30	A12			medium prismatic	low	soft	common
	30–60	C			medium prismatic	low	soft	common
8	0–3	A11	Eutric Leptosol	0.2 m	med. subang. blocky	low	soft	few
	3–15	A12			medium prismatic	low	soft	few
	>15	C			medium prismatic	low	soft	few
9	0–6	A	Eutric Leptosol	0.6 m	thick platy	low	soft	common
	6–60	AC			fine subang. blocky	low	soft	common

med. = medium; subang. = subangular.

TABLE 7.6 Chemical properties of the soils in the Rambla Honda catena

MA	Horizon (cm)	OM (%)	Total N (%)	pH (H₂O)	EC (S m⁻¹)	CaCO₃ (%)	CEC	Ca	Mg	Na	K
									(cmol kg⁻¹)		
1	0–4	1.57	0.18	7.3	0.016	1.7	2.85	7.06	0.69	0.25	0.28
	4–11	1.16	0.17	7.7	0.009	0.84	2.90	5.29	0.48	0.23	0.08
	11–42	0.77	0.14	7.8	n.a.	0.75	2.32	6.42	0.61	0.17	0.06
	42–73	n.a.	0.12	8.0	0.006	n.a.	2.25	4.33	0.54	0.23	0.06
	73–96	0.96	0.20	7.8	0.006	0.49	3.40	n.a.	n.a.	n.a.	n.a.
2	0–6	1.38	0.18	6.9	0.008	0.99	6.64	2.86	0.59	0.17	0.23
	3–34	0.68	0.11	7.0	0.003	0.42	2.52	1.65	0.39	0.17	0.06
	34–60	0.75	0.10	7.3	0.004	0.44	2.65	n.a.	n.a.	n.a.	n.a.
3	0–6	1.38	0.14	7.0	0.014	0.99	3.00	2.29	0.57	0.47	0.18
	6–21	0.93	0.14	7.6	0.005	0.48	n.a.	1.79	0.37	0.18	0.08
	21–38	0.71	0.11	7.6	0.004	0.61	2.50	1.52	3.35	0.15	0.05
	38–70	0.76	0.10	7.4	0.003	0.44	3.39	1.56	0.37	0.23	0.04
4	0–4	1.83	0.22	7.5	0.022	1.46	3.57	7.27	0.90	0.18	0.18
	4–20	0.76	0.10	7.2	0.004	0.47	4.12	3.90	0.49	0.26	0.05
	20–50	0.92	0.09	7.2	0.003	0.76	3.65	3.88	0.50	0.21	0.05
5	0–3	2.57	0.32	7.3	0.019	1.13	4.29	7.00	0.79	0.35	0.29
	3–20	1.35	0.15	7.3	0.007	0.42	2.92	4.48	0.46	0.18	0.08
	20–60	1.08	0.13	7.7	0.007	3.75	2.62	15.64	0.71	0.16	0.03
6	0–5	1.82	0.24	7.3	0.018	2.9	3.69	11.13	0.67	0.15	0.24
	5–16	1.28	0.16	7.8	0.010	3.02	3.04	12.40	0.51	0.15	0.08
	>16	1.42	0.20	7.6	0.016	9.85	4.39	17.59	0.52	0.17	0.04
7	0–5	5.2	0.33	6.9	0.004	2.01	9.45	12.96	1.45	0.37	0.51
	5–30	1.91	0.19	6.8	0.005	0.97	5.19	5.92	0.81	0.19	0.15
	30–60	1.89	0.17	6.5	0.004	1.12	7.30	5.29	0.97	0.18	0.06
8	0–3	4.8	0.34	6.9	0.016	1.9	4.74	9.77	1.32	0.19	0.51
	3–15	2.41	0.19	6.9	0.005	0.64	4.69	4.98	0.76	0.17	0.21
	>15	2.54	0.28	7.3	0.006	0.99	4.69	6.16	1.18	0.28	0.08
9	0–6	3.82	0.28	7.1	0.010	1.29	4.82	7.75	1.00	0.16	0.40
	6–60	1.67	0.17	7.2	0.004	0.86	4.35	5.14	0.88	0.27	0.05

n.a. = not available.
OM = organic matter; Total N = total nitrogen; EC = electrical conductivity; CEC = cation exchange capacity.

distributed and unoriented gravels, constitutes the whole soil. Soil structure, when present, is fine to medium crumby, very weakly developed and it arises only around and among roots. Bioturbation is not observed, so that sedimentary structures are well preserved.

Soils from the slope deposits also show a surface layer <10 mm thick of pure fine gravels; under the gravels a loamy layer of about 5 mm may have an origin similar to the one observed in alluvial fan soils. Under these two layers, a much less delineated layering, with alternate layers of coarse and fine textures, constitutes the whole soil mass. Bioturbation is present but rare.

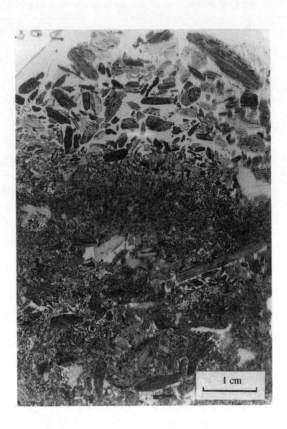

FIGURE 7.9   Thin section of the upper 7 cm of soil from MA 3, showing a washed fine gravel surface horizon overlying a loamy accumulation horizon

In general, in spite of sandiness, the water released from 0.01 to 1.53 MPa is relatively high, between 33 and 60%, of soil volume in surface horizons (Figure 7.10), as can be calculated from the moisture retention curves (Figure 7.11).

Infiltration, which was determined by simulating rainfall at 50 mm h^{-1} (Calvo et al., 1988), is high in the lower part of the alluvial fan, in *Retama* areas (24 to 41 mm h^{-1}), intermediate in the upper part of the alluvial fan, in *Anthyllis* areas (9 to 25 mm h^{-1}), and quite low in higher slopes, in *Stipa* areas (3.7 to 6.4 mm h^{-1}). Infiltration values under bushes are generally lower than under bare soil (Figure 7.12). These data are in agreement with the particular morphology of these soils.

There is a large variation in hydraulic conductivity (K), which is a consequence of local heterogeneity due to layering. The highest values appear in *Retama* soils, averaging from 1 to 2 m day^{-1}, and the lowest in the apex of the alluvial fan (MA 5), from less than 0.1 to 0.4 m day^{-1}. Under *Stipa* soils, K ranges from 0.3 to 1 m day^{-1}. In all cases, K is slightly higher in soils under or close to bushes than away from them. Figure 7.13 shows typical draw-down curves for the three main soils.

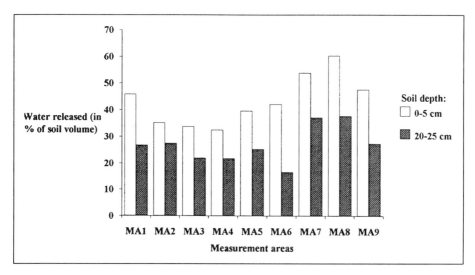

FIGURE 7.10   Water released between 0.01 and 1.5 MPa in soil profiles from the different MAs

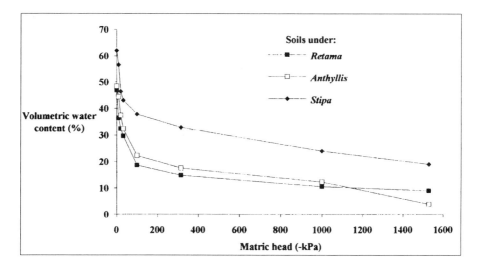

FIGURE 7.11   Water retention curves for the first mineral horizon (0–7 cm, excluding surface gravels) from representative soils in open areas with different vegetation types

7.2.5 Vegetation and land use

The natural vegetation includes bushes and thorny shrubs of the series *Rhamno lycioidi–Querceto cocciferae* sigmetum betic facies with *Ephedra fragilis* (Rivas-Martínez, 1987) but the remnants of this primeval matorral (*Rhamnus lycioides, Pistacia lentiscus,* etc.) are restricted to very small patches. A large part of the area is covered with alfa grass, *Stipa tenacissima,* with variable proportions of *Anthyllis cytisoides.*

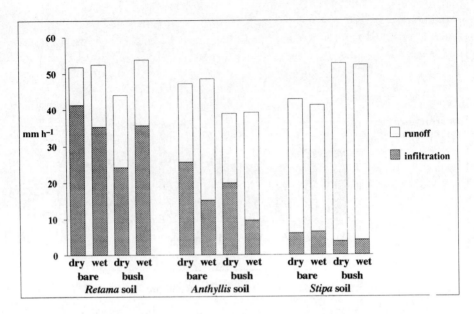

FIGURE 7.12   Runoff and infiltration rates after 30 minutes of rainfall simulation over circular plots on wet and dry soils both in open areas and above three vegetation types

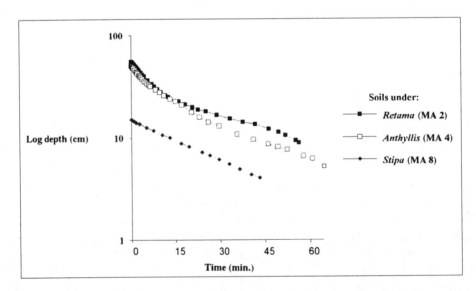

FIGURE 7.13   Draw-down curves for representative soils in open areas within the three vegetation types. These data were used to calculate saturated hydraulic conductivity

The lower sector of the catchment, where the field site is located, has an area of 5.28 km². Until the 1960s, three farms, with nine families, were exploiting the area. The total number of sheep and goats owned by the farmers was around 600. The mean grazing pressure may be estimated at around 1.13 sheep per ha. The alfa grass was harvested for cellulose and the rest of the area was cultivated for grain. Alfa grass and livestock accounted for more than 90% of the cash income in the traditional economy. Three types of fields have been identified: terraced plots on the slopes; fields without specific structures on the alluvial fans and along the drainage ways; and fields enclosed with stone walls, which were irrigated by flooding. In the first two, the main crop was barley, and in the third type, wheat and maize were often grown. The uncultivated floor of the rambla was covered with bushes of *Retama sphaerocarpa* which were cut for fire wood.

This traditional agricultural system was abandoned some 30 years ago. The steppe of *Stipa tenacissima* is encroaching with its own unharvested litter. The terraces and upper sectors of alluvial fans are being invaded by dense populations of *Anthyllis cytisoides*. The lower sectors, still occasionally cultivated in rainy years, are covered with scattered populations of *Retama sphaerocarpa* which sprout from root systems that remain in the fields after ploughing.

7.3 EXPERIMENTAL LAYOUT

7.3.1 Data acquisition system

A new low-speed digital radio network (LAN) field facility has been developed for this project. It consists of several remote microprocessor stations (based on Motorola MC68HC05/11 microcomputer families) and a central unit connected to an IBM AT computer, which has real-time overall control, monitoring, housekeeping, calibration and logging functions. The radio frequency link is bi-directional (half duplex), operating in the upper part of the VHF band, with a modified X.25 protocol, at 7200 baud with PCM-NRZI-BPSK modulation format. Multiple access has been implemented with a CSMA/CD (Carrier Sense Multiple Access Collision Detect) algorithm. For long-distance links or to avoid areas of radio shadow, the remote stations can also be programmed as zone relay stations (digipeaters) which have their own associated address group.

Each station which has a unique address according with zone–subzone–function topology, is polled cyclically by the host computer, as requested by the software, e.g. the Rambla Honda weather station has a polling period of 20 seconds. The host averages and stores data on hard disk in ASCII format at 10-minute intervals (5 minute for rainfall data). The data file can be accessed by other computers connected to the EEZA ETHERNET LAN for DBASE/EXCEL transformations and streamer backup.

Each station has telemetry and telecontrol functions and can be controlled through the network or locally by an RS-232 interface, using similar dialogue language for both paths. The system is actually managing the meteorological sensors but it is easily expandable to manage further instruments.

7.3.2 Weather data

An automatic weather station was installed in the Rambla Honda, at 640 m altitude. It is integrated in the EEZA radio network data acquisition system and it is provided with the following sensors listed in Table 7.7. One of two rainfall intensity gauges is located at the base of the slope, within the weather station, and the other, near the divide, at 750 m altitude. In addition, each runoff plot has been provided with a rain gauge for measuring total precipitation per event.

TABLE 7.7   Sensors in automatic weather station

Variable	Sensor model	Range	Resolution
Temperature	National Semicond. LM35CZ	-20 to $+80°C$	8 bits
Relative humidity	EE Electronik HC500	0 to 100%	8 bits
Wind speed	Health	0 to 25.5 m s^{-1}	8 bits
Wind direction	Health	360°	4 bits
Solar radiation	Kipp & Zonen CM6B	0 to 1200 W m^{-2}	8 bits
Rainfall (2 units)	Health	—	0.23 mm

7.3.3 Water and sediment yield

Surface runoff and sediment yield are estimated on the closed runoff plots described in Section 7.2.1. Each plot is provided with two collector tanks of 200 litres. When the first overflows, it conveys to the second 1/10 of the flow through a slot divisor. Readings are made after each rain event and samples are taken for the determination of coarse sediment (>2 mm) and fine sediment (<2 mm) bulked with total solutes by drying.

7.3.4 Vegetation

In the core field programme, biomass, above ground net primary productivity (AGNPP), leaf area (LAI), cover and litterfall are measured throughout the year in the nine MAs in sampling site 2. Plant cover, by functional types, and plant height are measured in 180 permanent quadrats (50×50 cm); observations, which started in October 1991, are made at 3-monthly intervals. Surface litter was sampled in November 1991 from 165 quadrats (50×50 cm). Litterfall is collected at monthly intervals from 105 traps installed in November 1991. In *Anthyllis*, five funnel traps for individual plants were installed per each MA. The distribution of traps and quadrats is at random within the main plant cover types.

Biomass and GNPP of shrubs are estimated through allometric relations based on branch dimensions which were established by field sampling in each MA during the winter of 1991/92. Later periodic sampling of twigs with leaves through the year provides estimates of seasonal variations of leaf and twig dry weight, leaf area, flower, fruit and standing dead dry weight.

In the case of *Stipa tenacissima*, biomass was estimated by allometric methods using the tussock dimensions as predictors. Litterfall, leaf mortality and AGNPP are estimated by measuring leaf numbers and length in samples of previously marked tillers. This operation, which started in November 1991, involves a total of 195 marked tillers in tussocks of different size and vigour.

Coarse root biomass was estimated through allometric relations based on root diameter. Fine root biomass (diameter <5 mm) was estimated in cubes of soil (10 cm^3), taken at each 10 cm in two columns per MA from the soil surface to the basal rock or up to a depth of 100 cm. Samples were collected in August 1991 and processed by means of dispersing agents (hexametaphosphate), by floatation and decantation.

The cellulolytic activity in the soil was estimated by the cotton strip assay (Sagar, 1988) as an index for soil biological activity. Six situations have been considered: *Retama*, *Anthyllis* and *Stipa* vegetation types and in each of them, under the canopy and in open areas. In each situation, five quadrats were installed and six cotton strips were exposed per quadrat, three in the ectoorganic horizon (soil surface in open areas) and three in the uppermost mineral horizon. The strips were exposed for 30 days and three runs have been completed: spring (17 March to 14 April 1992), summer (24 June to 22 July 1992), and autumn (21 October to 21 November 1992). After exposure, the strips are sent to the University of Bristol for the tensile strength determinations.

7.3.5 Plant ecophysiology

A total of five sets of measurements were completed and they were timed for the periods of minimum temperature (February 1992), maximum growth (May 1992), onset of drought (June 1992) and maximum drought (August and October 1992). From uniform stands of each of the three species characterizing the catena (*Retama sphaerocarpa*, *Anthyllis cytisoides*, *Stipa tenacissima*) eight randomly selected plants were chosen per species per set. One species was sampled per day, hence a set took 3 days. Measurements were carried out before dawn, a few minutes after sunrise, two hours after sunrise, at solar noon, midway between solar noon and sunset, and before dusk. For each measurement, a leaf, shoot or cladode was taken from each of the four cardinal positions on a plant. For relative water content (RWC), all leaves or cladodes per plant sampled at each time of the day were bulked. For water potential, the four leaves, cladodes or shoots per plant sampled at each time of the day were measured separately with a pressure chamber (Skye Instruments, UK). An infra-red gas analyser (LCA3, ADC, UK) in combination with a narrow leaf chamber (PCL-N, ADC, UK) measured stomatal conductance and net photosynthetic rate. Four leaves per shoot were measured simultaneously in the cuvette. The same sets of leaves were measured throughout the day. Leaf areas were measured from images obtained with a grey level scanner (Scanman, Logitek, USA). Fluorescence ratio rather than absolute fluorescence was measured with a fluorimeter (PEA, Hansatech, UK). Hourly meteorological data were obtained over the whole period from the weather station approximately 0.5 km from the sites of the *Retama* and *Stipa* stands. For sets two to four there was an on-site station recording 10-minute averages of micrometeorological variables.

7.4 HYDROLOGICAL RESPONSE TO RAINFALL EVENTS

7.4.1 Weather features in the period September 1991–August 1992

Total rainfall in the 1-year study period was 318 mm (Figure 7.4) and average temperature was 15.5°C. Both values are close to the average estimates in the area (see Section 7.2.2). Monthly maximum and minimum average temperatures range from 33°C in August to 3°C in January (Figure 7.14). Rainfall shows a large annual variability, both in its total amount and in its monthly distribution (Figure 7.4). Within the general trend of rainfall concentration in the cold season, the study period is characterized by a dry autumn, and a very dry spring, with only late and abnormal rains in June.

Average monthly values of vapour pressure deficit (VPD) are small, around 0.5 kPa, in winter, but in March they start to grow and, in August, they reach average values of 2.5 kPa and monthly averages of daily maxima of 5.5 kPa (Figure 7.15). This trend is only interrupted in the humid period which was recorded in June. The annual radiation curve (Figure 7.7) shows that during rainy periods (winter and June) there are both troughs and

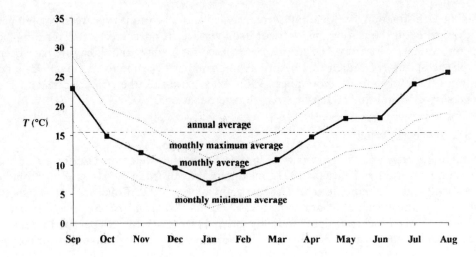

FIGURE 7.14 Distribution of monthly average temperatures (average, maxima and minima of daily values) at the Rambla Honda

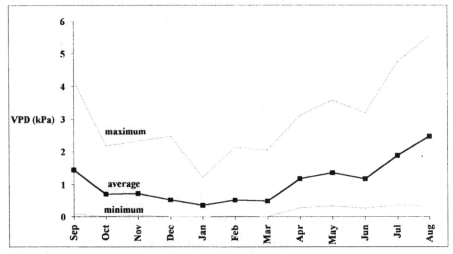

FIGURE 7.15 Distribution of monthly averages of vapour pressure deficits (VPD) (average, maxima and minima of daily values) at the Rambla Honda

peaks; the latter are probably due to cloud reflectance that may increase irradiance by up to 10% above the main cloudless trend.

7.4.2 Soil–water relationships

As a consequence of the origin of the parent material, layering as well as particle sorting across the slope are the main features in Rambla Honda soils that explain their morphological differences. When studied at a very detailed scale, they exhibit enough variations to account for their different hydrological behaviour. These differences refer to their surface morphology as well as to their horizon layering. Stones on and within the surface are much larger in slope soils than in fan soils. As a consequence the surface roughness coefficient is much higher over slopes than over the fan surface (Table 7.4). Surface roughness may influence surface runoff and infiltration at the slope scale.

It is also noticeable that most rock fragments larger than 100 mm are embedded in the soil surface; smaller sizes are usually found on the surface, and very seldom are they embedded. As slope soils contain larger rock fragments which are mostly embedded, and fan soils contain smaller rock fragments just on the top of the surface, the findings of several authors (Abrahams and Parsons, 1991; Poesen and Ingelmo-Sánchez, 1992) suggest that higher infiltration in fans and higher runoff in slopes could be expected. Runoff and infiltration in Rambla Honda confirm these expectations.

The surface gravel layer which characterizes the lower part of the alluvial fan, has a total porosity of 40% (calculated from bulk density measurements), which can be completely assigned to macroporosity, enabling the soil to drain internally. The layers just below the gravels, however, have macroporosities always less than 40% and as low as 20%. This means that in some cases (when rainfall intensities exceed a given threshold), subsurface flow (on the top of this lower macroporous layer) may be important in these soils. In fact, runoff data from sampling site 2 (after 39 rain events) show a significant and positive correlation between runoff and the percentage of gravel at the surface (Puigdefábregas et al., 1992). The latter may be interpreted as an indicator of a layered structure of contrasting macroporosity.

As infiltration (measured from a simulated rainfall of 50 mm h^{-1}) and saturated hydraulic conductivity are lower under *Stipa* soils than under *Anthyllis* and *Retama* soils

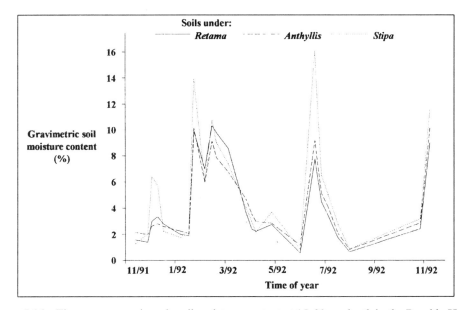

FIGURE 7.16   The average gravimetric soil moisture content at 15–20 cm depth in the Rambla Honda (mean values of the three MAs of each vegetation type in the catena) over a 12-month period

(Figure 7.13), it may be thought that the former soils get less water. This may be true, but in the first case, water is stored by the soil and, in the second, water is lost by internal drainage. The changes in soil moisture content during a 12-month period (Figure 7.16) confirm this.

A physical consequence of the observed dissimilarities in the different soils is that the water released between 0.01 and 1.53 MPa (Figure 7.10) in slope soils is higher than in fan soils (from 40% to 60% of the volume of the fine earth in slope soils; 33% to 40% in fan soils).

At soil water potentials of -1.5 MPa, the volumetric soil water content is still 0.1 LL^{-1} under *Retama* soils and 0.2 LL^{-1} under *Stipa* soils (Figure 7.11). But in desert ecosystems, water released between 0.01 and 1.5 MPa should not be considered as the only available water for plants. During the summer the soil water content is well below 0.07 LL^{-1} (1% gravimetric) (Figure 7.16), and measured leaf water potentials with psychrometric techniques during this season vary from -2 MPa in *Retama* to -5 MPa in *Stipa* (see 5.3.) (Pugnaire et al., 1996). This suggests that some water might still be available for plants at bulk soil water potentials lower than -1.5 MPa.

7.4.3 Water and sediment yield

The 10 runoff plots in sampling site 1 have provided data for runoff and sediment yield for a period of 2 years (1 December 1990 to 1 December 1992) in which 39 rainfall events took place. Results show a large variation between years (Table 7.8). Despite the fact that 1992 was drier than 1991, outputs in terms of runoff and sediment were two and three times higher respectively.

Significant runoff and sediment yield only occurs when total rainfall per event is larger than 20 mm (Figures 7.17 and 7.18). Beyond this threshold, the outputs are very variable because they are controlled both by the rainfall patterns and by the site characteristics, mainly the physical properties of the uppermost soil layer (Puigdefábregas et al., 1992).

The 18 runoff plots in sampling site 2, during the period 1 September 1991 to 1 December 1992, recorded 19 rainfall events but only four of them were significant in terms of runoff and sediment yield (>0.5 mm and >0.5 g m^{-2}). Three events which occurred in the cold season (November to March) produced 90% of the total runoff and sediment output during the period.

No permanent groundwater flux has been found in the Rambla Honda bore holes. Ephemeral groundwater has been recorded only on two occasions, in spring and autumn 1992, both in bore hole 2, which is drilled in a channel that trenches the sediment fill and

TABLE 7.8   Average runoff and erosion from the 10 runoff plots in sampling site 1

	Rainfall (mm)		Runoff (l m^{-2})		Runoff coeff. (%)		Erosion (g m^{-2})	
	Annual total	SE	Annual total	SE	Annual total	SE	Annual total	SE
1991	265.43	1.26	14.02	1.59	5.27	0.59	15.77	3.55
1992	215.00	16.00	27.23	2.67	12.88	1.25	39.59	11.23

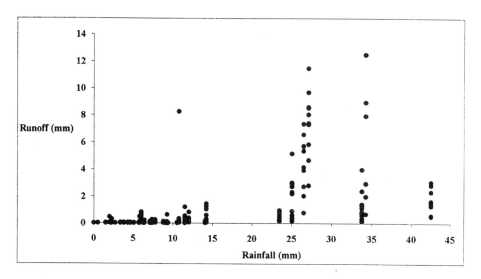

FIGURE 7.17   The relationship between runoff and total rainfall per event (1991/92) in the runoff
plots of sampling site 1

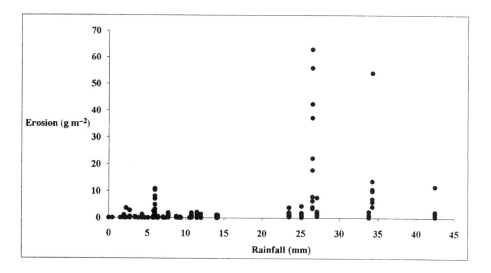

FIGURE 7.18   The relationship between erosion and total rainfall per event (1991/92) in the runoff
plots of sampling site 1

drains a slope sector of about 0.06 ha. Both occasions were preceded by a succession of
rain events, the time lag of the response was 1 to 4 days, and the maximum water level
above the micaschist basal rock was between 40 and 70 cm. These facts suggest that water
is provided by the top layers of the sedimentary body. These layers are more gravely and

may be connected in some way with the drainage net. The river channel (rambla) lacks a permanent saturated zone and the slope runoff produces only local and ephemeral perched water-tables at the outlets of gullies that become lost within the main sedimentary formation.

Available data are still too scarce to establish definitive comparisons between plots along the catena, because they do not allow the separation between effects of site and rainfall. Nevertheless, preliminary results (Table 7.9) show that in terms of runoff, the values from *Retama* plots (MA 1 to 3) are almost three times lower than those of *Anthyllis* (MA 4 to 6) and *Stipa* plots (MA 7 to 9). Densely vegetated plots tend to produce less runoff than the sparsely vegetated ones, particularly in *Retama* plots.

In terms of sediment yield, *Retama* plots produced four times less than those of *Anthyllis* and five times less than those of *Stipa*. The effect of plant cover density is conspicuous in the case of *Stipa*, where densely covered plots double the sediment yield of sparse ones. In the rest of vegetation types, the effect of plant cover is not apparent. On the whole, sediment yields (6–$50\,\mathrm{g\,m^{-2}\,yr^{-1}}$) are moderate in comparison with data from a similar climate (Romero Díaz et al., 1988).

Concerning the slope–fan relationships for sediment yield, the former (*Stipa* covered) have values two to five times greater than the latter (*Anthyllis* and *Retama* covered) (Table 7.9). This may be explained by a greater availability of fine sediment on the slopes, which is enhanced by higher weathering rates of the fresh micaschists compared to the more resistant residual gravels of the alluvial fans. In those rapidly weathering slopes, the tussocks of *Stipa tenacissima* are very effective in promoting spatial heterogeneity in the distribution of runoff which increases the storage of sediment, mainly of fine size classes. As this storage is related to the growth and decay patterns of the tussocks (Sánchez and Puigdefábregas, 1994), it is highly dynamic and, hence, the sediment yields remain high. Apart from the linear erosion by trenching of the alluvial fans, micaschist slopes show greater runoff, sediment yield and probably weathering rates, than the sedimentary bodies.

TABLE 7.9   Runoff and erosion in the three vegetation types of the Rambla Honda sampling site 2 in the study period (1 September 1991 to 31 August 1992)

Vegetation type	Runoff (mm)		Runoff coeff. (%)		Erosion ($\mathrm{g\,m^{-2}}$)	
	Average	SE	Average	SE	Average	SE
Retama dense ($n=3$)	3.84	2.58	1.33	0.91	6.25	1.60
Retama sparse ($n=3$)	7.99	5.98	2.78	2.10	5.91	1.29
Retama average ($n=6$)	5.91	4.42	2.06	1.55	6.08	1.30
Anthyllis dense ($n=3$)	13.73	4.95	4.50	1.86	17.95	0.12
Anthyllis sparse ($n=3$)	18.02	4.68	5.30	1.58	19.62	17.39
Anthyllis average ($n=6$)	15.88	4.62	5.76	1.65	22.38	12.90
Stipa dense ($n=3$)	13.13	3.80	4.94	1.40	48.02	16.55
Stipa sparse ($n=3$)	16.99	4.61	6.34	1.56	20.80	0.91
Stipa average ($n=6$)	15.06	4.06	5.64	1.43	34.41	14.87

coeff. = coefficient; SE = standard error.
Dense/sparse values are averages of the three runoff plots per vegetation type.
Data for vegetation type are average values of six runoff plots.

It is notable that plots from sampling site 1 (8 m long) produced higher runoff and sediment yield than longer runoff plots from sampling site 2 (10 m long). This fact may be interpreted as a result of re-infiltration along the slope which reduces overland flow as the length of the plot increases.

7.5 VEGETATION STRUCTURE AND DYNAMICS

7.5.1 Biomass, productivity and litterfall

The summary of the available results pools the data from the three MAs of each vegetation type (Table 7.10). The above-ground living biomass (AGB) of the *Stipa* stands (356 g m^{-2}) is between two and three times greater than that of the shrub stands, and its standing dead biomass is also greatest and about six times its AGB. Below-ground biomass (BGB) is very large in *Retama* stands and is 23 times greater than the AGB, whereas this ratio varies between three and five in the other vegetation types.

TABLE 7.10 Biomass, net primary productivity, litterfall and leaf area in the three vegetation types of the Rambla Honda catena

Plant community attribute	Plant community attribute	*Retama* MA (n = 3)		*Anthyllis* MA (n = 3)		*Stipa* MA (n = 3)	
		Average	SE	Average	SE	Average	SE
Canopy cover (%)	Perennial plants	14.03	3.29	28.89	2.00	34.91	2.33
AGB (g m^{-2})	Living	117.23	9.87	133.25	15.60	356.31	44.22
	Dead	44.42	1.42	119.28	29.08	2343.41	234.91
BGB (g m^{-2})	Stump	90.78	14.71	16.05	4.59		
	Coarse roots	2168.53	769.57	176.03	33.77		
	Fine roots	403.85	94.74	541.64	20.46	1143.62	40.12
	Total	2663.16	719.97	733.71	58.30	1143.62	40.12
LPOOL (g m^{-2} yr^{-1})		202.70	7.49	212.77	33.56	2343.11	234.91
AGNPP (g m^{-2} yr^{-1})	Wood/bark	33.41	4.10	16.32	3.32		
	Curr. twig & leaf	75.54	6.82	174.53	3.30	316.75	20.32
	Annual plants	79.33	7.96	47.89	4.05	9.54	2.23
	Total	188.28	14.47	238.74	9.36	326.29	18.12
LFALL (g m^{-2} yr^{-1})	Leaf (peren. + ann.)	134.76	9.46	107.31	2.26		
	Wood	5.88	1.10	1.23	0.12		
	Total	140.64	8.36	108.54	2.35	396.86	38.52
LAI MAX (m^2 m^{-2})	Perennials	0.16	0.01	0.44	0.01	0.98	0.08
LDW MAX (m^2 m^{-2})	Peren. + ann. plants	150.37	12.88	93.52	2.95	364.67	40.66

AGB = above-ground biomass in winter.
BGB = below-ground biomass.
LPOOL = litterpool on the soil surface.
AGNPP = above-ground net primary productivity.
LFALL = litterfall.
LAI MAX = maximum leaf area index in the growing period.
LDW MAX = maximum leaf dry weight in the growing period.
curr. = current year.
peren. = perennial.
ann. = annual.

Above-ground net primary productivity (AGNPP) is of the same order as AGB if annuals are included in the latter. Annual biomass increments (AGNPP−Litterfall) range from $131 \, \text{g m}^{-2}$ for *Anthyllis* to $-70 \, \text{g m}^{-2}$ for *Stipa*, although these figures should be considered with caution because of the short study period and the probable annual variability of the standing dead compartment.

The above-ground biomass distribution along the catena (Table 7.10) shows the contrast between the more stable *Stipa* community on the upper slope sector and the unstable formations of *Anthyllis* and *Retama* in the alluvial fan sector. The former seems to be near a steady state; it shows slightly negative biomass increments and large pools of dead debris. The *Anthyllis* and *Retama* communities are growing after the cessation of cropping; they show positive biomass increments, similar to those reported for other Mediterranean ecosystems (Specht, 1969; Schlesinger, 1980; Black, 1987). Biomass increments are particularly high in *Anthyllis*, probably due to the suppression of sheep grazing during the study period. Annual plants seem to suffer strong competition from perennials, since their biomass is inversely proportional to the cover of perennial plants (Table 7.10).

The two independent estimates of leaf dry weight (Table 7.10), the allometric one (LDW MAX) and the litterfall one (leaf LFALL), provide coherent figures, with differences that vary between 11 and 18%.

The huge below-ground biomass of *Retama* is explained by two facts: first, ploughing removed branches, but roots and root crowns able to sprout remained in the soil; second, the branches of *Retama* were cut for fire wood. Therefore, the root system of *Retama* is often much older than the aerial part. The higher fine-root biomass of the *Stipa* stands may be related to the finer textures of their soils, to their higher water-holding capacity and to the genetic adaptation of *Stipa* to use water from small rains at all times of the year.

Litter pool versus litterfall ratios (LPOOL/LFALL from Table 7.10) provide provisional figures of the litter turnover time: 1.44 years for *Retama*, 1.96 years for *Anthyllis* and 5.90 years for *Stipa* stands.

The longer turnover time of *Stipa* litter may be more related to the characteristics of its leaves, higher C/N ratios (Table 7.6) and richness in recalcitrant compounds, than to limiting soil moisture.

The cotton rotting rates (CRR) in Table 7.11, show that the largest differences among vegetation types occur in early summer, below canopies and in the mineral soil. These CRR rates seem to express well the decomposition rate of the litter, as they show a negative trend with LPOOL/LFALL ratios (Figure 7.19).

7.5.2 Phenology

The phenological stages through the year for the main species (Figure 7.20) show that vegetative growth is concentrated in autumn, winter and early spring, while in summer the reproductive stages and leaf fall prevail. The effects of the unusual rainfall recorded in June 1992 are shown by the rise of soil moisture content (Figure 7.16) and by the regrowth of the vegetation. This regrowth emphasizes the significance of the spring rains in this semi-arid environment. If they come early, they set up the soil moisture storage that has to support the larger part of the annual productivity. If they come too late, they produce short-lived extra growths that in general are useless to the plant.

TABLE 7.11 Cellulolytic activity of soils in the different vegetation types of the Rambla Honda catena. Values are expressed as cotton rotting rates (CRR per year)

			Time of year			
			18 Mar. to 15 Apr. 1992		22 Jan. to 24 Jan. 1992	
			Average $n = 5$	SE	Average $n = 5$	SE
Retama	Shade	Ectoorganic	6.496	1.792	8.052	1.354
		Mineral	18.454	1.137	31.41	4.405
	Open	Ectoorganic	6.855	0.624	6.006	0.635
		Mineral	12.94	2.075	15.91	2.084
Anthyllis	Shade	Ectoorganic	7.194	0.831	8.861	1.068
		Mineral	13.12	1.027	23.14	4.298
	Open	Ectoorganic	6.303	1.429	5.237	0.693
		Mineral	10.902	0.931	11.197	1.390
Stipa	Shade	Ectoorganic	8.54	1.802	7.778	0.411
		Mineral	14.09	2.545	15.85	3.600
	Open	Ectoorganic	8.873	1.206	6.064	0.647
		Mineral	8.886	2.384	8.697	1.132

Shade = below shrubs or tussocks.
Open = in open areas among shrubs.

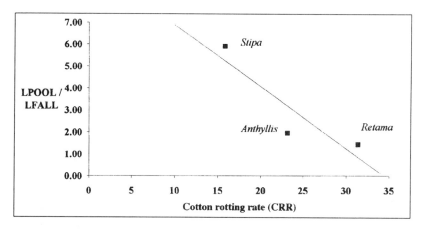

FIGURE 7.19 Cotton rotting rates (CRR) versus litter pool/litterfall ratio in the three vegetation types of the Rambla Honda catena

7.5.3 Ecophysiology

There were significant differences between species, times of the day and some plants in all sets for all attributes except orientation. In the case of total water potential of leaf or shoot (ψ_1), *Retama* had the highest (least negative) and *Stipa* the lowest values. These differences increased from February to September. There were no consistent trends between sets because of irregular weather conditions. For the relative water content (RWC) *Anthyllis* always had the lowest values. Diurnal differences in each run were small. There were significant differences between runs for *Stipa* and *Anthyllis* and some consistent trends over all runs for each species. For example *Anthyllis* had values of 0.6 to 0.8, which are lower than those for *Stipa* and *Retama*, 0.7 to 0.95. The relationship between ψ_1 and RWC differed significantly between species (Figure 7.21a). Stomatal conductances (g_s) in *Stipa* were the most uniform during the day, with the lowest g_s, whereas *Anthyllis* and *Retama* had high g_s at the start of the day which decreased rapidly as the day progressed (Figure 7.21b). In the case of net photosynthetic rate (P_n), *Retama* had the highest rates under most conditions. *Stipa* and *Anthyllis* had approximately similar and low rates. Fluorescence (F_v/F_m) ratio showed some significant diurnal and seasonal differences between and within species, with *Retama* the most uniform (Figure 7.21c).

It may be pointed out that the absence of any significant difference between orientations in the canopy suggests that diurnal changing factors nullify their effect. The significant difference in water relations ψ_1 between some replicate plants suggests a genetic and/or a soil moisture heterogeneity at the scale of the plants' root systems. The significant diurnal changes in ψ_1 are to be expected in a semi-arid environment. In the driest conditions there were measurable changes in water status during the day, suggesting that any summer dormancy state was partial. The failure of any species to regain full turgor by dawn suggests high plant and/or soil resistance to water movement in the rooting zone. The significant diurnal differences between the species is probably partly due to differences in the soil moisture regimes experienced. However, ψ_1 changes in *Stipa* more rapidly under all conditions, indicating a genetic difference. There were some consistent differences between species, notably the low potential of *Stipa* compared with *Retama*, and the increase in this difference as the season progressed. This divergence was probably due first,

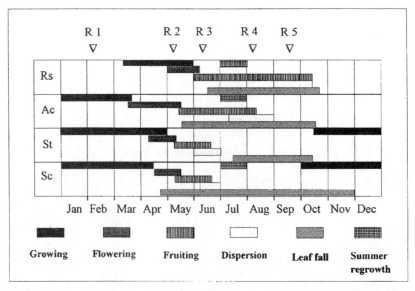

FIGURE 7.20 Plant phenology events in Rambla Honda during 1992. Rs = *Retama sphaerocarpa*, Ac = *Anthyllis cytisoides*, St = *Stipa tenacissima*, Sc = *Stipa capensis* (as a representative of winter annual plants). R1 to R5 show when ecophysiological measurements were carried out

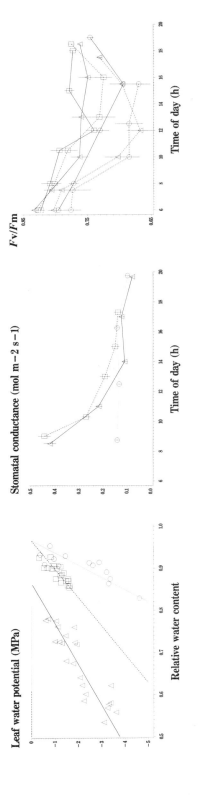

FIGURE 7.21 Some ecophysiological characteristics of the three species studied in Rambla Honda and their diurnal changes. (a) The relationship between leaf water potential and relative water content for *Retama* (□), *Anthyllis* (△) and *Stipa* (○) using data from sets 1 to 4. Lines show fitted linear regression lines. (b) Diurnal changes in stomatal conductances for *Retama* (□), *Anthyllis* (△) and *Stipa* (○) in June 1992; error bars represent SE. (c) Fluorescence ratio for *Retama* (□), *Anthyllis* (△) and *Stipa* (○) for set 3, well watered plants (- - - -) and set

to the *Retama* community's higher and more constant soil moisture content and second, to *Stipa*'s more rapid and greater response to changes in soil moisture content.

Retama differed least seasonally in relative water content. It maintained the highest RWC because *Retama* has an extensive root system that can effectively exploit soil water. RWC for *Anthyllis* varied significantly between runs, reflecting perhaps greater dependence on current precipitation than on stored water, because of its smaller root system. *Stipa* maintained a high and constant RWC, not because of access to readily available water but because of its low transpiration rate. In species with a high dry matter content, changes in absolute and not relative water contents may be more important physiologically. Thus a small decrease in RWC and hence in water content from *Stipa*'s near-constant 0.8–0.9 RWC throughout the year, although unimportant in its effect on water relations, could reduce the number of viable cells present and adversely affect protein functioning.

The relationships between ψ_1 and RWC (Figure 7.21a) suggest that *Stipa*'s cells have a higher bulk modulus of elasticity than *Anthyllis*'s and *Retama*'s. This could explain the more rapid diurnal changes in ψ_1 and RWC in *Stipa*.

Maximum stomatal conductances for *Retama* and *Anthyllis* were high (0.5 mol m^{-2} s^{-1}), but for *Stipa* (0.1 mol m^{-2} s^{-1}) were typical for species from similar habitats elsewhere. Stomata of *Retama* and *Anthyllis* responded rapidly to diurnal increases in vapour pressure deficit (VPD) bringing their g_s down to values more usually expected for this environment. *Stipa*'s response to VPD was not noticeable perhaps because folding of its leaves gave a long pathway for water vapour from inside the leaf to the leaf surface.

Rates of carbon assimilation in *Retama* were similar to those of other C3 plants (11–25 mmol m^{-2} s^{-1}). Rates were lower, around 10 mmol m^{-2} s^{-1}, in *Stipa* and *Anthyllis*. The values for *Anthyllis* may be atypically low as they were obtained when ψ_1 was low and when some of the leaves may have been in the initial stages of senescence.

A preliminary interpretation indicates that fluorescence is correlated with diurnal and seasonal changes in the other attributes and that it reveals marked differences between species regarding departures from optimum conditions for photosynthesis.

7.6 CONCLUSIONS

1. The slope–fan system on micaschists in Rambla Honda shows two contrasting types of soils. Slope soils are richer in fines and organic matter and show a lower saturated hydraulic conductivity, but higher water-holding capacity and sediment availability. Therefore, they produce higher outputs of runoff and sediments. The fan soils offer better opportunities to deep-rooting plants, but their water-holding capacities are less. They show a surface layer of washed gravel upon a thin very fine sandy layer of much lower macroporosity. This structure enhances subsurface runoff with light rainfalls and overland flow in large events. In the fan system, the internal drainage is high and no permanent saturated layer has been found. The phreatic levels are local and ephemeral and probably connected with the upper sector of the sedimentary body which shows a higher permeability.

2. The *Stipa* stands on the slopes are near a steady state in terms of biomass accumulation (300–400 g m^2) and LAI (0.9–1.0), but the storage in standing dead biomass is six times living biomass because of the cessation of harvesting 25 years ago.

The *Anthyllis* and *Retama* stands are unstable. They are growing and occupying the area after cessation of cropping. The turnover time of their litter is three to four times lower than for *Stipa*. *Anthyllis* and *Retama* stands show very high root/shoot ratios, especially *Retama* (20–25) in which the root system is much older than the shoots due to the traditional management.

3. *Retama* is least affected adversely by high temperatures and high VPDs. It closes its stomata rapidly in response to VPD and its narrow-diameter cladodes and open canopy facilitate loss of heat. Due to the large volume of soil exploited (the main roots can extend >15 m), high root/shoot ratios ($<16:1$), and the presence of relatively moist bands of soil in the rooting zone, water stress is least severe in *Retama* and its water relations are not closely related to current rainfall events. *Retama* is a drought avoider by habitat and a high-temperature avoider by morphology. *Anthyllis* is sensitive to VPD, resulting in high early morning photosynthetic rates but a short assimilating period over the day. Its summer deciduousness confers drought and heat avoidance. *Stipa* is a drought and high-temperature endurer, having low rates of photosynthesis and water loss due to low gas conductance. It can lower its leaf water potential substantially without appreciable loss of water and it keeps a high biomass of fine roots in the upper soil horizons. In all three species, root growth and arrangement are related opportunistically to the temporal and spatial distribution of soil water. Each species has a different adaptive syndrome for the same climate but a similar strategy for the different soil water regimes.

4. The variability of rainfall in terms of events, seasons and years has a strong hydrological and ecological significance. One or two events may account for 80–90% of the total annual runoff and sediment yields. Two years with similar total rainfall but different time distribution, may show very different amounts of runoff. The timing and amount of the spring rainfall set up the water storage for sustaining the largest part of the annual plant growth.

7.7 ACKNOWLEDGEMENTS

Miguel Angel Domene, Domingo Fernández, Juan Paris and Montse Guerrero are thanked for analysing soils, water and plants; Alfredo Durán, for indispensable technical assistance in field instrumentation; and the owners of field sites, J.J. Sánchez and R. Coca, for permission to carry out research and to exclude animals.

7.8 REFERENCES

Abrahams, A.D. and Parsons, A.J. (1991) Relation between infiltration rate and stone cover on a semiarid hillslope, Southern Arizona. *Journal of Hydrology*, **122**, 49–59.

Black, C.H. (1987) Biomass, nitrogen and phosphorus accumulation over a southern California fire cycle chronosequence. In Tenhunen, J.D. et al. (eds) *Plant Response to Stress*. Springer-Verlag, 445–458.

Calvo, A., Gisbert, J.M. Palau, E. and Romero, M. (1988) Un simulador de lluvia portátil de fácil construcción. In M. Sala and F. Gallart (eds) *Métodos y técnicas para la medición en el campo de procesos geomorfológicos*. Monografía No. 1, Sociedad Española de Geomorfología, Zaragoza.

FAO (1990) FAO-UNESCO soil map of the world. Revised legend. FAO, Rome.

Harvey, A.M. (1987) Patterns of Quaternary aggradational and dissectional landform development in the Almeria region, south east Spain: A dry-region, tectonically active landscape. *Die Erde*, **118**, 193–215.

Lázaro, R. and Rey, J.M. (1991) Sobre el clima de la provincia de Almeria (S.E. Ibérico): Primer ensayo de cartografia automática de medias anuales de temperatura y precipitación. *Suelo y Planta* **1** (1), 61–68.

Le Houérou, H.N. (1991) La Méditerranée en l'an 2050: Impacts respectifs d'une éventuelle évolution climatique et de la démographie sur la végétation, les écosystèmes et l'utilisation des terres. Etude prospective. *La Météorologie*, **36**, 4–37.

Poesen, J. and Ingelmo-Sánchez, F. (1992) Runoff and sediment yield from topsoils with different porosity as affected by different rock fragment cover and position. *Catena*, **19**, 451–474.

Pugnaire, F.I., Haase, P., Incoll, L.D. and Clark, S.C. (1966) Response of the tussock grass *Stipa tenacissima* to watering in a semi-arid environment. *Functional Ecology*, **10** (in press).

Puigdefábregas, J. (1992) Mitos y perspectivas sobre la desertificación. *Ecosistemas*, **3**, 18–22.

Puigdefábregas, J., Sole, A., Lazaro, R. and Nicolau, J.M. (1992) Factores que controlan la escorrentía en una zona semiarida sobre micaesquistos. In F. López-Bermúdez, C. Conesa and M.C. Romero (eds) *Estudios de Geomorfología en España*. Sociedad Española de Geomorfología, Murcia, 117–127.

Rivas-Martínez, S. (1987) *Memoria del mapa de series de vegetación de España*. ICONA, Madrid.

Romero Díaz, M.A., López-Bermúdez, F., Thornes, J.B., Francis, C.F. and Fisher, G.C. (1988) Variability of overland flow erosion rates in a semi-arid mediterranean environment under matorral cover, Murcia, Spain. *Catena Supplement*, **13**, 1–11.

Sagar, B.F. (1988) The Shirley Soil Burial test fabric and tensile testing as a measure of biological breakdown of textiles. In Harrison, A.F., Latter, P.M. and Walton, D.W.H. (eds) *Cotton Strip Assay: An Index of Decomposition in Soils*. ITE Symposium No. 24, Institute of Terrestrial Ecology, Grange-over-Sands, 11–16.

Sánchez, G. and Puigdefábregas, J. (1994) Interactions between plant growth and sediment movement in semi-arid slopes. *Geomorphology*, **9**, 243–260.

Schlesinger, W.H. (1980) Biomass, production and changes in the availability of light, water and nutrients during development of pure stands of the chaparral shrub, *Ceanotus megacarpus*, in the Santa Ynez Mountains, California. *Ecology*, **59**, 1256–1263.

Specht, R.L. (1969) A comparison of the sclerophyllous vegetation characteristic of mediterranean type climates in France, California and southern Australia. II Dry matter, energy and nutrient accumulation. *Australian Journal of Botany*, **17**, 293–308.

Walter, H. (1973) *Vegetationenszonen und Klima*, 2nd Edition. Verlag Eugen Ulmer, Stuttgart. 244 pp.

8

The El Ardal Field Site: Soil and Vegetation Cover

F. López-Bermúdez, A. Romero-Díaz, J. Martínez-Fernández
and J. Martínez-Fernández

Departamento de Geografía Física, Universidad de Murcia, Spain

8.1 INTRODUCTION

Soil erosion has been recognized as a problem in the Mediterranean belt since the earliest times. About 80% of land is potentially erodible at unacceptable rates and 50% of the area of Spain has had its productivity seriously reduced. In the semi-arid Mediterranean areas of southeast Spain, the driest in Europe, soil erosion is one of the most serious processes of land degradation, because of its serious agronomic and environmental implications. Erosion affects the physio-chemical properties of the soil and its productivity, structural deterioration and nutrient removal. It reduces reservoir capacity by sedimentation, and causes irreversible degradation and instability to vast tracts of arable and forested lands. In sensitive areas of clays, marls and gypsum, soil loss can be extremely heavy, rates often reaching $200\,m^3\,ha^{-1}\,yr^{-1}$. These high erosion values are, however, easily surpassed during the short, but intense, periods of heavy rain when over 200 mm can fall in 24 hours (López-Bermúdez, 1990). Nevertheless, the sediment production rate for reservoir catchments are lower, oscillating between 2.11 and $11.00\,t\,ha^{-1}\,yr^{-1}$ (Romero-Díaz et al., 1992).

8.2 GEOGRAPHIC SETTING

The southeastern part of the Iberian Peninsula experiences powerful climatic contrasts. It is the most dry and arid in the whole of Europe (Vilá-Valentí, 1961; Geiger, 1970; López-Bermúdez, 1972). The El Ardal field site is located within the Mula basin which extends for $647\,km^2$ (Figure 8.1). The climatic conditions are semi-arid with hot, dry summers and mild winters. The average annual rainfall is about 300 mm but it has a high inter-annual and spatial variability (Figure 8.2). There are commonly prolonged dry periods followed by heavy rainfall, particularly in the autumn. The rainfall is concentrated in a few days during the year. There is a lot of sunlight, 2900 hours of sunshine per year, and an average temperature of 18°C with more than 100 days when the temperature is greater than 30°C. Global radiation is about $20\,MJ\,m^{-2}\,yr^{-1}$.

As a result of these conditions potential evapotranspiration is about $1100–1200\,mm\,yr^{-1}$ and a water deficit of around $900\,mm\,yr^{-1}$ is estimated. However, within the area there are

Mediterranean Desertification and Land Use. Edited by C. Jane Brandt and John B. Thornes.

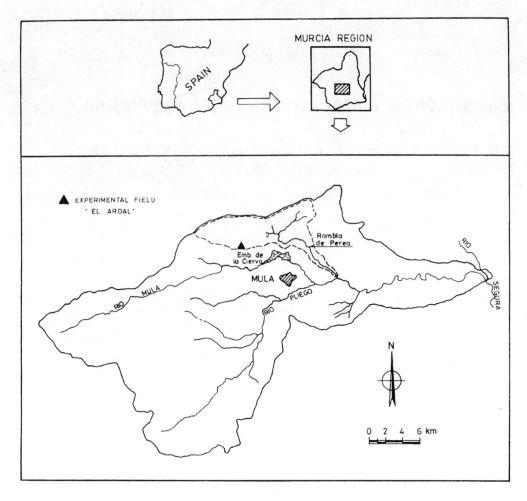

FIGURE 8.1   Rainfall series at La Cierva reservoir station (Murcia, Spain)

climatic differences which depend on topographic altitude, orientation and exposition. This gives rise to topoclimates or local climates whose conditions are mirrored by landscapes.

The El Ardal experimental field is on the northern side of the Mula basin, in the middle of the Region of Murcia (Figure 8.1). It is at an altitude of 550 m, on an average gradient of 20% and receives an average 320 mm of rain per year (although this is highly variable). It has 2870 sunshine hours per year, an average temperature of 14.5°C and daily global radiation and albedo of about 285 and 47 W m^{-2} respectively.

The site is within a micro-catchment with limestones outcropping on the upper slopes. These are covered with *Rosmarinus officinalis/Thymus vulgaris* mattoral scrubland with some patches of *Pinus halepensis*. There is a clear break of slope related to a change in lithology to marls. The middle section is occupied by abandoned fields and cereals and the lower zone with cereals, almonds, olives and viticulture. The soils are poorly developed,

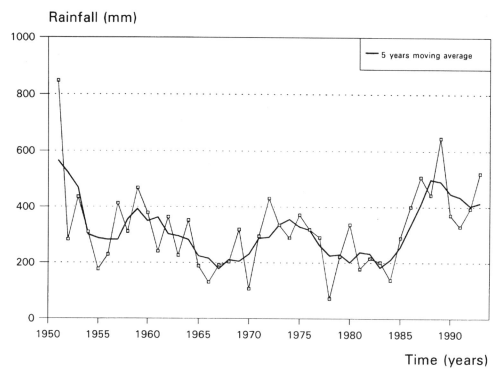

Mean = 320.1 mm   Standard deviation = 143.2 mm   CV = 44.74%

FIGURE 8.2   Location of study area

belonging to either the Xerorthent sub-order or the Xerollic palaeorthid group. Petrocalcic horizons at limited depths are typical in this area.

8.3 EXPERIMENTAL LAYOUT AND TECHNIQUES

8.3.1 Core programme site

The design of the experimental area is shown in Figure 8.3. There are 3 plots for vegetation monitoring, each of 20×30 m, and eight plots (of different sizes) to measure the impact of vegetation on interception and throughfall. There are 17 erosion/overland flow plots (12 of 8×2 m and 5 of 10×2 m). An automatic weather station is situated on the top the hillslope recording speed and direction of the wind, temperature, dew point, precipitation, global radiation, albedo, net radiation and air humidity. The data is stored in a data logger at 10-minute intervals (López-Bermúdez et al., 1991). There is an instrumented weir the mouth of the micro-catchment. In addition there is a network of 18 Hellman pluviometers covering the whole site, and 300 ground pluviometers, placed under the different plant

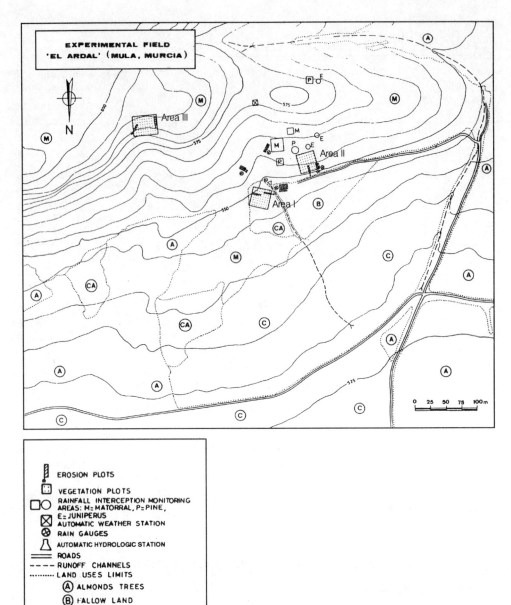

FIGURE 8.3   Experimental design of El Ardal site

species in order to determine their interception capacity (Belmonte-Serrato and Romero-Díaz, 1992).

Together with the Amsterdam team, rainfall simulations were undertaken and the results are reported in Bergkamp et al. (submitted). The simulator was a sprinkler type (van Mulligen, 1990) with Tee Jet 20 W nozzles (Spraying systems Co, USA), covering an area of $20\,m^2$ in a plot of 2×10 m. The intensity was kept at $70\,mm\,h^{-1}$ with a mimimum duration of 30 minutes (this has a recurrence interval of approximately 15 years at the El Ardal field site). Parameters such as runoff, soil moisture, *in-situ* pF curves and soil temperature, were recorded.

8.3.2 Soil analysis

To determine soil aggregate distribution, eight fractions were separated by sieving, and for aggregate stability the Drop Test was used (Farres and Cousen, 1985).

Water retention capacity was measured for pF 0 up to pF 2.3 using the sand box technique (Martínez-Fernández, 1990) and for pF 3.0 to pF 4.2 using the Richards pressure membrane (Richards, 1947).

The saturated hydraulic conductivity was measured in the laboratory using the constant head permeameter technique on undistorted soil samples. In the field the inverse auger hole method was used (Kessler and Oosterbaan, 1980). For soil moisture content the gravimetric method was used.

8.3.3 Vegetation analysis

For vegetation analysis three areas 30×30 m size were selected: Area I, on an abandoned field close to the shrubland; Area II, in the lower part of the hillslope, with a semi-natural shrubland with *Rosmarinus officinalis* and *Juniperus oxycedrus* as the main species; and finally Area III, on the upper part of the hillslope, covered by the semi-natural shrubland with the above species together with *Pinus halepensis* dominant.

Vegetation cover for shrubs was measured along 12 line transects in each area, each 10 m long. The cover of the herbaceous layer was measured using 20 point quadrats per measurement area, each 0.5×0.5 m. Semi-natural vegetation cover in the six erosion plots was also measured with 10 point quadrats per plot. The sampling was carried out every 3 months, in the middle of each season, and data are available from spring 1991 to autumn 1992.

Above-ground biomass was measured directly (destructive) for herbaceous plants and indirectly (using allometric relationships) for shrubby plants. Litter production has been measured using three different systems. Standard litter boxes 0.5×0.5 m were used for *P. halepensis* and shrubs. Individuals plants were enclosed in a mesh bag to measure total individual litter production. Newly designed, modified litter boxes were used to collect litter from the herbaceous layer. In these devices, litter was directly collected in square micro-plots (0.5×0.5 m) using a vacuum cleaner. The micro-plots were protected from losses by a small fence, 20 cm tall.

The growth of shoots of *Rosmarinus officinalis* was monitored by measuring the lengths of randomly selected twigs. Litter decomposition was estimated using the cotton strips method, expressed as values of CRR (cotton rooting rate). Finally, below-ground biomass was measured down the soil profile by taking soil samples of known volume.

TABLE 8.1   Experimental plot characteristics

Plot no.	Dimension (m)	Slope (%)	Orientation (N)	Treatment	Vegetation cover (%)	Date of start
1	2×8	7	N-3	1	0–90	2/89
2	2×8	7	N-3	2	10–30	2/89
3	2×8	7	N-3	3	10–30	2/89
4	2×8	7	N-3	4	0–10	2/89
5	2×8	28	N-120	5	10–30	2/89
6	2×8	28	N-120	6	10–70	2/89
7	2×8	22	N-25	6	65–70	2/89
8	2×8	22	N-25	5	10–20	2/89
9	2×8	11	N-350	6	65–70	2/89
10	2×8	11	N-350	5	10–20	2/89
11	2×8	7	N-3	7	5–20	10/89
12	2×8	7	N-3	8	0–90	10/89
13	2×10	20	N-110	2	10–60	9/90
14	2×10	20	N-115	2	30–60	4/91
15	2×10	11	N-350	6	20–30	4/91
16	2×10	22	N-190	6	45–50	4/91
17	2×10	22	N-40	6	10–20	4/91

Treatments: 1 = Crop barley, 2 = Fallow land with stones, 3 = Fallow land without stones, 4 = Ploughed down slope, 5 = Cut matorral, 6 = Natural matorral, 7 = Treated with polymers $(40\,\mathrm{g\,m^{-2}})$ and fallow land, 8 = Crop wheat.

8.3.4 Erosion

Soil losses were measured on 17 plots (Table 8.1). The plots are on different slopes (between 7 and 28%), with different orientations and treatments, and with different vegetation covers, changing according to the season. Six of the plots are arranged in pairs and the existing natural vegetation has been removed from three of them (plots 5, 8 and 10) in order to observe the effects of cover removal on soil loss. Barley and wheat were sown annually on plots 1 and 12. The surface stones were removed from plot 3. Plot 4 was ploughed downslope. Plot 11 was treated with polymers $(40\,\mathrm{g\,m^{-2}})$. The rest of the plots were not altered and remained with fallow or with their original vegetation.

In some plots (numbers 1, 2, 9, and 11), and over a 2-year period, 1990–91, samples from the A horizon were taken every three months in order to follow the seasonal variation of organic C, total N, available K, P, Fe, Cu, Mn and Zn (Alías et al., in press). These soil components and nutrient elements were also determined in the sediments caught in the runoff troughs, when the amount collected was sufficiently large. The clay fraction was obtained from the Ap horizon of fine earth using methods proposed by Martín Vivaldi and Rodríguez Gallego (1961), Brindley (1966), and Whitting and Allardice (1986) and the mineralogical composition was studied using X-ray diffraction. The clay fraction of the sediments was studied using the same methods.

Runoff was measured after every rainfall event. Water samples were analysed for soluble Na, K, NH_4-N, NO_3-N, and PO_4-P by using a Dionex Model 2000 i/SP Ion Chromatograph. An atomic absorption spectrophotometer was used to determine soluble

Mg and Ca. Electrical conductivity was also measured and the sodium adsorption ratio, SAR, was calculated.

8.4 MICRO-CLIMATE

The automatic weather station started in 1990, although rainfall data are available from 1989 (which was an atypically wet year). The years since then have been considerably drier, with values always under 300 mm.

Without doubt, the amount of rainfall is the most important of all meteorological parameters in determining the runoff coefficients. Therefore, the most rainy year (1989, 713 mm) is the one in which the highest coefficient (2.06%) was recorded, and the poorest one (1992, 315 mm) has the lowest coefficient (1.14%). Although the data series is short it has been possible to observe the following:

1. The spatial distribution of precipitation is irregular.
2. Rainfall differs according to the orientation of the slope.
3. A close relationship between the amount and kind of rainfall produced in each storm and the topographical configuration of the micro-basin has been verified.
4. There is a minimum threshold for precipitation (3–5 mm) below which runoff will not occur on any of the plots. However some plots, especially those with vegetation, need much bigger amounts to produce runoff.
5. The percentage interception is usually higher when the precipitation is lower although the speed and direction of the wind produce spatial differences. The highest interception percentages have been obtained in *Juniperus oxycedrus* and *Pinus halepensis* (40–50%). In areas with varied vegetation the percentage fluctuates between 25 and 30%.

8.5 SOIL CHARACTERISTICS

Seven soil profiles and a 2-km long soil transect have been studied, from the Lomo Herrero watershed to the valley bottom. Additional information has also been obtained from sampling at several other points on the hillslope. The soils of El Ardal are shallow (40–50 cm average) and only at the valley bottom surpass 1 m depth. On the upper slopes they lie over limestone or limestone conglomerate and on the other slopes over an horizontal, sometimes strongly cemented, petrocalcic layer. The latter has a high content of $CaCO_3$, which can exceed 50%.

Soil texture changes depending on the hillslope position because, while in the upper and lower slopes clay and silt prevail, in the whole mid-slope which is used for cultivation, sand is the most important fraction. The organic matter content is low in the cultivated area (about 1%), and higher in the shrublands (between 5 and 7%).

8.5.1 Physical characteristics

Aggregate distribution

The fine aggregates, from 0 to 20 mm, have been analysed separately from the large, bigger than 20 mm (Table 8.2). In the cultivated soils, aggregates smaller than 2 mm represent more than 50% of the fine fraction, as the result of the continuous fragmentation due to

TABLE 8.2 Distribution (%) of aggregate size (mm), soil organic matter (OM %) and land use*

Land use (and no. of samples)	OM	Aggregate size							
		0–1	1–2	2–4	4–5	5–8	8–10	10–20	>20
Cultivated (32)	1.70	37.9	11.4	13.3	4.2	9.7	5.21	8.4	29.9
	(0.86)	(14.7)	(2.0)	(3.8)	(1.1)	(2.8)	(1.2)	(8.3)	(22.3)
Scrubland (16)	5.38	14.3	12.2	22.1	6.5	15.4	7.3	22.1	45.0
	(2.00)	(6.9)	(2.9)	(3.7)	(1.0)	(2.9)	(2.0)	(7.4)	(14.5)
Cultivated									
cereal (12)	1.92	28.4	11.5	14.7	4.6	10.9	5.6	24.2	40.7
	(0.81)	(18.3)	(1.6)	(3.7)	(1.3)	(3.0)	(1.2)	(10.2)	(26.0)
vineyard (6)	1.24	46.6	11.0	11.1	3.3	7.2	5.3	15.6	15.5
	(0.33)	(2.4)	(0.9)	(2.1)	(0.3)	(0.8)	(0.7)	(3.0)	(11.5)
almonds (2)	1.17	46.9	11.1	11.9	3.6	7.5	4.6	14.3	9.6
	(0.14)	(1.3)	(0.5)	(0.4)	(0.1)	(0.7)	(0.1)	(0.7)	(5.1)
fallow (8)	1.13	45.6	9.8	10.5	3.8	8.7	5.3	16.3	37.8
	(0.21)	(8.2)	(1.2)	(2.3)	(0.7)	(1.6)	(1.5)	(3.5)	(15.4)
Scrubland									
under veg. (4)	6.30	12.6	14.7	22.8	6.9	13.5	7.3	22.2	57.6
	(1.07)	(6.2)	(3.5)	(2.6)	(0.1)	(0.8)	(2.8)	(8.5)	(4.4)
bare soil (12)	5.02	14.9	11.3	21.9	6.4	16.0	7.3	22.1	40.8
	(3.30)	(7.0)	(2.0)	(3.9)	(1.2)	(3.1)	(1.6)	(6.9)	(14.2)
Abandoned									
4 years (2)	2.46	28.7	14.3	19.5	5.7	12.3	3.4	16.1	17.2
	(0.67)	(4.6)	(0.9)	(1.2)	(0.6)	(0.4)	(0.6)	(4.8)	(3.6)
10 years (2)	3.80	37.8	16.1	17.1	4.5	13.2	4.6	6.6	10.1
	(0.71)	(9.5)	(1.0)	(4.1)	(1.1)	(0.3)	(0.1)	(3.5)	(3.1)

* Mean values and standard deviation (in parentheses). Aggregates from 0 to 20 mm size on 100% basis. >20 mm size on total sample basis.
Sampling depth: 0 to 20 cm.

ploughing and the low soil organic matter content. In the soils on the uncultivated hillslope the distribution is much more uniform. Nevertheless, in these soils the percentage of large aggregates is higher, especially in soils under vegetation. The reason for this macro-aggregation is the large amount of organic matter and the active role of the roots.

Observing the relation between aggregate size and the other variables, it can be noted that the fraction between 4 and 5 mm, which is most stable, has a close dependence on the organic matter content (Figure 8.4).

Aggregate stability

In all the cases analysed, an increase in stability was produced when the sample was taken from a dry state to a moist pF 1 (Table 8.3). This increase is remarkable, especially in the samples which had a lower stability in air dry state. The stability is greater in the natural hillslope soils, and even more so under vegetation. In cultivated soils the aggregates are

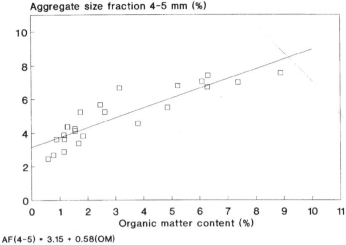

AF(4-5) = 3.15 + 0.58(OM)
$r^2 = 0.80, p < 0.001$

FIGURE 8.4   Aggregate size fraction of 4–5 mm and soil organic matter content relationship

TABLE 8.3   Aggregate stability* and land use

Land use (and no. of samples)	Air dry conditions			pF 1 moist conditions		
	Mean	SD	CV**	Mean	SD	CV
Crops (320)	10.2	4.2	40.7	14.5	11.2	77.0
Scrubland (160)	27.3	21.0	76.9	43.4	39.1	90.1
Crops						
cereal (120)	14.6	7.4	50.3	15.2	11.9	78.2
vineyards (60)	8.1	2.2	27.7	17.7	15.8	89.4
almonds (20)	7.3	1.7	23.0	10.8	8.0	73.7
fallow (80)	8.9	3.6	40.8	9.0	6.3	70.4
abandoned (40)	9.0	2.0	22.0	21.1	14.0	66.3
Scrubland						
under veg. (40)	53.8	42.7	79.4	62.1	48.0	77.3
bare soil (120)	18.4	13.7	74.5	37.2	36.2	97.2

* Number of drops (Drop test). Aggregate size: 4–5 mm. Sampling depth: 0 to 20 cm.
**CV Coefficient of variation (%).

much less stable. The analysis of aggregate morphology clearly shows two different types: those which have not been altered by cultivation, with a massive and compact morphology and small intergranular porosity, as a result of the active organic matter role in cementation; and the aggregates of cultivated soils which have an unconnected morphology, forming only simple groups of independent grains, with scarcely any trace of cementation agents, organic or inorganic.

Through multiple regression analysis, the relation between stability and the rest of the variables was studied. The result shows a very close and inverse relationship with bulk density and a direct relationship with the organic matter content and 4–5 mm aggregation fraction (the fraction used for the stability test). The combination of that fraction and the organic matter with the bulk density explains more than 75% of the variance.

8.5.2 Hydrological characteristics

Water retention capacity

The soils in El Ardal have low retention capacities because of the low organic matter content (even 1%) in some cases, and because of the high sand fraction (even more than 50%) in others. In the non-cultivated soils the characteristic curves show a high moisture content for low pF values, because of the abundance of organic matter and the balanced texture (silt loam). Sand loam predominates the intermediate sectors of the slope and, because they are cultivated, there is a low organic matter content.

The characteristic values of the pF curve have been studied, especially those more connected with the vegetation. The theoretical moisture content at field capacity (for these soils pF 2.3) fluctuates between 0.2 and 0.3 $cm^3 cm^{-3}$ (volume water/volume soil), although this is not often reached in natural conditions. The permanent wilting point (pF 4.2), is less variable, being almost always around 0.1 $cm^3 cm^{-3}$. Most of the time the soil moisture content is at or below permanent wilting point. Since the soils are shallow in all cases the available water for plants is scarce and difficult to uptake.

Saturated hydraulic conductivity

This is generally high or very high. Factors such as the sandy texture, high macroporosity and root content usually give values close to or higher than 100 $cm h^{-1}$. This characteristic was confirmed by experiments with rainfall when, in spite of using a high intensity (70 $mm h^{-1}$), no runoff was produced. Moreover the runoff coefficients in the erosion plots are always very low. The conductivity of the fallow soils (126.6 $cm h^{-1}$) is remarkably high compared with the rest of the cultivated soils (58.4 $cm h^{-1}$).

Under natural conditions there are even more striking differences between soils under vegetation which have a very high conductivity (197 $cm h^{-1}$) and those which are not

TABLE 8.4   Saturated hydraulic conductivity *in situ* (inverse auger hole method)

Profile	Land use	Depth (cm)	$cm h^{-1}$	$m day^{-1}$
1	cereal	0–25.2	6.15	1.48
2	scrubland	0–27.5	10.28	2.47
3	cereal	0–75.5	6.06	1.46
4	vineyard	0–57.0	8.15	1.96
5	fallow	0–60.7	5.55	1.33
6	abandoned field	0–18.5	11.99	2.88
7*		20.0–39.3	5.87	1.41

*Petrocalcic horizon.

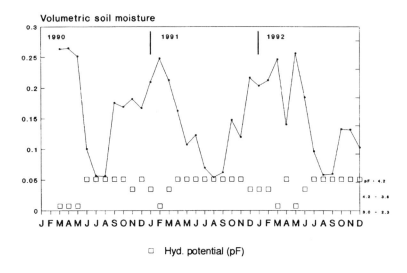

FIGURE 8.5   Soil moisture content (cm³ and cm⁻³) and hydraulic potential (pF) series in the El Ardal experimental site 1990–1992

covered (47.2 cm h⁻¹). It is thought that trampling between plants has reduced the soil's conductivity considerably.

Hydraulic conductivity profile

The cultivated soils all show similar conductivity values and in all of them a clear reduction is observed as the depth increases as the result of the compaction caused by ploughing. Table 8.4 shows the average values for the profiles under different land uses. In can be seen that the highest conductivities occur on soils which have been abandoned or which are under natural conditions.

Soil moisture evolution

The measurement period covers the 34 months from March 1990 to December 1992 (Figure 8.5). Generally the values are low all the time. The soil moisture content values have almost never been higher than $0.25\,cm^3\,cm^{-3}$ (volume water/volume soil basis). Only in the winter months and sometimes at the end of spring, have values above $0.20\,cm^3\,cm^{-3}$ been measured. The lowest soil moisture content appears in summer (July, August, about $0.06\,cm^3\,cm^{-3}$), although the critical period can be extended from May/June to October/November, as happened in 1991–92. These conditions provide only a small water content available for plants and not so much for the quantity but because of the difficulties of its extraction. In more than 60% of the months the hydraulic soil potential (in pF terms) was below the theoretical wilting point. This has happened also during long periods (in 1991, 8 months). This situation represents very precarious conditions for the vegetation cover.

8.6 VEGETATION STRUCTURE AND DYNAMICS

8.6.1 Species composition

The vegetation at the field site is a typical Mediterranean shrubland dominated by *Rosmarinus officinalis*. Other species present are *Juniperus oxycedrus, J. phoenicea, Genista scorpius, Rhamnus lycioides* and *G. valentina*, with isolated individuals of *Pinus halepensis* and chaemephytes like *Thymus vulgaris* and *Cistus clusii*. The species composition and the general structure indicate a relatively good state of conservation, despite the influence of traditional uses like grazing and hunting. In the abandoned field a different community appears, as a result of a secondary succession, with species like *Artemisia campestris, Teucrium capitatum, Medicago minima, Sonchus tenerrimus* and *Eryngium campestre*.

8.6.2 Seasonal and spatial dynamics

A total of 81 species have been identified. In the shrubland, 60% of species are shrubs, while in the abandoned field, annuals predominate. This generates a different structure and dynamic behaviour, both seasonally and in relation to environmental conditions. The greatest fluctuations in the vegetation cover occur in the abandoned field because of the presence of annuals. In the shrubland the vegetation is mostly formed by shrubs and perennial grasses such as *Brachypodium retusum*. *Rosmarinus officinalis* organizes the community into a spatial mosaic, in which other species, mainly herbaceous ones like *B. retusum*, show a close spatial dependence. This generates the same spatial pattern in several vegetation-related parameters, such as the litter pool and fine-root biomass. These in turn create the same spatial distribution of soil characteristics such as aggregate stability and water retention capacity. The final result is a hierarchical series of relationships between several parameters which is, ultimately, controlled by the spatial pattern of *R. officinalis*.

8.6.3 Vegetation, soil moisture and sediment yield

The closest relationship between vegetation cover and other physical factors, such as soil moisture, is found in the abandoned field. High correlations have been found between the cover of annuals plants and soil moisture ($r^2 = 0.72$; $p < 0.01$) and between total litterfall and soil moisture ($r^2 = -0.70$; $p < 0.01$) (Figure 8.6). The vegetation cover in the shrubland is also strongly related to soil moisture ($r^2 = 0.76$; $p < 0.006$), although it shows seasonal differences that are less marked than those of the abandoned field (Figure 8.7).

As expected, average values of runoff and sediment yield are poorly correlated with average cover, and each event was analysed separately. Significant correlations have been found between several cover variables and plot responses, in particular between herbaceous cover and runoff and between total litterfall and sediment yield (Figure 8.8) ($r^2 = 0.72$, $p < 0.05$). Nevertheless, no relationships have been found between vegetal cover and sediment yield. This might mean that litterfall plays a sufficiently important role in the control of sediment yield that it can be detected using averaged data. This could be explained by the fact that, at the plot scale, qualitative differences in vegetation cover (such as architecture) are very important in terms of its capability to protect against soil particle losses in high-intensity storms. On the other hand, the litter layer probably always effectively protects the soil regardless of rainfall intensity.

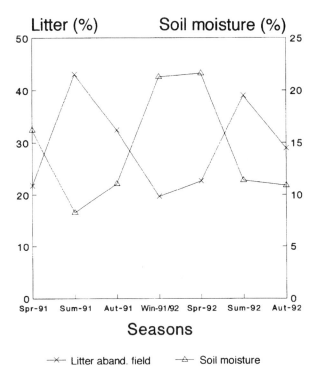

FIGURE 8.6 Litterfall in the abandoned field and soil moisture series (spring 1991 to autumn 1992)

8.6.4 Above-ground biomass, litterfall, productivity

The present biomass clearly separates the abandoned field from the shrubland. In the shrubland, there are also differences between the upper part of the hillslope, with more rocky outcrops (Area III), and the lower part, with greater depth of soil which increases water availability (Area II).

Destructive sampling was carried out in Area II to calculate total biomass per unit area in the herbaceous layer. The above-ground biomass of *R. officinalis* was also estimated using allometric relationships. In Area II, total biomass in the herbaceous layer is $1748\,\mathrm{g\,m^{-2}}$ and the estimated value of above-ground biomass of *R. officinalis* is $764\,\mathrm{g\,m^{-2}}$, giving an estimated total biomass of $2512\,\mathrm{g\,m^{-2}}$. In fact this value may be a little low, because shrubs of others species were not measured, but these generally represent less than 5% of the percentage cover. Despite this, the shrub biomass is about 10 times that of the abandoned field (between 200 and $250\,\mathrm{g\,m^{-2}}$). It is also between four and six times greater than the one communicated by Francis and Thornes (1990) with regard to other areas in Murcia, and is almost 10 times more than the biomass found in semi-arid communities, but it is similar to data from other Mediterranean areas without a semi-arid regime.

In Area II litter pool was also measured, and had an average value of $1359\,\mathrm{g\,m^{-2}}$, not far from that found in the biomass of the herbaceous layer.

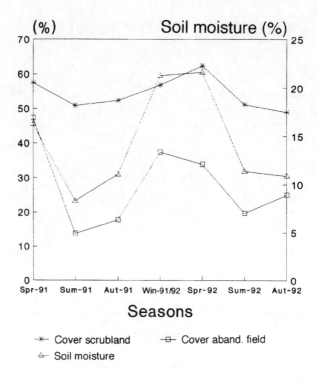

FIGURE 8.7   Vegetal cover in the shrubland and abandoned field and soil moisture evolution (spring 1991 to autumn 1992)

In Area III, on the rocky upper slopes, the above-ground biomass of *R. officinalis* is much lower, $246\,g\,m^{-2}$, as a result of the smaller size of the individuals plants (246 and 373 g per individual in Areas III and II respectively), and the lower density (around one and two individuals per square metre in Areas III and II respectively). Using several relationships with biomass, the leaf area index (LAI), was estimated, with values ranging between 1.1 and 3.4 (using not projected but total leaf area).

Litter production is significantly related to both vegetation cover and above-ground biomass (Figure 8.9). This has permitted the estimation of litter production in the different areas. The overall values are $540\,g\,m^{-2}\,yr^{-1}$ in the shrubland and $288\,g\,m^{-2}\,yr^{-1}$ in the abandoned field. Litter production due to *R. officinalis* $(122\,g\,m^{-2}\,yr^{-1})$, is not far from values found in species belonging to other non-semi-arid Mediterranean communities.

The rosemary growth data have been used to estimate net primary productivity and to assess the response of vegetation to the environmental changes, especially water availability. The growing period is clearly in spring. There is a high individual variability, but significant relationships ($p=0.001$) between growth and soil moisture with a 2 months delay period (probably the necessary time to obtain a physiological response) have been found (Figure 8.10). The net primary production of *R. officinalis* has been estimated using above-ground biomass, growth and litter production. The average values were around $300\,g\,m^{-2}$.

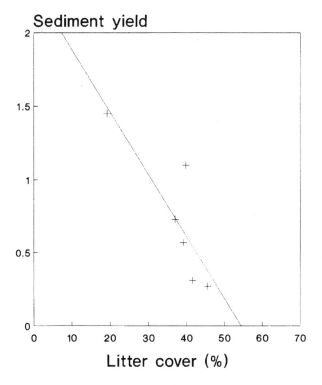

FIGURE 8.8   Litter cover (%) and sediment yield (g m^{-2}) relationships in the erosion plots ($r^2 = 0.72$)

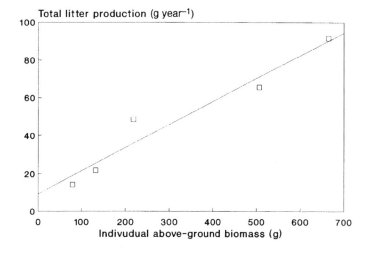

FIGURE 8.9   Individual above-ground biomass and total litter production of *Ròsmarinus officinalis* ($p < 0.006$)

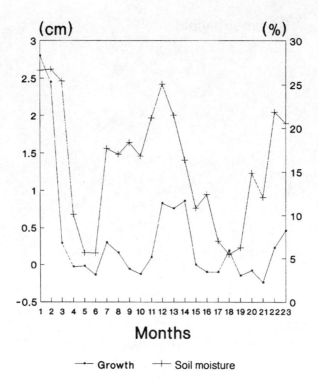

FIGURE 8.10   Growth of *Rosmarinus officinalis* twigs and soil moisture evolution (May 1990 to March 1992) in El Ardal experimental site

8.6.5 Below-ground biomass

The below-ground biomass in the top 10 cm of the soil in the shrubland is between four and six times greater than in the abandoned field (444.4 and 118.7 g m^{-2} respectively). In addition the smaller diameter roots, which are closely related to macro-aggregation, especially appear in the to 10 cm of the soil profile, and are closely related to the spatial distribution of *Brachypodium retusum* (a perennial grass found in the shrubland). The highest values of aggregate stability are also related to the presence of *B. retusum*, showing the same spatial pattern and confirming the expected results.

Decomposition of the cotton strips was significantly higher in the abandoned field than in the shrubland (7.66 and 4.98 CRR respectively), though the factors determining these results are not completely clear. In all cases the decomposition was higher in the mineral horizon than the ectorganic one, while there was no difference between the spring and summer periods.

8.6.6 Summary of vegetation

The structural and dynamic parameters of the shrubland are far from those of communities under high water stress, and indeed are quite similar to those of non-semi-arid Mediterranean shrublands. The northern exposure of the hillslope, the hydrological characteristics of the soil and the existence of additional water inputs from fogs, dew and hoarfrost may explain this situation.

8.7 EROSION

8.7.1 Soil losses and runoff

After running the experimental site for 4 years, it can be said that the soil loss rates in El Ardal are low. This can be explained by the soil type, vegetation cover and precipitation. The average erosion value for the period observed is $20 \, \mathrm{g \, m^{-2}}$. However, there are very marked differences due to the different years of study (Figure 8.11), different rain events and different land uses on the experimental plots.

An average soil loss of $100.52 \, \mathrm{g \, m^{-2}}$ was recorded in 1989 and was reduced to $18.42 \, \mathrm{g \, m^{-2}}$ in 1990; $4.92 \, \mathrm{g \, m^{-2}}$ in 1991, and $1.19 \, \mathrm{g \, m^{-2}}$ in 1992. The soil loss is more related to the rainfall intensity than to the total amount of rain; an example of this is the storm of 13 September 1990, which recorded a loss of $40 \, \mathrm{g \, m^{-2}}$, 90% of total soil loss that year.

It is important to point out that the plots with higher erosion and with the highest runoff coefficients are the most unprotected, those whose vegetation had been cut down, the one ploughed down slope, the one with the stones removed and those sown with wheat and barley (Figure 8.12). In the case of the last two plots, the most important events took place when the harvest had been collected. In contrast, the plots with natural vegetation, even if the cover is scarce or has been left fallow for a long time, have suffered less soil losses.

8.7.2 Fertility losses

The work on the seasonal variability of soil nutrient loss suggests a more efficient biogeochemical cycle under natural shrubby vegetation and a contrasting progressive loss of fertility on the arable plots.

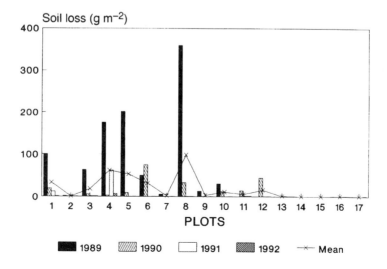

FIGURE 8.11   Soil loss in the erosion plots in El Ardal 1989–92

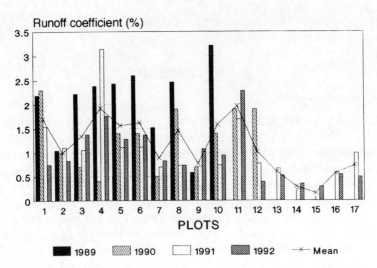

FIGURE 8.2   Runoff coefficient in the erosion plots in El Ardal 1989–92

The losses of nutrient elements associated with the sediment and runoff water are small. The plot cultivated for barley (plot 1) and that with cut shrubs (plot 10) have the highest values, whereas losses are extremely low for the plot covered in shrubs (plot 9) (Alias et al., in press).

8.8 CONCLUSIONS

The cultivation of a section of the hillslope has resulted in a substantial alteration to the environmental and soil conditions. Under the semi-natural shrubland the soil has better hydro-mechanical characteristics: more organic matter, finer texture, more aggregate stability and higher water retention capacity, higher soil moisture content and higher root density. In an intermediate stage between the cultivated and semi-natural areas are the fallow plots or abandoned sectors. The latter, which are located near to the natural hillslope, show evident signs of improvement in their characteristics, reverting to their state prior to cultivation. The El Ardal soils have an almost permanently low moisture content, and the water is strongly retained by the soil leading to unfavourable conditions for vegetation.

The hydrodynamic characteristics of these soils, especially their high conductivity, favour infiltration, so the runoff coefficients are generally low. Because of that, the subsurface flow must have great importance within the water balance of this system.

In this zone, the biological parameters of the semi-natural shrubs define the community as a non-semi-arid shrubland, with relatively high LAI and biomass values. Considering the great capacity of adaptation and the resistance which most of the species have to long droughts, possible changes such as a gradual precipitation decrease would probably produce no more than a local reorganization of the general vegetation structure. In contrast, different land uses and the field abandonment may have important effects on the soil properties. The results from the abandoned field indicate that there is a progressive recovery of the vegetation cover, litterfall, organic matter and aggregate stability, even though most of the variables are seasonally dynamic.

The erosion rates at El Ardal are very low due to the soil, climatic and vegetation-cover-type characteristics. The amount of soil loss depends on precipitation type and land use. The plots in which most erosion was recorded are those least protected by vegetation or stones and those which have been cultivated. On the contrary, the plots which have recorded least erosion are those uncovered or left fallow without any treatment. The fertility losses are also more important in cultivated plots and in those in which the cover has been removed and less important in plots with vegetation.

8.9 REFERENCES

Alías, L.J., López-Bermúdez, F., Marín-Sanleandro, P. et al. (1996) Clay minerals and soil fertility lost on Petric Calcisol under a semiarid mediterranean environment. *Soil Technology* (in press).

Belmonte-Serrato, F. and Romero-Díaz, M.A .(1992) Evaluación de la capacidad de interceptación de la lluvia por la vegetación y su relación con la erosión de los suelos en el SE semiárido español. Primeros resultados. In *Estudios de Geomorfología en España* (ed.) F. López-Bermúdez, C. Conesa-García and M.A. Romero-Díaz), Sociedad Española de Geomorfología, Murcia, 33–43.

Bergkamp, G., Cammeraat, L.H., Boer, M.M. (1996) Water movements and vegetation patterns on shrubland and abandoned fields in desertification threatened areas: results of large rainfall simulation experiments in Spain. *Earth Surface Processes and Landforms* (in press).

Brindley, G.W. (1966) Ethylenglycol and glycerol complexes of smectites and vermiculites. *Clay Min. Bull,* **6**, 119.

Farres, P.J. and Cousen, S.M. (1985) An improved method of aggregate stability measurement. *Earth Surface Processes and Landforms*, **10**, 321–329.

Francis, C.F. and Thornes, J.B. (1990) Runoff hydrographs from three Mediterranean vegetation cover types. In *Vegetation and Erosion. Processes and Environments*. (ed. J.B. Thornes), Wiley, Chichester, 363–384.

Geiger, F. (1970) *Die Aridität in Südostspanien*. Stuttgarter Geographische Studien. Stuttgart.

Kessler, J. and Oosterbaan, R.J. (1980) Determining hydraulic conductivity of saturated soils. In *Drainage Principles and Applications*, Vol. III, ILRI, 253–296.

López-Bermúdez, F. (1972) El agua en la Cuenca del Segura. *Papeles de Geografía*, Universidad de Murcia, Vol. 6, 9–24.

López-Bermúdez, F. (1990) Soil erosion by water on the desertification of a semi-arid Mediterranean fluvial basin: The Segura basin, Spain. *Agriculture, Ecosystems and Environment*, **33**, 129–145.

López-Bermúdez, F., Romero-Díaz, M.A. and Martínez-Fernández, J. (1991) Soil erosion in semi-arid mediterranean environment. The Ardal experimental field (Murcia, Spain). In *Soil Erosion Studies in Spain* (ed. J.M. García Ruiz, M. Sala and J.L. Rubio), Ediciones Geoforma, Logroño, 137–152.

Martín Vivaldi, J.L. and Rodríguez Gallego, M. (1961) Some problems in the identification of clay minerals in mixtures by X-ray diffraction. I. Chlorite–kaolinite mixtures. *Clays Min. Bull.*, **4**(26), 288–292.

Martínez-Fernández, J. (1990) Estudio de las propiedades hidrodinámicas de los suelos: El método del recipiente de tensión hídrica. *I Reunión Nacional de Geomorfología*. Sociedad Española de Geomorfología, Teruel. Vol. II, 699–708.

Richards, L.A. (1947) Pressure membrane apparatus: construction and use. *Agricultural Engineering*, **28**, 451–454.

Romero-Díaz, A., Cabezas, F. and López-Bermúdez, F. (1992) Erosion and fluvial Sedimentation in the River Segura Basin (Spain). *Catena.*, **19**(3/4), 379–392.

Van Mulligen, E.J. (1990) *Erosieonderzoek voor mais—suikerbieten—en aardappelteetl*. Rapport ROC-UVA, 17–25.

Vilá-Valentí, J. (1961) La lucha contra la sequía en el Sureste de España. *Estudios Geográficos, Madrid.* **XXII** (82), 25–49.
Whitting, L.D. and Allardice, W.R. (1986) X-ray diffraction techniques. In *Methods of Soil Analysis, Part 1. Physical and Mineralogical Methods* (ED. A. Klute), ASA and SSSA, Madison, 331–362.

9

The Rio Santa Lucia Site:
An Integrated Study of Desertification

A. ARU

Dipartimento Scienze della Terra, Università di Cagliari, Sardinia, Italy

9.1 INTRODUCTION

Desertification in Sardinia has formed the object of research for some years now. The term desertification in this context is intended to denote the depletion of non-renewable resources, or of resources whose recovery would take an extremely long time. Investigations have addressed large-scale problems such as forest fires and erosion, as well as localized problems such as land loss due to urban expansion and problems associated with salt-water intrusion in underground aquifers and salinization of soils irrigated with water tapped therefrom. The effects on desertification of extracting aggregate along water courses, afforestation with exotic species and of heavy metals pollution from mining and other industrial activities are poorly documented.

A number of field sites have been set up in various Mediterranean regions, in the framework of the MEDALUS I project, to investigate these problems. One such site is the Rio Santa Lucia catchment area. The choice of site hinged on the existence of large areas which have more or less preserved their peculiar natural characteristics, numerous forests planted with exotic species, a disused mine and extensive areas given over to quarrying, urban development and agriculture.

This variety of scenarios allows comparisons to be drawn with similar situations existing elsewhere in Sardinia and also in other Mediterranean countries. For example in watersheds in western Spain, erosion of slopes and salinization in the valley floors has resulted in badly desertified areas. In parts of Greece, and particularly in the islands (Mytilene, Chios, Crete for instance) anthropic pressures and erosion have led to desertification and reversal of the process will involve extremely long times. In southern Portugal and especially in the region between the Alentejo and the Algarve there are extensive degraded areas where erosion has been brought about by continuous tilling of thin soils. Another aspect concerns the cork-producing countries, Portugal, Spain and Sardinia, where continuous working of the soils, overgrazing and fires have had disastrous consequences on soil conservation and regeneration of cork forests creating adverse conditions for plant renewal. Many coastal areas in Sardinia, Spain and Portugal have been developed for housing, leading to the loss of agricultural soils as well as surface and ground waters. Greece appears to have been less severely affected as the phenomenon has been limited to areas near to urban centres.

The problem of pollution due to mining activities is circumscribed to areas in the vicinity of mines and processing plants of metallic ores. The problem has come to a head with the closure of mines, since management of waste dumps has ceased. The former mining district in the Sulcis-Iglesiente in southwest Sardinia and the Sao Domingo mines in southern Portugal are badly affected by heavy metal pollution.

9.2 PHYSIOGRAPHIC SETTING

The Rio Santa Lucia site is located in southern Sardinia (see Figure 9.1) and covers an area of some 110 km² extending from the Punta Sebera and Monte Lattias highlands in the west down to the Santa Gilla lagoon at the southernmost edge of the Campidano plain. The hilly and mountainous area consists chiefly of Palaeozoic formations, Plio-Quaternary sediments underlying the plains and coastal zone. The Palaeozoic basement is composed of a sequence of metasandstones, metaquarzites with alternations of quartz–metaconglomerates, siltites and metapelites, metalimestones and silicified metalimestones.

Two distinct tectonic units can be recognized: the lower unit consists of autochthonous Ordovician and Siluro–Devonian sedimentary formations, while the

FIGURE 9.1   Location of the Santa Lucia catchment, south Sardinia, Italy

upper unit consists of allochthonous Cambro–Ordovician ones. The two units are separated by an overthrust striking NW–SE and dipping SW.

During the late Hercynian, a granite intrusion came into contact with the two overlying tectonic units and formed a metamorphic aureola with a variety of ore deposits (San Leone mine). Granite rocks are represented by equigranular biotite leucogranites with medium-to-coarse grains of pink feldspars (orthoclase and microcline). The granite rocks, generally intensively fractured, are of the biotite leucogranite type with pink feldspars of medium-to-coarse grain.

The Plio-Quaternary sediments in the hilly area consist of screes and thin alluvial deposits. The foothills are bordered by thick deposits of Plio-Pleistocene debris, composed of Palaeozoic cobbles and gravels interbedded with layers of sands and clays, usually fairly well cemented. In the coastal plain, the Plio-Pleistocene-Holocene continental, lacustrine and marine sediments are made up of clays, silts, sands and to a lesser extent, gravels. Tyrrhenian sandstones, conglomerates and organogenous limestones are interlayered in marine sediments. Several fracture systems are present throughout the area related to the formation of the Sardinian graben of Hercynian and Alpine ages.

The geological and structural features of the area are such that it can be divided into two sectors. The first, the Palaeozoic basement is rough and hilly, with a high relief and deep valleys where sheet, rill and gully erosion are visible. In the valley heads and slopes unstable talus cones composed of loose mounds of cobble abound and embankments are frequent.

The evolution of the fluvial landforms has been influenced by tectonic events, which produced a subparallel, hydrographic drainage network in a NE–SW (Rio Gutturu Mannu and Rio Guttureddu), E–W (last segment of Rio Guttureddu), N–S and sometimes NW–SE direction. Torrent erosion is particularly active throughout the area and much coarse material is transported as a result of extreme rainfall events.

The second sector begins where the Rio Gutturu Mannu and the Rio Guttureddu join the Rio Santa Lucia which flows into the Campidano plain. This extensive area consists essentially of two alluvial formations. The oldest one, dating back to the Plio-Pleistocene, outcrops in the foothills and is composed of coarse, strongly weathered material that has undergone intense soil genesis. The younger formation (Holocene) consists of a large fan, that extends, gently sloping, down to the Santa Gilla lagoon and the Gulf of Cagliari.

9.3 DESCRIPTION OF SOILS

The choice of the Rio Santa Lucia site is particularly significant both for Sardinia and the Mediterranean as a whole. Numerous soil types are encountered. Their formation depends on the principal pedogenetic factors while their degradation and loss can be attributed chiefly to human causes. The catchment area provides a comprehensive picture of both the landscape's evolution and its desertification.

One of the major factors in soil formation is the parent material. The most common substrata in the area are granites, metamorphites, Plio-Pleistocene alluvia and glacis, Holocene alluvia, clays in the lagoons and dune sands. Landforms are related to the lithological types and with the vegetation play an important part in the development of soils; human activity is the major contributor to their degradation. Soils are described and defined taking into account both lithology and landforms. Soil in this context can be defined as the natural body of the earth's surface which supports or is able to support plant life. Its properties derive from the continued effects through time of both climate and living organisms on the parent rock, and are governed by the relief.

Soils have been classified according to the soil taxonomy developed by the USA Soil Survey Staff. This system comprises six categories of different degrees of generalization: orders, suborders, major groups, subgroups, families and series. Soils are classified on the basis of permanent data evaluated for each of the identified horizons, through field and laboratory measurements. Soil horizons are layers of soil approximately parallel to the surface with well-defined characteristics and are the result of soil formation processes. The series of horizons determines the soil profile.

In the area under study the most common soil orders are:

Alfisols: soils containing a diagnostic horizon with accumulation of clay with medium to high base saturation. The suborders Palexeralfs and Haploxeralfs are typical of areas of the Mediterranean on pre-Wurmian deposits.

Inceptisols: these soils are characterized by a cambic horizon which represents the beginning of soil formation. The cambic horizon can be distinguished from the surface horizons and the parent material by its colour and structure. The large group Xerochrepts is commonly found throughout the Mediterranean.

Entisols: these soils do not contain a diagnostic horizon. They are in the very early stages of formation either because the rock weathers poorly or because erosion phenomena have rejuvenated the soil profile. The large group Xerorthents is found in many Mediterranean areas.

9.3.1 Soils on metamorphites

Here the soil type is closely related to the type and extent of plant cover. On the hill tops, on the complex steep slopes and in the valley floors there are tracts of land where rock outcrops or is covered with a thin layer of argillic soil, highly skeletal even on the surface, which can be classified as Lithic Xerorthents or Lithic–Ruptic–Xerorthents. Plant cover is scanty or consists of low and degraded bush. Where the vegetation is more continuous, even on rough ground and steep slopes, the soils are more developed and belong to the Lithic Xerorthent, Dystric Xerorthents and Lithic Xerochrepts classes. In those areas where the morphology is not as rugged and where vegetation consists of developed bush or holm oak woods, mainly Lithic Xerochrepts and Typic Xerochrepts soils are found, shallow to average depth and typically sandy, subacid and acid. In the footslopes, deposits may be present having gentler form and slope than the rest of the landscape. The soil profile on these formations is well differentiated due to the presence of an argillic soil horizon (Bt) characterized by illuviation and stronger alteration. The catena in these formations consists from the bottom upwards, and depending on plant cover, of Alfisols (Palexeralfs), Inceptisols (Xerochrepts) and Entisols (Xerorthents). Entisols prevail in degraded and desertified areas while Inceptisols and Alfisols are found in the less degraded and well-preserved zones. In severely degraded landforms where the soil has disappeared for one reason or another, it cannot be restored; in less degraded environments soil regeneration is more likely.

These landscape types are commonly found in Sardinia and indeed in many parts of the Mediterranean where similar problems of evolution, degradation and management exist. Clearly the gravity of the degradation and the difficulties of recovery depend on the fact that the parent material weathers slowly and on the climate, characterized by xeric and ustic conditions. At any rate soil stability is strongly related to the organic matter contained in the upper soil horizons. In fact it is the organic matter that ensures the formation of stable aggregates, improves surface retention and enhances infiltration rates, all factors that contribute to reducing runoff and hence soil loss through erosion. The loss of organic matter accelerates desertification processes.

9.3.2 Soils on intrusive rocks

In spite of their different physico-chemical properties (particle size, base saturation, amount and distribution of organic matter) the soils found on granite rocks exhibit a topographic pattern similar to those overlying metamorphic rocks. Thus Lithic Xerorthents are found on the steeper slopes with sparse vegetation, Typic Xerorthents, Dystric Xerorthents, Lithic Xerochrepts in those areas of regular morphology and average slope with bush and holm oak woods. Typic Xerochrepts and Dystric Xerochrepts are present in the belts of debris on the less rugged ground and where the bush is tall, developed or intermingled with holm oak woods. Here too slope deposits are to be found having more mature soils characterized by a thicker vegetation cover.

Higher acidity and perhaps a different water regimen have created more favourable conditions for the cork oak which abounds in certain areas. The soil catena, as on the previous landforms, consists of Palexeralfs, Xerochrepts and Xerorthents with numerous subgroups such as dystric, lithic, cumulic, etc.

The study of desertification processes of soils and vegetation in landscapes of this or similar types is of particular significance in that it is on these soils that the majority of cork oak woods in Sardinia, and indeed in many other Mediterranean regions such as Spain and Portugal, grows.

9.3.3 Soils on screes

Screes, which have accumulated where the delta meets the hills, give rise to fairly well-developed, strongly weathered soils (Typic and Ultic Palexeralfs), which are deep, highly skeletal, and typical sandy or argillic. These soils are usually cultivated and fairly eroded. They are the most mature soils found in the catchment area and, apart from alteration, they are also characterized by illuviation, cementation by sesquioxides, and by low porosity and permeability.

In the past the plant cover consisted of mixed woods of evergreen sclerophyllous species, chiefly cork oaks. More recently these lands have been cultivated or used for grazing but with low economic returns. Such soils are fairly common on the older Pleistocene formations in a great many alluvial areas of the Mediterranean. In some cases they have been irrigated but this practice has not met with much success in economic terms.

9.3.4 Soils on Pleistocene alluvia

The soils most commonly found on this parent material can be fairly well developed (Typic and Aquic Palexeralfs) in the more ancient alluvial terraces or of average development (Typic Haploxeralfs and Typic Xerochrepts) in the more recent terraces. The variability of soils on Pleistocene alluvia depends not only on the geolithological features of the source area but also on the sediments' age and on the extent of alteration. The morphology of the soil profiles clearly shows these processes especially as far as alteration, illuviation and cementation are concerned. The most recent soils have high agricultural potential while the older ones are less suitable for farming purposes.

In the lower part of the Rio Santa Lucia watershed where land planning is totally lacking, these lands have been used for housing developments.

9.3.5 Soils on Holocene alluvia

The soils covering Holocene alluvia, confined to small areas near water courses, are poorly developed (Typic Xerofluvents), highly skeletal and drain freely. These soils have been affected by excavation of aggregates, earth movement and urban development. The soils with finer grain size represent one of the most important natural resources of the area as they have high agricultural potential and are planted with the major crops both in the open and in greenhouses. Favourable climate has encouraged the spread of greenhouses in this area. Much of this land, especially the soils with coarser grain size, has been lost due to extraction of aggregates for use in the building industry and for road construction.

9.3.6 Soils on marshy and coastal sediments of the Holocene

In the vicinity of the lagoons, covering muds and clays of marshy origin, are soils whose development is governed by the presence of surface salty groundwater. The soils concerned are Salorthids with high salt contents and typical halophytes.

9.4 DESCRIPTION OF THE VEGETATION

Vegetation in the Rio Santa Lucia catchment is characterized by thermophilous forests of evergreen sclerophyllous species which are present throughout the area. *Quercus ilex, Phillyrea latifollia, Phillyrea angustifolia, Myrtus communis, Pistacia lentiscus* and *Olea sylvestris* are the dominant species. *Juniperus oxycedrus* and *J. phoenicea* are prevalent in the most arid areas, characterized by steep slopes and shallower soils. On the rocky areas facing south are *Euphorbia dendroides* and other thermophilous species of the Mediterranean bush, especially in those zones particularly impoverished by grazing and fire.

Riparian formations largely consist of *Alnus glutinosa* and *Salix purpurea* along the rivers; where the riverbed broadens due to the deposits of sand and stone, *Nerium oleander* grows. Cork oaks, only grow in those areas where human activities have been carried out in the past, mainly in the low-lying and most accessible places, along valley floors and near sheep-folds. Garrigues are confined to sandy riverbeds, the rocky slopes and hill tops where *Teucrium arum* and *Euphorbia spinosa* are predominant.

Surveys in areas planted with *Eucalyptus* spp. reveal a considerable reduction in undergrowth species, due both to the layer of leaf litter and to the diminished availability of water in the soil. It is unlikely that this phenomenon can be attributed to shading, since no correlation has been found between decrease of heliophytes and increase of sciophilous species. Shrubs are also rare in the eucalyptus woods though they may be abundant nearby.

9.5 LAND USE AND ITS HISTORY

9.5.1 Population dynamics

Population trends in the catchment area can be summarized as follows: the population in the uplands gradually declined and they were eventually deserted. By contrast a large increase in population took place after the Second World War in lowland rural areas not far from Cagliari wherever there was no shortage of water, and wine and grain could be produced. In these most densely populated parts of the Santa Lucia watershed, continued growth has been accompanied by a number of housing developments. The territory extends from the coast—or from the valley floors—up to the central hilly area. As in the rest of Sardinia, the population has been increasing in the coastal areas. Today the resources offered by the hills (mines, game, charcoal, firewood) cannot compete with the attractions of the coast.

The population of the inland communes, in the valley of the river Cixerri in the lower Sulcis region, have remained stagnant, except for a slight increase between 1936 and 1961. As a whole, demographic growth in the nine communes concerned is higher than the regional average, especially from 1961 onwards and in particular in Capoterra, a dormitory town, and Sarroch where there is a large industrial complex. The reasons are numerous: high birth rate, only slight impact of the decline in mining and above all proximity to Cagliari which was, from the beginning, a market for farm and dairy produce as well as firewood. Cagliari later provided employment for workers in the building trade during the post-war reconstruction of the city and its subsequent straggly and rapid expansion.

The large increase in population, once small and stationary when the resources of the hills were more important, has coincided with a decline in economic activity as well as in the number of permanent residents in the hilly areas. In this peri-urban environment, the new attractions are the city and the coast.

9.5.2 Changes in land use in the catchment area

The hilly part of the catchment area

Mining activity fluctuated. The first concession dates back to 1862 (Compagnie des Hauts Fourneaux) but production soon halted. In 1906 the mine was taken over by Società Mediterranea which resumed mining operations after 16 years of inactivity. The mine changed hands several times and operations were halted and then resumed. Maximum production was achieved in 1957 with 426,000 tonnes of crude ore, but at the end of 1960 open cast mining was discontinued and soon after production ceased altogether.

The San Leone mine created an enormous blot on the landscape, especially the open cast excavations which caused deep cuts, landslides, and collapse of rock masses, changing the face of the whole area. In addition, dumps of aggregates and flotation muds were eroded by heavy rainfall. Consequently, the Bureau of Mines commissioned Kovisar SaS, the current proprietors, to draw up a project for reclaiming the area.

Land use in the area has a long and varied history. In feudal times the people of Assemini had common right to the land around the river Gutturu Mannu for grazing cattle, hunting and woodcutting. Animals were also allowed to graze in the oak forests. Some families lived off hunting so the forests were well frequented. Wood was cut and transported by boat across the lagoon to Cagliari to be sold. Cagliari was also the market for game. The forests were already being exploited to quite a large extent and partly degraded. Towards the end of the 19th century, the Compagnie des Hauts Fourneaux acquired by public auction part of the forests and set up a system for the rational management of about 12 000 of the 20 000 ha of forest, still considered among the best preserved in Sardinia.

The hills represented a major resource for the inhabitants of the nearby communes. The people of Capoterra gathered firewood and made charcoal to be sold in Cagliari and this ensured preservation of the woods; furthermore these activities were controlled by the proprietors. Game and fowl were sold in Cagliari and fowling was a deterrent against fires, because the fowlers had long-term rights and were careful to conserve the woods. Although prohibited by the regional law of 1978, wildfowling is still practised for commercial gain.

The years during the Second World War saw the end of charcoal exports from Sardinia. In 1951 the forests were sold to the Poli brothers of Pistoia and largely exploited for their wood. After several changes of hands they were eventually acquired by the State Forest Agency between 1982 and 1986 and became part of a single forest district.

The delta

The lower valley of the Rio Santa Lucia is divided into two distinct parts:

— The cone of Capoterra, on which stand the village of Capoterra and new housing developments. The ancient alluvia are cultivated with rain-fed crops, and also carob and almond trees. At the base of the cone is a riverbed from which gravel is excavated.
— The delta, composed of recent alluvia deposited by the river when it flooded over the surrounding area, has been developed in more recent times. An exception is the land east of the village, the Tuerra (marshy land) which is devoted to irrigated crops.

In the 1920s the whole area was included in the Capoterra land improvement project (about 1500 ha), and hydraulic engineering works were carried out on the riverbed and channels were dug. The fields in the area where improvements were carried out are larger and more regularly shaped, and vegetables, citrus and other fruits are cultivated.

The Rio Santa Lucia flows into the Santa Gilla lagoon which, with its bar, already existed in Roman times. Genoese, Pisan and Aragonese galleys could sail almost right up to Capoterra. This part of the lagoon was, up until the early 1800s, still deep enough for boats to sail across it to be loaded with wood from the nearby forests.

In Aragonese times there were only two entrances to the marshlands, one at La Scaffa in Cagliari and the other at the Capoterra end at Santa Maria Maddalena which remained important for a long time, being the first entrance on the west side of the lagoon.

At the beginning of the 19th century, 10 canals connected the lagoon and the sea to facilitate the passage of fish. Some of the canals were closed during the major alterations in the 1920s. At the moment there are four openings, one at the mouth of the Rio Santa Lucia, and the other three in the salt flats. There are entrances at the Cagliari end to the canal-port, the canal of the Santa Gilla lagoon and the branch sewer. The wetlands have diminished overall and particularly in the area around the mouth of the Rio Santa Lucia, as can be seen from a comparison of present conditions with old documents and place names (see Figures 9.2, 9.3 and 9.4).

Some important landowners collaborated to make major improvements in the lower delta. In the early 1800s an extensive 15 000 ha estate (Tanca di Nissa and Villa d'Orri) was colonized and later modernized. Subsequently 8000 olive trees were planted at Su Loi, recently developed for housing. The San Leone farm at Baccu d'Inghinu was also an example of farm improvements, 100 ha of wood and shrubland being converted into a model estate with vineyards, irrigated crops, extensive olive groves, and another 500 ha were planted with pine and carob and other trees.

Settlement spread out over the delta. Modern farmlands began to appear with small plantations of carob trees (typical of large estates), vineyards and olive groves on higher land, and orange groves in the low-lying, more fertile and irrigable land. This landscape has been partly overrun by the industrial complex at Sarroch and by peri-urban expansion.

Part of the land was sold at the end of the 19th century and, in the 1920s, the middle part of the lagoon was converted into salt flats. Colonization continued up to the 1950s, the continuing change of proprietors of farmland leading to an increase in the subdivision of estates accompanied by an irregular agricultural expansion.

The greatest alterations to the landscape are due to the land improvements made in the 1920s and the more recent construction of a canal port in the Santa Gilla lagoon as well as an industrial complex. Quarries for sand and gravel have been opened up along the bed of the Rio Santa Lucia. Groundwater is being tapped from the aquifer which surfaces at the fountains of Sa Tuerra to provide water for irrigation and domestic supplies with deeper bores supplying the industries. These developments have brought about major topographical and hydrographical changes.

The road running through the foothills of the Cixerri which is flanked by a thick embankment and by the pipes of the Rumianca, divides the region of the Rio Santa Lucia mouth from the rest of the wetlands. More roads are being built including one which will link the industrial zones of Macchiareddu and Sarroch.

The first areas to be developed by the expansion of Capoterra, to house people moving out of the town of Cagliari, were Poggio dei Pini in the hills, and La Maddalena Spiaggia by the sea. New housing developments built along the coast (Su Spantu, Torre degli Ulivi) affect the Santa Lucia river mouth. Housing for about 1000 people, complete with a marina, is also planned. All these will have a considerable impact on the environment in view of the fact that farmland will be occupied and the equilibrium between fresh and sea water will be upset. The development of the coastal area between Santa Margherita and Capoterra means that there will be more people seeking to spend their free time in the hills and in consequence a greater demand for the safeguarding of the natural heritage. Grazing and fire control, as well as a ban on hunting, are already some of the positive effects.

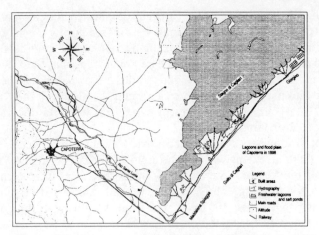

FIGURE 9.2   Lagoons and flood plain of Capoterra in 1898

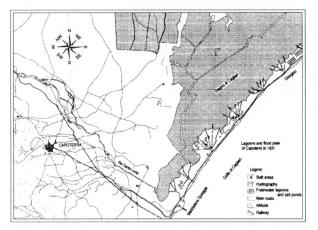

FIGURE 9.3   Lagoons and flood plain of Capoterra in 1931

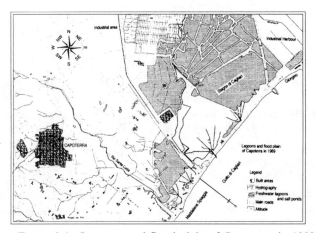

FIGURE 9.4   Lagoons and flood plain of Capoterra in 1989

Roads and railways

Among the first railways to be built in Sardinia for transporting minerals, was the San Leone to La Maddalena line, opened by the Company Petin et Gaudet in 1867. It turned out to be used more for transporting wood than for mining purposes. Another railway opened on the other side of the hills in the Sulcis region in 1890–91 connected Porto Botte to Santadi and Pantaleo. In Pantaleo a line also served a chemical factory which between 1915 and 1920 produced wood derivatives. This railway was used for transporting wood up until 1938. The engineer Angelo Vanini (1910) designed a link between the two sections to connect the Sulcis region with Cagliari. Today the only road which passes through the valley of the river Gutturu Mannu is the provincial road of the middle Sulcis which follows an old mule track, the shortest route from the Sulcis region to Cagliari. Hence access to the valley was not that difficult.

9.6 HYDROGEOLOGY: WATER DEVELOPMENT AND LAND USE

The geological formations in the Rio Santa Lucia catchment area consist of fractured granites and schists of the Palaeozoic in the mountainous part and alluvial and lacustrine deposits in the lowlands. Recent alluvia are multilayered and underground aquifers of little consistency occur between the layers. The deep aquifers are to be found at an average depth of 40–50 m beneath the ancient alluvia of different ages. Draw-down is brought about by pumping. In fact in the area concerned there are about 300 wells tapping the aquifer for domestic use, irrigation and industrial use. So far it has not been possible to quantify accurately the amount of water being extracted. However, salt-water intrusion has been observed and this phenomenon is likely to be aggravated as water requirements increase in the future.

Salt-water intrusion is commonly encountered throughout the Mediterranean in low coastal areas. Continuous monitoring indicates a rise in salinity in those areas where groundwater is overexploited. Salinity varies considerably from winter and spring to summer and autumn. In fact, values range from 2 to $7 \, g \, l^{-1}$, with sodium absorption rates ranging from 5–6 up to 20 mg kg^{-1} or more.

9.7 DIFFERENT KINDS OF DESERTIFICATION

The problems related to desertification in the Rio Santa Lucia catchment area are multifaceted and attributable to both physical and human causes. While the natural events are of negligible importance, desertification caused by human activities often has serious consequences. The major human causes of desertification in Sardinia appear to be the following.

9.7.1 Forest fires

A problem endemic to Sardinia, forest fires are one of the major causes of degradation and desertification. Some investigations have already been conducted into the causes and effects of forest fires, especially within the framework of the project 'Soil Conservation' of the Italian National Research Council. Nevertheless, it was thought appropriate to go into greater depth and two experimental plots were set up (Is Olias farm) in the catchment area and investigations were extended to other parts of the island. One clear fact that has emerged in all the regions examined is the desertification in those areas where several fires have occurred. The difficulties in vegetation regeneration as well as of soil formation are very slow especially where the parent materials (quartzites, sandstones, granites, etc.) are so little prone to weathering because of their high silica contents and slopes are rough and steep.

9.7.2 Reafforestation with exotic species

In the Rio Santa Lucia catchment area there are a number of plantations of exotic tree species, mainly eucalyptus. From the first investigations it emerged that:

— the soils are undergoing intense degradation due to erosion. In other sites in Sardinia and Portugal as well, soil profiles have been observed to be truncated due to erosion;
— tree growth rates are far lower than predicted. In eucalyptus plantations in Sardinia yields do not usually exceed $5\,m^3\,ha^{-1}\,yr^{-1}$. On thin soils this value decreases to as little as $2\,m^3$. Only in the valley floors with fertile soils and groundwater do growth rates exceed $15\,m^3\,ha^{-1}\,yr^{-1}$, but such soils are few and far between and in any case are used for intensive agriculture which is much more profitable than wood production.

These findings prompted the work group to extend the investigation to the whole of Sardinia and comparisons have also been drawn with other Mediterranean countries. Three hundred plantations have been counted throughout the island, chiefly of eucalyptus and *Pinus radiata*. The first checks confirmed the observations made above. The extent of this problem is such that at least in those regions with soil and climatic conditions similar to Sardinia, afforestation policies will have to be seriously re-examined or even drastically changed.

9.7.3 Loss of soils on recent alluvia due to sand and gravel extraction

Sand and gravel workings along the reach of the Rio Santa Lucia, that flows through the plain, began shortly after the Second World War. In fact construction works, modernization of existing ports and the building of new roads all contributed to the growing demand for aggregates.

The lithological nature of the hydrographic basin makes the riverbed material ideal for engineering uses. A large number of quarries are now operating and it is estimated that $500\,000\,m^3$ of material are removed each year. Virtually all the soils in recent alluvia have been destroyed by the extraction, especially in those areas where coarser material is to be found. Hence over 50% of recent alluvia of the Rio Santa Lucia, potential intensive agricultural land, has been destroyed.

The problem of quarries is not just peculiar to the Rio Santa Lucia catchment but concerns Sardinia as a whole and many other Mediterranean regions. It is now prohibited to extract aggregate from recent alluvia in Sardinia and future land planning envisages the closure of existing workings and their reinstatement.

9.7.4 Land loss through urban development

Urban development has affected areas around existing urban agglomerations, areas of industrial activity, areas in the vicinity of main highways and coastal zones. All these have been built up to a greater or lesser extent depending on the type of development, but the areas most affected are along the coast, within a belt no wider than 1 km, and industrial zones.

Housing and tourist developments along the coast have been increasing exponentially with the demand for housing outside the town of Cagliari and with the expansion of tourism. In these cases building complies with authorized planning and hence includes rational plans for infrastructures and services. The urban expansion has paid no regard, however, to the need for harmony with the surroundings and the need to minimize the destruction of good farmland and problems connected with overexploitation of groundwater resources. Numerous greenhouses for the production of vegetables and flowers and extensive fruit-growing areas are now being replaced by housing and tourist developments, services and small industries.

The industrial estate in the Rio Santa Lucia catchment where the Macchiareddu petrochemical complex has arisen, has been built on relatively poor farmland (Palexeralf soils). Many housing projects have been submitted for approval to the planning authorities, which would involve the further loss of soils throughout the entire catchment area and seriously undermine the future economy of the region.

9.7.5 Pollution caused by mining and industrial activity

Until a few decades ago, iron ore was extracted from the San Leone mine in the catchment area. The ore was treated in flotation plants and the tailings were stocked in dumps. With the closure of the mine, management of the dumps ceased and consequently they are presently undergoing severe erosion. Though pollution in the Rio Santa Lucia catchment area may seem to be of minor significance, the same cannot be said for other mining areas. Suffice it to say that in Sardinia 18 million cubic metres of muds are stocked and the same or larger amounts have eroded and have been transported into the sea, rivers and soils.

9.7.6 Erosion

Erosion is the most significant factor in desertification processes in Sardinia as well as in the Mediterranean as a whole. The causes of erosion are multifold, though the main ones are forest fires, tilling of thin soils and steep slopes, the introduction of exotic tree species ill-adapted to local conditions, and overgrazing. Of the above forms of degradation at least three are present in the Santa Lucia catchment area and are attributable chiefly to human causes.

For a better understanding of erosion and its importance, a number of experimental plots were set up in the catchment area at Is Olias near Capoterra (SW Sardinia). Three slope catenas were selected with different land uses, representative of large areas of the island, and indeed of other Mediterranean regions. Six runoff plots measuring $20 \, m^2$ were installed in each of the three catenas.

Slope catena 1 is on former grazing land, now abandoned and invaded by shrubs consisting mainly of *Cistus* spp. with different density (from 10 to 70%). Slope gradient varies from 17.6 to 12.2% and soils are Typic Xerochrepts formed on a Holocene slope deposit. Slope catena 2 is in macchia burned at the end of June 1991 where grazing is prohibited. Slope gradient ranges between 46.6 and 34.4% and soils are mainly Typic Xerochrepts on metasiltites. Slope catena 3 is in a eucalyptus stand planted about 15 years ago where soils are predominantly Alfisols on a Pleistocene slope deposit. The gradient varies from 31.5 to 12.2%.

Rainfall distribution from September 1992 to August 1993 is shown in Table 9.1.

Sediment yield (kg ha^{-1}) for slope catenas 1, 2 and 3 are given in Table 9.2. The most important erosion events observed for the three studied catenas are in October, following the first rainfall after the dry season. October gives 60% of the total eroded sediment during the hydrological year for slope catena 1, 64% for slope catena 2 and 80% for slope catena 3 for rainfall equivalent to 27% of the annual value but of high intensity.

Although the three slope catenas differ not only in land use but also in slope, soil and geology, the findings suggest that differences in sediment yield are largely affected by land use. In fact the preliminary results reported above clearly indicate that sediment yield is significantly higher in the woods planted with exotic species with rapid growth. However, in this area, and likewise throughout most of the island with similar soils and climate, these species do not find ideal soil and climatic conditions and in fact annual growth rates do not exceed 5 m^3 ha^{-1} This aspect is of prime importance in many Mediterranean countries where extensive areas are being planted with exotic species for wood production.

Irrational grazing, overgrazing and fires are commonplace in Sardinia with disastrous consequences almost everywhere. Soils are destroyed or impoverished, the productive capacity of pastures diminishes with the loss of edible species, the hydrological balance is upset resulting in an increase in runoff and decrease in concentration times, sedimentation

TABLE 9.1   Rainfall (mm) at Is Olias, September
1992 to August 1993

Month	Rainfall (mm)
Sep. 1992	0.8
Oct.	144.4
Nov.	34.8
Dec.	117.6
Jan. 1993	12.6
Feb.	71.2
Mar.	56.0
Apr.	52.8
May	43.9
Jun.	6.0
Jul.	0.6
Aug.	0.0
Total	540.7

TABLE 9.2 Sediment yield (kg ha^{-1}) in the three slope catenas September 1992 to August 1993

Month	Slope Catena 1	Slope Catena 2	Slope Catena 3
Sep. 1992	0.00	0.00	0.00
Oct.	804.74	791.39	3078.23
Nov.	5.71	10.86	15.76
Dec.	35.15	86.39	76.09
Jan. 1993	16.73	134.64	28.08
Feb.	283.51	53.24	248.12
Mar.	27.69	25.09	75.30
Apr.	141.69	29.83	229.34
May	17.92	103.08	90.00
Jun.	0.00	0.00	0.00
Jul.	0.00	0.00	0.00
Aug.	0.00	0.00	0.00
Total	1333.14	1234.52	3840.92

in the intensively farmed plains, destruction of bridges and houses through flooding and sometimes the loss of human life. There is a considerable imbalance between natural productivity of grazing land and number of livestock. For this reason extensive areas are burned in order to create more pasture but this causes endless damage to the woods and bush.

9.7.7 Salt-water intrusion and soil salinization

Salt-water intrusion and soil salinization are common to many coastal areas of the Mediterranean, like Italy, Spain, Greece and North Africa. The steady increase in water requirements has led to an ever growing demand for water supply, far in excess of that actually tapped from the aquifers. In the valleys of the Rio Santa Lucia catchment a number of users draw water from wells, including some petrochemical industries (SARAS and ENICHEM), the municipality of Capoterra for domestic use, and many agricultural concerns for greenhouses, fruit crops and market gardens.

Because water consumption exceeds aquifer recharge, salt-water intrusion is occurring in places where large amounts of water are tapped (industrial areas and farms). The majority of wells in the industrial areas are no longer used because of their high salinity (7–9 g l^{-1}). The first signs of soil salinization are beginning to emerge in farmland.

9.8 CONCLUSIONS

The process of desertification consists above all of the gradual impoverishment and hence the total loss of soil for productive purposes. For the time being, not enough data are available to show the effect of any climatic changes on desertification; however, the consequences of specific land uses are clear. The findings of the investigations carried out in the framework of MEDALUS I have enabled a comparison between desertified areas and other better preserved ones. The difficulties of recovering the degraded areas are evident, especially as far as soil and vegetation are concerned. The soil formation processes in the Mediterranean climate are extremely slow, estimated to be of the order of tens of millennia. The morphology and lithology of the hilly part of the Rio Santa Lucia watershed, where the rocks weather very slowly, are such that former ecosystems can only be restored in geological times. There are no signs of desertification in those areas where land use has allowed the soil–plant ecosystem to be maintained, but the reverse is true of altered ecosystems.

Desertification in the plain may be attributed to extraction of aggregate, salinization of the soils, extensive urban and industrial expansion and development of infrastructures. Particularly favourable soil and climatic conditions and the availability of good quality groundwater has resulted in intensive farming locally. The overexploitation of the aquifer system for a variety of uses (agricultural, industrial and domestic use) in conjunction with the conversion of extensive areas to salt flats and to sand and gravel workings, is causing the gradual intrusion of sea water, both in the upper and lower parts of the aquifer. Irrigation using water with a high salt content increases the salinity of the soil, rendering it unproductive. The good state of preservation of the forests seems to be attributable not so much to their isolation but more to good management, especially when owned by the Compagnie des Hauts Fourneaux. Hunting also helped to keep forest fires down. Human activities in the hilly areas began to decline in the 1930s when the railway line was closed and continued into the 1950s when the San Leone mine was shut down and inhabited areas were depopulated.

The growing population in the delta is explained by the desire of people living in towns to escape to the countryside. In 1973 a conservation order was placed on the forests of Capoterra and in 1975 on those of Gutturu Mannu where every year a mountain festival is held by the commune of Assemini. Today there is a WWF oasis (3000 ha) in the Guttureddu valley and ecological weekends are organized there. Lastly, law No. 31 of 7 June 1989 provided for a natural park of some 68 000 ha in the Sulcis region.

With the decline in hunting, and also because of the nature reserves and the end of industrial and forest production, wood cutting has virtually ceased and the occurrence of fires is now so rare in this forest that it is one of the few regionally owned forests where there are no fire breaks. The changing land uses have improved the environmental equilibrium which has led to the restocking of the forests.

The qualitative and quantitative data collected in the catchment area allow a better interpretation of situations observed elsewhere in Sardinia and indeed in the Mediterranean, where the phenomena are even more marked but cannot yet be directly correlated with time and past and present land use. The problem of desertification and environmental recovery depends undeniably on management which in any event must consider the principles of ecosystems. In this regard the Rio Santa Lucia watershed offers significant examples of both degradation and recovery.

9.9 BIBLIOGRAPHY

Aru A. et al. (1983) *Il consumo delle terre con l'espansione urbana di Cagliari e del suo hinterland.* C.N.R., P.F. Conservazione del Suolo.

Aru, A., Baldaccini, P, and Vacca, A. (1991) *Nota illustrativa alla carta dei suoli della Sardegna.* Dip. Scienze della Terra, Univ. Cagliari. Assessorato Progr. Bil. Ass. Terr., Centro Reg. Progr., Regione Autonoma della Sardegna.

Aru, A., Baldaccini, P. and Vacca, S. (1992) *Importanza degli studi pedologici nella pianificazione di bacino*. Convegno Soc. Geol. It. sui Piani di Bacini.

Lai, M.R. (1987) Analisi ambientali e valutazione del territorio in un'area della Sardegna meridionale (Margine Rosso, Capitana). *Procs. Cultura del paesaggio e metodi del territorio*, ed. Fernando Clemente, Ed. Gallizzi, Sassari, p. 102–119.

MEDALUS I Final Report (1993) Rio Santa Lucia, Sardinia, p. 534–559.

Fuddu, R. and Lai, M.R. (1994) Analisi pedologiche e geoambientali del bacino del S. Lucia (Sardegna meridionale). Progetto MEDALUS. *Geolica Romana, Vol. XXX*, 335–350.

USA Soil Survey Staff (1992) *Keys to Soil Taxonomy*. SHSS Technical Monograph No. 19, Pocahontas Press, Blacksburg, Virginia, 556 p.

10

The Spata Field Site:
I. The Impacts of Land use and Management on Soil Properties and Erosion.
II. The Effect of Reduced Moisture on Soil Properties and Wheat Production

C. S. KOSMAS, N. MOUSTAKAS, N. G. DANALATOS
and N. YASSOGLOU

Laboratory of Soils and Agricultural Chemistry,
Agricultural University of Athens, Greece

10.1 INTRODUCTION

Soils developed on Tertiary and Quaternary rolling surfaces (marl, conglomerates and shale–sandstones) in the semi-arid zone of Greece are cultivated mainly with vines and olives or rain-fed cereals. The soils of these areas usually have a restricted rooting depth, normally limited by subsurface layers (petrocalcic horizons, gravelly and stony layers, or bedrock) and high erodibility. These limitations combined with improper land use and management create favourable conditions for accelerated erosion and soil deterioration. Any further soil loss drastically reduces the effective soil depth, leading to a reduced productivity and finally desertification (Yassoglou, 1989).

In addition there are growing indications that the earth is experiencing a long dry and warm cycle with increasing global air temperature, and decreasing precipitation accompanied by changes in seasonal distribution. Such a climatic change will seriously affect the earth's vegetation cover both in gross production and the geographic distribution of crops (Maracchi, 1991).

The work at the MEDALUS field site in Spata, Greece, concentrates on these two important issues.

10.1.1 Effect of land use and management on soil properties and erosion

Large-scale deforestation followed by intensive cultivation with improper management in Thessaly (central Greece) has resulted in a dramatic decrease in the organic matter content

Mediterranean Desertification and Land Use. Edited by C. Jane Brandt and John B. Thornes.
© 1996 by John Wiley & Sons, Ltd.

and in accelerated erosion. Danalatos (1993) reported an average reduction rate of approximately 1.25% per annum in the last 40 years; mollic epipedons can hardly be found today on these areas which, not more than 60 years ago, were characterized as dark 'Rendzina' soils (Papoutsopulos and Svorikyn, 1932). In similar hilly areas in the St Genis basin (France), an average erosion rate of $190 t ha^{-1} year^{-1}$ has been reported by Bufalo and Nahon (1992).

Francis and Thornes (1990) have demonstrated that runoff and sediment yields decrease exponentially with increasing vegetation cover in a wide range of environments. Changes in crop management may cause soil degradation and particularly soil structure deterioration which is largely responsible for reduced crop yields (De Boodt et al., 1961; Perfect et al., 1990). Relatively few studies have characterized the changes in soil properties taking place along with the gradual removal of surface layers by natural erosion (Rhoton and Tyler, 1990). It is against this background that the MEDALUS core field programme was used to focus on the effects of land-use and management change on soil properties and erosion rates.

10.1.2 The effect of reduced rainfall on soil properties and wheat biomass production

Rainfall amount and distribution are the major determinants of wheat production on hilly lands in the semi-arid climatic zone of Greece. Areas under the same climatic conditions, but with soils formed in different parent materials, show different vulnerability to desertification. Kosmas et al. (1993b) demonstrated that stony soils on Tertiary shale–sandstone and conglomerate deposits may support considerable gross wheat production even in extremely dry years by conserving appreciable amounts of soil water from evaporation through surface mulching. In contrast hilly soils on marl deposits under annual cropping, despite their considerable depth and high productivity in normal and wet years, cannot support any vegetation in particularly dry years and therefore are very susceptible to desertification (Kosmas et al., 1993a).

The second line of investigation at the Spata site was to study the effects of diminishing soil moisture on soil properties and biomass production of wheat in an experiment where rainfall was partially excluded from plots of wheat. The results have been used to predict productivity of rain-fed wheat grown on similar soils under different weather conditions.

10.2 EXPERIMENTAL DESIGN

10.2.1 The study site

The Spata field site is situated about 30 km northeast of Athens at an average elevation of 140 m above sea-level (coordinates 37°58'N and 23°55'E) (Figure 10.1). The site is a sloping hillside draining down to a flat alluvial plain (Figure 10.2).

An automatic meteorological station was installed in the vicinity of the experimental site and data such as solar radiation, net radiation, dry and wet air temperatures, wind speed and direction were recorded hourly. Additionally, open pan evaporation was recorded every 6 hours, while rainfall was recorded every 5 minutes.

The climate is thermo-Mediterranean. The average annual air temperature is 17.8°C. The average annual precipitation is 496 mm with 71% of it falling in the period November through to April. Over the study period of 640 days (March 1991 to November 1992) the

FIGURE 10.1   Location of the Spata and Tanagra field sites in Greece

total rainfall was 724 mm (averaging to 467 mm per year). Four small storms occurred in 1991, on 10 March (46 mm), on 20 March (36.6 mm), on 29 May (30.5 mm), and on 9 November (41.8 mm). A large storm occurred on 6 December bringing a total of 106.4 mm of rain at a (maximum) intensity of 16.6 mm h^{-1}.

For more than 160 years, the major land use in the area was olives under semi-natural conditions. The natural understorey vegetation in the olive grove is dominated by winter-annual plants and shrubs of the following species: *Avena barbata, Vicia* sp., *Phagnalon graecum, Hordeum murinum, Muscari comossum* and *Thymus capitatus*. The overall maximum percentage of the natural vegetation cover under the olive trees is quite heterogeneous ranging from 44 to 95%. In a part of the study area, the olives have been removed and vines grown since 1979. The vines grow under rain-fed conditions. Fertilization and weed control are nearly optimal for vines. The (bare) soils are deeply ploughed in winter and once again in early spring before fertilization. The soils are tilled once again in late spring for weed control.

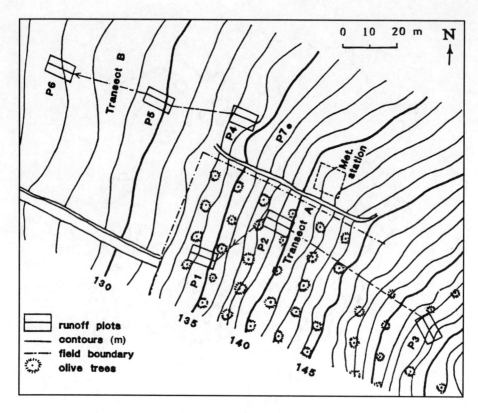

FIGURE 10.2   Map of the instrumented hillslope at Spata (sites P1, P2 and P3 correspond to transect A and sites P4, P5, P6 and P7 correspond to transect B)

10.2.2 Field measurements

The field site was designed according to the specifications of the core programme. Two transects were selected, extending from the top to the bottom of the hill (Figure 10.2). Transect A with an average slope of 19.0% was under the olives, while transect B with an average slope of 11.3% was under the vines. Within each transect were three experimental sites and six plots (two in each site) 3×10 m which were used for measuring runoff. Each plot was bounded with a trench to lead away runoff water from upslope. Runoff from the plots was drained into cement troughs 3 m long, and collected in large containers buried in the ground. The containers were periodically, and after each rainfall event, cleared of sediment and water. The soils and the natural vegetation along transect A were kept as undisturbed as possible, with the installation of the runoff troughs being the only disturbance. The sites along transect B were subjected to tillage, fertilization and harvesting operations, all typical for the general area. The field experiment was conducted over a period of 21 months (from March 1991 to November 1992).

As stated above, precipitation was measured by an automatically recording rain gauge at intervals of 5 minutes. Crust permeability was determined with the trickle irrigation method as described by Boiffin and Monnier (1985). Total above-ground biomass was measured in 1 m² quadrats by harvesting all the annual plants (above-ground parts) late in June. The below-ground biomass was determined in soil cores, 10 cm in diameter and 10 cm in height. The roots were separated from the soil by washing and sieving through a 0.5-mm sieve. The soil moisture content was measured gravimetrically from samples collected in triplicate from the soils adjacent to the runoff plots at depths of 5–7 cm, 10–15 cm and 20–25 cm usually weekly and after each rainfall event. A base topographic map was compiled at a scale of 1 : 250.

Finally, seven pedons located next to runoff plots (P1, P2, P3, P4, P5, P6 and P7; Figure 10.2) were described in detail and sampled for physical and chemical characterization in 1991. The soils of the study area had already been mapped and described in detail in 1979, before partly clearing the olive grove. Comparison of these data reflects a noticeable change in depth of the A horizon in both upper and lower hill sites and gives first evidence of the impact of land-use change on soil erosion.

The rainfall exclusion experiment was carried out very near the core field site on the footslope of the hill, with an average slope grade of 5%. The soil is very deep, well drained and fine-textured. It has a calcic horizon at a depth of 90 cm and is classified as Calcixerollic Xerochrept (Soil Survey Staff, 1975). Twelve plots (four treatments with three replicates each) 2.5×5.0 m in size were partially shielded by wave plexiglas strips (transparent to sunlight) 10 cm wide and 5.5 m long, installed 1.8 m above the soil surface in such a way that 35, 50, 70 and 100% of the rain was allowed to reach the ground. The whole experiment and the plots were again bounded by trenches to lead away any runoff from upslope and also runoff initiated between the plots. Rainfall under each treatment was measured by three rain gauges.

A wheat durum cultivar was sown in November of both years 1990 and 1991. Before sowing all plots received the same dressings of 75 kg N ha^{-1} as ammonium fertilizer and 140 kg P_2O_5 ha^{-1}. The leaf area index (LAI) was measured every 10 days with a LI-COR-LAI 2000 plant canopy analyser. At the end of the growing period, the plots were harvested and weighed for above-ground biomass production. Subsamples were dried at 65°C for moisture content correction.

Ammonia volatilization from the applied inorganic ammonium fertilizer was measured using a volatilization chamber 35 cm in diameter and 50 cm in height, covered by a plexiglas lid. The ammonia-charged air escaping from the soil surface was pumped and passed through a chemical trap (172 mm of 2% boric acid indicator solution) for 10 minutes, and the amount of trapped ammonia was titrated with standardized HCl (Kissel et al., 1977).

The volumetric soil moisture content was measured in each plot with a neutron and gamma probe at depths of 15, 45, 75 and 90 cm every week. Soil samples were collected from each plot and from each soil horizon for laboratory determinations.

At another site, Tanagra, 100 km to the north of Athens, biomass production of wheat was measured in triplicate at 15 sites (plots each of 1 m²) along hillslope catenas at the end of three successive growing periods (1989/90, 1990/91 and 1991/92). The Tanagra sites are located in hilly areas with rolling topography and soils formed in similar parent materials as at Spata. Daily meteorological data for this area were supplied by the nearby Meteorological Station of Tanagra (Greek Meteorological Service).

10.2.3 Soil analyses

The following analyses were carried out on the soils in the core programme plots.

Visible spectra of the surface soil were recorded using a Perkin Elmer spectrophotometer equipped with a diffuse reflectance accessory containing a ring collector. Powder amounts of dry soil samples collected from different sites along the study transects were prepared by passing the sample through a 0.25-mm sieve, backfilling the sample into a hole 24 by 24 mm in an aluminium disk (3 mm thick), then gently pressing the powder against unglazed paper. Reflectance measurements were made relative to a $BaSO_4$ standard from 800 to 360 nm. The CIE tristimulus values X, Y and Z were calculated using colour matching functions for the CIE 1931 Standard Colorimetric Observer and standard illuminant C (Wyszecki and Stiles, 1982). The tristimulus values were converted to Munsell notations using graphs and tables given by Wyszecki and Stiles (1982). Soil colour is usually described by comparing it with standard colour charts. The reflected visible spectrum of soil samples determines what we perceive as colour very accurately and they can be converted to any colorimetric system. The reflected spectrum, recorded with a spectrophotometer, can also be analysed in detail to yield specific information about the different soil constituents (Kosmas et al., 1986).

The wet sieving technique by Yoder (1936) was used for the determination of the mean weight diameter (MWD) of the soil aggregates. Prior to sieving, the soil aggregates were passed through a 8-mm sieve, air-dried and placed on the top of a nest of sieves consisting of 4 mm, 2 mm, 1 mm and 0.2 mm sieves. After sieving for 10 min and correcting the dry soil weight in each sieve for sand and gravel, the MWD was calculated by summing the product of mean diameter (x) times the fraction of the corrected for sand and gravel aggregate weight (F_i) in that class using the formula:

$$\text{MWD} = \sum_{i=1}^{5} x_i F_i, \qquad i = 1, 2 \ldots 5$$

The particle size distribution of the <2 mm fraction was determined by the Bouyoucos hydrometer method (Gee and Bauder, 1986). The organic carbon content was measured using the modified Walkley–Black wet oxidation procedure (Nelson and Sommers, 1982). Inorganic carbonates were determined by measuring the volume of CO_2 evolved in reaction with excess of HCl (Nelson, 1982). Total nitrogen was measured with salicylic–sulfuric acid digestion (Bremner and Mulvaney, 1982). Free iron oxides were measured in a Varian Techtron atomic absorption spectrophotometer after extraction with citrate–bicarbonate–dithionite solution (Mehra and Jackson, 1960).

In addition to the above, the following techniques were carried out on the soils in the rainfall exclusion experiment.

The mineralizable organic nitrogen (N_t) was determined at the end of the second growing period (July 1992), under laboratory conditions in an incubation experiment at 30°C for 28 weeks according to the mineralization procedure described by Stanford and Smith (1972). Then, the potentially mineralizable nitrogen (N_o) was calculated from the cumulative amount of nitrogen (N_t) mineralized at various times t using the exponential model: $N_t = N_o (1 - e^{kt})$ (where k is the specific rate of decomposition) (Stanford and

Smith, 1972). The nitrogen actually mineralized under field conditions was predicted from its calculated potential value (N_o) and corrected (k value) for the prevailing soil temperature using an Arrhenius-type equation (Addiscott, 1983), and for the measured soil moisture content using the Stanford and Epstein (1974) equation.

The soil on each plot was sampled to a depth of 90 cm at the end of the growing period (July 1992) and immediately analysed for NO_3-N and NH_4-N after extraction with 2 M KCl according to the cadmium reduction and indophenol blue methods respectively (Keeney and Nelson, 1984). The total nitrogen uptake by the crop was measured after digestion with salicylic–sulfuric acid (Bremner and Mulvaney, 1982).

Undisturbed 300 cm^3 soil cores were taken in triplicate from all sampled horizons and used to determine the soil moisture retention characteristics. The soil moisture content was determined in undisturbed samples for saturated soil and for suction levels up to -33 kPa (field capacity) in low-pressure chambers. Disturbed subsamples were placed in high-pressure chambers to determine the soil moisture content at pressure heads from -300 kPa to -1500 kPa (the latter referred to as the wilting point) (Klute, 1986). The bulk density was measured by drying the cores at 105°C until constant weight.

10.2.4 Evapotranspiration analyses

In the rainfall exclusion experiment, the potential evapotranspiration rate (ET_{pot}) was calculated from daily values of maximum and minimum air temperature, net radiation, air humidity and wind speed, according to Penman (1948; modified by Frere, 1979). The crop maximum evapotranspiration rate ET_{max} was estimated from the ET_{pot} and the crop leaf area coefficient (FAO, 1977, 1979). The actual crop water use ET_{act} was calculated by a simple soil water balance model which uses the momentary moisture content in the root zone. The algorithm is based on daily rainfall and evapotranspiration rates, measured soil parameters such as field capacity and wilting point, and crop parameters (coefficient of crop cover, effective rooting depth, depletion fraction of the total soil water storage, and the actual crop calendar).

10.3 EFFECT OF LAND USE AND MANAGEMENT ON SOIL PROPERTIES AND EROSION

10.3.1 Soil properties

The soils of the transects under study are moderately deep to deep (depth 52–120 cm), well-drained, medium to moderately fine-textured with prevailing fine and very fine sand fractions, slightly gravelly to gravelly (gravel content 5–29%; Table 10.1), and strongly calcareous (equivalent carbonates 15–66%). They are characterized by a calcic or a petrocalcic horizon and are classified as Typic or Calcic Xerochrepts according to the Soil Taxonomy (Soil Survey Staff, 1975).

The depth of the A horizon in all pedons along transect B was altered as a consequence of the 12 year of cultivation and erosion/deposition processes. The uppermost pedon, P7 (Figure 10.3), was severely eroded; pedon P4 was moderately eroded, while the lowermost pedons, P5 and P6 (Figure 10.3), were located at the erosional–depositional boundary of the sediment plain. Pedons P1, P2 and P3 along transect A (Figure 10.2) represent slightly eroded soils.

TABLE 10.1   Characterization of the soils along the transects under study (in parentheses data of 1979)

Property	Transect A (olives)			Transect B (vines)			
	P1	P2	P3	P4	P5	P6	P7
Altitude (m)	134.5	139.2	153.2	133.5	129.8	127.0	137.5
Slope (%)	17.2	18.0	22.3	12.4	9.3	7.0	16.7
Clay (%)	21.4	28.4	36.5	29.1	30.1	34.4	14.8*
	(23.4)	(27.9)	(37.5)	(28.6)	(27.1)	(35.9)	(19.7)
Silt (%)	39.3	32.0	35.4	48.9*	38.9*	31.2	40.2
	(42.1)	(35.6)	(34.6)	(28.1)	(29.5)	(36.4)	(37.6)
Sand (%)	39.3	39.6	28.1	22.0*	31.0*	34.4*	45.0
	(34.5)	(36.5)	(27.9)	(43.3)	(43.4)	(27.7)	(42.7)
Gravel (%)	29.0	20.9	34.2	19.0	9.6	10.9	5.2
CaCO$_3$ (%)	42.2	36.9	32.3	33.2	39.4	32.3	45.1
	(44.6)	(38.7)	(31.4)	(28.7)	(31.6)	(33.0)	(39.6)
Org. C (%)	1.22	1.57	1.82	1.06*	0.87*	0.92**	0.50***
	(1.28)	(1.66)	(1.75)	(1.22)	(1.10)	(1.24)	(1.73)
Total N (%)	0.11	0.13	0.15	0.09**	0.08*	0.09*	0.04***
	(0.12)	(0.14)	(0.16)	(0.12)	(0.10)	(0.11)	(0.15)
C/N	11.3	12.1	11.8	12.3*	10.4	10.6	11.9
	(11.6)	(11.8)	(10.9)	(10.2)	(11.0)	(11.3)	(11.5)
Fe$_2$O$_3$ (%)	1.43	1.50	1.02	1.43	1.58	1.62	1.21
Aggr. size (mm)	3.15	5.78	7.09	0.58	0.08	0.94	0.12
Runoff (mm)	2.31	2.62	0.32	12.25	4.87	8.88	—
Sedim. (g m^2)	0.80	0.45	0.06	441.91	67.06	83.83	—

Note: Asterisks denote significant differences of the properties per sampling site between the years 1991 and 1979 (*$a=0.05$, **$a=0.01$ and ***$a=0.001$).

10.3.2   Soil moisture

The distribution of the moisture content in the upper 20 cm of soil during 12 months (from May 1991 to May 1992) is presented in Figure 10.4. The moisture content reached rather low values during the dry period (early summer to mid-autumn 1991) which were significantly different between the two transects during most of this period (significant at $a=0.01$). Actually, the moisture content in the upper 20 cm soil layer fluctuated between 0.04 (air-dry) and 0.18 cm^3 cm^{-3} under olives (transect A), while under vines (transect B) it was higher, fluctuating between 0.12 and 0.18 cm^3 cm^{-3}. The difference is attributed mainly to the absence of transpiration by natural vegetation (weeds and shrubs) and to the mulching effect of tillage on transect B. No significant differences in moisture content were found between the two transects from mid-autumn to early winter. This is due to the lack of annual plant cover in the olive grove (no transpiration), whereas practically no water was transpired from the upper 20 cm soil layer in the deep-rooted vineyard. During the winter and early spring months, when rainfall inputs were frequent, soils in both transects had a high moisture content, 0.30–0.40 cm^3 cm^{-3} for transect B (vines) and 0.12–0.22 cm^3 cm^{-3} for transect A (olives). The moisture content of the soils under olives were lower

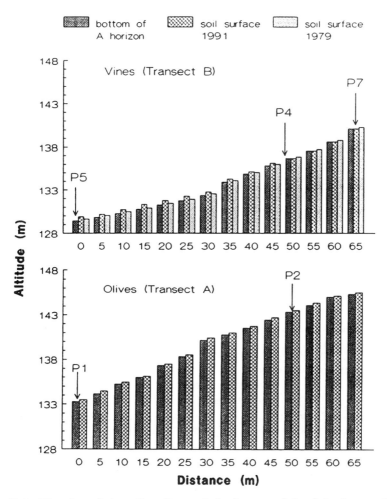

FIGURE 10.3 Elevation of the soil surface and the bottom of the A horizon at Spata along the transects as measured in 1979 and 1991 (the soil surface under olives in 1979 coincides with soil surface in 1991)

(significant at $a = 0.01$) due to the greater amount of rain and to transpiration of the annual vegetation. These differences were more pronounced in late spring when the leaf area index was measured at maximum. Although the bulk density of the surface layer did not practically differ (significant at $a = 0.01$) between the two types of land use (average values $1.43\,g\;cm^{-3}$ and $1.48\,g\;cm^{-3}$ on transects A and B, respectively), significant differences in crust permeability of the topsoil were found (significant at $a = 0.95$) in November ($4.2\,cm\;h^{-1}$ and $0.93\,cm\;h^{-1}$ on transect A and B, respectively). The average crust permeability of the topsoil under vines changed from 0.93 to $0.57\,cm$ h^{-1} during the winter and resulted in a greater overland flow. In contrast, the crust permeability of the (top)soils under olives remained almost constant at the high average levels of $4.2\,cm\;h^{-1}$.

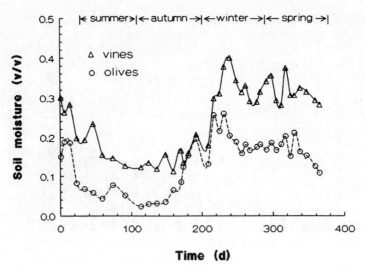

FIGURE 10.4   Soil water content in the upper 20 cm soil layer under olive trees with an understorey of annual vegetation (site P1), and under vines with weed control (site P4)

10.3.3 Biomass production and soil structure

The total biomass production of the annual vegetation under olives (379 g m^{-2}, SD = 128 g m^{-2}) was about eight times greater than under vines (48 g m^{-2}, SD = 13 g m^{-2}). Vine root mass decreased sharply from the plant row to the inter-row space because of cultivation. A relatively small portion of the rooting system of vines developed in the plough layer (35 cm), whereas in the olive grove the growing plants developed a dense root system in the upper 40 cm (Figure 10.5). It was found that 85.8%

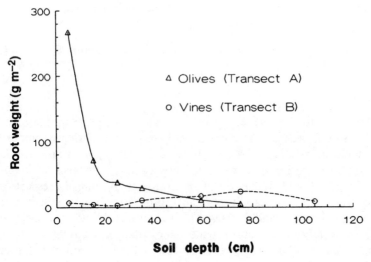

FIGURE 10.5   Mean root distribution (g m^{-2}) with soil depth (cm) for the transects under olives and vines

of the weight of the roots was present in the upper 30 cm along transect A. In contrast, only 20.3% of the roots (including roots of natural vegetation and vines) was measured in the same soil layer on transect B; most of the below-ground biomass was found in the deeper layers between 30 and 90 cm, corresponding to 77.3% of the total root weight (Figure 10.5).

Soil aggregation was affected by soil organic matter content and cultivation practices. Both these parameters were significantly different under the two types of land use. Data on the mean aggregate size of soil samples collected in November 1991 (Table 10.1) suggest that the shift from olive grove to vine cropping had a degrading effect on the size and stability of the soil aggregates. Thus cultivation over 12 years reduced the aggregate size by about 10 times in transect B compared with transect A (assuming equilibrium in soil aggregates under olives), increasing the erodibility of these soils.

A significant amount (43.7%) of the total roots in the olive grove was found in the uppermost 10 cm soil layer (Figure 10.5). As Tisdall and Oades (1982) mentioned, roots penetrating through the intra-aggregate pores stabilize macro-aggregates on transect A, while on transect B roots were rather scarce in the same soil layer. The organic carbon content on transect B decreased by about 33% during the 12 years of cultivation, while the C/N ratio remained almost constant (Table 10.1). No significant differences in organic carbon content were found between the years 1979 and 1991 along transect A. On transect B, the differences were significant (significant at $a = 0.01$) for the same years and are attributed mainly to the accelerated erosion as well as the reduced biomass incorporated each year into the soils on this transect. The difference in organic carbon content and total soil nitrogen between the study transects was significant (significant at $a = 0.01$) but only for the year 1991. Furthermore, the presence of winter-annual plants significantly decreased the soil moisture in the surface layer in the olive grove. The drying of these soils appears to be enhancing cementation of the organic fraction and soil particles into water-stable aggregates (Reid and Goss, 1982).

10.3.4 Soil colour

Reduction of the organic carbon content caused by accelerated erosion and decreasing biomass production is associated with an increase in the reflectance of the 400 to 800 nm wavelengths in all study soils (Figure 10.6). The less eroded soils P2 and P3 on transect A have a higher organic matter content and a comparatively lower light reflectance in all wavelengths. They also show a lower slope of the reflectance spectra with a convex to linear shape. The severely (pedon P7) and moderately (pedon P4) eroded soils along transect B are characterized by a change in the overall reflectance and the increase in slope occurring in the 400 to 580 nm wavelengths. As Table 10.1 shows, there are no great differences in the total iron oxides between the study transects. Therefore the difference in reflectance, which is significant particularly in all wavelengths greater than 530 nm (significant at $a = 0.10$), is attributed mainly to the amount of the organic matter present in the pedons.

Munsell colour value and chroma both increased with the degree of erosion. The surface horizon became progressively lighter as the organic matter content decreased (Figure 10.7), and features of the underlying B horizon appeared on the surface. The correlation

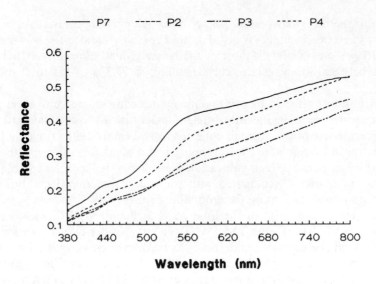

FIGURE 10.6   Visible reflection spectra for two slightly eroded soils (P2, P3) on transect A, and a moderately eroded (P4) and a severely eroded (P7) soil on transect B (reflectance is a dimensionless quantity)

coefficients for the linear regression between the parameters colour and chroma were 0.87 and 0.85 respectively.

10.3.5 Surface runoff and soil loss by erosion

The change in land use has significantly affected the rates of surface runoff and soil loss by affecting soil aggregation, water infiltration and the plant canopy. Soil survey data obtained in 1979 indicated that the depth of the A horizon in the lower part of transect B was only 23 cm. The deposition of new materials through erosion increased the depth of the Ap horizon to its present value of 48 cm, during the following 12 years (Figure 10.3). Note that the original surface horizon is still present but quite altered, mixed with the upper part (about 10 cm) of B horizon by tillage. Considering a bulk density of 1.4 g cm^{-3}, deposition of new material of at least 180 t ha^{-1} year^{-1} has occurred on this site.

Alternatively, the A horizon of the soils in the upper part of transect B (site P7) has been eroded away, and the parent material has been exposed to the surface. Considering the average thickness of the A horizon equal to 17.7 cm, measured in the upper part of this transect in 1979 (Figure 10.3), the soil loss in the last 12 years is estimated to be at least 205.8 t ha^{-1} year^{-1} as a result of the change in land use.

The great impact of land-use change on surface runoff is reflected by the runoff values measured at sites along the two transects over the entire observation period. Note that the slope gradients of the upper part of transect B under vines (sites P7 and P4) are similar with those of the lower part of transect A under olives (sites P1 and P2) (Table 10.1). The total runoff ranged from 0.32 to 2.62 mm on transect A, whereas on transect B, much higher values (4.87–12.25 mm; Table 1) were measured despite the lower slopes on this transect (site P4 versus sites P1 and P2). The annual vegetation and the plant residues on transect A prevented raindrop impact from creating a surface seal. Moreover, the

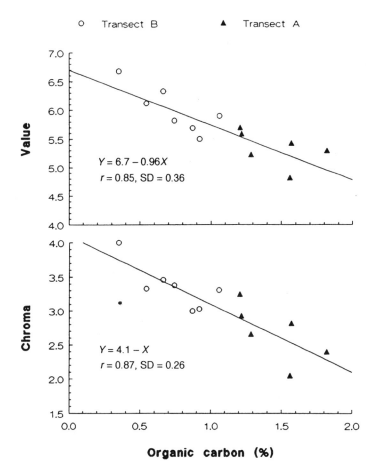

FIGURE 10.7   Relationship between Munsell colour value and chroma and organic carbon in selected sites along the transects

vegetation slowed down the velocity of runoff water. The root system of the vegetation increased the porosity of the upper horizon and hence increased its capacity for infiltration. Under prolonged rainfalls, hillslope seepage was considerable on the lower parts of both transects (sites P1 and P6) in at least three rainfall events. This occurred as a result of water moving initially downward through the surface layer and encountering subsurface layers with lower permeability. A layer of low permeability was found at depths usually deeper than 45 cm, overlaid by a high oxidized zone indicating a downslope water throughflow especially in the olive grove.

A high correlation between sediment loss (SL, in g m^{-2}) and runoff (D, in mm) was found on transect B ($SL = 7.3D^{1.44}$, $r = 0.85$, $n = 25$). This relation, however, was poor on transect A because of the very small amounts of sediment which were difficult to measure accurately. Indeed, the total amount of sediments on transect B ranged from 1 to 560 g m^{-2}, whereas on transect A, the sediment never exceeded 1 g m^{-2}.

The total soil losses per experimental plot recorded in the entire period of study are summarized in Table 10.1. It can be seen that total sediment loss varied according to land use, soil conditions and slope grade. Site P4, the uppermost runoff plot under fallow (transect B) having a low aggregate stability and a slope grade of 12.4%, produced the highest runoff and sediment losses (Table 10.1). The total average sediment measured during the period of 21 months was 0.44 g m^{-2} on transect A, while the extreme value of 441.9 g m^{-2} was measured on transect B (Table 10.1). However the total soil loss, estimated from the soil survey data, is considerably higher than the values measured during this 21-month study. Such a loss should therefore be attributed to the occurrence of few, but high-intensity rains which are typical in the Mediterranean region and are characterized by a rather high erosion capacity.

10.3.6 Conclusions

The data obtained from the MEDALUS core programme field site in Spata suggested that the change in land use from olive grove to vines had a degrading effect on the size and stability of soil aggregates. The organic carbon content was reduced by about 33% in 12 years; and this decrease was associated with an increase in Munsell colour value and chroma of the upper soil layer.

The depth of the A horizon was altered as a consequence of the 12 years of cultivation with vines. It was estimated that at least 205.8 t ha^{-1} year^{-1} have been eroded away from the upper part of the transect under vines. The total runoff in 21 months ranged from 0.32 to 2.62 mm under olives, in contrast to the much higher values measured under vines which ranged from 4.9 to 12.2 mm. The total average sediment output measured over the experimental period of 21 months was 0.004 t ha^{-1} in the olive grove, whereas an extreme of 4.4 t ha^{-1} was recorded in the vineyard.

All the above point to the great importance of land use and management on soil properties and land characteristics. The impact of land-use change on environmental conditions, particularly the shift from olives to vines in hilly areas under Mediterranean conditions, should be seriously considered in future land-use planning. In contrast to the olives, vines create favourable conditions for overland flow, erosion and desertification.

10.4 THE EFFECT OF REDUCED RAINFALL ON SOIL PROPERTIES AND WHEAT BIOMASS PRODUCTION

10.4.1 Climatic conditions

The total rainfall during the growing period (from 11 November 1991 to the end of May 1992) totalled 360.7 mm. This amount was reduced to 255.8 mm or 70.9%, 176.7 mm or 48.9%, and 122.9 mm or 34.1%, in the experimental plots which were partially shielded by plexiglas trips. Figure 10.8 schematically presents the 10-days rainfall (R) and calculated potential evapotranspiration (ET_{pot}) during the growing period. It is apparent that ET_{pot} largely exceeded rainfall after late December in all treatments (except for the control). In the control plots (100% applied rain), water surplus occurred in mid-winter and late March, but after that time rainfall lagged considerably behind ET_{pot} too. Note that the total ET_{pot} was 427 mm.

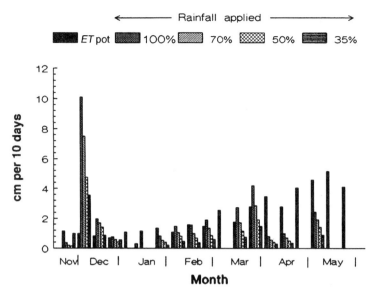

FIGURE 10.8 Rainfall and potential evapotranspiration in the partial rainfall exclusion experiment at Spata during the growing period of 1991/92

Figure 10.9 schematically presents the 10-day rainfall and ET_{pot} in the Tanagra sites for three successive growing periods. Based on the long-term average climatic data, the first growing period (1989/90) was extremely dry ($R=95$ mm vs. $R=370$ mm in an average year). The following growing period (1990/91) was particularly wet with rainfall ($R=663$ mm) exceeding ET_{pot} (345 mm) except at the end of the growing period when a slight water deficit occurred (Figure 10.9). The third growing period (1991/92) was a typical one for the area ($R=389$ mm, $ET_{pot}=397$ mm).

10.4.2 Soil properties

The change in soil moisture during the growing period 1991/92 is schematically presented in Figure 10.10 for all treatments and for two soil depths. It can be seen that the moisture content in the plots with 35% rain application was continuously lower than the permanent wilting point after mid-February. In contrast, the soil moisture in the plots receiving 100 and 70% of the incoming rain was sufficient for the growing plants in the early growing period, and only after the end of April (at the time of silking of the wheat) did severe water deficiency occur. The soil moisture was highly variable in the subsoil depending on the amount of applied rain (Figure 10.10). It can be further observed that the deviation of moisture content between the dry treatments (35 and 50% rain applied) and the wet treatments (75 and 100% rain applied) is greater in the subsoil than in the plough layer. This is due to the restricted downward movement of soil water in the dry treatments.

The potentially mineralized nitrogen (N_o) was not significantly different in the plots of 100 and 35% rain application ($N_o=69.7$ and 68.3 mg N kg^{-1} soil, respectively). The highest values of N_o were measured in the plots of 70 and 50% application ($N_o=96.3$ and 93.8 mg N kg^{-1} soil, respectively). As already known, the soil moisture content at field

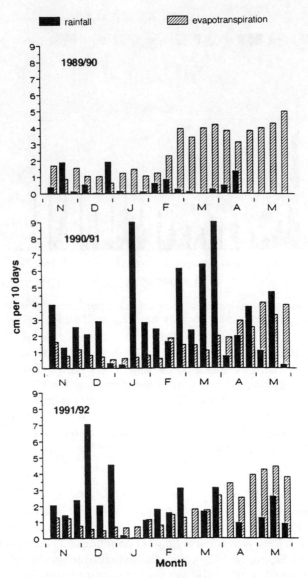

FIGURE 10.9   Rainfall and potential evapotranspiration in the Tanagra area during the three
growing periods of wheat (Kosmas et al., 1993a)

capacity is considered optimal for nitrogen mineralization (Stanford and Epstein, 1974).
Apparently, the plots receiving the maximum rainfall (100%) had adequate soil moisture
for nitrogen mineralization, and N_o was reduced due to biological decomposition of the
organic materials. The opposite might have occurred in the plots receiving the lowest
amount of rain, where only modest amounts of organic residues were produced and
incorporated in the soil. In the plots receiving 70 and 50% of the total rain, relatively high

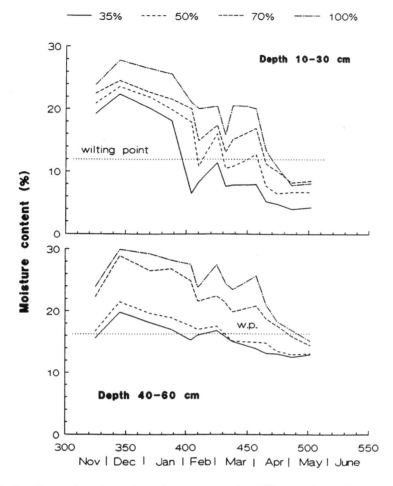

FIGURE 10.10   Change in volumetric moisture content for different rain applications and two soil depths at the Spata site during the growing period of 1991/92

amounts of plant residues were incorporated into the soil but soil moisture was not optimal for nitrogen mineralization.

Figure 10.11 illustrates the distribution of nitrates with soil depth at the end of the growing period (July 1992). At that time, the maximum concentration of nitrates was found in the topsoil in all treatments. Actually, the amount of nitrates in the topsoil followed the order of rain exclusion with the highest amounts measured in the treatment of lowest rain application. As Figure 10.10 suggests, leaching of nitrates below the root zone occurred during the rainy season at least in the wet treatments of 100 and 70% rain application. On the other hand, most of the applied nitrogen fertilizer was measured in the topsoil of the driest treatment (35% rain application), due to the restricted downward water movement. Finally, in the treatment of 50% rain application, an intermediate situation appeared, with a moderate nitrate movement through the soil profile and nitrates

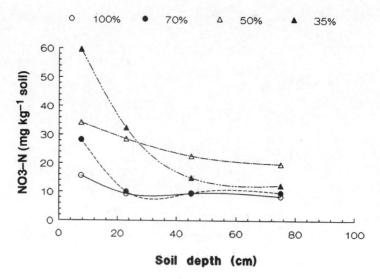

FIGURE 10.11   Distribution of NO₃-N with soil depth under conditions of different rainfall
treatments at the end of the growing period (July, 1992)

remaining in the subsoil in higher amounts, apparently due to the lower (or absent)
nitrogen uptake by the roots in the subsoil (Figure 10.11).

The nitrogen balance was calculated by the following equation:

$$N_{tot} = N_s + N_f + N_m + N_{at} - N_v - N_u - N_l,$$

where N_{tot} is the residual soil NO₃-N and NH₄-N measured at the end of the growth
period, Ns is the residual soil NO₃-N and NH₄-N measured before sowing, N_f is the
amount of applied N fertilizers, N_m is the mineralized N, N_{at} is the atmospheric deposition
of N, N_v is the volatilized ammonia, N_u is the plant uptake and N_l is the N leached by deep
percolation of soil water.

The results of these calculations for the upper 90 cm soil layer are summarized in Table
10.2. It was found that for the plots of 35, 50, 70 and 100% rain application, respective
NO₃-N amounts as much as 0, 6.7, 20.3 and 59.7 kg ha⁻¹ were lost below the root zone
(deeper than 90 cm) from late November to late March 1992 (Table 10.2). Since runoff did
not occur in the plots with 50 and 35% rain application, nitrogen leaching into the subsoil
or removal by runoff was limited, and the adverse consequences in the environment were
thus restricted under the prevailing weather conditions.

The mean organic matter content decreased with diminishing moisture content of the
topsoil from 1.4% in the plot of 100% rain application to 1.2, 1.1 and 0.9% for the plots
of 70, 50 and 35% rain application respectively at the time of measurements (July 1992).
Data on mean aggregate size demonstrated that diminishing soil moisture status
negatively affected the aggregate stability. The mean aggregate size was measured at
10.2, 5.3, 6.9 and 5.6mm for the plots of 100, 70, 50 and 35% rain application,
respectively. The effects of soil moisture reduction on aggregate stability are not clear as
yet because of the limited period of investigation.

TABLE 10.2 Nitrogen fluxes (kg ha^{-1}) in the upper 90 cm soil layer measured during the second growth period of wheat (1990/1991) for different rainfall applications

Rainfall (%)	Ns	N_f	N_m	N_{at}	N_v	N_u	N_l	N_{tot}
100	201.7	75	143.2	1.8	1.9	232.3	59.7	127.8
70	160.3	75	94.0	1.3	2.2	143.8	20.3	164.3
50	213.5	75	80.8	0.9	2.4	71.8	6.7	289.3
35	232.4	75	22.5	0.6	2.9	16.1	0.0	311.5

10.4.3 Biomass production

The change in soil moisture caused by the partial exclusion of rainfall in the experimental site affected the germination and maturation time. A delay of germination in the driest plots (35% rain application) by as long as 14 days can be observed in Figure 10.12. As expected, the leaf area index (LAI) of the crop was also greatly affected throughout the growing period (Figure 10.12). The maximum LAI values measured in the treatments of 100, 70, 50 and 35% rain application were 5.23, 3.73, 2.88 and 1.57 respectively (significant at $a = 0.005$).

The plants in the plots receiving 35% of the rain were completely dry 37 days earlier than the plants in the open field (Figure 10.12).

The total above-ground biomass production of wheat ($TAGBP$, in kg m^{-2}) was proportionally reduced with the fraction of rain applied (RI) and it was related with the following equation: $TAGBP = -0.46 + 1.64RI$ ($r = 0.97$). Reduction in biomass by 90, 71.4 and 53.4% was measured in the experimental plots in which rainfall was excluded by 65, 50 and 30% respectively.

In many studies (Rijtema and Aboukhaled, 1975; Slabbers, 1980) the ratio of actual evapotranspiration (ET_{act}) over maximum crop evapotranspiration (ET_{max}) is used to describe the reduction of biomass production. A logarithmic relation between $TAGBP$ and the ratio ET_{act}/ET_{max} was actually found in this study for soils both in the Spata and the Tanagra sites, the latter occurring on similar landscape positions and having similar parent materials and fertility status:

$$\log[TAGBP] = 13995 + 8263 \log[ET_{act}/ET_{max}] \quad (r = 0.97, \; n = 27)$$

When the maximum biomass production is known, the relative biomass production (the ratio of actual over maximum biomass production) can be used as an alternative to explain the production fluctuations under different weather conditions. The biomass productions harvested in both the Tanagra sites (Kosmas et al., 1993a) and in the Spata site (18.5 t ha^{-1}) in the wet year 1990/91, when water availability was practically at optimum throughout the growing period, are considered as the maximum. As Figure 10.13 demonstrates, the measured relative biomass production in both the rainfall exclusion experiments at Spata and the Tanagra sites in the successive years 1989/90, 1990/91 and 1991/92 were strongly related to the relative evapotranspiration rate (ET_{act}/ET_{max}; see Figure 10.13). The data can be further used to assess the reduction of $TAGBP$ in hilly areas on marl deposits under different weather conditions and to make predictions for future

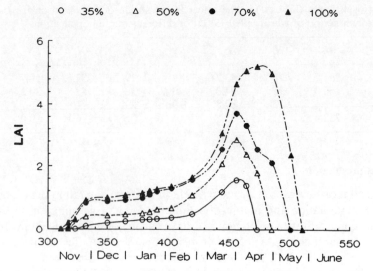

FIGURE 10.12   Change in the LAI of wheat grown in plots receiving 35, 50, 75 and 100% of the incoming rainfall in 1991/92

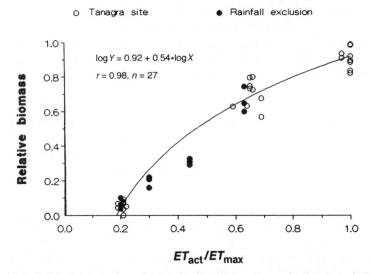

FIGURE 10.13   Relation between the calculated relative evapotranspiration rate (ET_{act}/ET_{max}) and the relative biomass production for all sites and for three growing periods

changes in biomass productivity of rain-fed wheat especially in areas vulnerable to desertification.

10.4.4 Conclusions

The effects of diminishing soil moisture on soil properties and biomass production of a widely used wheat durum cultivar were investigated under semi-arid conditions in a rainfall exclusion experiment at the Spata field site. The total biomass production of the same cultivar was additionally measured along a number of soil catenas in another semi-arid area (Tanagra) with a similar substratum of Tertiary marl but under quite different weather conditions.

A highly variable soil moisture content was found, depending on the amount of rain applied throughout the growing period. The distribution of nitrates in the soil profile was also greatly affected. The highest concentrations were found in the topsoil of the plots receiving the minimum amounts of rain (35% of total), whereas nitrates were leached below the root zone in the wet treatments (70 and 100% rain applied). Potentially mineralized nitrogen was affected by the amount of rain and the plant residues incorporated into the soil. Diminishing amount of soil moisture negatively influenced the organic matter content and the aggregate stability. The leaf area index and the total (above-ground) biomass production of the crop were almost proportionally reduced with the amount of rainfall excluded. In all cases, relative biomass production was strongly related to the relative evapotranspiration rate (ET_{act}/ ET_{max}). Such results can serve to reliably assess the biomass reduction of rain-fed wheat in similar areas vulnerable to desertification.

10.5 REFERENCES

Addiscott, T.M. (1983) Kinetics and temperature relationships of mineralization and nitrification in Rothamsted soils with different histories. *Journal of Soil Science*, **34**, 343–353.

Boiffin, J. and Monnier, G. (1985) Infiltration rate as affected by soil surface crusting caused by rainfall. In *Proceedings of Symposium on Assessment of Soil Surface Sealing and Crusting*, Ghent, 210–217.

Bremner, J.M. and Mulvaney, C.S. (1982) Salicylic acid–thiosulfate modification of Kjeldahl method to include nitrate and nitrite. In Page, A.L. et al. (eds) *Methods of Soil Analysis, Part 2*, 2nd edition. Agronomy Monograph 9. ASA and SSSA, Madison, WI, 595–624

Bufalo, M. and Nahon, D. (1992) Erosional processes of Mediterranean badlands: a new erosivity index for predicting sediment yield from gully erosion. *Geoderma*, **52**, 133–147.

Danalatos, N.G. (1993) Quantified analysis of selected land use systems in the Larissa region, Greece. PhD Thesis, Agricultural University of Wageningen, 370 pp.

De Boodt, M, De Leenheer, L. and Kirkham, D. (1961) Soil aggregate stability indexes and crop yields. *Soil Science*, **91**, 138–146.

FAO (1977) *Guidelines for Predicting Crop Water Requirements*. Irrigation and Drainage Paper 24. By Doorenbos, J. and Pruitt, W.O., Rome, 144 pp.

FAO (1979) *Yield Response to Water*. Irrigation and drainage Paper 33. By Doorenbos, J. and Kassam, A.H. Rome, 133 pp.

Francis, C.F. and Thornes, J.B. (1990) Runoff hydrographs from three Mediterranean vegetation cover types. In Thornes, J.B. (ed.) *Vegetation and Erosion: Processes and Environments*. Wiley, New York. 363–384.

Frere, M. (1979) *A Method for the Practical Application of the Penman Formula for the Estimation of Potential Evapotranspiration and Evaporation from a Free Water Surface*. FAO, AGP: Ecol./1979/1. Rome, 26 pp.

Gee, G.W. and Bauder, J.W. (1986) Particle size analysis. In Klute, A (ed.) *Methods of Soil Analysis, Part 1*, 2nd edition. Agronomy Monograph 9. ASA and SSSA, Madison, WI, 383–411.

Keeney, D.R. and Nelson, D.W. (1984) Nitrogen—inorganic forms. In Page, A.L. et al. (eds). *Methods of Soil Analysis, Part 2*, 2nd edition. Agronomy Monograph 9. ASA and SSSA, Madison, WI, 643–698.

Kissel, D.E., Brewer, H.L. and Arkin, G.F. (1977) Design and test of a field sampler for ammonia volatilization. *Soil Science Society of America Journal*, **41** 1133–1138.

Klute, A. (1986) Water retention: laboratory methods. In Klute, A. (ed.) *Methods of Soil Analysis, Part 1*, 2nd edition. Agronomy Monograph 9. ASA and SSSA, Madison, WI, 635–653.

Kosmas, C.S., Franzmeier, D.P. and Schulze, D.G. (1986) Relationship among derivative spectroscopy, color, crystallite dimensions, and Al substitution of synthetic goethites and hematites. *Clays and Clay Minerals*, **34**, 625–634.

Kosmas, C.S., Danalatos, N.G., Moustakas, N., Tsatiris, B., Kallianou, Ch and Yassoglou, N. (1993a) The impacts of drought on the wheat biomass production along catenas in the semi-arid zone of Greece. *Soil Technology*, **6**, 337–349.

Kosmas, C.S., Moustakas, N., Danalatos, N.G. and Yassoglou, N. (1993b) The effect of rock fragments on wheat biomass production under highly variable moisture conditions in Mediterranean environments. *Catena* **23**, 199–198.

Maracchi, G. (1991) Impacts of climatic change on crops. In Fantechi, R. et al. (ed.) *Climatic Change and Impacts: A General Introduction*. Commission of the European Communities, EUR 11943 EN, 343–350.

Mehra, O.P. and Jackson, M.L. (1960) Iron oxide removal from soils and clays by a dithionite–citrate–bicarbonate system buffered with sodium bicarbonate. *Clays and Clay Minerals*, **7**, 317–327.

Nelson, D.W. (1982) Carbonates and gypsum. In Page, A.L. et al. (eds) *Methods of Soil Analysis, Part 2*, 2nd edition. Agronomy Monograph 9. ASA and SSSA, Madison, WI, 181–187.

Nelson, D.W. and Sommers, L.E. (1982) Total carbon, organic carbon, and organic matter. In Page, A.L. et al. (eds) *Methods of Soil Analysis, Part 2*, 2nd edition. Agronomy Monograph 9. ASA and SSSA, Madison, WI, 539–579.

Papoutsopoulos, I.G. and Svorikyn, I. (1932) *Research on the Soils of the Prefecture of Larissa*. Ministry of Agriculture Report, Athens, 268 pp. (in Greek).

Penman, H.L. (1948) Natural evaporation from open water, bare soils and grass. *Proceedings of the Royal Society of London*, **193A**, 120–145.

Perfect, E., Kay, B.D., van Loon, W.K.P., Sheard, R.W. and Pojasok, T. (1990) Rates of change in soil structural stability under forages and corn. *Soil Science Society of America Journal*, **54**, 179–186.

Reid, J.B. and Goss, M.J. (1982) Interaction between soil drying due to plant water use and decreases in aggregate stability caused by maize roots. *Journal of Soil Science*, **33**, 47–53.

Rhoton, F.E. and Tyler, D.D. (1990) Erosion-induced changes in the properties of a fragipan soil. *Soil Science Society of America Journal*, **54**, 223–228.

Rijtema, P.E. and Aboukhaled, A. (1975) Crop water use. In Aboukhaled, A. et al. (eds) *Research on Crop Water Use, Salt Affected Soils and Drainage in the Arab Republic of Egypt*. FAO Regional Office for the Near East, 5–61.

Slabbers, B.J. (1980) Crop response to water under irrigated conditions. In *International Institute for Land Reclamation and Improvement, Publ. 27*, pp. 139–147.

Soil Survey Staff (1975) *Soil Taxonomy: A Basic System of Soil Classification for Making and Interpreting Soil Surveys*. USDA, CS Agric. Handb. US Government Printing Office, Washington, DC, 436 pp.

Stanford, G. and Epstein, E. (1974) Nitrogen mineralization–water relations in soils. *Soil Science Society of America Journal*, **38**, 103–107.

Stanford, G. and Smith, S.J. (1972) Nitrogen mineralization potentials of soils. *Soil Science Society of America Journal*, **36**, 465–472.

Tisdall, J.M. and Oades, L.M. (1982) Organic matter and water-stable aggregates in soils. *Journal of Soil Science*, **33**, 141–163.

Wyszecki, G. and Stiles, W.S. (1982) *Color Science: Concepts and Methods, Quantitative Data and Formulae, 2nd edition*. Wiley, New York, 950 pp.

Yassoglou, N. (1989) Desertification in Greece. In Rubio, J.L. and Ricon, R.J. (eds) *Strategies to Combat Desertification in Mediterranean Europe*. Commission of European Communities Report, EUR 11175 E/ES, 148–162.

Yoder, R.E. (1936) A direct method of aggregate analysis and a study of the physical nature of erosion losses. *Journal of the American Society of Agronomy*, **28**, 337–351.

11

The Petralona and Hortiatis Field Sites (Thessaloniki, Greece)

J. Diamantopoulos, J. Pantis, S. Sgardelis, G. Iatrou,
S. Pirintsos, E. Papatheodorou, A. Dalaka and G. P. Stamou

Aristotelian University, Thessaloniki, Greece

L. H. Cammeraat

V.F.G.B., Universiteit van Amsterdam, The Netherlands

and

C. Kosmas

Laboratory of Soils and Agriculture, Agricultural University of Athens

11.1 INTRODUCTION

The closed physiognomy of the Mediterranean shrublands does not make grazing easy. Historically, once natural pastures had been transformed into agricultural land, shrubs in the surrounding hills or mountain foothills were cleared for husbandry. The easiest and quickest method for clearance was by fire. This started the vicious cycle of erosion–vegetation degradation–erosion, leading to desertification.

The aim of the study at Thessaloniki within the framework of MEDALUS is to investigate relations between the two dominant land-use types (wheat cultivation and shrub grazing) and runoff–sediment yield, in an area where these two land-use types exist side by side. The sites at Hortiatis and Petralona were selected for this purpose (Figure 11.1).

11.2 GEOGRAPHICAL SETTING

11.2.1 Geology

The Petralona site lies on a limestone block of Kimmeridgian–Portlandian age (upper Jurassic) surrounded by Miocene–Pliocene (upper Tertiary) deposits at an altitude of 200–250 m (Figure 11.2). The Hortiatis site lies on Phyllitic metamorphic rocks of the Svoula group at an altitude of 400 m (Kauffman et al., 1976).

Mediterranean Desertification and Land Use. Edited by C. Jane Brandt and John B. Thornes.
© 1996 by John Wiley & Sons, Ltd.

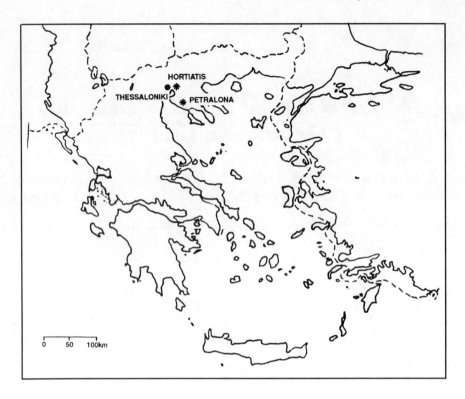

FIGURE 11.1   Location of the Petralona and Hortiatis sites, Greece

11.2.2 Soil

Soil properties at Petralona vary along the catena formed by the limestone hill (Figure 11.3). The upper part of the catena is on the broad shoulder of the hill (Figure 11.2). The soil is shallow and discontinuous. In the pockets between the bare karstified limestone rock, the soil is generally not more than 10 cm deep. The very shallow profiles are classified as lithic leptosols (FAO) or lithic xerothents (USDA) and have a gravelly and stony clay–loam to silt–loam texture. Shrinkage cracks are common. More than 40% of the surface is covered by bare solid rock and large stones, 18.9% by bare soil, 35.2% by of vegetation and 4.9% by litter and dead plant material (excluding beneath the canopy). Approximately one-third of the soil is covered by stones resting on the surface, another third is covered by stones situated in the soil and the rest is occupied by bare soil and shrinkage cracks.

   In the middle section of the catena bare rock is not common and the pattern of vegetated and bare areas is more regular. The effect of grazing is evident. These soils are better developed, although truncated. In the middle part of the slope the soil profile is most developed. Here the classification is different because the soil depth is the limiting factor. The texture is silt–loam to clay–loam, with gravel common. On top of the Ah horizon a discontinuous litter and humus layer is present. Further downslope, petrocalcic crusts and horizons start to appear, which continue towards the lower section of the catena.

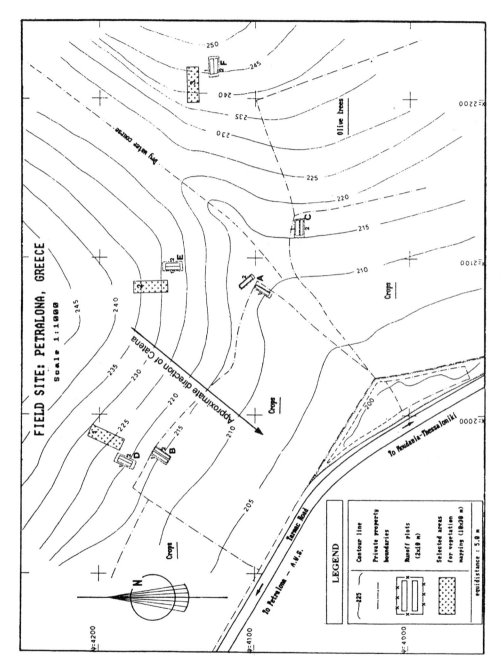

FIGURE 11. 2   The topographic diagram and runoff plots at the Petralona site

Soil profiles of Catena 1, Petralona

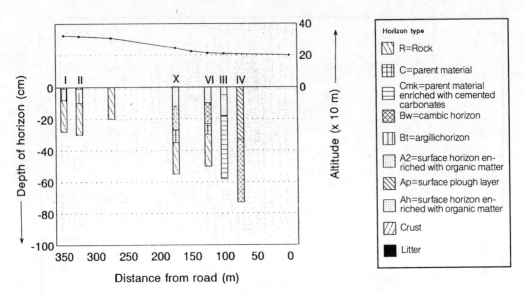

FIGURE 11.3   Soil profiles for the catena at the Petralona site

The lower section is characterized by a change in lithology (from limestone to marl) and land use (semi-natural, bushy rangelands to cultivated wheat). These changes are also reflected in the soil profiles. Two types of soil profiles can be distinguished, first an AC profile on the limestone–marl transition, with a petrocalcic horizon at shallow depth and a second profile developed on colluvial deposits and marls. The light-coloured soils with AC profile at the slope transition are characterized by their petrocalcic horizon, an indurated horizon consisting of a $CaCO_3$ concretion. The petrocalcic horizon might be related to unsaturated soil water flow and its evaporation at the slope transition, or to a change in parent material. The lower profiles are well-developed soils, characterized by their greater depths, a ploughed A and a cambic (weathering: Bw) horizon and by their $CaCO_3$ contents.

In general, the nitrogen and organic carbon content decrease with depth, which is to be expected as the input of these nutrients is related to litterfall or addition of manure. The organic carbon contents are relatively high (4.6%) in the semi-natural soils and lower (2–3%) in the agricultural soils. This will probably also be reflected in the aggregate stability. In general, the structural stability increases with organic carbon content. Except for the soils with a petrocalcic horizon, the soil aggregate size distribution peaks between 2 and 0.25 mm diameter, whereas the other size classes are variable in importance.

11.2.3 Climate

Mean monthly temperature and precipitation for the 1981–1990 decade at the nearest meteorological station to Petralona (Agios Mamas) are shown in Figure 11.4a. The climate is

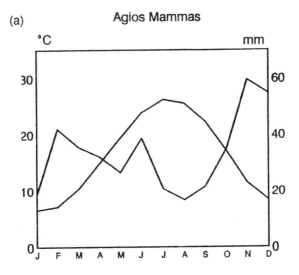

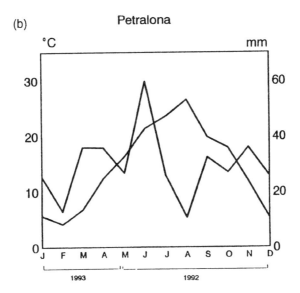

FIGURE 11.4   Meteorological data given in the form of an ombrothermic diagram (Klimadiagram) for Agios Mamas (a) and Petralona (b). 10°C corresponds to 20 mm rainfall

Mediterranean in character as shown by the fall in rainfall over the hot summer months. The driest months are July, August and September, although some rainfall is still probable.

11.2.4 Vegetation

The vegetation is maquis composed of evergreen, broad-leaved shrubs. Four species are dominant: *Quercus coccifera, Pistacia lentiscus, Juniperus oxycedrus* and *Phillyrea* sp. A quick census of the flora of both sites has produced a list of 50 plant species. The relative cover and position of each of the dominant species is shown in Figure 11.5.

11.3 EXPERIMENTAL DESIGN AND METHODS

The study areas were selected to include the two types of land use most common in Greece where maquis exists, grazing and winter wheat cultivation. In the maquis attention was focused on *Q. coccifera*, the species which is most commonly grazed and on two of the other dominant shrubs, *J. oxycedrus* and *Phillyrea* sp.

In the Petralona site 12 runoff plots were installed, six in the maquis and six in the wheat field (see Figure 11.2). The runoff plots were positioned along the slope in order to follow the soil catena of the limestone hill. Soil moisture was determined gravimetrically.

Decomposition rates were studied using both cotton and cellulose strips. With the cotton strips, the measurement of decomposition is related to the loss of tensile strength of the material. Three lots of 90 strips of 'Shirley Soil Burial Text Fabric' (each one 10×4 cm) were used at the Petralona site. Packs of 90 strips were wrapped together in aluminium foil, and sterilized in an oven for 15 minutes at $121°C$ in order to avoid contamination by alien organisms in the soil. The strips were placed under the two types of dominant vegetation and in the open, between canopies. At each measurement site, the strips were placed at two depths; at the base of the ectorganic horizon and at 3 cm within the mineral horizon. The strips were buried between the same dates as used by all the other MEDALUS groups (18 March 1992–15 April 1992, 24 June 1992–22 July, 21 October 1992–21 November 1992). Once they had been retrieved, the strips were washed with a jet of water. They were then soaked in 70% ethanol at room temperature for 4 hours and afterwards dried at room temperature before being tested for loss of tensile strength.

In addition to the cotton ones, cellulose strips were put in 10×10 cm bags with a 0.1 mm opening, and were inserted in the soil at 10 cm depth in both the maquis and the wheat field. The bags were retrieved on a seasonal basis. On each sampling occasion 12 bags from each site were brought to the laboratory, where they were dried to constant weight at $60°C$ and weighed.

The standing biomass was measured by clear harvesting three different categories of *Q. coccifera* shrubs (overgrazed, moderately grazed and ungrazed). The material harvested was separated into leaves, modules (stems from which growth starts, including new leaves and twigs) and wood. In order to evaluate the root biomass as well as its distribution in different horizons, a soil sampler (15×7 cm) was used. Soil cores were taken at three depths (0–15, 15–30, 30–45 cm), and in the same three categories of *Q. coccifera* shrubs. Cores were also taken from the open areas between the shrubs which are occupied by grass species. Roots were extracted from the soil samples by wet sieving and were classified according to their diameter, in three size categories (< 1, 1–3, 3–6 mm).

The leaf area index (LAI) was determined from the area/weight relationships obtained from representative samples of wheat and *Q. coccifera* in each sub-area. The material harvested for biomass estimation was also used for this purpose. LAI was seasonally determined for the three *Q. coccifera* shrub categories. Leaf area was determined using a leaf area meter.

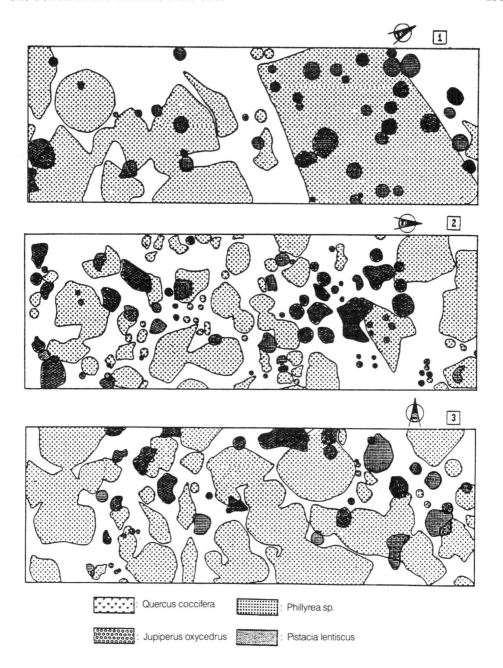

: Quercus coccifera   : Phillyrea sp.

: Jupiperus oxycedrus   : Pistacia lentiscus

FIGURE 11.5   Representation of the spatial distribution of the four dominant woody species at areas 1, 2 and 3 selected for vegetation mapping shown in Figure 11.2

In order to evaluate the leaf production and the stem elongation in the three categories of *Q. coccifera* shrubs, 10 stems were marked in ungrazed, moderately grazed and overgrazed shrubs. The number of leaves was counted and the length of each stem measured each month.

To measure stem elongation of the three dominant woody species (*Q. coccifera, J. oxycedrus, Phillyrea* sp.), 10 stems on plants of each species were marked and the length of each stem was recorded twice a month.

In order to estimate the litterfall from the *Q. coccifera* shrubs, funnels 20 cm in diameter were placed under randomly selected shrubs.

11.4 SOIL AND EROSION PROCESSES

11.4.1 Weather conditions

In Figure 11.4b the weather conditions prevailing in Petralona from May 1992 to April 1993 are given in the form of a Klimadiagram. This type of diagram shows the mean monthly temperature and rainfall and follows Walter and Lieth (1960). In the diagram, 10°C temperature corresponds to 20 mm of rain. The temperatures in 1992, especially after July, are similar to the decade averages, while for May and June 1992 and most of 1993, values are lower by approximately 2°C.

The variance coefficient for rainfall approaches 100% in some months. This trend is illustrated in the rainfall values recorded for February 1993 (about 13 mm) while the 1981–1990 decade average is over 40. What persists however in both graphs is the general trend, with peaks in the decade averages in June and November corresponding to peaks in the values recorded in the field.

11.4.2 Soil moisture

Mean values for the soil moisture at each of the depths sampled for the wet (winter, spring) and dry (summer, autumn) periods are shown in Figures 11.6 and 11.7. During summer and autumn in the surface layers, soil moisture was at its lowest (5%) in the wheat field but was considerably higher (13%) in the maquis. However at a depth of 25 cm the wheat field was wetter (16%) compared to 14% under the maquis vegetation. During winter and spring the wheat field was wetter than the maquis site.

11.4.3 Decomposition

The decomposition rate results are reported in Table 11.1 and Figure 11.8 and show that decomposition was greater in the wheat field than under the natural vegetation. These higher values are recorded despite the fact that soil moisture was higher during that period in the maquis than in the wheat field.

11.4.4 Runoff and sediment yield

Characteristics of the runoff plots are presented in Table 11.2 and the yearly total sediment and runoff yields are shown in Figure 11.9. Runoff versus precipitation is shown in Figures 11.10a and 11.10b. The higher runoff values occurred with rainfall events of about 30 mm. Similar results are reported also by Sala and Calvo (1990).

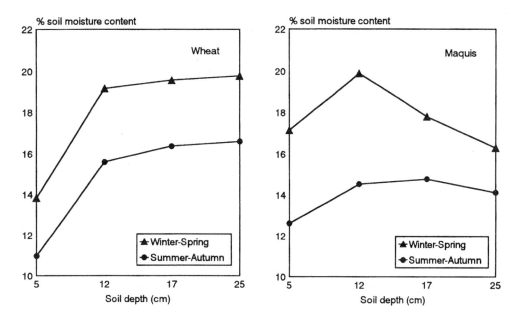

FIGURE 11.6   Soil moisture conditions during the measurement period at four depths (5, 12, 17 and 25 cm), (a) for the wheat field and (b) for the maquis

11.5 VEGETATION PROCESSES

Four woody species were dominant in the maquis, forming a heterogeneous mosaic of randomly distributed shrubs: *Quercus coccifera, Phillyrea* sp., *Juniperus oxycedrus* and *Pistacia lentiscus. Q. coccifera* dominates vegetation in the area with a mean cover of 56.6%, while the contribution of *P. lentiscus* is almost negligible. The overall vegetation cover was 67.4% with a mean vegetation height of 0.52 m.

The living leaf biomass is higher in the maquis sites than in the wheat field (Table 11.3). Nevertheless, the productivity in the wheat field is higher than in the maquis sites. In the ungrazed site the total biomass per square metre of *Q. coccifera* is higher than in the overgrazed site. The percentage contribution of leaves is the same in ungrazed and overgrazed sites and is higher in moderately grazed sites (Table 11.4). In addition, the percentage contribution of modules increases, while the percentage contribution of wood decreases with increasing grazing pressure (Papatheodorou et al., in press). From this it appears that grazing pressure affects the biomass allocation priorities of the *Q. coccifera* shrubs.

According to the results of an analysis of variance, the underground biomass of *Q. coccifera* is significantly affected by both the depth and the season, while it is independent of the grazing pressure. Thus, the root biomass is considerably higher in the upper soil layer (0–15 cm). In addition, the distribution of the fine-root biomass oscillates by statistically significantly amounts all the year round. In spring, the fine-root biomass is low, while in autumn it displays the highest values. The growth of fine-root biomass occurs in autumn when the climatic conditions are favourable (high temperature and rainfall).

Maquis sites

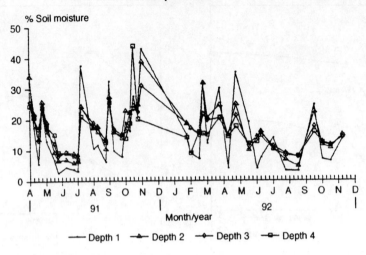

Wheat field

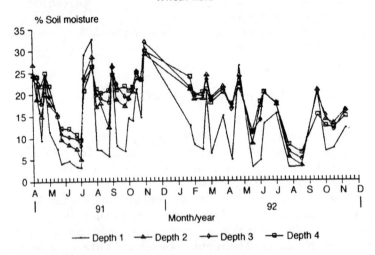

FIGURE 11.7   Gravimetric soil moisture at four depths for the maquis and wheat sites from April 1991 to December 1992

The wheat displays higher LAI during spring than *Q. coccifera* (Table 11.5). This is probably due to the different plant architecture, since wheat is characterized by a small number of long and broad leaves, while *Q. coccifera* shrubs are characterized by numerous small leaves. The ungrazed shrubs have the highest LAI in spring, the moderately grazed in summer, while the LAI in overgrazed shrubs seems to remain stable all the year round (Table 11.6). The emergence of new leaves occurs in April in the three categories of shrubs, independently of the grazing intensity (Figure 11.11).

TABLE 11.1   Cotton tensile strength loss (percentage)

Overstorey vegetation	Spring 1991	Summer 1991
Wheat (ectorganic)	73.58	73.17
Wheat (mineral)	85.08	81.00
Q. coccifera (ectorganic)	0.91	24.73
Q. coccifera (mineral)	15.29	29.55
Bare soil (ectorganic)	-2.79	48.26
Bare soil (mineral)	31.78	50.82

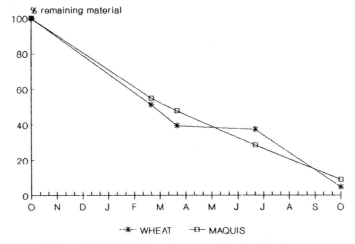

FIGURE 11.8   Cellulose decomposition in Petralona

TABLE 11.2   Vegetation cover (%) and slopes (°) of the runoff plots at Petralona, Greece

Plots	Herbaceous cover (%)	Shrub cover (%)	Slope (°)
A1	60	—	6.3
A2	100	—	7.3
B1	75	—	11.5
B2	75	—	11.5
C1	85	—	10.3
C2	70	—	10.3
D1	—	32.5	15.4
D2	—	54.5	15.4
E1	—	27.6	19.4
E2	—	30.0	19.4
F1	—	76.12	17.1
F2	—	73.15	17.1

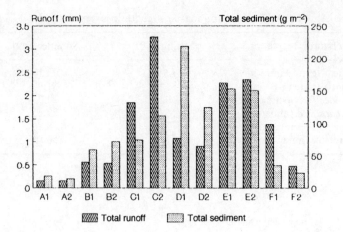

FIGURE 11.9   Total yearly runoff and sediment yield in Petralona in 1992

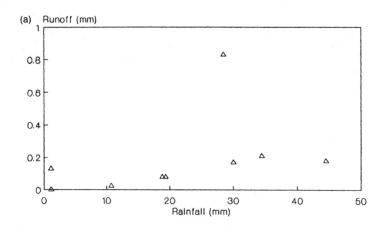

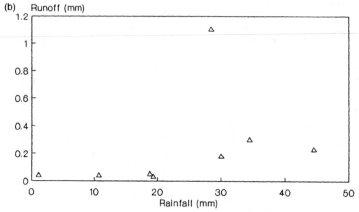

FIGURE 11.10   Runoff vs precipitation in (a) maquis and (b) a wheat field in 1992

TABLE 11.3 Above ground biomass of the runoff plots (g m^{-2}). Letters A, B and C represent wheat field sub-areas and letters D, E and F represent maquis sub-areas

Area	Living leaves		Living woods	
	Spring	Summer	Spring	Summer
A	1054.00	—	—	—
B	1055.32	—	—	—
C	1411.60	—	—	—
D	1250.1	1388.2	3134.7	4944.5
E	1852.4	2015.6	2559.8	6970.1
F	1667.2	2072.1	5345.5	7021.4

TABLE 11.4 Number of shoots and modules per square metre and dry biomass (g m^{-2}) of leaves, modules and wood at the three levels of grazing intensity in *Q. coccifera*. Percentage contribution to the total biomass is given in parentheses

	Overgrazed	Moderately grazed	Ungrazed
No. of shoots	19.6	17.6	27.3
No. of modules	1499.8	1990.0	3554.4
Leaf biomass	173.5	313.8	729.9
	(5.3)	(6.4)	(5.1)
Module biomass	760.1	1045.8	2639.4
	(23.3)	(21.8)	(18.5)
Wood biomass	2329.2	3514.5	1 0915.5
	(71.4)	(72.1)	(76.4)
Total biomass	3262.8	4874.1	1 4284.8

Nevertheless, in the ungrazed shrubs, leaves emerge and become larger during this period, while in the moderately grazed shrubs, leaves emerge at the end of spring and increase in size later in summer.

The emergence of new stems and the elongation of the old ones begins in April and the highest values of overall stem elongation are in May (Figure 11.12). The growth period of

TABLE 11.5 Leaf area index values for each sampling area during spring and summer 1991. Letters A, B, C, D, E and F represent the individual runoff plots

	Wheat field			Maquis sites		
	A	B	C	D	E	F
Spring 1991	3.85	6.15	4.68	4.11	2.82	3.27
Summer 1991	—	—	—	2.91	5.31	3.22

TABLE 11.6   Leaf area index of ungrazed moderately grazed and overgrazed *Q. coccifera* shrubs

	Autumn	Winter	Spring	Summer
Ungrazed	9.20	9.34	12.25	10.41
Moderately grazed	2.78	2.96	3.81	6.55
Overgrazed	2.33	2.51	3.73	3.84

Q. coccifera, and *J. oxycedrus* lasts from April to September, while growth of *Phillyrea* sp. ends in July. The phenology of *Q. coccifera*, *J. oxycedrus* and *Phillyrea* sp. shrubs follows the typical pattern exhibited by shrubby vegetation in the Greek Mediterranean ecosystems. Growth occurs in late spring and early summer when climatic conditions are favourable. Grazing pressure does not affect the temporal pattern of growth of *Q. coccifera*, but it affects stem elongation and leaf production (Figure 11.13). Thus the ungrazed stems display the higher values of biomass, while the moderately grazed have the higher percentage of leaf biomass. Apparently severe grazing has a negative effect on the growth of stems, while moderate grazing seems to stimulate the growth of stems as well as the production of new leaves.

Litterfall occurs all the year round but is highest in spring and summer (Figure 11.14). Thus, leaf fall occurs during the same period as leaf emergence. A hypothesis concerning nutrients translocation will be examined in the future.

11.6 CONCLUSIONS

New biomass in the maquis represents about 20% of that produced in the wheat field. This finding is in accordance with Lewis (1969, cited by Bartolome, 1993) that primary productivity of the rangelands is lower than croplands or forestlands.

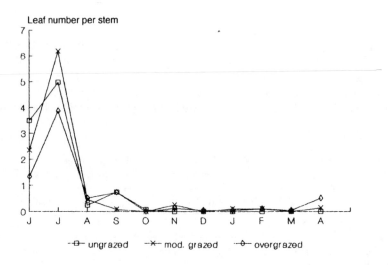

FIGURE 11.11   Leaf emergence in *Q. coccifera* in 1991/92

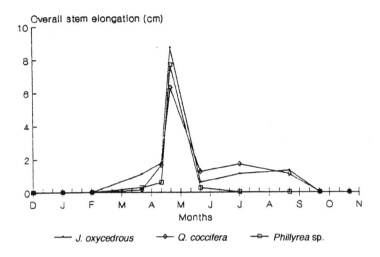

FIGURE 11.12   New stem production in the dominant shrubs in 1992

Evergreen shrubs in the natural state form a thick interwoven fabric, leaving just a few openings in places where there are outcrops of parent rock. This structure obviously keeps out grazing animals, unlike other ecosystems which seem made to be grazed, like tropical savannahs and temperate grasslands. If the maquis-covered areas were to be used for other purposes, such as cereal cultivation, a complete destruction of the shrubby vegetation cover would be required. The evergreen broad-leaved Mediterranean type shrublands are compatible with only minor use.

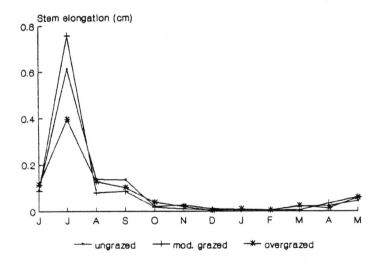

FIGURE 11.13   Stem elongation in *Q. coccifera* in 1991/92

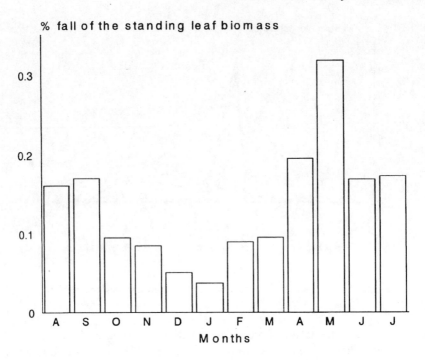

FIGURE 11.14   Litter production as a percentage of leaf standing biomass in *Q. coccifera* in 1991/92

In cases where grazing results in shrubs being left, the species persisting are the most resistant ones such as *Quercus coccifera*, while other woody species are eliminated. In the openings some unpalatable perennial or annual grasses and herbs become established but, once grazing is stopped, vegetation regrowth is vigorous. Grazing does not seem to affect phenological events in the *Q. coccifera* shrubs (Papatheodorou et al., 1993).

The incompatibility between the evergreen broad-leaved maquis and heavy human use, leads to its destruction when income-generating activities are introduced. The replacement of the natural vegetation leads to lower organic matter content, a dry soil surface which is prone to erosion and finally to a desertified landscape. Desertification in Greece can therefore be attributed to the complete incompatibility between natural ecosystems and human use. Whether the income gained by rural populations under present land use makes any economic sense, or whether it can be substituted from other sources, is a question that cannot be answered at present.

11.7 REFERENCES

Bartolome, J.W. 1993. Application of herbivore optimization theory to rangelands of the western United States. *Ecological Application*, **3** (1), 27–27.

Kauffman, G., Kockel, F. and Mollat, H. 1976. Notes on the stratigraphic and paleogeographic position of the Svoula Formation in the Innermost Zone of the Hellenides (Northern Greece). *Bulletin de la Société Géologique de France*, **18**, 225–230.

Papatheodorou, E., Pantis, J.D. and Stamou, G.P. The effect of grazing on growth, spatial pattern and age structure of *Quercus coccifera*. *Acta Oecologia* (in press).

Sala, M. and Calvo, A. 1990. Response of four different Mediterranean vegetation types to runoff and erosion. In Thornes, J.B. (ed.) *Vegetation and Erosion*. John Wiley and Sons, Chichester.

Walter, H. and Lieth, H. 1960. *Klimdiagamm Weltatlas*. Jena Fischer.

12

The Effects of Rock Fragments on Desertification Processes in Mediterranean Environments

J. POESEN* and K. BUNTE

Laboratory for Experimental Geomorphology, Katholieke Universiteit Leuven, Belgium

12.1 INTRODUCTION

Soils containing significant amounts of rock fragments (mineral particles 2 mm or larger in diameter, including all sizes that have horizontal dimensions less than the size of a pedon, Miller and Guthrie, 1984) in their top layers, are widespread in the Mediterranean belt. Except for France these soils occupy more than 60% of the land area in the countries listed in Table 12.1 (Poesen, 1990). A combination of various factors (such as geology, climatology, topography and land use) has led to the formation of shallow soils with a high rock fragment content in the Mediterranean region. Large areas are occupied by lithologies such as limestones, producing shallow soils under Mediterranean climatic conditions. Furthermore, a high percentage of the Mediterranean area consists of irregular terrain with steep slopes. In addition, long periods of land (mis)use (extensive deforestation, intensive cultivation and overgrazing since ancient times) as well as the frequent occurrence of fires in the Mediterranean has led to the progressive formation of skeletal soils through the inability of the vegetation and soils to regenerate themselves (e.g. Kosmas and Danalatos, 1994). In some cases, a high rock fragment content in the topsoil can solely be attributed to a particular land use. For example, since the 19th century soils in southern Portugal (Baixo Alentejo) have been heavily used for wheat production (Cutileiro, 1971). This overexploitation has led to intense water and tillage erosion and has therefore increased the areal extent of lithosols (i.e. soils with hard rock occurring at less than 10 cm depth; Figure 12.1). Although the geology of the area is similar on both sides of the political border between Portugal and Spain, lithosols are absent on the Spanish side because of a different land-use history.

Global circulation models predict a 3–4°C increase in average temperature and a decrease of mean annual precipitation (up to 300 mm for some parts of the Mediterranean region) before the middle of the 21st century (Brouwer and Chadwick, 1991). Such a temperature would lead to an increasing aridity of the Mediterranean, even if the annual

---

* National Fund for Scientific Research, Belgium

Mediterranean Desertification and Land Use. Edited by C. Jane Brandt and John B. Thornes.
© 1996 by John Wiley & Sons, Ltd.

TABLE 12.1 Areal distribution of soils containing a significant amount of rock fragments in the top layers for some Mediterranean countries based on the soil map of the European Communities and CORINE data-base (CEC, 1985)

Country	Portugal (%)	Spain (%)	France (%)	Italy (%)	Greece (%)
Lithosols	21.1	2.4	3.1	4.2	36.2
Phases:					
Lithic	2.0	20.2	0.4	23.2	—
Lithic/stony	—	0	10.0	8.6	—
Lithic/gravelly	—	—	0	8.4	—
Stony	11.5	—	9.6	7.4	46.2
Stony/gravelly	—	—	—	5.0	—
Gravelly	35.2	37.8	12.7	4.4	0
Concretionary	0	0	—	4.6	—
Total (%)	69.8	60.5	35.8	65.8	82.4
Land area (1000 km^2)	92.1	504.8	544.0	301.3	132.0

Lithosols: hard rock occurs at less than 10 cm depth.
Lithic phase: continuous coherent and hard rock occurs within 50 cm of the surface.
Stony phase: presence of stones (rock fragments with a diameter larger than 7.5 cm), boulders or rock outcrops in the surface layer or at the surface.
Gravelly phase: presence of more than 35% of gravels (rock fragments with a diameter up to 7.5 cm) in the surface layer.
Concretionary phase: presence of more than 35% of pedogenic concretions (with a diameter up to 7.5 cm) in or near the surface layer.

rainfall amount were to remain the same. The higher aridity will affect the intensity of various desertification processes such as the degradation of the vegetation cover and physical degradation (surface sealing, compaction), as well as soil erosion by water (De Ploey et al., 1991). Despite their vast areal extent, soils containing considerable amounts of rock fragments have received relatively little attention with respect to desertification processes. Therefore, this chapter reviews the literature regarding field measurements and laboratory experiments of how rock fragments at the surface or in the soil profile qualitatively or quantitatively modify the intensity of the most important desertification processes in the European Mediterranean belt (degradation of the vegetative cover, physical soil degradation and soil erosion by water). Before tackling these issues the effects of rock fragments on some important processes of the hydrological cycle as well as on plant growth are discussed.

12.2 EFFECTS OF ROCK FRAGMENTS ON SOME KEY HYDROLOGICAL PROCESSES

When discussing the role of rock fragments in hydrological processes, a distinction should be made between rock fragments at the soil surface and rock fragments below the soil surface (Figure 12.2). Rock fragments resting on the soil surface or partly incorporated in the top layer affect the partitioning of rainfall into interception, rock flow (runoff generated by the rock surface), infiltration, overland flow and evaporation (Figure 12.2). Rock fragments situated below the soil surface affect the water percolation rate and thus also the infiltration rate and runoff generation.

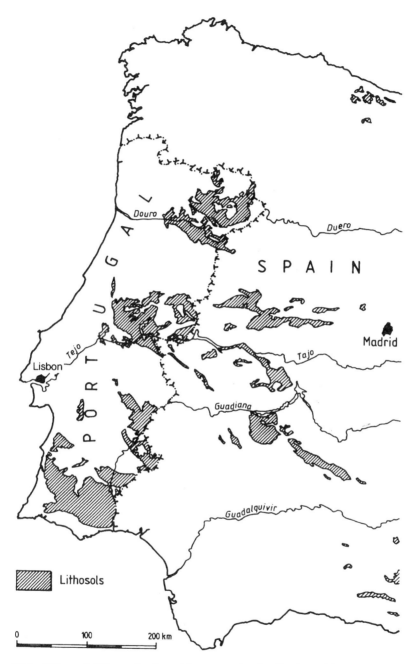

FIGURE 12.1   Effect of different land-use history on the distribution of lithosols, i.e. soils with hard rock occurring at less than 10 cm depth, along the southern Portugese–Spanish border (based on the Soil Map of the European Communities and CORINE data-base, CEC, 1985). Although the geology is identical on both sides of the political border in southern Portugal, lithosols are widespread in southeastern Portugal because of overexploitation of the soil for wheat production leading to intense water and tillage erosion, while no lithosols have been mapped in southwestern Spain

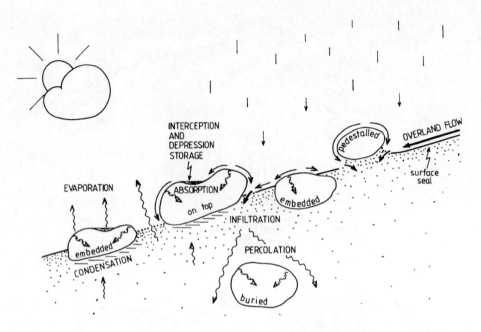

FIGURE 12.2 Sketch to illustrate possible effects of rock fragments in different vertical positions (pedestalled, on top, embedded and buried) on some key hydrological processes (interception, depression storage, absorption, infiltration, percolation, overland flow generation and evaporation)

12.2.1 Interception and storage

Rock fragments at the soil surface intercept raindrops. The volume of rainwater intercepted depends on the cover percentage and the microtopography of the rock fragments as well as on the angle of incidence of raindrops in relation to slope angle and aspect. Part of this rainwater is stored at the rock surface (depression storage), the remaining part will either penetrate the rock fragment (absorption) or will evaporate (Figure 12.2). The water-holding capacity of rock fragments (*MC*, expressed as a gravimetric moisture content at saturation) is determined by rock porosity which is a function of rock type and weathering state. While flintstone, for instance, has a low *MC* of 0.2%, it can be as high as 91.7% for chalk (Poesen and Lavee, 1994). Since smaller rock fragments are usually more weathered and hence more porous than larger fragments in natural soils (Childs and Flint, 1990), smaller rock fragments can absorb larger water quantities per unit of rock mass than larger ones (Poesen and Lavee, 1994). Once rock fragments at the soil surface are saturated, rainfall excess will flow from the rock surfaces as rock flow. Rock flow discharge is then determined by rainfall intensity, fragment size (Poesen and Lavee, 1991) and inclination of the rock surface.

12.2.2 Infiltration

A rock fragment cover can affect water infiltration rate in different ways. On the one hand, rocks prevent direct infiltration into the soil of the rain they intercept. On the other

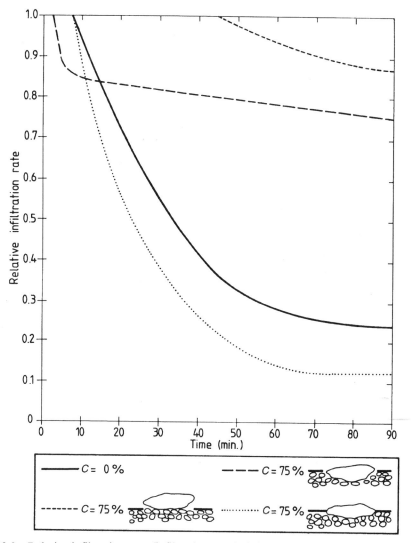

FIGURE 12.3   Relative infiltration rate (infiltration rate/rainfall intensity = 36 mm/h) versus time for a bare soil surface (C=0%) and soil surfaces covered with rock fragments (C=75%) having different vertical positions in the soil top layer: i.e. 'on top', 'embedded in a well structured top layer' and 'embedded in a completely sealed surface'. C = rock fragment cover. The soil is a silt loam which had initially a fine seedbed structure (median of dry aggregate size distribution ranges between 0.5 and 1.0 cm). After Poesen and Ingelmo-Sanchez (1992)

hand, however, rock fragments increase water intake rates by protecting the soil surface against raindrop impact forces which prevents soil surface sealing. Whether the total volume of infiltration is finally increased or decreased by the presence of rock fragments depends on various factors such as position, size and cover of rock fragments, as well as structure of the fine earth (Yair and Lavee, 1976; Poesen, 1986; Poesen et al., 1990, Poesen and Lavee, 1991; Poesen and Ingelmo-Sanchez, 1992).

Figure 12.3 (after Poesen and Ingelmo-Sanchez, 1992) illustrates how different rock fragment positions and topsoil structure affect relative infiltration rate into a silt loam soil that is very susceptible to surface sealing. Highest infiltration rates occur where the rock fragments lie on top of the soil surface. Rock fragments embedded in a well-structured top layer cause somewhat lower infiltration rates, though infiltration is on average still more rapid than on a bare soil surface. However, when rock fragments are well embedded in a completely sealed topsoil, infiltration rates are less than those for a bare soil. The effects of various rock fragment positions on infiltration rates were found to increase with rock fragment cover (Poesen et al., 1990; Poesen and Ingelmo-Sanchez, 1992).

Poesen and Lavee (1991) studied the effect of rock fragment size on infiltration rates. For a given rock cover, infiltration rates on a loamy sand soil are highest when covered by small rock fragments, because large rock fragments produce larger rock flow discharges per unit of rock fragment perimeter than smaller ones. During high intensity storms, this rock flow discharge can exceed the infiltration capacity of the fine earth surrounding the rock fragments. This produces overland flow and a lower overall infiltration rate. Surfaces with smaller rock fragments have a reduced potential for this mechanism to operate. The effects of various rock fragment sizes on infiltration rates increase with increasing rock fragment cover.

12.2.3 Runoff production

Because rock fragments have a variety of contrasting effects on infiltration, overland flow generation on soils covered by rock fragments is quite complex. Nevertheless, some trends can be indicated. Poesen and Ingelmo-Sanchez (1992) proposed a structural model to explain discrepancies in the relations between rock fragment cover and runoff production from field plots. If rock fragments are well incorporated in a sealed top layer (i.e. a top layer with only textural porosity), rock fragment cover reduces the infiltration rate and increases runoff production. On the other hand, if rock fragments rest on the soil surface or if they are partly incorporated in a soil top layer having structural pores, a rock fragment cover will increase the infiltration rate and, hence, reduce runoff coefficients.

The effects of a rock fragment cover on runoff production for a given soil can change with time because topsoil structures are highly dynamic. A switch from a negative to a positive effect occurs in the case of physical degradation (such as that due to land abandonment) while topsoil cultivation, for example, can lead to an instantaneous reduction of runoff production. Another factor that makes it difficult to predict the effects of rock fragments on runoff production is the high spatial variation of rock fragment content, size and position in undisturbed and cultivated soils. A high runoff discharge generated on areas with embedded, large rock fragments may flow towards areas with either smaller rock fragments or with rock fragments resting on the soil surface where higher infiltration rates reduce runoff discharge again. Hence, during a rainfall event a variety of contrasting effects may take place at the soil surface over short distances leading to considerable compensating effects on runoff production.

12.2.4 Percolation

Gras (1972b) investigated the effects of simulated porous and non-porous rock fragments with various shapes and sizes, on water transmission in a sand layer. He concluded that non-porous rock fragments reduced permeability and that the reduction was proportional to the rock fragment cross-sectional area perpendicular to the flow direction. Since then, various equations have been proposed to predict the negative effects of rock fragment content on saturated hydraulic conductivity. Peck and Watson (1979) proposed the following equation for rock fragments with negligible hydraulic conductivity

$$K_x = 2(1 - R_v)K_{fe}/(2 + R_v) \tag{1}$$

while Brakensiek et al. (1986) found the following equation using published experimental data

$$K_x = (1 - R_m)K_{fe} \tag{2}$$

where K_x = saturated hydraulic conductivity of a soil containing $x\%$ rock fragments by volume or by mass; K_{fe} = saturated hydraulic conductivity of the fine earth matrix; R_v = rock fragment content by volume (fraction of total volume); and R_m = rock fragment content by mass (fraction of total mass).

Both equations, only valid for soils with textural porosity (Poesen and Ruiz-Flano, in preperation), predict a decrease of K_x with increasing R_v or R_m. Therefore, an increasing rock fragment content in a topsoil matrix with essentially textural pores also increases the potential for runoff generation. In the case of soils with structural porosity however, the opposite effect of rock fragment content on K_x is to be expected (Poesen and Ruiz-Flano, in preperation). For such soils, rock fragment content has a negative effect on bulk density of the fine earth (BD_{fe} = mass of fine earth/(volume of fine earth + volume of textural and structural pores), see Figure 12.4) and hence a positive effect on macroporosity. Rock fragments in these soils are likely to enhance by-pass flow or short-circuiting if macropores are interconnected.

Rock fragments below the soil surface also affect soil water distribution within the profile. Hillel and Tadmor (1962) observed in the Negev highlands of Israel that a given quantity of water penetrates to a greater depth in rocky soils compared with non-rocky soils of similar fine earth texture, provided that the rock fragments themselves absorb only small or negligible quantities of water. This mechanism is particularly important in arid regions where the deeper penetration of a limited amount of precipitation reduces water loss by evaporation (Munn et al., 1987).

12.2.5 Evaporation

A layer of rock fragments at the soil surface acts as a mulch. It may change the soil's radiation balance, water balance and temperature regime among other factors. Various studies have demonstrated the reduction in evaporation rate by a surface layer of rock fragments (e.g. Corey and Kemper, 1968; Unger, 1971; Ingelmo-Sanchez et al., 1980). The reduced evaporation rate was attributed to the very low unsaturated hydraulic conductivity of a pure rock fragment layer at low suctions. This decreases the amount of water that can be transported to the surface by capillary rise. As a result, water transport upward through the rock fragment layer is by vapour diffusion only. Because this quantity is usually low, total evaporation losses are correspondingly low. As with other mulch materials, a rock fragment cover has its greatest inhibiting effect upon evaporation during the first few days after a rainfall event. The long-term beneficial effect of such a mulch depends largely upon the frequency and amount of rainfall. A rock mulch is less effective in climatic conditions where only very small amounts of rain occur at long intervals, because most of this water may evaporate irrespective of the mulch (Corey and Kemper, 1968).

12.3 EFFECTS OF ROCK FRAGMENTS ON PLANT GROWTH

12.3.1 Effects of rock fragments on soil properties affecting plant growth

Before discussing the effects of rock fragments on plant growth, we review the literature concerned with the effects of rock fragments on soil properties important for plant growth. When assessing the effects of rock fragments on soil quality, the prevailing assumption is that rock fragments do not improve the soil quality significantly because they decrease the total volume of fine earth (i.e. the effective soil). This is reflected by the concept of effective soil depth (Childs and Flint, 1990):

$$D_{eff} = D_{tot}(1 - R_v) \tag{3}$$

where D_{eff} = effective soil depth, D_{tot} = total soil depth, and R_v = rock fragment content by volume.

Though in some instances this concept may be valid, it is questionable whether a decreased effective soil depth is always deleterious to plant growth. Rock fragments influence processes in a manner which can be beneficial to plant growth. Gras (1972a) observed in an area of the Rhône valley, where annual evaporation exceeds annual precipitation by 250 mm, that no significant difference in peach tree productivity occurred between sites with no rock fragments and sites having up to 60% of rock fragments by mass. He referred to similar observations in other regions in France that also have soils containing significant amounts of rock fragments. From this striking observation Gras (1972a) concluded that rock fragments are not simply inert material diluting the fine earth volume but that rock fragments contribute to the water-holding capacity of a soil.

Various researchers, listed by Poesen and Lavee (1994), have reported that the water contained in rock fragments can contribute to plant available water. Available water for plants is expressed by available water capacity (AWC, gravimetric %) and represents the mass of water stored between a 'field capacity' value and a dry or 'unavailable water' volume. *AWC* of rock fragments varies with rock type, degree of weathering and rock fragment size. *AWC* can range between 0% for flintstone and 91–124% for pumice (Poesen and Lavee, 1994). *AWC* usually increases with decreasing rock fragment size. Hence, although rock fragments reduce the overall water-holding capacity of the soil, they can still contribute to available soil water because of the water contained within the rocks.

Additionally, rock fragments in soils have also been reported to contribute significantly to the nutrient content and cation exchange capacity (Munn et al., 1987). This is noteworthy given the current view which attributes the cation exchange capacity of soils to their clay and organic matter fractions. Rock fragments not only directly provide the fine earth with matter dissolved from the rocks by weathering, but rock fragments also affect the constituents of the fine earth in an indirect way. Again, the negative effect of rock fragments on 'effective soil depth' is counteracted by benevolent effects: inputs to the soil (such as decaying organic matter, fertilizer, water) are concentrated in the fine earth fraction (Childs and Flint, 1990). This then will affect other soil properties, soil development and soil productivity. For instance, rock fragment content affects fine earth bulk density (Figure 12.4) and hence macroporosity, which, in turn, improves root development.

One has to conclude that the rock component of soils may contribute, in varying degrees, to those soil properties conducive to plant growth.

12.3.2 Effects of rock fragment content on plant growth

Few systematic studies have been devoted to the effects of rock fragments in the surface and subsurface soil horizons on plant productivity. The scarce results indicate that the effects of rock fragment content on plant growth are quite complex and vary with fine earth properties, vegetation type and climate.

Fine earth properties

Lutz and Chandler (1946) reported that 'reasonable amounts' of rock fragments in heavy-textured soils have to be regarded as favourable for tree growth, while in sandy soils rock fragments appear to have an unfavourable effect. Babalola and Lal (1977b) found that the inhibitory effect of rock fragments on root development of maize (*Zea mays* L.) seedlings was more pronounced for sandy fine earth than for a sandy loam or a clay fine earth.

Vegetation type

A negative effect of rock fragment content on plant growth has been attributed to the decrease in soil volume for nutrient supply, which in turn decreases the productivity of forest soils (Childs and Flint, 1990). Voiculescu et al. (1983) found that walnut growth in Romania was restricted in soils containing more than 20% rock fragments. Research on irrigated reclaimed sites in Colorado (USA) showed a negative effect of rock fragment content on grass production (Munn et al., 1987). Babalola and Lal (1977b) observed root development of maize seedlings to be adversely affected by rock fragments when rock fragment content by mass (R_m) exceeded 10–20%. Adams (1967) found that a rock mulch in Texas increased the temperature next to plant stems enough to adversely affect sorghum yields.

Several studies have also demonstrated a positive effect of rock fragment content on soil productivity. Experiments conducted by Jackson et al. (1972) indicate that soil associations with a high rock fragment content provide the conditions most favourable to the emergence, growth and development of blueberry seedlings (*Vaccinium angustifolium* Ait.) because in these soils a greater amount of root branching was possible than in similar rock-free soil. A study on the effects of rock fragment content on reclaimed sites in Colorado on the growth of shrubs, forbs and grasses revealed that the highest productivity of fourwing saltbush (*Atriplex canescens*) occurred on a site with 85% rocks in the surface horizon (Munn et al., 1987). Albaladejo (1990) reported for southeastern Spain that thyme bushes (*Thymus*) prefer soils with a high rock fragment content in the top layer. The beneficial effects of rock mulches on field crops such as sorghum (*Sorghum vulgare* and *Sorghum bicolor* L.M.), tomatoes (*Lycopersicon* sp.) and soybean (*Glycine max* L.M.) was attributed both to soil water conservation and to higher soil temperatures that promoted early season growth (Unger, 1971; Fairbourn, 1973). The amenable soil moisture and temperature regime found in soils containing rock fragments is also known to produce high-quality grapes for wine-making (Seguin, 1971).

Contrasting effects of rock fragment content on plant productivity have also been reported by a number of scientists. Rutherford (1983) found that surface rock cover in South Africa could increase or decrease herbaceous standing crop in non-wooded parts of savannah and in desert grassland depending on subsurface rock weathering patterns. Some researchers observed a positive effect of rock fragment content on plant growth up to an optimal rock fragment content (*R*.OPT). For rock fragment contents larger than the *R*.OPT-value the trend was reversed. Wollny (1897–98) was the first to report that R_v.OPT (expressed as a volume %) equalled 10–20% for a number of crops (cereals, root crops, vegetables) in southern Germany. He attributed the ambivalent effects of rock fragment content on productivity to the soil moisture and the soil temperature regime. Lutz and Chandler (1946) reported a R_v.OPT-value of 20% for tree growth. Above this value unfavourable effects such as restricted root space, excessive temperature extremes and decreased field capacity of the soil body began to outweigh the favourable ones. Saini and Grant (1980) found a R_v.OPT-value of 12% for potato (*Solanum tuberosum* L.) yield. The positive effect of rock fragment content on soil productivity below this value was attributed to higher soil temperature, higher soil moisture content, less soil compaction and reduced water erosion. Babalola and Lal (1977a, b) concluded from their experiments that R_m.OPT (expressed as mass %) equalled 10–20% for the growth and development of maize roots. Magier and Ravina (1984) reported a R_v.OPT-value of 25–30% for apple tree (*Pyrus Malus*) development and yield in Israel. Finally, Gras (1972a) did not observe any significant effect of rock fragment content ($0\% < R_m < 60\%$) in soils on the growth of peach trees in southern France.

Climate

From an extensive literature review, Munn et al. (1987) concluded that rock fragment content effects on soil productivity varied along a moisture gradient from humid to arid climates. In humid climates plant productivity is generally higher on fine-textured, non-rocky soils. As precipitation declines, the relationship generally continues to hold, but the difference in productivity potential diminishes and is eventually reversed in areas of very low precipitation (< 300 mm). In arid and semi-arid regions coarse-textured soils and soils containing rock fragments in surface horizons produce deeper penetration of limited precipitation. As a consequence, these soils are often more productive than finer textured soils in comparable upland topographic positions. Kadmon et al. (1989) reported a positive effect of rock fragment content on the growth of woody perennials in the northern Negev (170 mm annual rainfall) and attributed this to the favourable effects of rock fragment content on water availability. Similar results were recently reported by Kosmas et al. (1994) in Greece. In a wet year, with 725 mm of rain during the growing season, they observed that biomass production of rain-fed wheat along catenas on shale–sandstone soils containing 40 to 65% of rock fragments was only 60 to 80% of the biomass production on marl soils which had no rock fragments at all. However, in a dry year, with only 95 mm of rain during the growing season, biomass production on the shale–sandstone soils was 5 to 10 times the biomass production on marl soils. The different behaviour of these two soil types could be attributed, at least partly, to differences in rock fragment content. Along the same lines, Yair and Shachak (1987) came to the conclusion that in arid areas an increased ratio of bare bedrock outcrop to soil cover decreases the ecological aridity of the area. Rocky slopes were found to maintain a more favourable environment for plant growth than non-rocky slopes because rock outcrops increased runoff frequency and magnitude (and therefore water availability) as well as soil desalination in rock-free patches.

From this review we conclude that the relation between rock fragment content and plant productivity is complex. Some general statements, however, can be made.

1. Rock fragments seem to be more beneficial for plant growth in clay soils than in sandy soils.
2. Some shrubby deep-rooting plants seem to be better adapted to soils containing rock fragments than shallow-rooting grassy plants or some trees.
3. Moderate rock fragment contents can beneficially affect the moisture and temperature regime of soils. However, beyond an optimal rock fragment content varying between 10 and 30%, the abundance of rock fragments begins to adversely affect plant productivity by restricting rooting space and the nutritional capacity of the soil and by increasing soil temperature extremes above plant-tolerable values.
4. In dry climatic conditions rock fragments seem to create favourable conditions for plant growth. Consequently, degradation of the vegetative cover due to climatic change might be less severe on soils rich in rock fragments compared to rock-free soils.

12.4 EFFECTS OF ROCK FRAGMENTS ON PHYSICAL DEGRADATION

Physical degradation of the soil refers to adverse changes in soil physical properties, including porosity, permeability, bulk density and structural stability (FAO, 1979). The most important degradational processes are surface sealing, crusting and compaction.

12.4.1 Surface rock fragments

Surface rock fragments protect the soil against aggregate breakdown, surface sealing and crusting in so far as these processes result from physical dispersion of soil aggregates by raindrop impact and filtration of dispersed soil particles transported in infiltrating water. Rock fragments tend to increase infiltration rate as shown in Figure 12.3 (after Poesen and Ingelmo-Sanchez, 1992), which leads to a decrease in runoff rate. On the other hand, a removal of rock fragments from upland areas has been shown to increase runoff (Grant and Struchtemeyer, 1959; Jung, 1960; Evenari et al., 1982). However, when rock fragments are well embedded in a seal or crust, they can even enhance sealing and activate runoff (see Section 12.2.3) and interrill sediment production (see Section 12.5.2) (Poesen, 1986; Poesen et al., 1990; Poesen and Ingelmo-Sanchez, 1992). If sealing and crusting result from essentially chemical dispersion, it is expected that the effects of rock fragments on these processes of physical degradation will be less important.

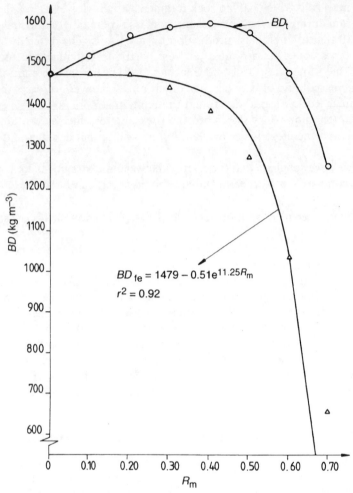

FIGURE 12.4   Relation between limestone rock fragment content by mass (R_m) in a Brown Soil developed on a river terrace (Ebro basin, agricultural land) and total bulk density (BD_t) as well as fine earth bulk density (BD_{fe}). Data on BD_{fe} were extracted from a scatter diagram published by Alberto (1971). BD_t was calculated as $BD_t = 1/[(1 - R_m)/BD_{fe} + R_m/BD_{rf}]$, where BD_{rf} = bulk density of the limestone rock fragments ($= 2080\ kg/m^3$)

12.4.2 Subsurface rock fragments

Two density values are commonly required for the interpretation of the physical behaviour of a soil containing rock fragments: total bulk density of the soil (BD_t) and bulk density of the fine earth (BD_{fe} = mass of fine earth/(total volume − volume of rock fragments) or BD_{fe} = mass of fine earth/(volume of fine earth + volume of textural and structural pores)). Figure 12.4 illustrates the relation between rock fragment content by mass (R_m) and BD_t as well as between R_m and BD_{fe}. With increasing rock fragment content, total bulk density increases to reach a maximum at 40% beyond which it decreases. In contrast to BD_t, BD_{fe} decreases monotonically with increasing R_m following the equation:

$$BD_{fe} = a - b(e^{cR_m})$$ (4)

where a, b and c are coefficients.

There are a number of possible reasons why a negative relationship between BD_{fe} and R_m might occur.

1. Insufficient fine earth to fill the voids in between the rock fragments at high rock fragment contents.
2. In a mixture of two particle size grades the smaller particles cannot pack as closely to the larger particles as they can with each other (Stewart et al., 1970).
3. The presence of rocks in the soil changes the nature of the fine earth fraction. Decaying organic matter, fertilizer inputs, rainwater, etc. become concentrated in a decreasing mass of fine earth when rock fragment content increases (Childs and Flint, 1990). This affects other soil properties such as soil structure. An increase in organic matter content of the fine earth fraction (corresponding to an increase in R_m) will lead to a decrease of the BD_{fe} value since the average bulk density of organic matter equals 0.224 g/cm^3 (Rawls, 1983). In addition to this effect, an increase in organic matter content may also lead to a higher porosity and a more stable structure of the fine earth fraction.

Figure 12.4 illustrates that even at high total soil bulk density, according to traditional standards, fine soil bulk densities are not excessive. This has important implications for plant growth: if plant growth is related to the soil physical properties of the fine soil fraction, high total soil bulk density in soils rich in rock fragments does not necessarily indicate a poor root-growth environment.

Rock fragments below the soil surface support the existing soil structure and, hence, diminish the susceptibility of the soil to compaction (compactibility). Saini and Grant (1980) reported that, when applying a dynamic load to a loamy soil, the presence of rock fragments reduced compaction of the fine earth. For a given rock fragment content, the smallest fragments were most effective in reducing compactibility of the fine earth fraction. Ravina and Magier (1984) found in laboratory experiments that the rock fragment content in clay soils had a positive effect on the resistance of the soil to compaction. Larger volumes of large pores were found after compaction of soils containing increased rock fragments. These results are along the same lines as the relation between rock fragment content and bulk density of the fine earth fraction (see Figure 12.4).

In conclusion, rock fragments help preserve favourable soil structures either at the soil surface by acting as a mulch or in the soil by acting as a skeleton which prevents the soil from being compacted. Hence, the presence of rock fragments tends to reduce the intensity of physical degradation in fine-textured soils.

12.5 EFFECTS OF SURFACE ROCK FRAGMENTS ON SOIL EROSION BY WATER

Poesen et al. (1994) reviewed the various effects of rock fragments on soil erosion by water and found these effects to be dependent on the size of the observation area. These effects

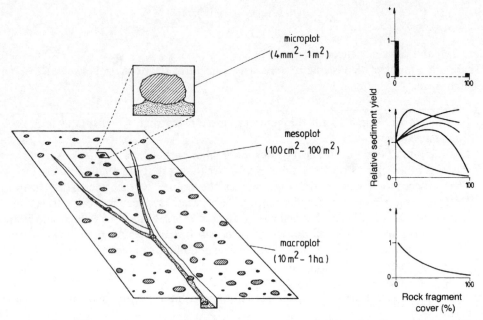

FIGURE 12.5   Effects of surface rock fragment cover on soil erosion by water (expressed by sediment yield relative to that of a bare, uncovered soil surface) at three different nested spatial scales (after Poesen et al. 1994). At the microplot scale, relative sediment yield reaches a maximum value inbetween individual rock fragments (0% rock fragment cover), and becomes zero below a rock fragment (100% cover). At the mesoplot scale (i.e. interrill areas), negative, positive and even convex upward relationships have been observed. Finally, at the macroplot scale (i.e. interrill and rill areas), relative sediment yield decreases exponentially with rock fragment cover

and the corresponding process mechanisms will be discussed at three nested spatial scales: the micro-, the meso- and the macroplot (Figure 12.5). Finally, some observations on the spatial variability of rock fragment cover at the megaplot scale will be reported and the implication for soil erosion rates along catenas in the Mediterranean belt will be discussed.

12.5.1 The microplot scale

The microplot is defined as an interrill soil surface which is covered by a single rock fragment and its area usually varies between 4×10^{-6} and $10^0 \, \text{m}^2$. Obviously, a rock fragment will protect the underlying soil surface against erosion by raindrop impact and surface flow. The mass of sediment detached by raindrop impact on a bare interrill soil surface which is partly covered by rock fragments can be estimated by a modified splash detachment model (Poesen, 1985):

$$SD = (1 - R_c)KE \, R^{-1} \tag{5}$$

where $SD =$ mass of sediment being detached per unit area and per unit time ($\text{kg} \, \text{m}^{-2}$ time^{-1}); $R_c =$ rock fragment cover (fraction of total surface); $KE =$ kinetic rainfall energy per unit area and per unit time $(J\,m^{-2}\,time^{-1})$; and R = resistance of bare fine earth soil surface to detachment $(J\,kg^{-1})$ (Poesen, 1985; Poesen and Torri, 1988).

Due to the protective effect of rock fragments at the microplot scale, splash pedestals may develop by the combined action of raindrop splash and rainwash (see Figure 12.2). Figure 12.5 illustrates this protective effect. Compared with an uncovered microplot (cover = 0%) having a sediment yield equalling unity, a microplot covered by a rock fragment (cover = 100%) will have a zero sediment yield.

12.5.2 The mesoplot scale

The mesoplot scale is defined by the size of an interrill area (Figure 12.5) which typically ranges between 10^{-2} and $10^2\,m^2$. Poesen et al. (1994) reported positive as well as negative relations between rock fragment cover and interrill sediment yield depending on the fine earth structure, on the vertical position and size of rock fragments and on surface slope, as well as on the occurrence of horseshoe-shaped vortex erosion (Bunte and Poesen, 1993, 1994).

Poesen and Ingelmo-Sanchez (1992) found a positive effect on interrill erosion for rock fragments well-embedded in a surface seal (i.e. a top layer with textural porosity). The surface seal reduces infiltration rate and does so more effectively with increasing cover of embedded rock fragments (see Section 12.2.2). This leads to increasingly reticular flow (Thornes et al., 1990) which becomes deeper and/or faster as the flow concentrates between the rock fragments. This flow concentration increases soil detachment and transporting capacity and sediment yields are high despite the reduced sediment supply by raindrop detachment. For rock fragments resting on the soil surface or partly embedded in a topsoil with structural porosity, Poesen and Ingelmo-Sanchez (1992) observed a negative exponential relationship between rock fragment cover and sediment yield. The effects of

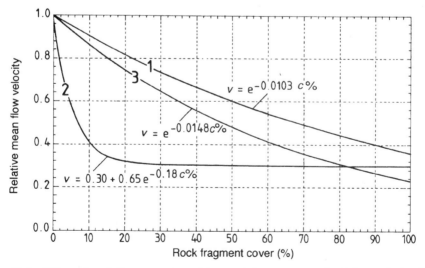

FIGURE 12.6 Effect of rock fragment cover (c) on relative mean overland flow velocity (v) for (1) cobble covers ($D50 = 8.6\,cm$) (after Bunte and Poesen 1994); (2) pebble covers ($D50 = 1.5\,cm$) (after Poesen et al. 1994); and (3) pebble covers ($0.6\,cm < D < 3.8\,cm$) (after Poesen et al. 1994)

rock fragment size on interrill sediment yield were studied by Poesen and Lavee (1991) who reported that large rock fragments were less efficient in reducing interrill soil loss than smaller ones. De Ploey (1981) observed an exponential decay of interrill sediment yield with increasing rock fragment cover on a 3.5% slope. For a steep 28.7% slope, however, interrill sediment yield was non-monotonically related to rock fragment cover. Bunte and Poesen (1994) investigated interrill sediment yield due to afterflow on a highly erodible sandy loam and its relation with rock fragment size and cover. Although an increasing cover of both large cobbles and small pebbles decreased average flow velocity (Figure 12.6), relative sediment yield was, for all rock fragment covers, still higher than for the same flow over a bare soil surface. This was attributed to local vortex erosion induced by the presence of rock fragments.

12.5.3 The macroplot scale

The macroplot scale denotes the size of upland areas where both interrill and rill erosion and, eventually, (ephemeral) gully erosion takes place (Figure 12.5). Its size typically ranges between 10^1 and $10^4 \, \text{m}^2$. An analysis of published data revealed an overall negative effect of rock fragment cover on relative interrill and rill sediment yield. For rock fragment covers exceeding 10% the mean trend could be represented by (Poesen et al., 1994):

$$IRR = e^{-0.04(R_c - 10)} \tag{6}$$

where IRR = mean relative interrill and rill sediment yield; R_c = rock fragment cover (%) with $10\% < R_c < 100\%$; and 0.04 = mean rate of decay of IRR with increasing R_c.

The overall negative effect of rock fragment cover on IRR is attributed to soil erosion reduction (sub)processes listed below that usually override those (sub)processes that promote soil erosion at the mesoplot scale (Poesen et al., 1994):

— protection against raindrop and flow detachment (see Section 12.5.1);
— reduction of topsoil physical degradation leading to a decrease of runoff production (see Section 12.4);
— retardation of overland flow velocity (see Figure 12.6) which leads to a reduction of the detaching and transporting capacity of the overland flow.

Perhaps more important than the mean trend is the scatter of the data (Poesen et al., 1994). This indicates that a given rock fragment cover can reduce interrill and rill sediment yield to various degrees depending on other site factors. These factors control the relative contributions of the various subprocesses that affect interrill and rill sediment yields.

At the macroplot scale, rock fragments in the soil top layer reduce not only interrill and rill erosion, but also gully erosion. Donker and Damen (1984) observed that gully erosion risk in northern Spain (Zaragoza province) decreased when the rock fragment content of the topsoil increased. They attributed this to several mechanisms, explained above, and to the depletion of fine earth in the top layer over time by selective erosion, resulting in a 'surface armour' (erosion pavement).

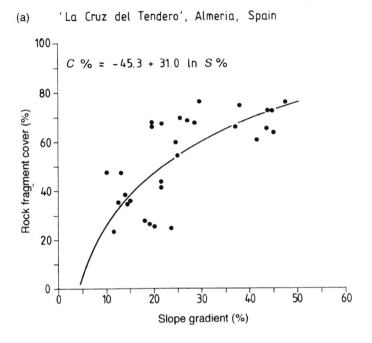

(a) 'La Cruz del Tendero', Almeria, Spain

$C\% = -45.3 + 31.0 \ln S\%$

Rock fragment cover (%)

Slope gradient (%)

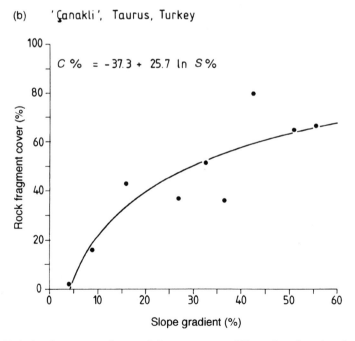

(b) 'Çanakli', Taurus, Turkey

$C\% = -37.3 + 25.7 \ln S\%$

Rock fragment cover (%)

Slope gradient (%)

FIGURE 12.7 Relation between surface rock fragment cover (*C*) on abandoned agricultural land and local slope gradient (*S*) for two catenas in the Mediterranean belt: (a) a southerly facing catena on micashist bedrock (SE Spain), (b) a northerly facing catena on limestone bedrock (SW Turkey). Both based on unpublished data collected by the authors

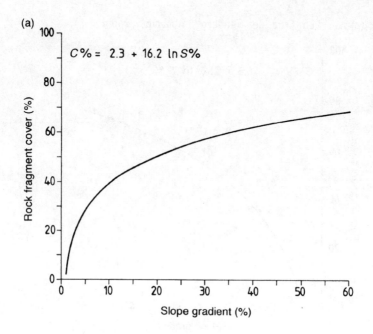

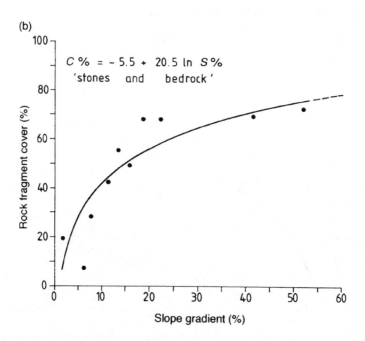

FIGURE 12.8   Relation between surface rock fragment cover (*C*) on natural land and local slope gradient (*S*) for (a) a semi-arid area (Walnut Gulch, Arizona, USA; after Simanton et al., 1994) and (b) an arid area (Avdat, northern Negev, Israel; data extracted from Lee, 1988)

12.5.4 The megaplot scale

The megaplot scale is defined as an area encompassing a complete catena and its size usually exceeds $10^4 \, m^2$. Field observations in the Mediterranean belt reveal that very often rock fragment cover on abandoned agricultural land increases non-linearly with hillslope gradient along catenas. Figure 12.7 indicates a logarithmic increase of surface rock fragment cover with surface slope for two catenas sampled in southeast-Spain and in southwest Turkey respectively. These relations are very similar to those found for catenas in semi-arid southwest USA (Simanton et al., 1994) and in arid southern Israel (Lee, 1988) (Figure 12.8).

The decreasing rock fragment cover with decreasing hillslope gradient can be explained as follows. The selective removal of fine earth material on the steeper slope sections leads to a coarsening of the soil surface. The deposition of these fines occurs in downslope positions along the catena and causes a fining of the surface material. Also, smaller rock fragments themselves can be eroded from upslope positions by concentrated overland flow in rills and in gullies (Poesen, 1987), and transported and redistributed over footslopes that usually receive fine material.

The implication of this systematic spatial variability of rock fragment cover on abandoned agricultural land along Mediterranean catenas is that soil erosion intensity will not necessarily be the highest on the steepest slope sections, since these are covered by a high rock fragment cover. At the macroplot scale (see Section 12.5.3), rock fragment cover will reduce soil loss exponentially. Consequently, the high rock fragment cover on steep slope sections will counteract the positive effect of hillslope gradient on soil erosion rate. Catenas in central Europe where soils contain little or no rock fragments have monotonically increasing sediment yield (power function) with hillslope gradient. It is highly probable that for sparsely vegetated catenas in the Mediterranean belt where rock fragment cover increases with hillslope gradient that the hillslope gradient–sediment yield relation will be different and can even be convex upward, as shown for debris-covered hillslopes in Arizona (Abrahams and Parsons, 1991). The relation between hillslope gradient and sediment yield must therefore be reconsidered for most Mediterranean environments.

In conclusion, at the mesoscale (i.e. on interrills) rock fragments at the soil surface can have negative as well as positive effects on soil erosion intensity. At the microscale and at the macroscale (upland areas where both interrill, rill and (ephemeral) gully erosion takes place), however, rock fragments at the soil surface were found to have a negative effect on soil erosion rate and must therefore be considered as natural soil surface stabilizers. Because at the megascale (i.e. an entire catena) rock fragment cover often increases (logarithmically) with hillslope gradient in the Mediterranean belt, this cover of soil surface stabilizers will counteract the positive effect of hillslope gradient on soil erosion rate.

12.6 CONCLUSIONS

This study outlines the various effects of rock fragments on processes explaining or contributing to desertification in Mediterranean environments. It also provides a basis for modelling the effects of rock fragments both at the soil surface and in the soil profile on fine earth bulk density, on organic matter and moisture concentration in the fine earth fraction, on infiltration and percolation rates, on overland flow production and velocity, and on soil erosion by water at the micro-, the meso-, the macro- and the megascale. Most of these effects are incorporated in the MEDALUS slope catena model (Kirkby et al., this volume, Chapter 14). The main conclusion is that the role played by rock fragments at the soil surface or in the soil profile is quite complex because of contrasting effects which can lead to compensations. Nevertheless, some general trends emerge which have been discussed in the various sections.

As an illustration of how the information collected in this study can be used, an assessment is made of what might happen on soils containing rock fragments in Mediterranean environments under a scenario predicting desertification. The higher evapotranspiration rates of water caused by higher mean temperatures will affect plant growth and, hence, the vegetative cover. How exactly plants growing on soils containing rock fragments will respond to such a change, in comparison with those growing on rock-free soils, will depend on the balance between the various physical effects of rock fragments. Negative effects of rock fragments include a reduction in rooting volume, a decline in total water-holding capacity, a reduction in total soil nutrients and increased surface temperatures as a result of surface rock coverage. Positive attributes of rock fragments, in the soil or at the surface, include an increase in rate and depth of infiltration, a decrease in evaporation rate, an increased heat capacity and a reduced susceptibility of the soil to physical degradation and water erosion. The net effect will also depend on the size, type and quantity of rock fragments, the characteristics of the fine earth fraction, the plant community and the climatic conditions. Some shrubby plant species, such as saltbush (Munn et al., 1987) or thyme (Albaladejo, 1990), have a preference for soils rich in rock fragments. In arid and semi-arid regions, rock fragments often make the soils more productive than soils that have a finer texture in comparable upland topographic positions. Therefore, degradation of the vegetation cover due to climatic change may be less severe on soils rich in rock fragments when compared to rock-free soils.

After degradation of the vegetation (by fire, e.g. Ruiz-Flano et al., 1992, overgrazing or cultivation practices) the beneficial effects of rock fragments on physical degradation and soil erosion by water will be significant. As shown by several experimental studies, a significant rock fragment content in the soil surface layers will conserve and might even improve fine earth structure. Therefore, rock fragments will often help to counteract physical soil degradation. Furthermore, soils containing significant amounts of rock fragments will be less vulnerable to erosion by water compared to rock-free soils. Although at the mesoscale (interrill areas), rock fragments might activate soil erosion, this trend will often be only temporary, since an increased surface erosion will often expose more rock fragments, leading to the formation of an erosion pavement. At the microscale and at the macroscale (interrill and rill areas) rock fragments at the soil surface act as natural soil surface stabilizers. Since at the megascale (an entire catena), rock fragment cover is often positively related to hillslope gradient on abandoned agricultural land in the Mediterranean, the high rock fragment cover on the steepest slope sections will counteract the positive effect of hillslope gradient on soil erosion intensity. Hence, because of their higher surface stability, soils containing rock fragments will better conserve organic matter in the topsoil and will also offer a less hostile environment for plant growth under drier conditions.

Based on the previous analysis we conclude that the presence of rock fragments at the soil surface or in the topsoil will reduce the intensity of the most important desertification processes in Mediterranean environments.

12.7 ACKNOWLEDGEMENTS

The authors acknowledge the help of Dr M. Jamagne and of Dr Y. Le Bissonais (INRA, Orléans, France) in providing data on the areal distribution of soils containing rock fragments in the European Communities. They would also like to thank Dr J. Puigdefábregas and Dr A. Solé (Estación Experimental de Zonas Aridas, CSIC, Almeria, Spain) who gave freely of their time and expertise and who supported our fieldwork in Almeria in every possible way. Fieldwork in Turkey was supported by the Belgian Programme on Interuniversity Poles of Attraction (IUAP 28) initiated by the Belgian State, Prime Minister's Office, Science Policy Programming. Dr Bas van Wesemael and an unknown reviewer are thanked for their comments on an earlier draft of this paper.

12.8 REFERENCES

Abrahams, A.D. and Parsons, A.J. (1991). Relation between sediment yield and gradient on debris-covered hillslopes, Walnut Gulch, Arizona. *Geological Society of America Bulletin*, **103**, 1109–1113.

Adams, J.E. (1967). Effect of mulches on soil temperature and grain soybean development. *Agronomy Journal*, **57**, 471–474.

Albaladejo, J. (1990). Impact of degradation processes on soil quality in arid Mediterranean environments. In Rubio, J.L. and Rickson, R.J. (eds) *Strategies to Combat Desertification in Mediterranean Europe*. Commission of the European Communities, Report EUR 11175, 193–214.

Alberto, F. (1971). Considérations sur la pierrosité des sols bruns à croûte calcaire du bassin de l'Ebre. *Bull. Rech. Agron. Gembloux*, **6**, 180–185.

Babalola, O. and Lal, R. (1977a). Subsoil gravel horizon and maize root growth. I. Gravel concentration and bulk density effects. *Plant and Soil*, **46**, 337–346.

Babalola, O. and Lal, R. (1977b). Subsoil gravel horizon and maize root growth. II. Effects of gravel size, inter- gravel texture and natural gravel horizon. *Plant and Soil*, **46**, 347–357.

Brakensiek, D.L., Rawls, W.J. and Stephenson, G.R. (1986). Determining the saturated hydraulic conductivity of a soil containing rock fragments. *Soil Society of America, Journal*, **50**, 834–835.

Brouwer, F.M. and Chadwick, M.J. (1991). Future land use patterns in Europe. In Brouwer, F.M., Thomas, A.J. and M.J. Chadwick (eds) *Land Use Changes in Europe*. Kluwer Academic Publishers, Dordrecht, 49–78.

Bunte, K. and Poesen, J. (1993). Effects of rock fragment covers on erosion and transport of noncohesive sediment by shallow overland flow. *Water Resources Research*, **29** (5), 1415–1424.

Bunte, K. and Poesen, J. (1994). Effects of rock fragment size and cover on overland flow hydraulics and sediment yield on an erodible soil surface. *Earth Surface Processes and Landforms*, **19**, 115–135.

Childs, S.W. and Flint, A.L. (1990). Physical properties of forest soils containing rock fragments. In Gessel, S.P., Lacate, D.S., Weetman, G.F. and Powers, R.F. (eds) *Sustained Productivity of Forest Soils*. University of British Columbia, Faculty of Forestry Publ., Vancouver, 95–121.

CEC (Commission of the European Communities) (1985). *Soil Map of the European Communities 1:1.000.000*. Directorate-General for Agriculture, 124 pp. + 7 sheets.

Corey, A.T. and Kemper, W.D. (1968). Conservation of soil water by gravel mulches. *Colorado State University Hydrology Paper* 30, 23 pp.

Cutileiro, J. (1971). *A Portugese Rural Society*. Clarendon Press, Oxford.

De Ploey, J. (1981). The ambivalent effects of some factors of erosion. *Mém. Inst. Géol. Univ. Louvain*, **31**, 171–181.

De Ploey, J., Imeson, A. and Oldeman, L.R. (1991). Soil erosion, soil degradation and climatic change. In Brouwer, F.M., Thomas, A.J. and Chadwick, M.J. (eds) *Land Use Changes in Europe*. Kluwer Academic Publishers, Dordrecht, 275–292.

Donker, N.H. and Damen, M.C. (1984). Gully system development and an assessment of gully initiation risk in Miocene deposits near Daroca—Spain. *Zeitschrift für Geomorphologie, Suppl.*, **49**, 37–50.

Evenari, M., Shanan, L. and Tadmor, N. (1982). *The Negev. The Challenges of a Desert*. Harvard University Press, 322–323.

Fairbourn, M.L. (1973). Effect of gravel mulch on crop yields. *Agronomy Journal*, **65**, 925–928.

FAO (1979). *A Provisional Methodology for Soil Degradation Assessment*. Rome, 84 pp.

Grant, W. and Struchtemeyer, R. (1959). Influence of the coarse fraction in two Maine potato soils on infiltration, runoff and erosion. *Soil Sci. Soc. Am. Proc.*, **23**, 391–394.

Gras, R. (1972a). Effets des éléments grossiers sur la dynamique de l'eau dans un sol sableux. I. Comportement des éléments grossiers poreux vis-à-vis de l'eau. *Ann. Agron.*, **23**, 197–239.

Gras, R. (1972 b). Effets des éléments grossiers sur la dynamique de l'eau dans un sol sableux. II. Dynamique de l'eau dans le système terre fine-éléments grossiers. *Ann. Agron.*, **23**, 247–316.

Hillel, D. and Tadmor, N. (1962). Water regime and vegetation in the central Negev highlands. *Ecology*, **43**, 33–41.

Ingelmo-Sanchez, F., Cuadrado, S. and Blanco, A. (1980). Evaporacion de agua en suelos de distinta textura. *Anu. Centr. Edafol. Biol. Apl. Salamanca*, **6**, 255–280.

Jackson, L.P., Hall, I.V. and Aalders, L.E. (1972). Lowbush blueberry seedling growth as affected by soil type. *Can. J. Soil Sci.*, **52**, 113–115.

Jung, L. (1960). The influence of the stone cover on runoff and erosion on slate soil. *IAHS Publ.*, **53**, 143–153.

Kadmon, R., Yair, A. and Danin, A. (1989). Relationship between soil properties, soil moisture, and vegetation along loess-covered hillslopes, northern Negev, Israel. *Catena Supplement*, **14**, 43–57.

Kosmas, C. and Danalatos, N.G. (1994). Climate change, desertification and the Mediterranean Region. In P. Loveland and M. Rounsevell (eds) *Proceedings of a NATO Advanced Research Workshop 'Soil Responses to Climate Change: Implications for Natural and Managed Ecosystems'*. Springer-Verlag, Berlin, pp. 26–37.

Kosmas, C., Moustakas, N., Danalatos, N.G. and Yassoglou, N. (1994). The effect of rock fragments on wheat biomass production under highly variable moisture conditions in Mediterranean environments. *Catena*, **23**: 191–198.

Lee, M.D. (1988). The development of a distributed computer simulation model of a reconstructed ancient water-harvesting system, Avdat, Israel. In Whitehead, E.E., Hutchinson, C.F., Timmermann, B.N. & Varady, R.G. (eds) *Arid Lands Today and Tomorrow*. Westview Press, Boulder, CO, 919–934.

Lutz, H.J. and Chandler, R.F. (1946). *Forest Soils*. Wiley, New York, 501 pp.

Magier, J. and Ravina, I. (1984). Rock fragments and soil depth as factors in land evaluation of terra rossa. *Soil Science Society of America Special Publication*, **13**, 13–30.

Miller, F.T. and Guthrie, R.L. (1984). Classification and distribution of soils containing rock fragments in the United States. *Soil Science Society of America Special Publication*, **13**, 1–6.

Munn, L., Harrington, N. and McGirr, D.R. (1987). Rock fragments. In Williams, R.D. and Schuman, G.E. (eds) *Reclaiming Mine Soils and Overburden in the Western United States. Analytic Parameters and Procedures*. Soil Conservation Society of America, 259–282.

Peck, A.J. and Watson, J.D. (1979). Hydraulic conductivity and flow in non-uniform soil. Paper presented at the Workshop on Soil Physics and Field Heterogeneity, Canberra, Australia, 31–36.

Poesen, J. (1985). An improved splash transport model. *Zeitschrift für Geomorphologie*, **29**, 193–211.

Poesen, J. (1986). Surface sealing as influenced by slope angle and position of simulated stones in the top layer of loose sediments. *Earth Surface Processes and Landforms*, **11**, 1–10.

Poesen, J. (1987). Transport of rock fragments by rill flow—a field study. *Catena Supplement*, **8**, 35–54.

Poesen, J. (1990). Erosion process research in relation to soil erodibility and some implications for improving soil quality. In Albaladejo, J., Stocking, M.A. and Diaz, E. (eds) *Soil Degradation and Rehabilitation in Mediterranean Environmental Conditions*, CSIC, Murcia, 159–170.

Poesen, J. and Ingelmo-Sanchez, F. (1992). Interrill runoff and sediment yield from topsoils with different structure as affected by rock fragment cover and position. *Catena*, **19**, 451–474.

Poesen, J. and Lavee, H. (1991). Effects of size and incorporation of synthetic mulch on runoff and sediment yield from interrills in a laboratory study with simulated rainfall. *Soil and Tillage Research*, 21, 209–223.

Poesen, J. and Lavee, H. (1994). Rock fragments in topsoils: significance and processes. *Catena*, 23: 1–28.

Poesen, J. and Ruiz-Flano, P. (in preparation). Effects of rock fragments on saturated hydraulic conductivity.

Poesen, J. and Torri, D. (1988). The effect of cup size on splash detachment and transport measurements. Part I: Field measurements. *Catena Supplement*, 12, 113–126.

Poesen, J., Ingelmo-Sanchez, F. and Mücher, H. (1990). The hydrological response of soil surfaces to rainfall as affected by cover and position of rock fragments in the top layer. *Earth Surface Processes and Landforms*, 15, 653–671.

Poesen, J., Torri, D. and Bunte, K. (1994). Effects of rock fragments on soil erosion by water at different spatial scales: a review. *Catena*, 23: 141–166.

Ravina, I. and Magier, J. (1984). Hydraulic conductivity and water retention of clay soils containing rock fragments. *Soil Science Society of America Journal*, 48, 736–740.

Rawls, W.J. (1983). Estimating soil bulk density from particle size analysis and organic matter content. *Soil Science*, 135 123–125.

Ruiz-Flano, P., Garcia-Ruiz, J.M. and Ortigosa, L. (1992). Geomorphological evolution of abandoned fields. A case study in the central Pyrenees. *Catena*, 19, 301–308.

Rutherford, M.C. (1983). Herbaceous standing crop in relation to surface and subsurface rockiness. *Bothalia*, 14, 259–264.

Saini, G.R. and Grant, W.J. (1980). Long-term effects of intensive cultivation on soil quality in the potato-growing areas of New Brunswick (Canada) and Maine (U.S.A.). *Canadian Journal of Soil Science*, 60, 421–428.

Seguin, G. (1971). Influence des facteurs naturels sur les caractères des vins. In Ribereau-Gayon, J. and Peynaud, E. (eds) *Sciences et Techniques de la Vigne*. Dunod, Paris, Vol. 1, 671–725.

Simanton, J.R., Renard, K.G., Christaensen, C.M. and Lane, L.J. (1994). Spatial distribution of surface rock fragments along catenas in semiarid Arizona and Nevada, USA. *Catena*, 23: 29–48.

Stewart, V.I., Adams, W.A. and Abdullah, H.H. (1970). Quantitative pedological studies on soils derived from Silurian mudstones. II. The relationship between stone content and the apparent density of the fine earth. *Journal of Soil Science*, 21, 248–255.

Thornes, J.B., Francis, C.F., Lopez Bermudez, F. and Romero-Diaz, A. (1990). Reticular overland flow with coarse particles and vegetation roughness under Mediterranean conditions. In Rubio, J.L. and Rickson, R.J. (eds) *Strategies to Combat Desertification in Mediterranean Europe*. Commission of the European Communities, Report EUR 11175, 228–243.

Unger, P. (1971). Soil profile gravel layers: I. Effect on water storage, distribution, and evaporation. *Soil Science Society of America Proceedings*, 35, 631–634.

Voiculescu, N., Craioveanu, G. and Popescu, I. (1983). The restrictive effect of soil skeletal material on walnut growth at Birsesti-Gorj. *Analele Institutului de Cercetari pentru Pedologie si Agrochimie*, 45, 145–151.

Wollny, E. (1897–98). Untersuchungen über den Einfluss der Steine auf die Fruchtbarkeit des Bodens. *Forschungen a.d. Gebiete d. Agrikultur-Physik*, 20, 363–395.

Yair, A. and Lavee, H. (1976). Runoff generative process and runoff yield from arid talus mantles slopes. *Earth Surface Processes*, 1, 235–247.

Yair, A. and Shachak, M. (1987). Studies in watershed ecology of an arid area. In Berkofsky, L. and Wurtele, M.G. (eds) *Progress in Desert Research*. Rowman & Littlefield Publishers, 145–193.

13

Mediterranean Ecology and an Ecological Synthesis of the Field Sites

S. C. CLARK

Department of Pure and Applied Biology, University of Leeds, UK

13.1 SOME ECOLOGICAL ASPECTS OF THE MEDITERRANEAN BASIN

The following overview of the tectonic, climatic, pedological and ecological history of the Mediterranean Basin is presented here so that the ecological research carried out within MEDALUS, often specialized and generated by local conditions, can be assessed in the context of the whole Mediterranean.

13.1.1 Origin

This account is based on those of Thrower and Bradbury (1973) and Windley (1984). The Mediterranean Basin is the latest in a series of basins that have occupied the same general position in relation to the adjacent continents probably since before the break-up of the 'supercontinent' Pangaea in the Early Jurassic period approximately 200 million years ago. The extent and position of these marine basins has varied because of continental movement. There is evidence that at least during the early part of their existence they formed an important barrier to plant and animal migration between continents. For example, they prevented the intermingling of the distinctive Glossopteris (southern) and Laurasian (northern) floras during the Carboniferous period.

Tectonic activity associated with the Alpine fold belt was continuous from the Jurassic onwards in the area and continues today, witness the current volcanic activity and frequent earthquakes in the region. Movement in the Pliocene and Pleistocene resulted in the Basin's present configuration. Thus there has been tectonic activity in what is now the Mediterranean region for 2 million years. This has resulted in '... a constantly evolving interconnecting network of mid-oceanic ridges, continental margins,

Mediterranean Desertification and Land Use. Edited by C. Jane Brandt and John B. Thornes.
© 1996 by John Wiley & Sons, Ltd.

island arcs ...' (Windley, 1984). It is not surprising that associated marine deposition of sediments, metamorphism and vulcanism have produced a complex lithology.

13.1.2 Ecological history

Raven (1973), Pons (1981), and Pons and Quezel (1985) have reviewed the ecological and plant evolutionary history of the Mediterranean Basin and this account is based on their work. The early history of the Mediterranean Basin is important in that habitats have been available for colonization there since the emergence of marine sediments to form dry land in the Mid-Cretaceous. There may therefore have been continuous evolution of flowering plants, in particular evergreen perennials, in the Mediterranean Basin from that time until the present with no major phylogenetic break. If so, there would have been ample time for the assemblage of adaptive gene complexes that, given the semi-arid climate in which the evergreens evolved, would have conferred pre-adaptation to the present Mediterranean environment. Also of current ecological relevance is that the geological and tectonic history of the Basin, resulting in the development of a wide range of substrata and altitudes, has aided evolutionary divergence and explains in part the species richness of the area.

Over the last million years the landscape has been much influenced by the destabilizing effects of the climatic changes of the Pleistocene with, in the wetter periods, the deposition and erosion of sedimentary features such as riverine terraces and alluvial fans (Bradbury, 1981). Although the area had been subjected to climatically generated erosion cycles throughout and before the Pleistocene, man's activities over the last 10 000 years and in particular since the Roman period, are thought to have exacerbated the instability owing to deforestation and possibly grazing.

Ecologically, the instability created a heterogeneous and dynamic habitat mosaic. Diversity increased as the vegetation was destroyed. This is because the contact and merging of previously genetically isolated species assemblages has been a powerful evolutionary force creating novel opportunities for recombination and selection. This process is demonstrated by the richness of the winter annual floras in the eastern Mediterranean and by the co-evolution in the same area of cereals and many of the most successful arable weeds of Mediterranean and temperate climates throughout the world. Given the unpredictable annual amount and seasonal distribution of the rainfall, opportunism must have been strongly advantageous before man enhanced the instability of the area. Hence, although climax (i.e. relatively static) communities were reduced in area as a result of man's activities, the disturbances of the Pleistocene meant that the vegetation was pre-adapted to the environmental pressures induced by man. The changes were in degree not in kind.

13.1.3 Soils

Climate and not rock or soil type is the main determinant of community type in the Mediterranean area. Mediterranean climate soils are similar throughout the world, with only slight profile development, and relict soils are widespread (Bradbury, 1981). They comprise a mosaic of different ages. Local variation is often in the form of catenas, for example Red Mediterranean Earths at low elevations, Brown Earths at medium elevations and Lithosols at high elevations. The effects of soil type on the vegetation, in particular soil properties related to pH, are not as marked as in cooler leaching climates. Thus maquis and garrigue[1] communities occur on both calcareous and non-calcareous soils in the eastern Mediterranean. Extremely acid soils (pH < 4.0) are virtually absent. However, calcicole and calcifuge species do occur although they are not so tightly confined to well-defined pH ranges as are, for example, many species of northwest Europe.

Soil type can, however, be locally important in differentiating communities in the same climatic regime. Thus in northern Israel, in a relatively cool, wet Mediterranean climate (500–1100 mm precipitation per year), Rabinovitch-Vin (1983) found that the type of vegetation was controlled primarily by the concentrations and proportions of the major nutrients. In addition to the mineralogical properties of the rocks and associated soils, physical features, for example the nature of the clay minerals and rock and soil porosity, were also involved because of their effects on soil–water relations.

In Mediterranean soils, fertility ranges from high to extremely low. A brown calcareous stony argillaceous soil over limestone in southern France supporting *Quercus coccifera* scrub contained as t ha^{-1} total nutrients in the top 300 mm, 5–7 N, 0.5–0.6 P, 16–22 K, 31–113 Ca, 9–13 Mg, 64–89 Fe, 2–3 Mn and 6–9 Na, and as t ha^{-1} available nutrients to the same depth, 0.06 P, 0.7 K, 34 Ca, 0.4 Mg and 0.05 Na (Rapp and Lossaint, 1981). Fire is a major cause of short-term change in soil nutrient status in the Mediterranean region as it results in up to a 50% increase in available nutrients.

Species have different responses to nutrients and many have become obligately associated with a particular nutrient level and this can determine their habitat range. Hence the large and sudden changes in soil nutrient status following fire are important in influencing the succession of colonizing species (Kruger, 1987). There is considerable variation in the nutrient content of species. This can be related to habitat. Thus in the context of regeneration after fire, species with high nutrient contents are mostly fire ephemeral shrubs and herbs, while species with lower concentrations are mainly longer-lived species that occur later in the succession. The short-lived pioneer species are opportunist and they tolerate low soil nutrient levels in the absence of competition and respond markedly as resources increase. The late-successional species have physiological and morphological adaptations to get nutrients from the low concentrations prevailing in the soil at that stage. These species, unlike the pioneers involved in regeneration after fire and other plants of fertile habitats, are unable to alter their rates of nutrient uptake and there may be an actual decrease in growth above certain concentrations.

Nutrients do not operate in isolation and soil–water relations are strongly involved in determining nutrient availability and also the soil volume available for nutrient movement by mass flow to the plant roots. Two types, then, of nutrient response have been selected for in the Mediterranean environment. One occurs mainly in perennial shrubs and trees of long-term stable habitats, for example those that develop during the later stages of regeneration after fire. This type is characterized by slow nutrient uptake and unresponsiveness to change in nutrient concentrations. The other is opportunist, plastic, highly responsive to changes in soil nutrient concentrations, and occurs in short-term, open, unstable habitats such as the initial stages of successions. Plants of this type are able to exploit transient supplies of nutrients as, for example, during the first few months after a fire.

---

[1]Maquis and garrigue are physiognomic terms describing respectively, dense, predominately sclerophyllous (evergreen) vegetation above 2 m in height and shrub–dominated vegetation 0.6–2 m in height (Tomaselli, 1981)

13.1.4 Climate

This account is based on the work of Aschmann (1973a), Nahal (1981) and Quezel (1985). The Mediterranean climate, broadly characterized by 'cool wet winters and hot dry summers', is unpredictable regarding the amount and timing of its rainfall, but there is always a dry season. This is of varying duration, again unpredictable but is always in summer. The rain falls typically as heavy, isolated storms. Less than 1% of the earth's surface experiences this climate and, of this area, more than half is in the Mediterranean Basin. It occurs mainly in belts north and south of the equator in latitudes 32 to 40°. The summer dry period occurs because of an extension polewards of the subtropical highs while the winter rainy season occurs because of an extension southwards of mid-latitude cyclones of mild oceanic air. Defining the Mediterranean climate is difficult due, ironically, to one of its most characteristic traits, namely the unpredictability of its rainfall regarding not only the annual total but also its distribution within the year. There have been many attempts to define it, including the use of indicator species; for example in the Mediterranean Basin itself, the climate has been equated with the range of the domestic olive, *Olea europaea*. Using climatic not biotic indicators, it has been defined as having 65% precipitation in winter with its lower rainfall limit at around 275 mm yr^{-1}; a winter temperature regime of at least one month below 15°C; and a minimum temperature not below freezing point for more than 3% of the year. The Mediterranean Basin's climate is generally drier in the east and wetter in the west where the summers are not rainless and where the dry period may be shortened by heavy rains in late spring and early autumn. However, there are many exceptions to this trend. For example the Almeria area of southern Spain, towards the western end of the Basin, is the driest in Europe because it is located in the rain-shadow of the Sierra Nevada to the west. The northern edge of the Basin is at the polewards limit of the Mediterranean climate, hence dry seasons there are relatively short and often interrupted by rainfall. The length of the dry season along the northern margin of the Basin, as for example in the northern Adriatic, is further reduced and seasonal irregularity increased by the effects of the nearby high mountains on the local climate.

The most important feature of the climate affecting the evolution of biota in the Basin and its ecology is probably the unpredictability between and within years in the amount and timing of the rainfall. For example the annual precipitation was only average for 14-years out of a 100-year period at Marseilles (Nahal, 1981). Precipitation does however tend to be maximum in autumn, winter or spring. A consequence of the rainy season being at the cooler time of year is that there has been selection in Mediterranean species for the ability to grow at winter temperatures. This is more noticeable in the winter annual than in the evergreen component of the vegetation. However the evergreens are never winter dormant and they rapidly reach high rates of photosynthesis in early spring. Of obvious ecological as well as evolutionary importance is the summer dry period: the one certain feature of the Mediterranean climate. Although an annual and hence predictable event, its length, severity of the associated drought, temperature regime and probability of being interrupted by rainstorms differ markedly over the whole region. The effect on the plants of a particular dry season regime will depend on the soil moisture reserves, the ability of the root systems to utilize those reserves and the rapidity with which they are depleted.

The physiological effects of the dry season environment on plant function and its effects on relationships between the species and individuals comprising communities are not precisely known because the combined and separate effects of water deficiency and high temperatures are difficult to disentangle. Most ecophysiological work on the effect of the dry season on plant function has been on water relations. Further work is needed on the role of high temperatures in plant communities. It is probable that high temperatures in themselves, by adversely affecting protein functioning in leaves and meristems, are important in determining the habitat limits of particular species and floristic differences between communities not only at the geographical but also at the local topographical scale.

13.1.5 History of the vegetation

This account is based on the work of Axelrod (1973), Raven (1973), Pons and Quezel (1985), and Quezel (1985). The Mediterranean Basin vegetation has two main components, evergreen sclerophyllous[1] broad-leaved shrubs and deciduous winter annuals. Their evolutionary history until the last few thousand years have been very different. The sclerophyllous component evolved from the Mid-Cretaceous onwards in a warm, moist, possibly subtropical environment with summer rains and a winter dry season. By the Late Tertiary, an oak–laurel flora had developed, containing genera (e.g. *Acer, Arbutus* and *Quercus*) that are still widespread today in the region and also in North America which had land connections with the Old World at the time these genera were evolving. Other contemporary genera with the same evolutionary history but absent from the New World are *Chamaerops, Myrtus* and *Olea*.

Uplift in the Oligo–Miocene resulted in greater extremes of temperature, a reduction in precipitation but with the rains still in summer, and in the spread of more drought-resistant species. These had apparently evolved in localized dry habitats that existed in the generally moist environments of the Tertiary, perhaps as well drained but never extensive rocky outcrops with shallow soils. There is evidence that the sclerophyllous species emanating from these limited perhaps edaphically controlled habitats, were important components of the vegetation during the higher rainfall phases of the climatic fluctuations in the Late Tertiary. By the Early Pleistocene there were some winter rains. Thus in a Villefranchian age site in Tunisia, there was both winter and summer rain with a diverse flora, part tropical, part temperate.

During the Quaternary, rainfall decreased and the sclerophyllous vegetation became restricted to its present areas in the Mediterranean Basin and there it was only important during the relatively brief periods of favourable climate in the interglacials. During the cold glacial periods, sclerophyllous arboreal and shrub communities survived in refugia, which in the eastern Mediterranean, were probably near the present coastal zones.

In the early Post-Glacial, on account of melting ice at the end of the Pleistocene glaciation, the oceans became cooler and as a consequence the air masses over the region became more stable in summer, which resulted in the development of the summer dry

---

[1]Sclerophyllous vegetation comprises species with sclerophyllous leaves; these are stiff, evergreen, with xeromorphic features (e.g. thick cuticles, sunken stomata, small cells and much sclerenchymatous tissue). Typical examples are *Olea europaea, Quercus ilex* and *Myrtus communis*.

period that is such a feature of present Mediterranean climates. At the end of the Glacial period, Mediterranean-type pine forest and juniper scrub were recorded from the northwest of the region and forest from the southeast. Evergreen sclerophyllous vegetation apparently only occurred at this time in unfavourable habitats liable to drought, with possibly the summer drought regime favouring its survival. However, it did not gain its present importance until man's disturbances favoured its spread during the last 10 000 years. It is probable that the present predominance of sclerophyllous species in the flora is due primarily to man's influence and not to an optimization of the climate for the sclerophyllous habit. Thus *Quercus ilex*, long assumed to be an important climax species, is now thought to have gained its importance in the Mediterranean vegetation as a result of man's activities in the postglacial. In contrast to the long evolutionary history of the sclerophyllous vegetation of the Mediterranean, the winter annual flora cannot have evolved before the Early Pleistocene by which time the Mediterranean climate had largely evolved. The forebears of this winter annual flora probably originated further east than the sclerophyllous component and possibly in a drier climate.

13.1.6 Past effects of man on the vegetation

This account is based on the work of Aschmann (1973b), Le Houerou (1981), and Pons and Quezel (1985). Until the Neolithic period (12 000 to 4000 years ago) man had little effect on the vegetation. Localized disturbances in the form of small clearances for crops commenced in the Paleolithic. Clearance had intensified by the end of the Neolithic, taking place some 4000 years later in the western than in the eastern parts of the Basin. Clearance and cereal production were widespread by 5000 BP and by 4500 BP there was increased anthropogenic effect on the vegetation. Further increases in the area of pasture and cultivated land occurred over the next 1500 years. Increasing change from natural to managed landscapes took place during the Medieval period, with the introduction of tree crops, including olive (*Olea europaea*), sweet chestnut (*Castanea sativa*) and walnut (*Juglans regia*). Man's effect on the vegetation appears from the pollen and macrofossil record to have been to bring about a decrease in the extent of stable communities dominated by long-lived arboreal and shrubby species. This was associated with an increase in communities of short-lived species including annuals, more tolerant of the disturbed habitats resulting from man's activities. In reality clearance was almost certainly haphazard, disjunct and often eliminated by recolonization with the trend towards clearance irregular and at times reversed, reflecting the changing activities of contemporary society. The more abundant evidence from the Roman period onwards supports this view. The natural vegetation over the last 10 000 years has been reduced to one-third its original area. Intensity of destruction over the period depended on the demands for timber products and land for agriculture, with the period of Roman–Byzantine domination from the 2nd to the 7th centuries AD and the period following the Second World War, the worst. On a smaller scale, tourist-orientated development has made inroads into coastal vegetation. Man's influence on the Mediterranean environment was greatly intensified in the 19th century owing to population increase and new agricultural techniques. Grazing intensity and fire frequency also increased.

   The main non-destructive anthropomorphic effects on the vegetation until the present have been the appearance of ruderal and nitrophilous species and the selection and spread of domesticated races of native and exotic species, the disappearance or decline of deciduous oak species, for example *Quercus pubescens*, and the corresponding increase in sclerophyllous oaks, notably *Q. ilex* and *Q. suber*. Other important forest trees also declined or vanished during this period, although some pine species increased. Macrofossil evidence shows that shrubby species, as a percentage of the flora, have increased over the period of man's influence in the Mediterranean. The shrubby sclerophyllous vegetation associated with man's activities apparently existed in some small areas of the Basin before man's presence there, for example in western coastal regions and in the east away from the coast, indicating that the propensity to form the types of vegetation associated with man already existed. Thus thorny xerophytic[1] assemblages were established as climax formations on the high mountains of the Basin by the end of the Late Glacial and extended, after the destruction of forests by man, to adjacent areas. Until recently it was thought that forest was the climax formation in the Basin. There is now evidence that, not only in arid and semi-arid parts of the Basin where forest only occasionally occurs but also in humid areas such as southern Anatolia and Corsica, maquis vegetation was the climax. Thus in Turkey and Corsica soil development and the age of the shrubs indicate four or five centuries of presence at least.

13.1.7 Current vegetation

This account is based on Quezel (1981). The vegetation of the Basin currently consists of a mosaic of transient vegetation types at various scales in space and time, resulting from sporadic cultivation, deforestation, fire, overgrazing, and tourist developments. The matorral[2] vegetation of the Basin is still important in itself and as a precursor of many of the communities resulting from man's activities there. It forms a narrow belt around the Mediterranean with greater development in the west than the east. Apart from man's effects on its distribution, there are natural constraints. Thus low winter temperatures, in particular the average minimum of the coldest month, explain its northern and altitudinal limit. Less than 200 mm annual rainfall prevents its development and precludes it from many areas at low altitudes. Above 2000 mm it is replaced by deciduous formations. Timing, as distinct from the amount of rain, is unimportant suggesting that opportunistic traits are present.

Soil characteristics affect distribution in that in the drier parts of the Basin, on the shallower soils, the formation fades out at a higher rainfall than it does on deeper soils with the same amount of rain. Also, the formation reaches higher altitudes on calcareous substrata than on metamorphic ones. Floristically, the mattoral is extremely poor compared with the grasslands, especially the annual grasslands, that occur in the same areas. In the Thermo-Mediterranean Zone, forest is uncommon and arborescent mattoral and garrigue predominate. In the Mediterranean Zone, forest is the theoretical climax although now rare because of man's activities. In the western Mediterranean soil type has

---

[1]Xerophytic: containing morphological and anatomical structures thought to be associated with reducing water loss (e.g. sunken stomata and thick cuticles).
[2]Matorral: a general term for the broad-leaved evergreen shrubby formations of the Mediterranean Basin, including maquis (high matorral) and garrigue (middle mattoral). Matorral vegetation varies in height and density and characteristically consists of shrubs that are not clearly differentiated into trunk and canopy because many branches arise from the same level at the base of the plant (Tomaselli, 1981).

an important effect on the vegetation. Thus in the arborescent communities *Q. ilex* and *Q. coccifera* are dominant on calcareous substrata, while *Q. suber* largely replaces *Q. ilex* on silieous soils. The shrub-dominated communities in the west form garrigue and maquis vegetation. As with the forests, the communities differ according to soil type. On calcareous substrata *Rosmarinus officinalis* is prevalent and on non-calcareous soils *Cistus* species are often dominant. Phrygana, consisting of thorny, cushion, drought-deciduous chamaephytes, is rare in the western Mediterranean. In the eastern Mediterranean, *Quercus ilex* is also the predominant forest tree. Unlike in the west, the forest and shrub communities on calcareous and non-calcareous substrata are similar. Phrygana is important in the east. In the Supra-Mediterranean Zone, sclerophyllous vegetation rarely develops and a dwarf shrub, open rock-heath predominates.

Functionally the Mediterranean vegetation has as its main elements winter annuals and two types of shrub, one drought-deciduous and one evergreen or sclerophyllous. The conventional view of the Mediterranean vegetation is that it consists of fragile unstable degenerative stages, derived ultimately from stable climax woodland whose coherence, owing to man's disruptive influence over millennia in the region, is small. A different interpretation can be argued in that the groups of transient and often ill-defined regressive and successional stages that comprise the vegetation in most parts of the region can be thought of as parts of 'mega communities' with closer floristic, temporal, and functional affinities than successional stages have in, for example, temperate climates. The evergreen and the winter annual floras have different evolutionary histories and phenologies. These elements of the same community also have different breeding systems, hence different population structures and widely disparate adaptive syndromes and responses to perturbations. This means that in spite of their current impoverishment, the Mediterranean two or three component communities comprising widely differing but compatible elements are peculiarly resilient in unfavourable environments. Thus the communities have the necessary variability to 'make the best' of post-perturbation environments and the proportions of each functional type at a particular stage following disturbance will vary in response to the prevailing conditions.

A number of factors contribute to the resilience of the composite community of shrubs and winter annuals in response to perturbations, whether these are relatively predictable as are, for example, grazing and fire, or unpredictable such as sporadic cultivation (Mooney, 1983). The annuals are opportunist and take up nutrients rapidly. The perennials grow slowly with slow rates of nutrient uptake. The annual leaf is a lavish user of water but is a rapid fixer of carbon dioxide, hence its water use efficiency (WUE mols CO_2 fixed/mols H_2O lost) can be high. The drought-deciduous shrub is the same and both types are drought avoiders. The sclerophyllous leaf is a drought-enduring organ tolerant of high temperature. It also has a high water use efficiency but this is a consequence of its low rates of water loss not because of rapid carbon gain. The elements of the winter annual shrub communities may be mutualistic in that the shrubs provide a favourable micro-environment for the annuals. Conversely the cover of annuals beneath the shrubs may extend the period during which nutrient acquisition by the shrubs can take place. This is because first, and most importantly, the annuals improve soil structure, thereby increasing the proportion of rain that infiltrates into, as opposed to runs off, the soil. Secondly soil surface temperatures are lower because of the shading effect of the annuals. This means that rates of evaporation from the upper layers of soil will be lower. It is possible then, that in spite of the transpirational demands of the annuals, their presence results in an increased reserve of soil water that is depleted less rapidly than when they are absent.

There is a tendency for most species to be unpalatable for at least part of the year with a low density of palatable plants from a wide range of species available at any one time. As a consequence of their evolutionary history the constituent species of the Mediterranean vegetation appear to be pre-adapted to the temperature and drought aspects of global warming but importantly not necessarily to the consequences of CO_2 enrichment, particularly as to its effect on competitive relationships.

13.1.8 Man's present effects on the vegetation

This section is based on the work of Quezel (1981) and Pons and Quezel (1985) and the inherently variable Mediterranean environment regarding climate and soils means that regressive and successional vegetational change have been a feature of the region since terrestrial vegetation first evolved there. The vegetation has apparently been regressing since the Neolithic; rapidly during periods of human population expansion and economic prosperity; slowly if at all and with some regeneration during periods of economic instability and war. Man's activities have in many cases merely exacerbated existing instability and many of the communities were pre-adapted to these changes. Forest is thought to be the theoretical climax over much of the region and man's current effects on the vegetation are often assessed as the extent of departure from forest communities. Thus arborescent matorral and maquis generally succeed sclerophyllous oak forest, and maquis and garrigue succeed deciduous oak. Coniferous woodland becomes arborescent mattoral and then garrigue. In the Supra-Mediterranean zone, dwarf shrub heath communities rather than sclerophyllous shrub vegetation succeed the forest. Generally all degradational change eventually leads to the formation of grass-dominated communities. In some countries, for example Spain, rural depopulation during the last few decades has in some areas resulted in regeneration of the vegetation where grazing pressure is low.

Most contemporary communities are short-lived stages resulting from abandoned cultivation, grazing, fire and other effects of man's activities. Definite and repeating sequences occur in response to the main perturbations of fire, cultivation and grazing. Communities rich in winter annuals are dominant in the 2 to 3 years following perturbation and are again important if the grazing pressure is too high for the regeneration of woody species.

A combination of climate and vegetation type make the vegetation of Mediterranean environments extremely susceptible to burning (Le Houerou, 1981; Kruger, 1983). The most susceptible types of vegetation are conifers and shrubby species containing inflammable oils. Thus, in response to the 6000 or more years the vegetation has been subjected to fire, there has been selection for specialized features such as thick, fire-resistant bark, and vegetative growth and seed germination that are stimulated by fire. Approximately 2000 km² of vegetation are destroyed in the Mediterranean Basin each year by fire. '

Burning has been used by man as a management tool for 2500–4000 years in the region. Fire, even as a rare event, has influenced the vegetation over most of the region probably since before the presence of man there (Trabaud, 1981). This long-term influence and its association with other factors, notably grazing, make it difficult to assess its effects on the vegetation other than in the few years following a burn. Fire in moderation can be beneficial by releasing nutrients from undecomposed plant material and destroying organic compounds, for example terpenes and phenolics, that are produced by one species and may affect the germination or growth of others. However, burning reduces species diversity and can enhance runoff.

A definite 'burn succession', that is rich in annuals initially with herbaceous and shrubby perennials coming in later, is a feature of regeneration following fire. The nature of this succession depends on the frequency and intensity of burning. Thus species are precluded if the interval between burns is too short for their successful reproduction or if an occasional high-temperature burn destroys shrubs that successfully regenerate vegetatively after normal low-temperature burns. Biomass generally reaches a maximum at some stage in the succession following burning and then declines to an equilibrium level with an associated decrease in the leaf:wood ratio. Observations in Portugal on the long-term effects of burns (Merino and Vicente, 1981) showed that biomass generally increased for about 20–30 years after the fire and then declined. There is evidence from this work that the decline terminated at an equilibrium because in the oldest stands, 80 years since the last fire, the ratio of productivity to litter production was one.

The effects of grazing and burning are closely related as burning temporarily increases the extent and palatability of the pasture while grazing, because of its effect on biomass, can affect the frequency of burning. By destroying seedlings in the areas made accessible by fire, grazing can also affect the type of vegetation that develops following a burn and reduce the rate of change.

Almost all the vegetation in the Basin, apart from a few forests, is grazed to some extent. The high stocking rates, the pastoral techniques employed and the frequent fires, have led over many centuries to a decline in pasture quality and productivity and an increase in the proportion of unpalatable species. Year-round grazing may be one of the main causes of degradation coupled with ineffective or non-existent rotational grazing. The vegetation of the region has been subjected to grazing since man domesticated grazing animals, so there has been time for the selection of features that favour survival. The pastures contain large numbers of species that are not grazed, for example spiny *Compositae* and aromatic *Umbelliferae*.

Grazing consists of browsing the palatable evergreen shrubs and searching in a matrix of unpalatable species for the palatable forage produced by small numbers of individuals of a wide range of species, mainly *Gramineae* and *Leguminosae*, that occur at low frequencies and cover in these pastures. This low probability of lethal overgrazing of individual plants and of extinction of particular species, suggests that these communities are resilient in the face of intensive grazing pressure. The stability of the Deheza, Montado and Argania savannahs of western Spain, southeast Portugal and southwest Morocco respectively, supports this view. Goats are the animal best adapted to many types of overgrazed pasture, too unpalatable for sheep, as they will eat the residual woody vegetation.

13.1.9 Production

The Mediterranean Basin, because of its inherent and human-induced environmental instability, supports for the most part short-lived and unstable communities where change is usually unidirectional and regeneration is often prevented by disturbance. The few relatively stable or 'climax' communities that are currently present (i.e. where change is cyclical and where any unidirectional change is gradual and in response to long-term environmental or endogenous changes) comprise sclerophyllous shrubs and trees. However, as a consequence of climatic differences within the Basin, there are variations in the life-form composition of these communities. Thus with decrease in rainfall the proportion of summer deciduous species increases and with decreasing winter temperatures winter deciduous trees and shrubs become important. Thus production varies not only because of direct climatic effects but because there is a change in the inherent productivities of the species present (Mooney, 1981, 1983).

Because of water and nutrient limits and genetically based attributes, Mediterranean vegetation has a low carbon fixing capacity. Evergreen *Quercus ilex* woodland, occurring in the less arid habitats, has a leaf area index of above 4, less than 3% of the biomass is leaves and the root weight ratio (RWR)[1] is approximately 0.17. By contrast dwarf shrub Greek phrygana, containing a high proportion of drought-deciduous species of extremely arid sites, has a leaf area index of less than 2 and a RWR of 0.6. In spring, leaves comprise nearly 20% of the biomass.

Evergreen forest accumulates almost twice as much woody material and produces approximately twice as much litter as do evergreen shrub communities. Above-ground biomass accumulation in evergreen shrub communities varies between 100 and 200 g m^{-2} yr^{-1}. Although the productive structures differ markedly between evergreen shrub and drought-deciduous communities, annual above-ground productivity (i.e. biomass accumulation plus litterfall) is about the same at around 400 g m^{-2} yr^{-1}. This is because the deciduous leaves are less costly to produce and photosynthesize more rapidly than do the evergreen ones, thereby compensating for the shorter annual period for carbon fixation in the drought-deciduous species (Mooney, 1983). Above-ground biomass accumulation in evergreen woodland is one-third higher than in shrub communities (Mooney, 1981).

Merino and Vicente (1981) showed the marked effects of soil–water relations on biomass in matorral vegetation after a fire along a moisture gradient in southern Spain. Biomass was inversely related to depth of the water-table and ranged from 0.5 kg^{-2} at 3.65 m depth of water-table on the ridges, to 2.2 kg m^{-2} at 0.9 m depth of water-table in the hollows. The differences in standing biomass were apparently a consequence of lack of available water because the plant water potential values in summer on the ridges were as low as -8 MPa while in the hollows they were around -0.8 MPa and never lower than -2.2 MPa (Merino et al., 1976).

Papanastasis (1981) showed in three grasslands at different altitudes in northern Greece that the important non-woody component of Mediterranean vegetation, the winter annual, was important in lowland grasslands, decreased with altitude and was absent from the highest sites at 1500 m. Phenologically, there was one group of species that was active in the warm season but not in winter and a second group that was active in the cool season but not in summer. Each group contained winter annuals and perennials. The perennials had a longer growing season than the annuals owing to the 3–4 month summer dormancy of the latter. Despite the floristic richness of the communities at all sites, only a few species

---

[1]$RWR = R/(R + S)$ where RWR =root weight ratio, R =root weight and S =shoot weight.

accounted for the bulk of production. Thus the three dominants at each altitude made up more than 50% of the biomass. There were the expected differences in the proportion of the total biomass contributed by each species at different times of year and there were also differences between years for the same sampling date resulting from changes in the structure of the communities between seasons and between years. Thus at the lowland site peak biomass in July 1975 was ($227\,g\,m^{-2}$), in June 1976 it was ($233\,g\,m^{-2}$), and in May 1977 it was ($112\,g\,m^{-2}$).

13.1.10 Carbon allocation

It is difficult to make realistic generalities about carbon allocation when there is no large body of information on the fate of photosynthate in species of the Mediterranean area. In addition there are, almost certainly, marked differences between the allocation patterns of different species even with the same life-form. However those species where growth, whether seasonal or opportunist, is rapid apparently need a readily available reserve to support the initial growth, particularly if new leaf formation is involved.

Times of carbon acquisition and growth are related to the availability of water but use of the photosynthate gained, for example for reproduction, may not be. The timing of carbon acquisition is not wholly determined by exogenous factors. Thus many sclerophyllous species fix carbon, albeit at low rates, all the year round, while drought-deciduous species, especially annuals, photosynthesize at extremely high rates but only during the short periods when soil water potential is high.

New growth is apparently supported by reserves accumulated during the previous year in many drough-deciduous and evergreen species. The nature and location of longer-term reserves, needed to support initial growth after one or more years of drought when no growth occurs, needs investigating.

Information on the allocation pattern in the sclerophyllous oak *Quercus agrifolia* in a Mediterranean climate and habitat in California (Mauffette and Oechel, 1987) is given here, with the assumption that this pattern is broadly similar to those of many evergreen species in the Basin. This study concerned the above-ground parts only. At monthly intervals leafy branches, still attached to the trees, were enclosed in polythene bags. ^{14}C-rich air was then introduced into the bags. Four weeks after labelling the branches were harvested. Greatest retention of label was in new stems in June (56%) and in new leaves in July (60%). This represented photosynthate produced in May before bud break and then translocated to the growing points. Photosynthate produced in July was used for current leaf development and little was exported. There was a larger percentage of incorporation into lipids in old than in young leaves in summer but the same amount in both ages at other times. Phenolics were higher in young than in old leaves. Incorporation into starch and sugars was more variable in young leaves (6–34%) compared with old (27–39%) but the seasonal pattern of high allocation in winter, decreasing in summer was the same in both age groups. This high allocation in winter was used to supplement current photosynthesis for early spring growth. Allocations to nitrogen compounds were the same for old and young leaves with no seasonal differences but this may have obscured differences between classes of nitrogenous compounds seasonally or in leaves of different age. Structural compounds, that is, lignin, hemicellulose and cellulose, had the highest amount of label incorporated in July.

Diamantoglou and Meletion-Christou (1981) found different allocation patterns in three sclerophyllous species in Greece. Thus in *Quercus coccifera* and *Pistacea lentiscus*, leaf content of sugars rose in summer and early autumn and decreased during the spring growing period, with lipid contents broadly mirroring sucrose. There was, however, the opposite pattern in *Ceratonia siliqua*.

Carbon allocation to below-ground parts is obviously important and in many Mediterranean species with high root:shoot ratios the root system can be the dominant sink. Photosynthate is needed by root systems for length and thickness increase of the larger roots, growth of rootlets and respiration for growth and maintenance. Growth and maintenance respiration of the root system alone can take 20–30% of the photosynthate.

The major root biomass of typical 25-year old mixed Californian chaparral was 565 g m^{-2} with an approximate 10% increase each year (Kummerow, 1981). There were marked seasonal differences in rootlet density; 800 g m^{-2} in spring, 1400 g m^{-2} in summer and 400 g m^{-2} in winter. The difference of 1000 g m^{-2} between minimum and maximum values represented annual rootlet production. Litterfall is generally higher below than above ground in chaparral, (240–355 g m^{-2}). Cultivation experiments involving three species of sclerophyllous shrub from a chaparral community gave biomass root weight ratios of 0.3–0.4 and rootlet:(leaf + rootlet) biomass ratios of 0.24–0.36. Biomass allocation, means of the three species, was 12% rootlets, 19% larger roots, 28% leaves and 41% stems (Kummerow, 1981).

13.1.11 Nutrient cycling

This involves cyclical processes in both soil and plant. In soils of low nutrient status, such as most of those in the Mediterranean Basin, the rate of cycling affects production as the more rapid the turnover the more effective is a particular size of nutrient pool in supporting growth (i.e. rapidity of turnover can substitute for concentration). The size of the nutrient pools in Mediterranean, compared with other environments, is generally small. Southern hemisphere Mediterranean heaths represent the extremes of impoverishment for the Mediterranean environment with pools 10–20 times smaller than in woodland in northern Europe. This small pool size, despite plant adaptation to it, means that rates of nutrient cycling, in particular litter breakdown, are important as a factor in nutrient supply.

In general, the content of lignin and the climate determine rates of decomposition (Read and Mitchell, 1983). Nitrogen content is also positively correlated with rate. Of the nutrients, potassium and calcium are released first and nitrogen later. The rate of phosphorus mineralization depends on the type of litter, carbon:phosphorus ratio, secondary metabolites in the litter and pH. The phosphorus content of litter is small because of recycling in the plant. Of the phosphorus in the litter the amount available may be as low as 1.5% of the total because of poor mineralization, rapid plant uptake or precipitation as insoluble phosphorus. The greatest amount of the total phosphorus pool is in the soil. Of that, only approximately 0.01% is available to plants. The availability of phosphorus in the soil is affected importantly by fluctuating soil moisture levels, resulting in 'pulses' of available phosphorus that are presumably utilized by the opportunist growth of feeding roots. Inputs of phosphorus to most natural Mediterranean ecosystems are at low or trace levels. As with phosphorus the greatest proportion of the pool of total nitrogen in low nutrient status Mediterranean soils in the southern hemisphere is in the rocks and soil (Groves, 1983). Unlike phosphorus, nitrogen in rain is an important input to the nitrogen pool in Mediterranean environments.

The soil fauna and microflora enhance the rate of litter breakdown (Read and Mitchell, 1983). Microfloral activity depends on temperature and the availability of water. Mycorrhizal enhancement of the uptake of phosphorus from low concentrations occurs on some soils and symbiotic nitrogen fixation is important on others. Sometimes the two processes occur in the same soil (Lamont, 1983). In general, phosphorus limits nitrogen fixation but not vice versa, as phosphorus is needed for nitrogen reduction to ammonia. Nitrogen fixers contain two to four times more nitrogen than non-fixers. The importance of free-living nitrogen fixers in the soils of Mediterranean environments is not known.

Sclerophyllous litter from parts of South Africa with a Mediterranean climate decomposed at approximately 15–24% yr^{-1}. Nitrogen turnover time was 2.5–4.6 years. Litter in Mediterranean southeast Australia from a similar habitat contained 0.06 $g\,m^{-2}$ phosphorus at the time of fall. This level remained constant for 2 years and the phosphorus was released in the third year (Groves, 1983). In the low nutrient status South African and Australian Mediterranean communities, there is efficient use of scarce nutrients in that 90% of the phosphorus and 30% of the nitrogen in leaves was internally recycled (Groves, 1983). Thus non-senescing leaves of a *Banksia* species contained 3.1 mg P per 100 leaves, while recently fallen litter contained only 0.4 mg P per 100 leaves (Groves, 1983). In *Quercus ilex* in Spain, (Escarre et al., 1987) nutrients were withdrawn from 1-year-old leaves into 1-year-old branches from where they passed to the young leaves of the current year. Of the nitrogen used in young structures 48% was retranslocated from older ones (i.e. 18% from 1-year-old leaves, 17% from dead leaves and 12% from branches). With phosphorus, 17% of that used in new structures came from dead leaves and 14% from 1-year-old leaves. None came from branches.

In 30-year-old stands of *Q. coccifera* in southern France, Rapp and Lossaint (1981) found that calcium had the highest cycling rate between soil and plant followed by nitrogen, potassium and magnesium in that order. Leaf fall occurred in April–June but normally there was little decomposition until the onset of the autumn rains. Hence time of leaf fall and timing of the rainfall largely determined the length of time that nutrients were locked up in the litter and also their rates of release. Considering the 820 $kg\,ha^{-1}$ of the major elements contained in the plant biomass above-ground, of the major elements together, approximately 30 $kg\,ha^{-1}\,yr^{-1}$ were retained in the plants and 102 $kg\,ha^{-1}\,yr^{-1}$ were returned to the soil in litter, mainly as leaf material (60% of the total) with flowers next in importance and wood least. The main elements returned were calcium (50%) nitrogen (30%) and potassium (11%). There was an input in the precipitation of 50 $kg\,ha^{-1}\,yr^{-1}$, mainly as sodium (40%) nitrogen (27%) and calcium (22%). Annual uptake into the above-ground parts was 132 $kg\,ha^{-1}\,yr^{-1}$. Of the above-ground nutrient mass 80% was in the wood. As a consequence, over a 17 year period, 629 $kg\,ha^{-1}$ were immobilized.

13.1.12 Water relations and gas exchange

The three main components of the vegetation of the Mediterranean Basin, the sclerophyllous tree and shrub, the drought-deciduous shrub/subshrub and the winter annual, have different origins and responses to the environment. The main factors affecting their evolution in the Mediterranean Basin in the long term have been the irregular and often inadequate supply of water, the length of the dry season and perhaps, more recently, fire and grazing (Bazzaz and Morse, 1991; Rundel, 1991).

Two major types of leaf have evolved: first, deciduous and drought avoiding with a large photosynthesizing capacity but no resistance to desiccation; and secondly, drought enduring, evergreen (sclerophyllous) with low rates of photosynthesis and a life span of up to 5 years but usually less than half this. Photosynthetic rates are maximum in annuals, equal or slightly lower in drought-deciduous species, less again in evergreens and least in cladodes (Mooney, 1983). Evergreens contain less nitrogen on a weight basis per leaf and use it less efficiently than do deciduous species, leading to low rates of photosynthesis. The low leaf nitrogen content also results in low light saturation. However, photosynthesis is less affected by drought than it is in deciduous species and the sclerophyllous leaf has a wide temperature tolerance of 16–29C (Mooney, 1983).

Merino (1987) has shown that, mainly because of their small size, sclerophyllous leaves are, as a group, more costly to produce than deciduous ones (i.e. 1.6 g glucose g^{-1} dwt compared with 1.5). Maintenance costs, however, are lower in evergreen leaves, 0.013 g glucose g^{-1} dwt day^{-1} compared with 0.015 g. Fossil evidence suggests that the sclerophyllous leaf did not evolve in response to the Mediterranean climate, but to low-nutrient status regimes before the winter rains/summer drought regime of the Mediterranean climate had developed. Thus the supposedly drought-resistant xeromorphic features of this type of leaf, except for cuticle thickness, did not vary consistently in response to increasing aridity in a range of Mediterranean climate sclerophyllous species (Kummerow, 1973). More probably it is the tolerance to a wide range of environmental conditions and the plastic responses to them of the sclerophyllous leaf type, with its opportunist and drought-enduring features, that have enabled it with little obvious evolutionary change to remain an effective type of leaf when the shift from summer to winter rains took place as the Mediterranean climate developed.

There are often large differences in the same community, both between species of the same life-form and between individuals of the same species, in the acquisition of resources and in the physiological response to the environment (Westman, 1983; Kruger, 1987). Some of these differences are inherent while others are a consequence of local habitat heterogeneity. Thus nutrient uptake characteristics and requirements may differ between species in the same habitat. Overall concentrations may be the same but requirements and contents of individual elements may differ substantially. There can also be large differences in water relations and photosynthesis in different species of the same community. Accumulation of nutrients occurs and hence uptake and use are not necessarily synchronized. Phosphorus is probably the most limiting major element.

Water is the main limiting resource in the Mediterranean not only because of an absolute lack of soil water in the dry season but also because of high vapour pressure deficits for substantial periods of the day over a far longer period than that of the dry season itself. Drought results in decreases in net photosynthesis, transpiration rates and leaf mesophyll photosynthetic capacity and in lower quantum efficiencies and conductances. For a given shortage of water these changes are not manifest to the same extent in all species and changes in one feature can be independent of change in others (Correia et al., 1987). Plants respond to the soil water and temperature regimes diurnally and seasonally. Plant water potential is usually maximum at dawn, but in the low humidities of the Mediterranean climate it is usually well below full turgor at that time, even in early spring, the most favourable growing season. Stomatal conductance will be maximum as soon as irradiance is sufficient to open the stomata fully.

Maximum rate of photosynthesis is, for many species, in early morning while vapour pressure deficit (VPD) is minimum. Stomatal conductance will decrease with VPD later in the day with midday closure occurring in some species. This is an attribute that conserves water but at the cost of a rise in leaf temperature. Thus over much of the day partial or complete stomatal closure limits the rate of carbon dioxide acquisition and rapid carbon uptake only occurs when conductances are maximum during the favourable early morning period (Poole et al., 1981). This is a characteristic of many drought-deciduous and Mediterranean annual species. The trait is less marked in the evergreen sclerophylls where photosynthesis occurs, although at slower rates, over longer periods of the day.

Reduction of the diurnal effects of high temperatures and drought consists then of regulating the absorptance properties of leaves, leaf orientation and leaf angle to minimize the absorbance of heat (Ehleringer and Comstock, 1987). Changes in one or more of these attributes in response to environmental change, in particular with regard to gradients of temperature and precipitation, can be intergeneric, with replacement of related species along the gradient, and occasionally within a single species. Changes within a single species can also occur in response to seasonal changes in temperature and water relations (Ehleringer and Werk, 1986). In addition water loss is minimized while maintaining sufficient CO_2 input by maximizing the use of the most favourable time of day for photosynthesis. This is achieved by rapid stomatal response to leaf water potential, to atmospheric humidity or to both.

Seasonal changes in temperature and soil– plant water relations offer more opportunity to the plant for regulating gas exchange. Thus in addition to the traits involved in diurnal change there are a number of others. High temperatures, by affecting water loss and by structurally damaging protein, are an important factor affecting carbon acquisition. High temperature acclimation in photosynthesis raises the optimum temperature for one or more of the component processes or at minimum reduces the rate of decline with rise in temperature.

The effects of the seasonal decline in soil water content can be reduced by increasing cell wall moduli of elasticity and by osmotic adjustment which can be 0.5–0.8 MPa (Larcher et al., 1981). Reflective cuticles or leaves reduce the heat load on the leaf as do changes in leaf orientation. More drastic seasonal responses include leaf and branch shedding and leaf dimorphism in drought-deciduous shrubs, and related to this an increase in the root weight ratio. Recent evidence indicates that hormones, through their effects on stomatal movement, are involved in plant–soil water relations. In *Arbutus unedo* and *Quercus coccifera*, high rates of CO_2 uptake were maintained in summer in the absence of water shortage over a wide temperature range with optima between 25 and 30°C, indicating that there is no endogenously induced summer dormancy in these species and that, in the field, low rates of CO_2 uptake are a consequence of the effects of exogenous factors (Tenhunen et al., 1987).

Canopy geometry affects stand productivity, because of its effect on rates of water loss (Miller, 1983). In evergreen canopies differentiation of sun and shade leaves is advantageous in terms of water use efficiency (WUE). This is because in shade leaves, experiencing low amounts of solar radiation within the canopy, stomatal conductance is lower without a matching decrease in carbon acquisition. Thus in 1–2-m tall *Quercus coccifera*, a leaf area index above 2.3 made shade-acclimation and differentiation into sun and shade leaves down the canopy advantageous as CO_2 uptake and WUE were enhanced (Meister et al., 1987). Old leaves become acclimated to shade as they become overtopped by younger leaves.

Reduced leaf area index (LAI) is the main response to increased aridity. This reduces transpirational water loss and is usually accompanied by shifts in the pattern of carbon allocation, for example towards higher root : shoot ratios. Any factor, such as severe drought, that causes a reduction in leaf area index may be beneficial in the short term as it reduces transpiration, but it will increase the probability of enhanced erosion when rain eventually falls.

13.2 SOME ECOLOGICAL ASPECTS OF THE MEDALUS PROJECT

13.2.1 Comparisons of the climate at the MEDALUS sites

The eight sites used in MEDALUS I spanned the northern Mediterranean and so experienced between them much of the climatological variation of that region. The Mediterranean climate is characterized by unpredictability regarding the amount and timing of the rainfall and by a summer dry season of varying length and timing. Data from the MEDALUS sites provide good examples of this irregularity and further details are given in this volume in the field site chapters.

At Vale Formoso, Portugal, annual rainfall for 1987–92 ranged from 277 to 044 mm, with 95% distributed approximately equally between autumn, winter and spring and the remaining 5% occurring in the dry period from June to August. The irregularities in the precipitation regime both between and within years probably do not affect the vegetation dynamics and directional changes in the semi-natural vegetation appreciably as the annuals have seed banks and the perennials internal reserves. Prolonged extremes, for example the exceptionally dry year 1991/92 or the exceptionally wet summer of 1938/39, with 104 mm rain, can affect successional changes. The effects of the timing of rainfall on wheat production are discussed in Chapter 6.

The site in the Rambla Honda, Spain, in the rain-shadow of the Sierra Nevada, is in the driest part of Europe and its climate differs from the markedly less arid but still Mediterranean climate of Vale Formoso. Here average annual precipitation over a 25-year period was 220 mm and the climate is an extreme Mediterranean one regarding aridity, high temperatures and low humidities. Because of the pronounced unpredictability of its precipitation regime the climate, especially during sequences of low rainfall years, tends to be arid. There is no definite wet season and any month from September to June has an equal probability of receiving the same amount of rain. There is a dry period in July and August in which there is often no precipitation. There is a marked difference between years as at Vale Formoso in total precipitation and in its distribution within years. Humidities are around 60–70% in winter decreasing to around 20% in summer with absolute minima of around 5%. Mean monthly air temperatures based on the period 1967–88 are highest in July and August, around 26.6°C, and lowest from December to February, around 11.0°C. Winter temperatures do not fall below 7°C. However, ground frosts do occur.

The El Ardal site in Murcia, Spain, although still experiencing a dry Mediterranean climate, is markedly less arid than Rambla Honda. Thus the 1940–90 annual average precipitation was

320 mm compared with 220 mm for a comparable period from Rambla Honda. Over this 50-year period, the regime consisted of an irregular series of alternating wet and dry periods, with most years departing widely from the mean. Extremes during this period were 850 mm in 1951 and approximately 50 mm in 1945 and 1978. Recording at the MEDALUS site at El Ardal began in 1989, when 700 mm rain were recorded. Values in later years were around 400 mm. Monthly values were also irregularly distributed between years. The summer dry period in which precipitation is low or absent, less than 10 mm during July and August, is the only certain feature of the precipitation regime.

The sites at Réart and En Ferran in Roussillon, France, have a higher rainfall and usually a shorter dry season than do the two Spanish sites. Rio Santa Lucia, Sardinia, experiences the characteristic variation between and within years as to the amount and timing of the rainfall and a dry season of varying length sometime during the period late June to early September.

In Petralona, north Greece, there is no obvious dry season and there is an exceptionally high summer rainfall. The distribution of rainfall ranged from 3 to 65 mm per month.

At the Spata site near Athens, average annual precipitation is 495 mm with 66% falling in the period November to April. Typically much of this rain falls as storms of varying degrees of intensity. The irregularity of the precipitation regime is apparent in that in the dry year 1988/89 evapotranspiration exceeded precipitation while in 1989/90, a very wet year, the opposite occurred. In 1990/91, which was a normal year, evapotranspiration again exceeded precipitation but only from spring onwards.

Ecologically important, and perhaps even more so in its effects on evolution, are not so much the seasonal characteristics of the Mediterranean climate but its irregularity in terms of the distribution and amount of precipitation both between and within years. Thus even though the occurrence of a period of high temperatures and drought is virtually certain each year, its timing and duration during the summer are unpredictable.

Similarly, the wet season can occur at any time from September until June and may last for a few weeks or several months. Consequently, there are interactive effects with temperature and other climatic factors that may differ from year to year. This unpredictability indicates that there should be an opportunist element in the behaviour of those plants that are unable to exploit deep water sources but rely on current precipitation. There is evidence from some of the MEDALUS sites, for example at Rambla Honda, that such behaviour is superimposed on seasonal activity in some species. However, opportunist growth in response to exceptional rainfall events just before the onset of the dry season or before it ends can result in increased and sometimes lethal water stress. This suggests that in order to maximize survival, responses to temperature and photoperiod or to some other reliable predictors of the seasons have been selected for, to initiate late winter and spring growth for example, when conditions are usually favourable for these activities. If this is correct the species concerned presumably evolved in a seasonal climate. Their opportunist behaviour will have evolved more recently, with as yet no obvious advantage, in response to the increasingly random element in the amount and timing of the rainfall as the Mediterranean climate developed.

13.2.2 Comparisons of the soils at the MEDALUS sites

An example of how the data from the MEDALUS core programme can be used to make detailed inter-site comparisons has already been given in this volume (Chapter 5,

Section 9) where aggregation is compared at a number of sites. The following section deals with much broader comparisons and looks especially at the relationship between soils and vegetation.

The soils at Vale Formoso, derived from Carboniferous schists and Pleistocene sedimentary deposits, are typical red and yellow Mediterranean soils with some profile development. They are shallow, hence liable to drought, with the B–C horizon approximately 0.5 m deep. Their permeability is poor especially after drying out. Hence because of runoff, they are liable to erosion.

At the Rambla Honda site soil vegetation relations were investigated down the western slopes of the valley. Here *Stipa* is dominant on the upper slopes, *Anthyllis* on the fans below and *Retama* on the floor of the valley. The rocks are strongly bedded and steeply dipping, and the direction of dip affects the soil–water relations of the valley. All soils, whatever their position in the valley, are stony loamy sands or sandy loams, with little profile development. When structure is present it is associated with plant roots. Infiltration rates varied with geomorphological position and with type of vegetation. Thus they increased down slope from 4–6 mm h^{-1} in *Stipa tenacissima* vegetation at the top of the fans, to 9–25 mm h^{-1} in *Anthyllis vulneraria* midway down the fan, to 24–41 mm h^{-1} in *Retama sphaerocarpus* on the valley floor. Hydraulic conductivity varied widely within particular types of vegetation and positions along the slope. Organic matter content was variable and related to vegetation type, *Retama* having the lowest, 1.4–1.6% and *Stipa* the highest, 3.8–5.2%. Total porosity and water-holding capacities of all soils were low. Values for stands of *Stipa* were slightly higher than for those of *Retama* and *Anthyllis*. Sediment loss was higher on the slopes than on the fans. However, where *Stipa* tussocks were present on the slopes, there was increased spatial heterogeneity in the pattern of runoff with increased storage of sediment beneath or adjacent to the tussocks compared with in the areas between.

At El Ardal the study plots were situated along a slope with undisturbed pine–shrub vegetation at the top, a zone of sporadic cultivation with abandoned fields midway down and uncultivated shrubby vegetation at the bottom. Soil moisture content at field capacity, rarely attained in the field, was 0.2–0.3 cm^3 cm^{-3}. Available water for plants was low with levels around permanent wilting point for most of the year. Infiltration rates were high, approximately 1 m h^{-1}, especially under natural vegetation. On arable fields, saturated conductivities decreased from 100–150 mm h^{-1} at the surface to 55–60 mm h^{-1} at 500–600 mm depth, the position of the plough layer. The soil moisture regime demonstrates the beneficial effects of high infiltration rates in maximizing the storage of any precipitation that does occur. The type of vegetation, both qualitatively and quantitatively, was affected by the soil moisture regime. Hydraulic conductivities, by affecting the proportion of the precipitation that enters the soil, affect the type of vegetation present within a particular climatic regime. As at Vale Formoso, cultivation reduces soil permeability and so reduces the water available to plants and increases the tendency for erosion to occur. The effects of sporadic cultivation depend to a large extent on the hydraulic properties of the soils involved. Thus on coarse-textured soils compacting and loss of hydraulic conductivity as a result of cultivation will be minimal and natural regeneration can be rapid. On fine-textured soils, however, structural damage can persist for many years after cultivation ceases, thereby retarding natural regeneration.

At Réart, southern France, the experimental plots are in abandoned and currently cultivated vineyards and abandoned olive groves. The soils have good profiles that have developed since cultivation ceased around 1985. Ecologically these soils appear more benign than many Mediterranean ones owing to their ability to absorb and store precipitation. However this ability depends on an absence of erosion, as the accumulation of gravel from the runoff and the resulting impervious pavements may reduce rates of infiltration and hence the chance of plants establishing. Low nutrient levels, especially of phosphorus, coupled with the presence of relatively high concentrations of sodium, may more than counteract the relatively favourable soil moisture regime. Hence establishment on abandoned land and regeneration of the natural vegetation could be slow on these soils.

The En Ferran and Réart sites are similar, the main difference being that the En Ferran plots are in cultivated rather than abandoned vineyards. As a consequence of the active cultivation regime, erosion is more severe at the En Ferran sites, with little profile development over parts of the site because of truncation.

The study area at Rio Santa Lucia in Sardinia is larger than the other sites and comprises an area of approximately 110 km. Its geology is extremely varied. Soil type in the area is closely related to plant cover. Thus on the tops of the hills, where there is only scant vegetation cover, there is a thin highly skeletal argillaceous soil. On the same terrain but with more vegetation cover, the soils are better developed with profiles. On less rugged terrain where, importantly from the erosion point of view, the slopes are less steep than on the tops and where scrub or woodland occurs, still deeper sandy soils develop. The runoff plots demonstrate, as at other MEDALUS sites, that microtopography and vegetation cover markedly affect runoff and sediment yield and that sediment yield is also closely related to rainfall.

The Petralona site in northeastern Greece is on an isolated uplifted limestone block surrounded by marls. It is horizontally bedded and massive. At the top of the catena the soils are shallow and form pockets between the limestone outcrops. Profiles are truncated with no A horizon. Rock and stone covers more than 40% of the surface with the pockets of soil themselves approximately 60% stone covered. Midway down the catena the soils are better developed and deeper although the profiles are still truncated. The lower soils are derived from the limestone marls that occur below the limestone outcrop. Nutrient levels decreased with depth on all soils. Organic matter content was high at the semi-natural sites (4–6%), but less (2–3%) on the agricultural soils. Structural stability was greatest in the semi-natural soils. Runoff and sediment yield differed markedly between the runoff plots, with values of 9.2 to 3.2 mm m^{-2} and 14 to 219 g m^{-2} respectively. Soil type and not slope or cover of vegetation was the most important determinant of sediment yield and hence of the intensity of erosion. This is because cover has to vary by large amounts to affect sediment yield appreciably.

At the study sites at Spata near Athens, the main underlying rocks are limestones while the surface strata are sandstone or conglomerate marls. The soils are argillaceous. Intensive cultivation over the past century has resulted in the removal of most of this material from the steep slopes leaving a residue of stony gravelly soil. The soils derived from the conglomerates are less prone to erosion because of their high content of stones and their greater permeability. Those derived from the marls are calcareous, finer textured, deeper, more fertile, less permeable and with a lower stone content than those derived from the conglomerates. Despite their susceptibility to erosion this is not severe because they occur on shallow slopes.

The following conclusions concerning Mediterranean soils arise from the work of the MEDALUS groups and others.

1. Profiles are generally poorly developed and often disrupted, e.g. truncated by erosion or cultivation.
2. Vegetation improves the size and strength of structural units thereby increasing infiltration rates and hence the proportion of the rainfall reaching the soil surface that enters the soil.
3. Cultivation by destroying structure and creating pans reduces water input, increases runoff and hence increases desertification.
4. The harmful effects of cultivation persist after abandonment due to the slow recovery of structure and hence the continued low permeability and consequential adverse soil water regime.
5. The liability to erosion and, related to this, soil–water relations, depends on a number of factors. These include the amount of precipitation, steepness of slope and the nature of the cover of vegetation. It also depends on the permeability of the soil. That in turn depends on textural and structural properties, the degree of profile development and the hydraulic conductivities of the horizons.
6. The heterogeneity in soil texture, with its associated differences in conductivity characteristics, together with the irregular rainfall regime with long periods between storms, probably explains the observed differences in water relations between plants and within stands. The consequences for establishment of the interactive effects of climatic and soil heterogeneity need exploring to increase understanding of plant fitness in the Mediterranean environment.
7. At the El Ardal and Rambla Honda sites observations indicated that the water relations of plant and soil need further investigation, in particular the mechanisms and dynamics of water movement to the roots, that enable turgor to be maintained and some growth to take place when the soil water is at the lower limits of availability for plants (i.e. at potentials more negative than the notional -1.5 MPa permanent wilting point). High root : shoot ratios, by reducing the rate of flow needed to satisfy a given demand, could be important in this context, as could the osmotic concentration of the cell sap.

13.2.3 Comparison of vegetation at the MEDALUS sites

In the Vale Formoso area the predominant type of semi-natural vegetation is open savannah with annual rich grassland and scattered evergreen trees, mainly *Quercus suber*. The density of the tree cover can be as low as two to three per hectare. The density depends on management as does the species composition of the grassland. The landscape is a heavily managed one with cork, olives, wheat and milk, meat and wool from sheep the main products. Where cultivation is impossible on steep slopes a dense evergreen shrub and tree community develops.

Agricultural activities are the main cause of environmental degradation in the area because of the loss of soil and nutrients in the runoff resulting from the widespread wheat–fallow agriculture system practised. Excessive grazing is also involved but to a lesser extent. These activities also prevent the vegetation from developing into the more stable woody communities that occur on the uncultivatable areas. In the longer term, arable cropping is resulting in a gradual decline in the ability of the Vale Formoso area to support agriculture or the regeneration of shrub–woodland communities.

The savannah vegetation, apart from the scattered tree component, comprises mainly annual species. These form a stable community when not disrupted by erosion brought about by excessive grazing or too frequent or too prolonged periods of cultivation. As already mentioned, environmental deterioration is occurring in the Vale Formoso area because agricultural and pastoral exploitation of the savannah are too intensive for the maintenance of the savannah community. However, the long-term stability of this community in other parts of southeastern Portugal and western Spain shows that systems of management can be developed in the Mediterranean environment that do not result in appreciable environmental degradation.

The vegetation of the Rambla Honda site, dominated by grassland and dwarf shrub communities, is physiognomically very different from that of the Vale Formoso area in Portugal. However there are similarities in that the types of vegetation are largely determined by man's activities and also that winter annuals, as in the savannah at Vale Formoso, are an important component of the vegetation. However, the community dynamics are markedly different since there is a greater tendency, suppressed by management practices, for more rapid successional change at Vale Formoso. By contrast at Rambla Honda the constraints of aridity rather than man's influence prevent rapid successional change. Also the absence of sporadic cultivation and a grazing regime that has not destroyed the perennial vegetation, has allowed the development of communities on the slopes and valley bottoms that are sufficiently stable for mosaic regeneration to operate.

The El Ardal, Murcia, site is situated in an area of often steeply rolling hills on which shrubby vegetation predominates although it is discontinuous and interspersed with abandoned arable fields. Scrub often develops on long-abandoned fields. The occasional valleys, too steep-sided for cultivation, support shrubby evergreen vegetation. The shrub vegetation forms a discontinuous and irregular canopy, often dense in places, with *Rosmarinus officinalis*, *Juniperus oxycedrus* and *J. phoenicia* dominant, with *Pinus halapensis* also occurring on the upper slopes. Some 60% of the species in this shrub-dominated vegetation are shrubs, while on the abandoned fields annuals predominate. Some of the hills are cultivated for almonds. The shallower slopes are sporadically cultivated for winter wheat depending on the amount of winter rain and prices. Hence the fallow or 'abandoned land' phase is of varying length. The whole area is grazed by sheep and some goats in an extensive but managed way.

The vegetation then, depending on steepness of slope, soil depth and time since last cultivated, ranges in type from open pastures rich in winter annuals, through dwarf shrubs to scrub of irregular density. Thus despite the disturbance resulting from cultivation, successional stages towards woodland are present. Notwithstanding the irregular rainfall regime at El Ardal there were consistent effects on the one hand of slope and orientation on the amount of precipitation received and on the other of the type of vegetation on the amount of precipitation intercepted. Both of these affected soil moisture. Thus *Juniperus oxycedrus* communities intercepted 40–50% and other types of vegetation 25–30% of the precipitation.

Over much of the area, man's activity prevents the development of the relatively stable communities of the later stages of succession or of equilibrium situations with the development of cyclical regeneration. Erosion and runoff were inversely correlated with vegetation cover. However, some types of vegetation provide better protection than others against erosion. Thus *Cistus ladaniferus* dominated vegetation has bare areas between the bushes, once the annuals die in spring, that are susceptible to runoff. Well-developed annual communities on the other hand, that form a cover of litter on death of the vegetation, provide a superior protection against runoff. Abandoned fields with their relatively sparse vegetation had the highest runoff. The site average was $20\,\mathrm{g\,m^{-2}\,yr^{-1}}$ while in wheat and barley fields the figure was $30\,\mathrm{g\,m^{-2}\,yr^{-1}}$. Soil losses were more closely related to storm intensity than to total amount of rain.

The main anthropomorphic effect on the vegetation was not serious degradation because of soil and nutrient losses in runoff but the effect on the type of vegetation present. In the absence of man's activities, evergreen woodland, probably *Quercus ilex* dominated, would predominate on the lower and middle slopes and *Pinus halapensis* at higher altitudes. Today, however, the vegetation on the shallower slopes forms a series of unstable successional stages whose duration depends on the economics of cereal and of almond production. The uncultivatable slopes are also successional stages and not in equilibrium because of timber extraction and grazing. Despite man's extensive interference over millennia, the soils of the El Ardal area appear to be fully capable of maintaining woodland, or other stable woody vegetation, if the grazing pressure was reduced and sporadic cultivation ceased.

At Réart, Roussillon, two of the study plots are on vineyards abandoned approximately 5 years before the start of the MEDALUS project and the third is in a former olive grove. At the other site in Roussillon, En Ferran, the plots are in productive vineyards. There is much sheet and gully erosion in the cultivated and recently abandoned vineyards and olive groves at both sites and unstable colonizing communities predominate there. The types of vegetation are largely determined by the condition of the vineyards. Thus apart from the vines themselves, vegetation is virtually absent in the fully cultivated vineyards apart from isolated plants of herbicide-resistant weeds. Long-abandoned vineyards, however, are reverting to shrubby vegetation and there is a sequence of intermediate types between these two extremes. At Réart the colonizing flora and vegetation have reached the shrub stage in some areas, with *Quercus ilex*, *Q. pubescens* and *Erica arborea*. The floras are similar but the dominants differ between the sites. Available information suggests that although there are similarities between the 'mosaics of abandonment' at Réart and En Ardal, the vegetation is more resilient at El Ardal.

The study area in southwestern Sardinia comprises the Rio Santa Lucia catchment and extends from a mountainous hinterland, maximum altitude 1113 m, to coastal plains and lagoons. It embraces a wide range of communities and types of vegetation. Thus the upland vegetation is mainly evergreen, often dense, sclerophyllous woodland with *Quercus ilex*. Certain species have characteristic habitats there, for example *Phillyrea latifolia*, *Juniperus oxycedrus* and *J. phoenicia* occur on steep slopes and shallow soils and *Euphorbia dendroides* and other thermophilous species occur on steep south-facing slopes. At the highest altitudes on the steeper slopes open rock-heath communities occur. These are dominated by *Teucrium marum* and *Euphorbia spinosa*. The woodland and maquis vegetation is replaced by garrigue and dwarf shrub communities in the sandy valley bottoms and on rock faces.

At lower altitudes there is some *Q. suber* woodland and plantations of *Eucalyptus* spp. The *Eucalyptus* plantations, compared with the other types of woodland in the area, are notably lacking in vegetation beneath the tree layer. *Q. suber* only occurs in habitats strongly influenced by man, in particular in low-lying accessible sites in the valley bottoms and near sheep-folds. Along the lower river course, there is gallery woodland with *Alnus glutinosa, Salix purpurea* and *Nerium oleander* and there are salt-marsh communities adjacent to the lagoon.

The vegetation of the Petralona site in northeast Greece is typical of 20% of Greece's vegetation. The site consists of a flat-bedded block of limestone with pockets of maquis vegetation separated by bare limestone on the top. At the base of this block, there is a break of slope. Here the soils become deeper, more marly and there is maquis vegetation interspersed with cultivated areas.

Maquis is the dominant vegetation type at Petralona, with *Quercus coccifera, Phillyrea* spp., *Juniperus oxycedrus* and *Pistacea lentiscus* the dominant species. *Q. coccifera* is the predominant shrub, with 57% cover. Total cover averaged 67% in 1991/92. As in other parts of the Mediterranean region, the denser the vegetation the greater is the development and stability of the structure of the associated soils, the greater the permeability of those soils and the less is runoff and erosion. As observed on other MEDALUS sites, the wheat fields here have a poorer structure and a higher yield of sediment than do the areas of semi-natural vegetation. Cultivation then, leads to instability. By contrast, a moderate grazing pressure may actually be beneficial to the functioning of the maquis vegetation by stimulating growth, thereby speeding nutrient cycling and preventing the build-up of dead material with its attendant fire risk.

The overall conclusion from the work at Petralona is that the plants are well adapted to the present climate and are pre-adapted to the changes predicted as a result of global warming. Hence there are unlikely to be substantial changes in the vegetation there as a consequence of global warming. Rather, it is man's disruption and destruction of the vegetation, not climatic change, that is more likely to lead to irreversible degradational change and desertification in the future.

Most of the area around the site at Spata, near Athens, is cultivated for rain-fed annual and perennial crops, mainly cereals and olives. In 1979, some of the long-established (for more than 160 years) olive groves were cleared and vines planted. The study plots are on a hillside; half are in vineyards and half in olive groves. 'Natural' vegetation of winter annuals and shrubs has developed under the olives. Here shrub height was mainly 0.3–1 m and cover varied from 44 to 95% at the time of maximum growth. The climate, sloping topography and shallow soils make the plots susceptible to erosion and runoff. The depth of soil exploitable by plants at this site is limited because of the presence near the surface of bedrock, a petrocalcic horizon or gravelly/stony layers. Consequently, the vegetation is highly susceptible to desiccation. This can be lethal, especially if erosion reduces the depth of soil available for exploitation. Biomass above and below ground of the natural vegetation in the olive groves in July 1991 was 380 and 440 g m^{-2} respectively. It was approximately four times less in the vineyards.

What can be inferred of general relevance to the Mediterranean Basin from the information about the vegetation of the various MEDALUS sites?

1. That although most of the communities are short-lived regenerative or regressive successional stages due largely to man's direct or indirect activities, stable

communities do exist for example in the savannahs of southeastern Portugal and western Spain, the woodlands of the southern Sardinian uplands and the *Stipa* grasslands of eastern Andalusia. Moreover, such communities can be created and managed to maintain stability.

2. That the response of the Mediterranean vegetation to disruption and its resilience in the face of frequent perturbations, is largely because it is a two-component perennial shrub and winter annual system. Thus depending on the type and frequency of the disturbance, the vegetation becomes increasingly winter-annual or shrub (and eventually tree) dominant; or it may stabilize at an intermediate stage. The MEDALUS study sites encompass most of these phases. Thus resilience in the form of the persistent regeneration of successional stages back towards woodland in response to a long history of sporadic cultivation occurs at El Ardal. Stability at the two extremes occurs at Vale Formoso (winter annual dominant) and in the Santa Lucia uplands (evergreen shrub and tree dominant). The *Stipa* and *Retama* communities at Rambla Honda are relatively stable intermediates.

    The widespread occurrence of this two-component system is probably a consequence primarily of grazing pressure rather than fire. Thus grazing enhances canopy discontinuity and creates habitats between the shrubs that are favourable for the annuals. The dependence on grazing for the maintenance of the inter-shrub habitat is demonstrated at the Rambla Honda site, where protection from grazing has resulted in a reduction in the physiognomic distinction between the shrub and inter-shrub components. In the post-burn situation, however, annuals are pioneers and their maintenance in the succeeding shrubby vegetation will largely depend on the grazing regime.

3. The findings at all sites confirm the overriding importance of vegetation in reducing runoff and loss of sediment. They also demonstrate that soil type, in particular soil texture, affects the liability of a particular area to erosion. Thus at El Ardal the soils are relatively resistant and man's effect on the landscape is primarily to alter the type of vegetation. However, where the soil is highly impermeable, as at Réart and Spata, greatly enhanced runoff and erosion are the main consequences of reducing the cover of vegetation. Agriculture initiates and stimulates erosion primarily because of the soil disturbance involved and because the type of vegetation produced by crops is ineffective in reducing runoff.

    An obvious implication from these observations is that environmental degradation can be reduced and reversed in the Mediterranean Basin by creating conditions for the establishment and maintenance of an effective cover of vegetation. The MEDALUS sites also demonstrate the importance of grazing directly and indirectly in determining the type of vegetation and hence the amount of erosion. Thus the stability of the savannah vegetation and the re-creation of the annual-rich grassland after occasional wheat cultivation, as at Vale Formoso, depends on controlled grazing. Grazing, too, as shown at Rambla Honda, affects the structure of the shrub and winter annual communities, enhancing the pattern and thereby helping to create habitats between the shrubs favourable for the annuals.

4. Vegetation is more resilient in the face of degradational change than are soils and it regenerates more rapidly. Soil genesis from parent rocks that are resistant to weathering may take thousands of years in the Mediterranean climate, for example in

the uplands of the Santa Lucia basin. The rapidity with which plant communities develop will depend on the type of substrate and on other environmental factors, for example grazing intensity or the frequency of burning. Current ecosystems may be permanently lost because the ones developing as soil genesis proceeds will be different owing to climatic changes over the period of regeneration. The resilience of the Mediterranean vegetation is demonstrated, in that where cultivation has ceased and grazing intensity becomes less severe as consequences of rural depopulation, successional processes begin to operate.

5.   Observations at a number of the MEDALUS sites, for example Petralona and El Ardal, based on the response of the vegetation to differences in climate between years, suggest that first the vegetation will be able to tolerate and adapt to the predicted climatic changes resulting from global warming. Secondly, it is man's disruptive activities in the landscape that are the greatest threat to the existing vegetation and to the development of stable communities.

13.2.4 Productivity and vegetation dynamics studies at the Rambla Honda, El Ardal, Petralona and Spata sites

Studies specifically relating to the productivity and dynamics of the vegetation were carried out at the Rambla Honda (Chapter 7, Section 5), El Ardal (Chapter 8, Section 6), Petralona (Chapter 11, Section 5) and Spata (Chapter 10, Section 4) sites. The work is reported in detail in the relevant chapters, but is summarized here.

Rainfall was the main determinant of biomass at all sites as exemplified by the experiments on wheat at the Spata site. Here plant size was the variable most responsive to rainfall. Elsewhere, for example at the El Ardal site, plant density was the main determinant of biomass in a particular rainfall regime. The effect of rainfall was strongly mediated by other factors, for example aspect, slope and soil type, in particular permeability and stoniness. Vegetation itself also influenced the effectiveness of a particular rainfall regime to support growth, by first intercepting a proportion of the rain and secondly by influencing the proportion of throughfall that entered the soil as opposed to being lost as runoff.

The widespread practice of sporadic cultivation, with its associated abandoned fields, meant that where soil type and slope made cultivation practicable, as at the El Ardal site, the vegetation and associated soils consisted of widely differing short-lived successional stages developing towards shrubby communities but usually not achieving an equilibrium before the next perturbation. At some sites, as in parts of the Rambla Honda site, some of the communities were stable, for example the *Stipa* and *Retama* communities, with evidence for cyclical regeneration. Evidence from the MEDALUS sites suggests that the vegetation reaches a state of equilibrium within a few decades of the last perturbation and that despite frequent disturbance, cyclical regeneration is an important feature of the dynamics of Mediterranean vegetation. Observations on the study sites also indicate that cultivation, unless preserving or creating a cover of vegetation or litter, always initiates degradational soil processes in that runoff and erosion increase often by orders of magnitude in association with it.

Grazing, unlike cultivation, does not normally have a sudden disruptive effect on the vegetation but the types of community and their physiognomy are profoundly influenced by it even if not obviously so. Thus, as the studies in Petralona on *Q. coccifera* and on *Stipa* at Tabernas show, it can affect not only biomass but also phenology and resource allocation.

The adaptation of the vegetation to low levels of available water and its resilience in response to local catastrophes such as the severe drought on the shallow soils at El Ardal, suggest that any gradual decrease in water availability as a consequence of the climatic changes that may occur as a result of global warming is not likely to have any appreciable effect on the stability of the vegetation and soils, compared with the effects of a changed regime of disturbance initiated by man. This conclusion assumes that increased temperatures and decreased rainfall would be the changes likely to have the greatest effect on the vegetation. However, rain-exclusion experiments at Spata suggest that decreased rainfall could affect soil nutrient status and, perhaps more importantly, reduce structural stability. Both these could affect the vegetation.

The work on vegetation dynamics and production revealed several areas where further research is needed. Thus, what is the cause of the marked heterogeneity in the water relations of individual plants within a stand? How, as reported from El Ardal and suspected elsewhere, do some shrubby species use more water than is apparently available to them in the soil? Related to this problem is the apparent ability of *Retama* to extract more liquid-phase water than is available in the root run. Investigations of the effects of soil heterogeneity on plant–soil water relations are obviously needed. Finally, the degree of obligate seasonality as compared to opportunism in phenological and physiological activity and in growth needs investigating as this could affect response to climatic change. It would involve, among other approaches, studies of carbon allocation to determine the extent to which the resources for spring growth come from reserves or from current photosynthesis.

13.2.5 Final synthesis

The following are a series of propositions. Much of the evidence for these has been derived from the investigations carried out at the MEDALUS sites.

1. That the evergreen woody species have had a long and continuous history of evolution in similar environments to that of the present Mediterranean Basin. That, because of their long history, these species had time to acquire the adaptive gene complexes that pre-adapted them to the current Mediterranean climate as it developed during and after the Pleistocene.
2. That the winter annual's life cycle has evolved in response to climates with winter rains and a summer dry season. Hence the prevalence of annuals with this type of life cycle in the Mediterranean Basin is a consequence of the change there from summer to winter rains as a Mediterranean type of climate developed.
3. That before man became an important factor in the Mediterranean Basin, winter annuals and shrubs (since they had both evolved before the advent of man) probably formed mixed open communities in the limited drought liable shallow soil habitats available.
4. That this shrub–annual community extended as man created suitable habitats. Thus it is possible that much of the current vegetation of the Mediterranean Basin consists of

communities that had integrated before man appeared and whose constituent species were already adapted to environmental instability. If this is correct, then much of the current non-arboreal vegetation consists of long-established temporally stable assemblages and not transient remnants of more complex communities that supposedly developed in response to more stable environmental conditions.

5. That this long history of selection for disturbance has produced, because of the resilience of the constituent species, a community resilience with the three functional types, evergreen and drought-deciduous shrubs and winter annuals, not temporally and spatially overlapping members of different (now vanished) communities but integral parts of the same assemblage, although each component is often temporally separated from the others.

6. That, unlike seral stages that are separated by time in a unidirectional successional sequence, this community does not embrace any permanent directional shift. Instead, there is a change in the proportion of the components, depending on the current environment. For example, the savannahs of southwest Spain and southeast Portugal contain tree and winter annual elements that are at the opposite ends of what could be a successional series. However, management or other environmental change could result in the establishment of intermediate shrubby species that would merely be one of the currently absent components of the same community and not a successional stage. Again, depending on subsequent environmental (including man-induced) changes, the proportions of the various components would also change. These changes would be in response primarily to exogenous factors and not to the endogenous community induced changes involved in succession.

7. That the winter annual shrub community of the Mediterranean Basin is a resilient, tolerant assemblage that has developed in response to a 'constant instability', where the periodicity of environmental change is often less than the time needed for the development of stable communities. This contrasts with the successional situation, where the periodicity of environmental change is longer than the time needed for the vegetation to stabilize. Although the mixed annual shrub community almost certainly evolved before man became an important environmental factor, man, by creating favourable habitats throughout the Basin, has been largely responsible for its current prevalence.

8. The activities of both woody and annual species are closely related to the seasonal changes of the Mediterranean climate. These species also respond opportunistically to the typically unpredictable rainfall within and between years of the Mediterranean climate. As unseasonal activity in response to an exceptional rainfall event can apparently be disadvantageous, because the rain to which the plants respond may not be sufficient to support the growth evoked by the rain, selection for opportunism is either not important or has as yet not been as effective as has selection in response to seasonal events.

9. That the communities of evergreen and drought-deciduous shrubs and winter annuals of the Mediterranean Basin owe their present extent to the creation of favourable habitats by man. That, although currently damaged and disrupted by human activity, these communities are inherently resilient and would rapidly recover if the pressure of disturbance was reduced. Spontaneous recovery of the arboreal vegetation is more doubtful.

13.3 REFERENCES

Aschmann, H. (1973a) Distribution and peculiarity of Mediterranean ecosystems. In di Castri, F. and Mooney, H.A. (eds) *Mediterranean Type Ecosystems*. Chapman & Hall, London/Springer-Verlag, Berlin, 11–36.

Aschmann, H. (1973b) Man's impact on the several regions with Mediterranean Climates. In di Castri, F. and Mooney, H.A. (eds) *Mediterranean Type Ecosystems*. Chapman & Hall, London/Springer-Verlag, Berlin, 363–371.

Axelrod, D.I. (1973) History of the Mediterranean ecosystem in California. In di Castri, F. and Mooney, H.A. (eds) *Mediterranean Type Ecosystems*. Chapman & Hall, London/Springer-Verlag, Berlin, 225–277.

Bazzaz, F.A. and Morse, S.R.(1991) Annual plants: potential responses to multiple stresses. In Mooney, H. A., Winner, W.E. and Pell, E.J. (eds) *Responses of Plants to Multiple Stresses*. Academic Press, San Diego, 283–305.

Bradbury, D. E.(1981) The physical geography of the Mediterranean lands. In di Castri, F. and Goodall, D.W. (eds) *Ecosystems of the World 11. Mediterranean-Type Shrublands*. Elsevier Scientific Publishing Company, Amsterdam, 53–62.

Correia, O., Catarino, F., Tenhunen, J.D. and Lange, O.L. (1987) Regulation of water use by four species of *Cistus* in the scrub vegetation of the Serra da Arrabida, Portugal. In Tenhunen, J.D., Catarno, F.M., Lange, O.T. and Oechel, W. C. (eds) *Plant Response to Stress*. Springer-Verlag, Berlin, 247–257.

Diamantoglou, S. and Meletiou-Christou, M.S. (1981) Changes of storage lipids, fatty acids and carbohydrates in vegetative parts of Mediterranean evergreen sclerophylls during one year. In Margaris, N.S. and Mooney, H.A. (eds) *Components of Productivity of Mediterranean-Climate Regions: Basic and Applied Aspects*. Dr. W. Junk Publishers, The Hague, 121–127.

Ehleringer, J.R. and Comstock, J. (1987) Leaf absorptance and leaf angle: mechanisms for stress avoidance. In Tenhunen, J.D., Catarno, F.M., Lange, O.T. and Oechel, W. C. (eds) *Plant Response to Stress*. Springer-Verlag, Berlin, 55–76.

Ehleringer, J.R. and Werk, K.S. (1986) Modifications of solar-radiation absorption patterns and implications for carbon gain at the leaf level. In Givnish, T.J. (ed) *On the Economy of Plant Form and Function*. Cambridge University Press, Cambridge, 57—82.

Escarre, A., Ferres, Ll., Lopez, R., Martin, J., Roda, F. and Terrades, J. (1987) Nutrient use strategy by evergreen oak (*Quercus ilex* ssp *ilex*) in NE Spain. In Tenhunen, J.D., Catarino, F.M., Lange, O. T. and Oechel, W. C. (eds) *Plant Response to Stress*. Springer-Verlag, Berlin, 429–458.

Groves, R.H. (1983) Nutrient cycling in Australian heath and South African fynbos. In Billings, W.D. (eds) *Mediterranean-Type Ecosystems: The Role of Nutrients*. Springer-Verlag, Berlin, 179–207.

Kruger, F.J. (1983) Plant community diversity and dynamics in relation to fire. In Billings, W.D. (eds) *Mediterranean-Type Ecosystems: The Role of Nutrients*. Springer-Verlag, Berlin, 446–472.

Kruger, F.J. (1987) Responses of plants to nutrient supply in Mediterranean-type ecosystems. In Tenhunen, J.D., Catarino, F.M., Lange, O.T. and Oechel, W.C. (eds) *Plant Response to Stress*. Springer-Verlag, Berlin, 415–427.

Kummerow, J. (1973) Comparative anatomy of sclerophylls of Mediterranean climatic areas. In di Castri, F. and Mooney, H.A. (eds) *Mediterranean Type Ecosystems*. Chapman & Hall, London/Springer-Verlag, Berlin, 157–167.

Kummerow, J. (1981) Carbon allocation to root systems in Mediterranean sclerophylls. In Margaris, N.S. and Mooney, H.A. (eds) *Components of Productivity of Mediterranean-Climate Regions: Basic and Applied Aspects*. Dr. W. Junk Publishers, The Hague, 115–120.

Lamont, B.B. (1983) Strategies for maximising nutrient uptake in two Mediterranean ecosystems of low nutrient status. In Billings, W.D., Golley, F., Lange, O.L., Olson, J.S. and Remmert, H. (eds) *Mediterranean-Type Ecosystems: The Role of Nutrients*. Springer-Verlag,Berlin, 245–273.

Larcher, W., de Moreas, J.A.P.V. and Bauer, H. (1981) Adaptive responses to leaf water potential, CO_2-gas exchange and water use efficiency of *Olea europaea* during drying and rewatering. In Margaris, N.S. and Mooney, A. (eds) *Components of Productivity of Mediterranean-Climate Regions: Basic and Applied Aspects*. Dr. W. Junk Publishers, The Hague, 77–84.

Le Houerou, H.N. (1981) Impact of man and his animals on Mediterranean vegetation. In di Castri, F. and Goodall, D.W. (eds) *Ecosystems of the World 11 Mediterranean-Type Shrublands*. Elsevier Scientific Publishing Company, Amsterdam, 479–521.

Maufette, Y. and Oechel, W.C. (1987) Seasonal photosynthate allocation of the Californian coast live oak *Quercus agrifolia*. In Tenhunen, J.D., Catarino, F.M., Lange, O.L. and Oechel, W. C. (eds) *Plant Response to Stress*. Springer-Verlag, Berlin, 437–444.

Meister, H.P., Caldwell, M.M., Tenhunen, J.D. and Lange, O.L. (1987) Ecological implications of sun/shade-leaf differentiation in sclerophyllous canopies: Assessment by canopy modeling. In Tenhunen, J.D., Catarino, F.M., Lange, O.L. and Oechel, W. C. (eds) *Plant Response to Stress*. Springer-Verlag, Berlin, 401–411.

Merino, J. (1987) The cost of growing and maintaining leaves of Mediterranean plants. In Tenhunen, J.D., Catarino, F.M., Lange, O.L. and Oechel, W. C. (eds) *Plant Response to Stress*. Springer-Verlag, Berlin, 553–564.

Merino, J. and Vicente, A.M. (1981) Biomass, productivity and succession in the scrub of the Donana biological reserve in southwest Spain. In Margaris, N.S. and Mooney, H.A. (eds) *Components of Productivity of Mediterranean-Climate Regions: Basic and Applied Aspects*. Dr. W. Junk Publishers, The Hague, 197–203.

Merino, J., Garcia Novo, F. and Sanchez-Diaz, M. (1976) Annual fluctuation of water potential in the xerophytic shrub of the Donana Biological Reserve (Spain). *Oecologia Plantarum*, **11** (1), 1–11.

Miller, P.C. (1983) Canopy structure of Mediterranean-type shrubs in relation to heat and moisture. In Billings, W.D., Golley, F., Lange, O.L., Olson, J.S. and Remmert, H. (eds) *Mediterranean-Type Ecosystems: The Role of Nutrients*. Springer-Verlag, Berlin, 134–166.

Mooney, H.A. (1981) Primary production in mediterranean-climate regions. In di Castri, F. and Goodall, D.W. (eds) *Ecosystems of the World 11. Mediterranean Type Shrublands*. Elsevier Scientific Publishing Company, Amsterdam, 249–255.

Mooney, H.A. (1983) Carbon-gaining capacity and allocation patterns of Mediterranean-climate plants. In Billings,W.D., Golley, F., Lange, O.L., Olson, J.S. and Remmert, H. (eds) *Mediterranean-Type Ecosystems: The Role of Nutrients*. Springer-Verlag, Berlin, 103–119.

Nahal, I. (1981) The Mediterranean climate from a biological point of view. In di Castri, F and Goodall, D.W. (eds) *Ecosystems of the World 11 Mediterranean-Type Shrublands*. Elsevier Scientific Publishing Company, Amsterdam, 63–86.

Papanastasis, V.P. (1981) Species structure and productivity in grasslands of northern Greece. In Margaris N.S. and Mooney, H.A. (eds) *Components of Productivity of Mediterranean-Climate Regions: Basic and Applied Aspects*. Dr. W. Junk Publishers, The Hague, 205–217.

Pons, A. (1981) The history of the Mediterranean shrublands. In di Castri, F. and Goodall, D.W. (eds) *Ecosystems of the World 11 Mediterranean-Type Shrublands*. Elsevier Scientific Publishing Company, Amsterdam, 131–138.

Pons, A. and Quezel, P. (1985) The history of the flora and vegetation and past and present human disturbance in the Mediterranean region. In Gomez-Campo, C. (ed.) *Plant Conservation in the Mediterranean Area*. Dr. W. Junk Publishers, Dordrecht, 25—43.

Poole, D.K., Roberts, S.W. and Miller, P.C (1981) Water Utilization. In Miller, P.C. (ed.) *Resource Use by Chaparral and Matorral*. Springer-Verlag, New York, 123–149.

Quezel, P. (1981) Floristic composition and phytosociological structure of sclerophyllous matorral around the Mediterranean. In di Castri, F. and Goodall, D.W. (eds) *Ecosystems of the World 11. Mediterranean-Type Shrublands*. Elsevier Scientific Publishing Company, Amsterdam, 107–121

Quezel, P. (1985) Definition of the Mediterranean region and the origin of its flora. In Gomez-Campo, C. (ed.) *Plant Conservation in the Mediterranean Area*. Dr. W. Junk Publishers, Dordrecht, 9–24.

Rabinovitch-Vin, A. (1983) Influence of nutrients on the composition and distribution of plant communities. In Billings, W.D., Golley, F., Lange, O.L., Olson, J.S. and Remmert, H. (eds) *Mediterranean-Type Ecosystems: The Role of Nutrients*. Springer-Verlag, Berlin, 74–85.

Rapp, M. and Lossaint, P. (1981) Some aspects of mineral cycling in the garrigue of southern France. In di Castri, F and Goodall, D.W. (eds) *Ecosystems of the World 11. Mediterranean-Type Shrublands*. Elsevier Scientific Publishing Company, Amsterdam, 289–301.

Raven, P.H. (1973) The evolution of Mediterranean floras. In di Castri, F. and Mooney, H.A. (eds) *Mediterranean Type Ecosystems*. Chapman & Hall, London/Springer-Verlag, Berlin, 213–224.

Read, D.J. and Mitchell, D.T. (1983) Decomposition and mineralization processes in Mediterranean-Type ecosystems and in heathlands of similar structure. In Billings, W.D., Golley, F., Lange, O.L., Olson, J.S. and Remmert, H.(eds) *Mediterranean-Type Ecosystems: The Role of Nutrients*. Springer-Verlag, Berlin, 208–232.

Rundel, P.W. (1991) Shrub life-forms. In Mooney, H.A., Winner, W.E. and Pell, E.J. (eds) *Responses of Plants to Multiple Stresses*. Academic Press, San Diego, 345–370.

Tenhunen, J.D., Beyschlag, W., Lange, O.L. and Harley, P.C. (1987) Changes during summer drought in leaf CO_2 uptake rates of macchia shrubs growing in Portugal: Limitations to photosynthetic capacity, carboxylation efficiency, and stomatal conductance. In Tenhunen, J.D., Catarino, F.M., Lange, O.L. and Oechel, W. C. (eds) *Plant Response to Stress*. Springer-Verlag, Berlin, 306–327.

Thrower, N.J.W. and Bradbury, D.E. (1973) The physiography of the Mediterranean lands with special emphasis on California and Chile. In di Castri, F. and Mooney, H.A. (eds) *Mediterranean Type Ecosystems*. Chapman & Hall, London/Springer-Verlag, Berlin, 37–52.

Tomaselli, R. (1981) Main physiognomic types and geographic distribution of shrub systems related to Mediterranean climates. In di Castri, F. and Goodall, D.W. (eds) *Ecosystems of the World 11. Mediterranean-Type Shrublands*. Elsevier Scientific Publishing Company, Amsterdam, 95–106.

Trabaud, L. (1981) Man and fire: impacts on Mediterranean vegetation. In di Castri, F. and Goodall, D.W. (eds) *Ecosystems of the World 11. Mediterranean-Type Shrublands*. Elsevier Scientific Publishing Company, Amsterdam, 479–521.

Westman, W.E. (1983) Plant community structure—spatial partitioning of resources. In Billings, W.D., Golley, F., Lange, O.L., Olson, J.S. and Remmert, H. (eds) *Mediterranean-Type Ecosystems: The Role of Nutrients*. Springer-Verlag, Berlin, 418–445.

Windley, B.A. (1984) *The Evolving Continents*, 2nd edition. Wiley, Chichester.

14

The MEDALUS Slope Catena Model: A Physically Based Process Model for Hydrology, Ecology and Land Degradation Interactions

M. J. KIRKBY[2], A. J. BAIRD[1], S. M. DIAMOND[3], J. G. LOCKWOOD[2], M. L. McMAHON[2], P. L. MITCHELL[3], J. SHAO[5], J. E. SHEEHY[4], J. B. THORNES[5] and F. I. WOODWARD[3]

[1]*Department of Geography, University of Sheffield, UK*
[2]*School of Geography, University of Leeds, UK*
[3]*Department of Animal and Plant Sciences, University of Sheffield, UK*
[4]*Creative Scientific Solutions, Marlow Bottom, UK*
[5]*Department of Geography, King's College, London, UK*

14.1 INTRODUCTION AND OVERVIEW

14.1.1 Overall aims and structure

The overall objective of the Physically-based Modelling Group (PMG) is to construct and validate an explanatory model which forecasts hillslope vegetation, hydrology and soil changes from known climate and land-use data. The model largely relies on established knowledge of individual processes. The main theoretical contribution is to identify more fully the dynamic interactions between processes. By so doing, the model will improve existing forecasts of both the direct and indirect responses to externally imposed changes in climate and/or land use. Within the overall framework of the MEDALUS project, the process model is intended to meet two main objectives.

(i) To extend existing physically based models to provide explicit forecasts of the short-term (0–10 year) and medium-term (10–100 year) impact of expected climate change and imposed land-use change on ecology, hydrology and sediment yield.
(ii) To contribute to the design of field experiments which effectively validate and calibrate the models developed.

Mediterranean Desertification and Land Use. Edited by C. Jane Brandt and John B. Thornes.
© 1996 by John Wiley & Sons, Ltd.

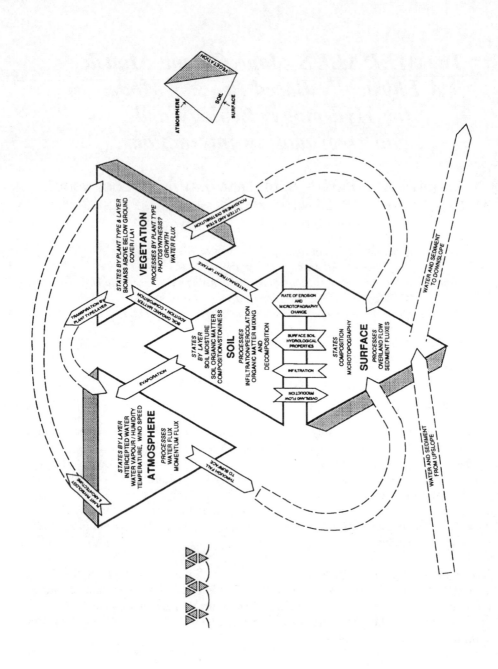

FIGURE 14.1 The main interactions between the four components of the MEDALUS model: Atmosphere, Vegetation, Soil and Surface

The overall focus of the model is the individual hillslope, with the potential for aggregating to small catchments, which would allow direct comparison with field experiments. Although the final objective is to forecast changes in average conditions, these averages must be built up from individual storm events, again to allow validation from the field experiments (Thornes, 1989). Estimates of average changes can then be made by explicit aggregation over the frequency distribution of storm events, taking account of changes in this distribution over time.

The model consists of four submodels which interact strongly and dynamically. The atmosphere model forecasts evapotranspiration; the vegetation model simulates plant growth for up to four functional types; the surface model simulates infiltration, overland flow and erosion; and the soil model simulates subsurface water movement, and changes in physical properties arising from erosion and organic matter addition, mixing and decomposition. The overall model structure is summarized as a flow diagram in Figure 14.1.

14.1.2 General need for a dynamic model

Global warming is expected to have strong direct impacts on the terrestrial biosphere. Indirectly, there will also be substantial changes in hydrology and sediment yield, and these are likely to be most severe in semi-arid areas. The focus of this work is primarily on hydrology and soil erosion, but vegetation cover and type has such a strong influence on hydrology and erosion that the effects of climatic change can only be forecast by including a substantial and explicit vegetation growth and ecological component in the model. These components are built into an integrated model, for which Figure 14.1 provides a broad framework.

Climatic change resulting from increasing greenhouse gas concentrations in the atmosphere may have a number of both direct and indirect impacts on the land biosphere, hydrology and sediment yields. Firstly there is the direct effect of increasing atmospheric carbon dioxide on vegetation growth and regional transpiration. Only half of the projected climatic changes are due to increasing atmospheric CO_2, and the rest are due to other trace gases. The influence of increasing CO_2 on vegetation may be relatively fast and closely linked to CO_2 concentrations. Other climatic responses to increasing greenhouse gases are more highly lagged. The most significant effects are the following:

(i) Changes in short- and long-wave radiation inputs, mainly due to changes in cloud distribution. Short-wave radiation influences vegetation through its control of stomatal conductance and as a forcing mechanism for surface evaporation. Long-wave radiation exerts a powerful control on surface energy balance.

(ii) Changes in rainfall distribution, particularly duration, frequency and intensity, influence the amount of water which passes through the vegetation canopy and into the soil. Changes in soil moisture, induced by precipitation/evaporation changes, feed back strongly into vegetation growth and structure.

(iii) Change in temperature averages and in particular temperature extremes influence biosphere activity. Changes in frequencies of very low (frosts) and high temperatures may be important.

(iv) Changes in extreme wind speeds may influence gap generation in forests.
(v) Changes in ultra-violet radiation, due to increasing ozone, may influence vegetation.

Land use is generally imposed or strongly influenced by man, in response to regional economic constraints, which have changed rapidly over the last few decades, and continue to do so. Although not all changes can be foreseen, a model which can explicitly respond to an imposed land-use history provides a valuable management tool. The physical responses to land use can be summarized, as in many existing soil erosion models, by imposing three elements.

 (i) A crop cycle for germination, and for irrigation where applicable. Growth is then able to respond additionally to the natural sequence of weather.
(ii) A harvesting regime, which responds dynamically to the quality of the crop, either through grazing or through direct harvesting for market or use.
(iii) A soil cultivation regime, which may be related to the calendar or to the onset of rains at the start of winter (for a Mediterranean climate). Tillage influences surface soil properties, rates of organic matter decomposition and vertical mixing.

14.1.3 Model components

For the atmosphere, an evapotranspiration (E-T) model is required, related to the layering of the vegetation canopy, which forecasts evaporative loss from the soil and from water intercepted on the canopy, and transpiration through leaf stomata. E-T cannot readily be explicitly calculated in the vegetation submodel without re-assessing parameter values obtained from previously validated studies, but must implicitly show similar responses to atmospheric and soil factors. The principles of E-T modelling are widely accepted, deriving from the work of Penman, Monteith and others. For present purposes we need formulations which are appropriate for sparse vegetation covers, and these rely on a modified version of the Shuttleworth and Wallace (1985) model.

A number of one-dimensional soil–plant–atmosphere models have been produced for use with atmospheric general circulation models. The best known of these are the American BATS (Biosphere–Atmosphere Transfer Scheme) and SiB (Simple Biosphere) models. Other similar models are SWATRE (Winand Staning Centre for Integrated Land, Soil and Water Research, Wageningen) and the UK Meteorological Office one-dimensional model. None of these models are suitable for this research project because they contain no vegetation dynamics. The preferred vegetation model must be multilayered in the vertical plane, and forecast plant growth in response to light, nutrients, water and CO_2; and losses due to leaf fall, grazing, death, etc. The model must explicitly estimate growth rates and net accumulation of biomass in the plant and the organic soil, using a mass and/or energy balance framework. There are, however, severe shortages of relevant data for plant growth, where many of the problems for Mediterranean species have previously been largely neglected.

The soil hydrological balance, and in particular infiltration rates, provide forecasts which are used to estimate overland flow runoff and grain size selective sediment transport from individual hillslopes. Soil is lost by water and wind erosion, the former in relation to overland flow, and the latter influenced by soil surface moisture content and plant cover. In severe cases, these losses may amount to several millimetres per year. Over a period of a few decades, and sometimes in a single storm, there can be serious on- and off-site effects, including effects which influence the performance of the evapotranspiration and vegetation growth submodels. Because of the episodic nature of erosion, it is vital to include a significant stochastic component in long-term sediment yield forecasts, and to note that field data generally show highly variable rates (Romero-Díaz et al., 1988).

For hydrology and sediment yields, the key unit is the single hillslope catena, allowing rational routing of water and sediment into streams across a defined topography. This appears to be a realistic scale, and one which links into specific site studies and into modelling on the catchment scale, which is important for many of the off-site effects. It also appears to be important to consider the partition of flow and sediment transport across microtopographic roughnesses which represent rills at one extreme and vegetation mounds at the other. Previous data show that the degree of flow concentration greatly increases erosion. The model must therefore have the capacity to forecast changes in the microtopography through net erosion and sedimentation.

For the plant a water balance is required between root uptake and transpiration loss. For the soil, a hydrological model is required to forecast moisture tension at the plant roots, as well as drainage losses through downward percolation, lateral subsurface flow and overland flow. There are a number of current models available for the soil hydrology, and it is essential that the parameters of the hydrology model are made to depend to a significant extent both on the organic soil, which is a major store of available soil moisture, and on soil stoniness, which responds to surface erosion and armouring.

The consistency of each individual component of the model is critical to the success of the whole. In addition, it is important, and critical to the success of the model as a whole that the interactions between components are represented in a dynamic way. Previous work (Kirkby and Neale, 1987) has shown the importance of these long-term interactions, and demonstrated the feasibility of a simple integrated model.

For example, a sequence of dry years will generally reduce plant production, and initially increase and then decrease additions of organic matter to the soil. As this material is incorporated into the soil by bioturbation and decomposes, the moisture storage capacity of the soil is reduced through the reduction in organic matter. Over several years, the reduction in storage capacity leads to greater erosion and less capacity for plant growth, beginning a possible spiral of positive feedbacks towards desertification. To simulate the dynamics of change, it is therefore vital to represent these linkages in a dynamic way, so that soil hydraulic parameters, for example, are not taken as fixed but are seen to depend functionally on organic matter and soil composition, which may be gradually modified through erosion or deposition.

Similarly surface microtopography actively controls the patterns and average rate of sediment transport, is in turn modified by erosion or deposition, and controls the spatial distribution of infiltration into the soil. In consequence, the pattern and average amount of soil water available to plants and evapotranspiration changes gradually in response to changes in surface flow and erosion.

14.1.4 Programming

The model is written for MSDOS machines, using TURBO Pascal for Windows. This provides executable programs suitable for a 386 PC system with VGA graphics or better. For each point on a slope catena, the main components are linked together by a shell which stores and updates shared variables and passes control to individual processes at time intervals defined by their needs. In this way, processes that are changing the variables only slowly do not waste computer run-times. For example, overland flow processes need to be updated at very short intervals during and immediately after rainfall, but only infrequently at other times, whereas evapotranspiration varies significantly throughout the day as long as there is any appreciable soil moisture.

The size of the final model is limited by a number of practical constraints, related to the use of MSDOS computers, the availability of data and potential for application as a planning tool.

(i) Our work shows the limitations of the chosen MSDOS micro-computer platforms for developing a model as complex as MEDALUS, even though its scope is clearly limited compared to very large models such as GCMs. The critical reason for using desktop computers remains the requirement to make the model widely available, both to the MEDALUS field groups and to other users.

(ii) The useful complexity of MEDALUS, or any model, is limited by the availability and detail of field data-sets. For example, if soil type data are only available for two positions in the soil profile, there are limited advantages in subdividing the soil into a very large number of layers for modelling purposes.

(iii) Application of the MEDALUS model to sites other than the project field areas will generally be constrained by even more restricted data-sets, so that a simple structure, with a limited number of required parameters, is essential for effectiveness.

(iv) Within the MEDALUS II project, we need to extract a simplified core from the present model which can be used for a large area model at a coarser resolution.

For all these reasons, one of the key stages in assembling the final model has been a careful assessment of the relative importances of individual model components in any particular context. Integration involves a clear understanding of the precise nature of the necessary links between components. In most cases this requires new science and novel concepts, making demands which could not have been clearly defined at the start of the project.

14.2 MODELLING CONCEPTS

Vegetation growth is forecast separately for trees, shrubs and herbs. It is computed on the basis of solar radiation, atmospheric humidity, CO_2, water availability and current plant morphology for each functional type (Jones, 1983). Evapotranspiration is estimated separately for each vegetation group, and for bare soil between plants. Forecasts are based on plant type and size, weather conditions as they change through the day, and the availability of soil moisture.

Surface runoff is estimated from the distribution of flow depths across the microtopography of natural surfaces, and includes both rill and inter-rill components.

Flow velocity is estimated from water depth and skin roughness as a basis for kinematic routing of the overland flow downslope. Flow depths also provide forecasts of tractive shear stress, detachment rates for soil and its size-selective transport downslope. The routing of overland flow and sediment provides the strongest interaction between successive points down the slope/soil catena (Baird et al., 1992).

Infiltration and soil water movement are calculated for a range of elevation classes within the local microtopography. Infiltration and bare surface evaporation interact strongly with the overland flow during and immediately after rainfall. Other soil processes considered include the incorporation and breakdown of organic matter and the evolution of near-surface grain size composition through selective erosional winnowing of fine materials.

We regard the interactions between the four submodels to be the main technical contribution within this modelling study. Although there have been some significant innovations within each of the submodels, it is only through their dynamic interactions that long-term simulations of soil and vegetation change can adequately forecast even the main changes involved in the complex process of desertification.

The assembled model is being calibrated and validated against meteorological, soil and vegetation data from the field sites, although this stage is far from complete. It is continuing, as planned, during the second phase of the MEDALUS project. There is also scope to examine more fully the sensitivity of model forecasts to uncertainties in field measurements and to climatic and land-use change scenarios for the immediate future. Processes modelled within each of the four groups are set out below.

14.2.1 Atmospheric evapotranspiration

The fixed physiology of the four plant types, together with their current morphology, determine the stomatal response of each type, and of the canopy as a whole. A modified version of the Shuttleworth and Wallace (1985) sparse vegetation model distinguishes the transpiration loss from each plant type, from bare soil beneath and between plants, and from free water on the surface, in response to current meteorological conditions and soil moisture availability. The standard model is being modified to take account of the interaction between above and below-ground conditions, based on the choice of a dynamic threshold soil moisture tension. This threshold controls plant stomatal opening and so distributes moisture loss to the plants between the various soil layers, providing an important bridge between aerodynamic/radiative demand and soil water availability.

Although the course of vegetation growth may be forecast sensibly on a daily basis, it is clear that radiation, photosynthesis, rainfall and evapotranspiration operate on a much shorter time span, so that these submodels require many iterations within each day. Their integrated effects may then be used within a daily growth model.

14.2.2 Vegetation growth and change model

The vegetation and evapotranspiration components in the MEDALUS model play an important part in predicting erosion. They are driven by climatic variables, exchange current values of several variables and produce output required elsewhere by the model (Figure 14.2). Although not shown in the figure, the vegetation model also requires input from the component of the model that simulates the decay of organic matter in the soil

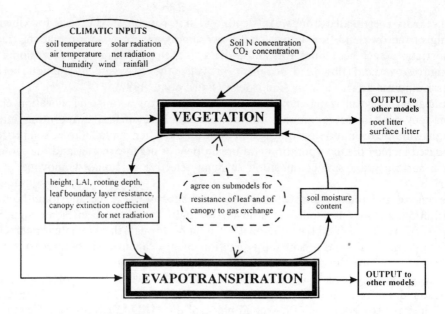

FIGURE 14.2   The relationships of the vegetation and evapotranspiration models in the MEDALUS
model

and the release of nitrogen. Where vegetation, or the leaf litter it produces, covers the soil
it reduces the erosive energy of raindrops. In the course of time leaf litter is incorporated
into the soil and decays where it joins the organic matter contributed by the death of fine
roots. The input of litter above and below the soil surface adds to the organic matter
content of the soil and affects its properties relevant to erodibility.

The main requirement of the vegetation model is to simulate the production of litter
accurately. The standing biomass is not needed provided that vegetation height, leaf area
index (LAI) and rooting depth can be predicted. However, biomass is of use in validating
the model against field data.

Given the diversity of types and productivity of Mediterranean vegetation it was
impracticable to construct a generalized model. Instead the important functional types
were identified and separate models were designed for each one. The distinction between
herbaceous, primarily grassy, vegetation and woody types is important in modelling. The
main grass model is for perennial grass from which a model for annual grasses could
eventually be derived by including reproduction followed by death of the plant, and by
modelling the survival and germination of the seeds. The main model for woody
vegetation is for an evergreen shrub. Modifications to leaf demography produce a
drought-deciduous shrub. A tree could be simulated by altering the allocation of woody
material to the stem fraction of the plant to make a single trunk as well as branches;
however, this is not a trivial change and would require substantial development.

The model for perennial grass was constructed first to make use of the comparative
simplicity of the structure of the plant. This facilitated the development of models for
processes such as photosynthesis, partitioning of assimilate, and in particular death of
parts of the plant. This model has been completed and incorporated into the MEDALUS

model. The evergreen shrub model is in preparation. It uses many features of the process modelling in the perennial grass model but the morphology of the plant and its phenology are more complex.

Within the scope of the MEDALUS I model, the area covered by each vegetation category remains fixed over time, while the density of the canopy, measured by LAI or percentage cover, is able to vary over time, so that a changing proportion of the area under each canopy type is accessible to direct rainfall impact. For the annual herbs only, the percentage cover may fall to zero in dry periods. Within each plant type, growth rates are integrated over each day, and total growth is then partitioned between roots, stems and leaves depending on vegetation type and size.

14.2.3 Surface water and sediment transport model

Rainfall, reaching the surface through and between plants, accumulates as a dynamic flowing network, which is subject to losses from free water evaporation and infiltration. A distribution of elevations represents the microtopography which determines the distribution of flow depths, and the rapidly changing ponded area on the surface. This ponded area is generally found between large plants, rather than directly beneath them. Sediment transport detaches material through its flow power and this is added to material detached by raindrop impact. Detachment is viewed as responding to mean surface grain size, but as not being size selective. Transport of detached material is highly size selective, and may produce some armouring of the surface. Flow detachment and transport are non-linear in flow depth, so that the distribution of depths is critical. Flowing water and transported material are budgeted through the successive slope segments which form the hillslope catena which is the target of the MEDALUS I model. Net deposition or erosion of surface material modifies the depth and composition of the surface armour layer. If there is erosion for any grain size class, material of all sizes is transferred to the armour layer from the underlying soil (of known composition).

Transport is calculated for each grain size, using an algorithm which allows transport of fines to be limited by availability, and transport of gravel to be limited by transporting capacity. An armour layer is allowed to develop, depending on transport conditions and the composition of the underlying soil. Erosion is then able to both sort the surface material and allow the microtopography to change, with its relief tending to increase at high rates of erosion and vice versa.

14.2.4 Soil water and organic matter model

The soil area is divided into five domains, which change dynamically in response to surface flow conditions. The first three are the areas directly under perennial plants, which are generally only slowly changing and are not generally ponded. The fourth domain is for the area of annual plants, including areas which are currently unoccupied. The fifth domain, which commonly (but not invariably) lies within the area of annuals, is the dynamically varying area of surface ponding. Infiltration and vapour transfer (to supply bare soil evaporation) are calculated separately for each partition, solving the unsaturated flow equations for the known field soil properties, defined by moisture retention curves and grain size distributions. The soil layering and properties change dynamically (though slowly) in response to changes in the armour layer (which is the top soil layer for infiltration) and incorporation of decomposing organic matter. The way in which the hydraulic properties of the soil change with these variables relies on empirical extensions from the field data. Surface crusting is modelled as a thin film which modifies the effective soil moisture tension in parts of the area unshielded by overhanging vegetation or large stones. Within the soil, downward movement of water within the five domains is modified by a diffusional lateral mixing between them. Water is abstracted by plant roots across all three partitions, in relation to the dynamic tension threshold established for plant transpiration, and the distribution of root depths by plant type.

14.3 EVAPOTRANSPIRATION

14.3.1 Conceptual basis

The standard unilayer Penman–Monteith equation does not directly yield information on transpiration and soil surface evaporation as separate entities. Since evaporation from the soil surface can dominate the total evaporative loss from very sparse vegetation cover after rainfall, it is essential that values for transpiration, soil surface evaporation and interception loss are available separately (Lockwood, 1993). Therefore the evapotranspiration model is a modified version of that used in Lockwood (1992), which in turn is derived from Shuttleworth and Wallace (1985), and estimates transpiration, soil surface evaporation and interception loss as separate quantities. It is a one-dimensional model that provides a physically plausible description of the atmospheric turbulent fluxes across the transition between a bare substrate and a closed vegetation canopy. The equations are expressed in terms of conceptual resistances, using Ohm's law in electricity as a direct analogue, now familiar to micro-meteorologists and plant physiologists. The elements of which the model is composed (e.g. LAI, energy fluxes, etc.) are defined as horizontal averages over scales in which persistent features occur in sufficient numbers to allow averaging.

14.3.2 Numerical model

The model is divided into a series of subschemes which describe aerodynamic resistances within and above the canopy, leaf boundary layer resistances, net radiation absorption within the canopy, heat conduction in the soil, stomatal conductance, bulk canopy resistance, bulk soil surface resistance, vegetation transpiration, soil surface evaporation, rainfall interception by canopy, interception loss, wet canopy drainage, soil moisture tension and soil water balance. Input to the model consists of hourly averages or totals of solar radiation, net radiation, air temperature (at 2 m above the ground), wet bulb temperature (at 2 m, and ventilated if possible), wind speed (at 2 m) and rainfall. If possible soil heat flux measurements should be available. Vegetation parameters required are daily values of LAI, vegetation height and rooting depth. Soil parameters required are information on moisture (initial soil moisture content of profile, total available soil moisture content), thermal properties, and aerodynamic roughness of the soil surface (vertical dimensions of surface litter, stones, etc.). The general aerodynamic roughness of the surrounding countryside is also needed.

Stomatal conductance varies with solar radiation, atmospheric water vapour deficit, temperature and soil moisture. The relevant equations need to be determined from fieldwork or the literature for each vegetation type. Typical default values are supplied with the model. The stomatal and cuticular resistance of the leaf and the LAI are combined (allowing for Mediterranean vegetation in general having stomata on the lower leaf surface only) to obtain bulk canopy resistance. Boundary layer resistance assumes a characteristic leaf dimension of 2 cm. The soil moisture function in the stomatal conductance scheme restricts transpiration when soil moisture is limiting by increasing the bulk canopy resistance; the other functions adjust the bulk canopy resistance so that the transpiration is limited to the lowest value which is consistent with the canopy energy balance under prevailing meteorological conditions (i.e. there is a great range of conditions when even with moist soils the transpiration is insignificant). Rooting depth is assumed to be 0.3 m for grasses and 1.0 m for shrubs and the maximum soil moisture available to the plant is set accordingly. A soil surface resistance is calculated, using the soil moisture model, which is analogous to the bulk canopy resistance for vegetation. Bulk canopy and soil surface resistances range from nearly zero (wet) to 10 000 s m^{-1} or above (very dry). The aerodynamic resistances are calculated from wind speed, vegetation height, LAI, soil surface roughness and, where appropriate, general landscape roughness. The net radiation when the sun is above the horizon is determined primarily by the direct solar radiation, and therefore the net radiation reaching the soil surface after attenuation by the canopy is calculated using a Beer's law relationship. The extinction coefficient of the canopy for net radiation is arbitrarily set at 0.7 but this could be in error in the case of an actively transpiring crop above a dry soil surface. Interception loss from the canopy was determined using a Rutter-type drainage/evaporation expression (Rutter et al., 1971).

An example of the use of this type of model is contained in Lockwood (1992). It provides a plausible description of a vegetation cover without being computationally expensive. It should be noted that this particular model does *not* contain a numerical description of the atmospheric mixed boundary layer which would allow the investigation of the water balance of a large variety of vegetation covers under any given meteorological conditions. Such models are computationally expensive and resources were not available within the MEDALUS project.

Investigations of transpiration using sparse vegetation models suggest that at low LAIs transpiration varies with, but not directly proportionally to LAI, but at moderate to high LAI values, transpiration is largely independent of LAI. They also suggest that with low values of LAI the total evapotranspiration can be dominated by the evaporative loss from the soil surface.

The stomatal conductance, G_s, (in mm s^{-1}), is estimated from solar radiation, specific humidity deficit, average moisture volume fraction in soil and soil moisture deficit (Lockwood et al., 1989). Stomatal conductance is not particularly sensitive to changes in solar radiation greater than 100 W m^{-2}, or soil moisture contents above about 20% of field capacity. A soil surface resistance is calculated which is analogous to a bulk canopy resistance for vegetation. In the MEDALUS model, soil resistance is related to pore water tension at the soil surface, which is taken to be the key evaporating surface by analogy with leaf stomata. The values are also modified at very high tensions to allow for the associated reduction in vapour pressure. Within the soil surface evaporation is balanced by upward flows of water and vapour within the soil. These are obtained directly from the available air-filled porosity for vapour diffusion, and from tension gradients using the Richards' equation for liquid movement.

The aerodynamic resistances were calculated from crop height, wind speed and LAI. Net radiation profiles within the canopy are a function of LAI. Examples of the use of this type of model may be found in Lockwood et al. (1989) and Wallace et al. (1990).

Interception loss from the canopy is described by a Rutter-type drainage/evaporation expression with the assumption that drainage occurs when the rainfall storage on the vegetation exceeds 0.2 mm per LAI (Rutter et al., 1971). Rainfall interception by the vegetation canopy follows the basic description used by Thompson et al. (1981), and is a non-linear function of LAI.

The input data required for the numerical models are hourly averages of solar and net radiation, air temperature, wet bulb temperature, wind speed and hourly totals of rainfall, all preferably from an automatic weather station. LAI and vegetation height are at present provided daily by a demographic growth scheme included within the models. Output consists of components of the evaporative loss, surface resistances, aerodynamic resistances within and above the canopy, and soil moisture deficits. The models were integrated using hourly time steps except when the canopy was wet when the time step for estimating interception loss was 1 minute.

14.4 VEGETATION GROWTH

14.4.1 Introduction

The place of the vegetation component in the MEDALUS model is summarized in Figure 14.2. This shows the environmental driving variables, the output to other components of the overall model and the close relationship with the evapotranspiration model. The vegetation and evapotranspiration models exchange current values of several variables. The primary objective of the vegetation model is to simulate the production of litter and the changes in height, LAI and rooting depth. The standing biomass is not needed except as means of validating the model against field data.

The vegetation model has been made as mechanistic as possible. This is because empirical models or those based on correlation of vegetation properties with climatic or soil factors cannot be used with confidence outside the restricted circumstances in which they were derived. Consequently, the vegetation model is larger and more complex, perhaps by an order of magnitude, than the other components which deal with well-understood, physical processes of water movement and erosion, and where equations can sometimes be solved analytically.

Given the diversity of types and productivity of Mediterranean vegetation it was impracticable to construct a generalized model. Instead, two basic models have been produced: for an evergreen, sclerophyll shrub and for vegetative perennial grass. This allows for future elaboration and development to produce versions for drought-deciduous shrubs and for trees, and for incorporation of reproduction into the grass model and derivation of an annual grass.

14.4.2 General features of the models

Conventions

The models are constructed from conceptual submodels which are self-contained modules as far as possible. Each can be replaced by an alternative or improved version without affecting the overall running of the model. Submodels are a convenient way of summarizing the assumptions and parameters in the model. The larger submodels correspond with a subroutine or unit in the Pascal code; the smaller ones are a simple equation which may be replaced by a single variable as a default. Most submodels are equations of some kind; some are sets of logical statements or pathways. The larger ones may incorporate one or more smaller submodels. There are 10 groups of submodels, including the six basic processes used in the models of Sheehy et al. (1979, 1980).

1. Setting up.
2. Phenology—the timing of growth and death.
3. Photosynthesis to produce assimilates.
4. Respiration which reduces the amount of assimilate available for growth.
5. Uptake of nitrogen (other elements are assumed to follow in the proportion needed).
6. Partitioning of assimilate and nitrogen for growth among parts of the plant.
7. Transformation of new growth into parts of the plant with functional attributes such as leaf area, and updating properties of the whole plant such as height.
8. Death of parts of the plant.
9. Modifications for a suboptimal environment, principally shortages of water or nitrogen.
10. Leaf and canopy properties for transpiration.

Several basic conventions are adopted. The vegetation model is driven by incoming photosynthetically active radiation (PAR, wavelengths 400–700 nm), in energy (not quantum) units and usually as daily totals ($J\,m^{-2}\,day^{-1}$). For all practical purposes PAR can be taken as half the amount of solar radiation (300–3000 nm) in energy terms (Szeicz, 1974). All calculations of photosynthesis, respiration and growth are carried out in weights of carbohydrate, i.e. notional CH_2O units. For grass this is equivalent to dry weight because 40% of grass dry weight is carbon, the same proportion as in CH_2O (Sheehy, et al., 1979). For vegetation where this proportion differs an adjustment will be required.

Each model is computed for $1\,m^2$ of ground area covered by the canopy. This means that if, on the larger scale, cover is less than 100% for grass or shrub the model output per square metre should be reduced proportionally. Shrubs, in particular, often occur as bushes or clumps. It is important to remember that leaf area index (LAI) is defined at a field or 1 hectare scale averaging over the ground between bushes as well as their crown area. Thus if the LAI is 6 and shrub cover is 50% then the cover area index is 12, that is the average number of layers of leaves above the area of ground covered by a vertical projection of the bush or clump canopy. What is referred to as LAI over $1\,m^2$ in this account of the model is in fact cover area index.

If cover is less than 100% then root spread may exceed crown spread so that water is gathered from a larger area. This is well known for desert shrubs (Rundel and Nobel, 1991) where root area can be 10–15 times crown area. Maximum rooting depth is assumed to be $0.3\,m$ for grasses and $1.0\,m$ for shrubs.

The negative sign for water potential is ignored in coding the model in Pascal. Several parts of the model are conceived as arrays even though they may be coded as linked lists in Pascal to make more efficient use of memory.

Outline of the models

Block diagrams of the shrub and grass models are given in Figures 14.3 and 14.4 (following a pattern suggested by Dr A. C. Terry). Driving variables such as climatic inputs or soil properties are shown in ellipses. The main ones are solar radiation and air temperature. Other important driving variables are soil water potential and the concentration of nitrogen in the soil, which are the outputs of other major component models in the MEDALUS model that are themselves driven by the climatic variables. The main outputs of the models are the litter and the current canopy properties (height, LAI, resistances to water movement) required by the evapotranspiration model.

The plant is divided into parts for which the dry weights are the main state variables, shown enclosed in a double box. These are roots, stems and leaves in the grass model, with leaves divided into photosynthetic mechanism—the components involved in the leaf's biochemistry—and leaf structure. The shrub model has in addition reproductive structures (flowers and fruits, referred to as fruits hereafter for brevity), and a store. This is the carbohydrate reserve material laid down in the ray cells of woody stems and roots which is called upon when photosynthesis is low or has stopped completely in unfavourable periods, or to recover from complete defoliation. Roots and stems are each divided into structure and maintainable components. The structure component comprises dead cells in wood whereas the maintainable component is the living cells which need assimilate for respiration. In this way woody plants can build up a large standing crop but keep the demands of respiration in bounds.

The models use arrays to keep track of each portion of root, stem, leaf or fruit produced each day. Each portion of tissue looses weight by daily respiration for maintenance. Portions of leaf have a specific leaf area and photosynthetic capacity assigned to them on the day of formation. The state variables are totals from the relevant arrays to provide the biomass or canopy properties.

The models need to run for a period to remove any effects of the particular set of starting values for state variables used to initiate the model. For the shrub model this period is a couple of years; for grass a few months of growth followed by removal of almost all above-ground parts simulating grazing or cutting is sufficient. In order to simulate growth from low starting values, as for re-growth of semi-natural vegetation on abandoned farmland, the models are being modified to remove some of this sensitivity to starting values. The run starts with empty arrays and starting values as variables. The arrays are filled day by day and the starting values become a decreasing proportion of the total. At the end of the year the arrays are summed to provide new starting values without any age structure, the arrays are emptied and then refilled day by day.

The similarity of the block diagrams for functional types of vegetation as different as grass and shrubs is because several fundamental aspects of plant growth are universal: photosynthesis, respiration and nitrogen uptake, for example. In these submodels the differences, if any, tend to be in the values given to parameters. These processes are comparatively well understood and mechanistic models are well developed in these areas. The different sizes and forms of the two functional types arise from the details of phenology, partitioning and modifications for suboptimal environment, how these occur in the plant and how they are simulated in the model. Here the submodels themselves tend to differ between grass and shrubs so that there may not be comparable parameters. It is

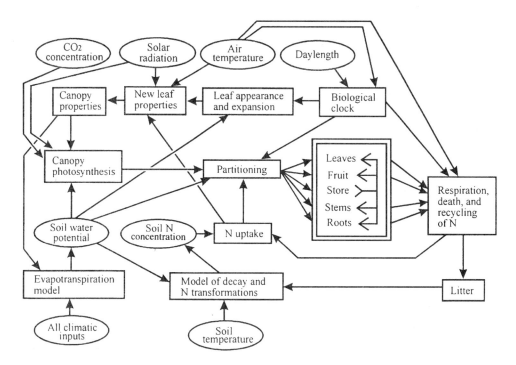

FIGURE 14.3   Block diagram of the shrub model. The environmental inputs are shown in ellipses and the main state variables in the double rectangle. Arrows represent influence or the movement of material

more difficult to make these submodels mechanistic because less is known about these aspects of plant growth.

The main difference between the two models is the greater role for the biological clock in the shrub, driven by day length, to control the phenology of growth and death of all parts of the plant. In the grass model the clock regulates the appearance of new leaves only. Perennial grass is perennial because there are always leaves, stems and roots present. But they are not the same ones: new ones appear and grow, and old ones die, faster or slower according to the weather. Thus phenology is determined by the environment, i.e. growth becomes slow or stops below certain temperatures or soil water potentials, and death of these ephemeral parts occurs when a fraction of the initial weight has been respired away. For shrubs, phenology has to be specified in the model because there are more plant parts and reproduction is included. There are submodels for the seasonal patterns of partitioning and of death and fruit drop.

There are fewer parts in the grass model because there are no fruits or store. Regrowth after cutting or grazing is funded initially by transferring some dry weight from the stem to produce new leaves and thereafter from the rapidly increasing photosynthesis of the new leaf canopy that develops.

Table 14.1 is another way of summarizing the model so as to show how the environmental variables drive the model. Under each environmental input is listed the submodels which it drives. The comparatively few entry points for environmental inputs

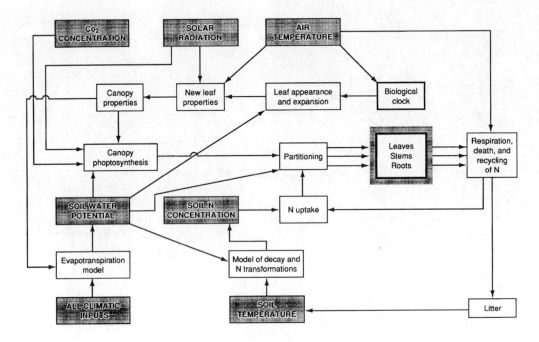

FIGURE 14.4   Block diagram of the grass model; details as in Figure 14.3

TABLE 14.1   Submodels driven by the climatic inputs and other environmental
factors in the vegetation models

Solar radiation
    Canopy photosynthesis
    Canopy attenuation
    Specific leaf area of new leaf
    Maximum rate of photosynthesis for new leaf
Temperature (air and soil assumed to be the same)
    Effect of temperature on respiration
    Specific leaf area of new leaf
    Maximum rate of photosynthesis for new leaf
Nitrogen concentration in the soil
    Nitrogen supply from the soil
Carbon dioxide concentration in the atmosphere
    Effects of elevated CO_2 concentration
Day length (treated as a climatic input for a given latitude)
    Clock (shrub)
Soil water potential
    Effects of shortage of water on leaf dynamics
    Effects of shortage of water on photosynthesis
    Effects of shortage of water on partitioning

indicate the mechanistic nature of the model: the internal workings of the plant are simulated so as to avoid the need for crude correlations of growth, reproduction, LAI, litter production, etc. with climatic variables.

14.4.3 Details of the models

(1) *Setting up*

Two submodels are concerned with setting up the model. One, the soil moisture characteristic curve, is the relationship between soil moisture content and soil water potential. The empirical equation given by Gregson et al. (1987) is used to convert the moisture content supplied by the evapotranspiration model to water potential which drives certain parts of the vegetation model. Another submodel sets up the distribution of roots exponentially down the profile so that 99.9% of root weight is accounted for in the fixed rooting depth. This submodel is run each day to distribute the current weight of root. The exponential pattern is commonly observed in crop plants (Gerwitz and Page, 1974) and is convenient for modelling; it is thought to be applicable to woody plants as well (Landsberg, 1986).

In the grass model the starting values are roots 400, stems 100, leaves 20, all in $g\,m^{-2}$ dry weight for ground area covered by the canopy. The values set at the start of the shrub model run are as follows: roots, 1400; stems, 1900; leaves, 1400; fruits, 0; store, 0. These were obtained from data in Kummerow (1981) for American evergreen sclerophyll shrubs from Mediterranean-type environments grown for 2 years in containers. Data from the MEDALUS field programme have not yet been made available.

(2) *Phenology*

The phenology of growth is simulated by a submodel of a biological clock. In the grass model the clock is simple because it controls only the appearance of leaves. The clock is set at 1.0 when a new leaf appears and runs down at the rate set for maintenance respiration. At the threshold value of 0.65 (after 29 days at a constant 15°C) a new leaf appears and the clock is reset. Together with the submodel for the death of ephemeral parts which is applied to leaves this ensures that there are usually three live leaves present in the model, the fourth almost completely dead when the newest leaf appears. This accords with the observation of the number of leaves on a tiller of perennial rye-grass. Since all parts of the grass plant are short-lived no phenology for death is needed.

In the shrub model the clock monitors the day length and triggers growth, death and the shedding of mature fruit at threshold day lengths. Day length is read in with the climate data day by day, or can be computed from the equations of solar geometry given the latitude. Threshold day lengths are specific to latitude. The clock determines which of six cases of assimilate partitioning is operating at a particular time, each case being a combination of parts of the plant being supplied with assimilates for growth. Roots are always capable of growth; stems, leaves and fruits grow at certain seasons; and the store is replenished when other, more important parts are not growing. Stems grow when leaves are growing because new twigs have to support the leaves. But stem growth also occurs when leaves are not growing as woody stems increase in thickness. Leaves grow from February to early May, stems grow from February to September, fruits grow from

April to early December, and the store increases in December and January. Field observations of phenology (e.g. Diamantoglou and Mitrakos, 1981) are used where possible. The general timing of growth of parts, relative to each other in partitioning, is a compromise of conflicting observations (Arianoutsou-Faraggitaki et al., 1984; Lo Gullo and Salleo, 1988; Arianoutsou, 1989; Salleo and Lo Gullo, 1990) which arises because evergreen sclerophyll shrubs, though an easily recognizable structural group, are not a uniform functional type of vegetation.

Shrub roots are treated as short-lived organs although a structure component, representing woody roots, accumulates continuously. The production of stem litter is simulated by one-tenth of all stem material dying on 31 December. Leaves die during the period of leaf growth so that the evergreen canopy maintains a near-constant LAI. Fruits drop in mid-March, about 4 weeks before new reproductive growth begins.

(3) *Photosynthesis*

Photosynthesis is modelled as gross photosynthesis, i.e. net photosynthesis (as would be measured with a gas analyser) plus the simultaneous respiration of the leaves (assumed to be the same as their respiration in the dark), because respiration is dealt with separately for all parts of the plant (Sheehy et al., 1979). Photosynthesis is modelled at the canopy scale as a hyperbolic curve against PAR (Sheehy et al., 1980). The initial slope is the apparent photochemical efficiency for photosynthesis by the C3 pathway, taken as a constant for a given CO_2 concentration. The asymptote is the maximum rate of photosynthesis, P_{max}, which is calculated for each part of the canopy, i.e. each portion of leaf tissue formed on a particular day. The effects of changes in CO_2 concentration are taken into account by a submodel which adjusts the quantum efficiency and P_{max} (Thornley et al., 1991).

(4) *Respiration*

The conventional division between maintenance respiration and respiration associated with growth and synthesis of new plant matter is adopted (Sheehy et al., 1979). When assimilate is allocated for growth 25% is subtracted for growth respiration. All live parts, so all biomass except the structural components of woody parts, require assimilate for maintenance respiration at the rate of 1.5% of their weight each day at the reference temperature of 15°C. The rate of maintenance respiration is affected by temperature, of the air for aerial parts and of the soil (if available) for roots. The current submodel of temperature sensitivity is exponential. Q_{10} is the factor, here 1.5, by which the rate is multiplied each 10°C increase in temperature above 15°C.

(3) *Uptake of nitrogen*

The uptake of nitrogen from the soil depends on the nitrogen density ($g\,N\,m^{-3}$ in rooting zone) which is provided from a model of nitrogen release during decomposition of organic matter such as that of Thornley and Verberne (1989). If the amount of nitrogen taken up, plus the nitrogen recycled from dead tissue, is insufficient to provide nitrogen at optimal concentration for the weight of new growth then a revised concentration is calculated. The optimal concentration varies. For grass it is 3% throughout the plant. In the shrub model it is 1% for roots and stems, 5% for the photosynthetic mechanism component of leaves, 1.25% for leaf structure and 4% for fruits.

(6) *Partitioning*

The assimilate produced each day by photosynthesis has to be partitioned to the parts of the plant for growth. The simplest model of partitioning is where fixed proportions of the assimilate available are allocated to parts of the plant that are growing. Although very simple to implement in the model, the result is vegetation which grows with the parts in rigid proportions, and which cannot respond to the environment or to the internal condition of the plant. For simplicity, fixed proportion partitioning is applied to four of the six cases, defined by phenology, in the shrub model.

An alternative is dynamic partitioning where the amount allocated is proportional to the amount of daily photosynthesis following the model of Sheehy et al. (1980). Graphically this is represented by a rectangle of height 0–100% of CH_2O to allocate and length 0–100 g CH_2O m^{-2} day^{-1} gross canopy photosynthesis. Inside the rectangle straight lines are drawn to separate areas for the plant parts. The partitioning on any day is read off a vertical line from that day's amount of photosynthesis. The lines are drawn teleonomically in the sense of Thornley and Johnson (1990), i.e. by guessing the goals sought by the plant although Sheehy and Johnson (1988) also derived this scheme from a mechanistic model of partitioning. Arithmetically, dynamic partitioning is represented by the equations for the lines with intercepts and slopes, the partitioning coefficients. The coefficients for all the parts must sum to zero. The signs and relative sizes of the coefficients reflect the importance of allocation to particular parts given the amount of assimilate produced on that day. Thus, for example, a small amount of assimilate may be indicative of a small leaf canopy so that leaf growth is favoured over roots and stems; if the daily production of assimilate then increases, less is allocated to leaves and more roots so as to restore the functional balance of the plant. Dynamic partitioning responds to the internal condition of the plant and to the environment through the effects on the daily output of assimilates by photosynthesis.

The growth of perennial grass is sufficiently well understood for dynamic partitioning to be used (Sheehy et al., 1980). In the shrub model dynamic partitioning is used in the cases when leaves are growing, with or without fruits, and fixed proportions for the rest of the year.

(7) *Transformations*

Submodels classed as transformations include the differentiation of woody tissue, the specific leaf area (SLA) and maximum rate of photosynthesis (P_{max}) of new leaves, and growth in height.

The vast majority of cells in woody tissue are dead and require no respiration for maintenance. The living cells are the cambia including any as yet undifferentiated derivatives, the functional phloem, and parenchymatous cells in the rays through the xylem and phloem. In the model this division into structure and maintainable components for stems and woody roots is initiated at setting up and continued by daily transfers from maintainable to structure of a fraction of the new growth: this represents differentiation of xylem vessels and tracheids, fibres throughout the xylem and phloem, and phellem (the true bark) produced by the bark cambium. All these cells are dead after differentiation is completed. Each day, 25% of the day's new growth of root becomes structure, requiring no further maintenance respiration. For stems, 10% of the maintainable weight is converted to structure each day. This submodel is not, of course, needed for grass where there is no woody tissue.

New leaves, that is each day's portion of leaf tissue, acquire SLA and P_{max} according to a relationship with the amount of PAR received on that day and its average temperature (Sheehy et al., 1980). There are slight differences in the parameter values between the models, and shrub leaves have their P_{max} scaled by the concentration of nitrogen in the leaf relative to an optimal 4%. In addition, in the grass model the photosynthetic capacity of a portion of leaf declines as it ages, simulated by a relationship involving the current weight of the leaf, which is determined by the daily loss through maintenance respiration and so is dependent on temperature.

Growth in height of the shrub is computed from the weight of new stems added each day assuming that 5000 g stem occupy 1 m^3 of crown volume and that the crown is a semi-ellipsoid with height equal to basal diameter. In the grass model, the height of the sward is taken as 5 cm for the pseudostem of leaf sheaths plus the length of the youngest leaf which is assumed to be vertical. The length of the leaf is calculated from its area assuming a constant width of 3 mm. Once the leaf has grown long enough to bend over, a younger leaf has appeared which is still short enough to be vertical.

(8) *Death of parts*

For short-lived plant organs it is possible to simulate the life span in a simple way. The model keeps track of each portion of leaf, stem or root produced on a particular day as an element in an array. On allocation of assimilate each portion loses 25% of its weight by respiration associated with tissue synthesis and then loses weight day by day through respiration for maintenance. When it reaches a critical fraction of its initial weight it dies. The life span is determined by the critical fraction at which death occurs, the rate of respiration and its temperature relationship, and the actual daily temperatures experienced. At a constant 15°C with maintenance respiration of 1.5% day^{-1}, a critical fraction of 0.7, 0.5, 0.3 or 0.1 produces a life span for a piece of tissue of 6, 28, 62 or 135 days. This method of simulating death of ephemeral parts is used for all parts in the grass model. For roots in the shrub model a slight modification is made to allow for the woody part of shrub roots which is very long-lived: the loss of weight by maintenance respiration applies only to the fraction of root tissue which is maintainable. Throughout the assumption is made that short-lived organs are made with a certain amount of assimilate and then receive no more. They respire away the materials that can be used, primarily soluble sugars and storage carbohydrates (Sheehy et al., 1980; Marshall and Waring, 1985).

When a portion of tissue dies, 80% of its nitrogen content is recycled and becomes available with the day's uptake of nitrogen from the soil for use anywhere in the plant. The dead tissue is added to the appropriate litter pool: leaf, stem or root. Leaf litter has an area as well as weight. No nitrogen is recycled from mature fruits that are shed but do not die in the usual way.

(9) *Modifications for sub-optimal environment*

The basic processes in the model are for conditions where water and nitrogen are freely available. When this is not so, measurable as a decrease in soil water potential or nitrogen concentration in the soil, then modifications are made to the standard patterns of photosynthesis, leaf expansion and partitioning.

Threshold water potentials are defined between which there is linear change in the process modelled. For example, in the shrub model, canopy photosynthesis is unaffected by soil water potentials from 0 (saturated) to -1.5 MPa. Between -1.5 and -2.0 MPa photosynthesis is reduced linearly with water potential from 100 to 25% of its normal value. From -2 to -3 MPa photosynthesis is scaled from 25% to zero and below -3 MPa photosynthesis stops. The threshold water potentials differ between grass and shrubs and vary with the process involved. Leaf expansion is affected before photosynthesis, i.e. at higher water potential, in accordance with observations (Terry et al., 1983; Hsiao et al., 1985). The modifications to partitioning are generally to direct assimilates to the roots, at the expense of stems and then of leaves, since only new root growth can tap further volumes of soil for water held at potentials high enough for movement into the plant.

Shortage of nitrogen arises from a low soil nitrogen concentration or small amount of roots through which uptake can occur, or both. The result is that, even with recycled nitrogen from dead tissue, there is not enough nitrogen to maintain the optimal concentration in new tissue being produced. In the grass model shortage of nitrogen affects partitioning to favour root growth. In the shrub model the P_{max} of new leaves is decreased in proportion to the actual nitrogen concentration in the new leaf as a fraction of the optimal concentration of 4% overall.

(10) *Leaf and canopy properties for transpiration*

The stomatal resistance of leaves influences the loss of water through transpiration and is controlled by several environmental factors. The evapotranspiration model uses the stomatal responses given in Lockwood et al. (1989) as a default. The resistance to water loss of the canopy as a whole is computed from the LAI, the stomatal resistance, and the cuticular resistance which is assumed to be constant. A small modification of the equation in Shuttleworth and Wallace (1985) has been made to allow for the fact that Mediterranean plants, certainly the shrubs, are expected to have stomata on one side of the leaf only.

Leaf boundary layer resistance and the canopy extinction coefficient for net radiation are shown as variables updated by the vegetation model in Figure 14.3. At the moment these are treated as constants; neither influences the evapotranspiration model greatly. The boundary layer resistance for a leaf (one side) is set at $25 \, s \, m^{-1}$. Using the formula in Jones (1983, p. 53) this is equivalent to a leaf characteristic dimension of 2 cm at a wind speed of $1.3 \, m \, s^{-1}$ or 6 cm at $3.9 \, m \, s^{-1}$ so is reasonable for many plants and wind speeds. The extinction coefficient is 0.7 for net radiation. This value is higher than the extinction coefficient for PAR (0.6) because PAR is absorbed by leaves more strongly.

14.4.4 Specific assumptions and simplifications

In the shrub model two simplifications are made to achieve a functioning model which stays within realistic limits: these are a maximum LAI of 7 and a maximum weight of

$10\,000\,\mathrm{g\,m^{-2}}$ for stems. Frequently the onset of summer drought in the model curtails the period of leaf growth so that LAI as high as 7 is not attained. However, if it is, then setting the area of dead leaves each day to equal the area of new leaves produced maintains the LAI at 7. If dead leaf area reaches $6.9\,\mathrm{m^2\,m^{-2}}$ ground area in a growing season then the growth of new leaves stops as this indicates complete turnover of the leaf canopy. Normally this does not happen and leaves last for 2 to 3 years. The maximum weight of stems is only likely to be approached after long runs with favourable growing conditions. If this occurs then dead stem litter is produced at the same rate as new stems.

Rooting depth is not modelled at present and is assumed to be 1.0 m for shrubs with root weight distributed exponentially from the surface through this depth. However, a further assumption is made that root spread exceeds crown spread by a factor of three. To keep to the existing practice of simulating shrub growth for one square metre of ground, this is handled in the model by trebling the amount of water available in the soil under the nominal 1 m². This achieves the result that the shrub on 1 m² exploits 3 m³ of soil without making the model more complicated.

There are different starting conventions for the grass model. There are assumed to be 10 000 tillers per square metre. The leaf sheaths have an area index initially of 1.0 (cf. Sheehy et al., 1979). The sheath area index is not constant but below a critical stem weight is adjusted in line with the weight of stem. The P_{max} of sheaths is set initially at 10% that of leaves created at that time; thereafter it is 5% of the average value for the whole canopy. Every day there is a 3% loss of weight of leaves and stem. This is based on field observations (Sheehy et al., 1979) that 3–4% of tillers die each day; here it is not intended to model the tiller population but to account for losses through grazing by insects and molluscs, diseases, competition for light, etc. It is weighted for temperature like maintenance respiration to simulate seasonal variation.

14.4.5 Examples of model output

Figure 14.5 shows the results of running the shrub model alone for 4 years with four combinations of temperature and atmospheric CO_2 concentration. The basic climate was that of Malaga, Spain, for which rainfall averages 51 mm a month from September to May, but totals only 10 mm for June to August. The elevated temperature was the current value plus 3°C. Current CO_2 concentration was 350 ppm by volume and the elevated concentration was 560 ppmv. Not surprisingly the model takes a year or two to settle down, partly because all of the initial root weight was assumed to be ephemeral so that woody roots took time to build up.

Increasing the CO_2 concentration had surprisingly little effect except that fruit production became negligible. The main effect of elevated temperature was to increase the standing weight of stems and, especially, of woody roots. Leaf weight was much the same in all runs and corresponded to a LAI of about 6. Although detailed interpretation of the results is not warranted at this stage, these results do show that the shrub model can run for some years under a variety of climatic conditions.

14.4.6 Conclusions

Modelling vegetation so that it responds sensitively to the environment, interacts with the evapotranspiration model, and accurately simulates litter production is a demanding task.

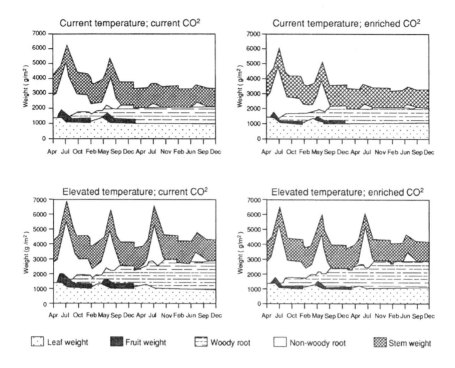

FIGURE 14.5   The standing weights of parts of the plant produced by the shrub model during a run of 4 years under four combinations of temperature and CO_2 concentration. The parts are shown in each diagram in the order leaves, fruits, woody roots, non-woody roots and stems from the bottom upwards

The two basic models developed run reliably and produce believable outputs. Fine tuning will be possible as data become available from the core programme of fieldwork in MEDALUS I. In addition results from specific experimental work and field observations will be used to remedy the gaps in our knowledge that modelling inevitably discloses.

14.5 SURFACE RUNOFF AND EROSION

14.5.1 Introduction

For the purposes of overland flow routing and erosion modelling, semi-arid slope surfaces can be described physically in two ways. In the first they can be considered as having a microtopography of rills and inter-rills (Kirkby, 1990). When we route overland flow on these surfaces, the roughness elements in either the rills or the inter-rill areas are described by a channel resistance factor (such as Chezy's C). In the second (Baird et al., 1992), the small roughness elements are modelled explicitly as a distribution of depths and flow is routed along a number of partitions, each partition representing a single roughness element. For each element the roughness is described by the grain/skin friction. The number of partitions chosen for any simulation will, of course, depend on the resolution required from the model and the physical characteristics of the roughness elements on the slope surface.

On slopes with generally small roughness elements the former description is probably the more appropriate, whereas on slopes which have a wide range of small and large roughness elements the latter description is probably better. To reflect these two approaches to describing the roughness of semi-arid surfaces, two models were initially developed. Subsequent work on the scales and magnitude of microtopography at the field sites has led us to prefer the second model.

14.5.2 Description of the model

In order to avoid the formal dichotomy between rill and inter-rill, we have adopted a series of depth categories which provide a continuum, related to the available microtopographic relief on each site. We envisage an initially rough surface with stones, plant mounds, rill and gullies. The roughness elements are represented by a series of cross-slope depths above or below the mean plane which locally defines the surface. This idea is very similar to the conceptual infiltration model of Moore and Clarke (1978). To avoid using a map to give the depth of each strip, the depths can be obtained from known depth distributions. The depths for micro-roughnesses are generally provided from a normal distribution, which has been found to give a good fit to field distributions of roughness elements. Initially the variance of this distribution is derived from direct field measurements.

With input rainfall in a given reach, the cross-profile is filled with water from the bottom upwards, the lowest hollows in each part of the cross section being filled first. Any excess is distributed to neighbouring hollows. In contrast, water routed from the section upslope is used to fill the deepest parts of the whole profile. In this way the rills and the deeper flow elements can be functioning when there is little or no flow across the upstanding roughness elements.

For routing, velocities are obtained from slopes determined from successive profiles of depths at each cross section. In this way, the velocities are adjusted to the configuration of individual roughness elements and their continuation, or otherwise, in a downslope direction. Velocities are determined using the Manning equation. Water is routed between successive sections as a kinematic cascade, using methods first developed by Kibler and Woolhiser (1970). The nodes separating slope elements are spaced at intervals of the order of 10 m, so that each catena site is subdivided into three to ten sections for overland flow routing.

The surface water and erosion model is based on the kinematic wave equations which consist of the unsteady continuity equation:

$$\frac{\partial A}{\partial t} + \frac{\partial Q}{\partial x} = i \tag{1}$$

where A is the flow cross-sectional area in cm^2, Q is the discharge in cm^3 s^{-1}, t is the time in seconds and i is the lateral inflow (cm^2 s^{-1}) at distance x; and a uniform flow equation (the Chezy equation):

$$V = U_c C \sqrt{R\Lambda} \tag{2}$$

where V is mean velocity (cm s^{-1}), R is hydraulic radius (cm), $U_c = 5.52$ cm$^{1/2}$s^{-1}, C is the dimensionless Chezy roughness and Λ is the tangent slope gradient.

Initially the Chezy coefficient was regarded as a constant; but review of the literature now suggests that it varies enormously when vegetation is involved (Palmer, 1946; Parsons, 1949; Petryk and Bosmajian, 1975). The following Chezy values are used in the current model:

Hillslope plant cover (%)	0	20	40	80	99
Chezy C	200	70	40	30	15

These values are typical for silty slopes. There will be slight differences between them and values for sandy and clayey slopes.

These equations are solved numerically in the model using a one-dimensional explicit finite difference scheme.

For any time step the dynamic partitioning of water is as follows. Residual water left on each strip in each cell provides a dynamic water surface topography, shown as *temporary depth* in Figure 14.6a. The relationship between depth and volume for this temporary surface is then calculated. Since the total flux arriving from upslope for the previous time step is known, the fill depth can be calculated. The water arriving from an upslope reach (*active volume*) is then added to every strip below the fill depth. Rainfall additions and infiltration losses are then allocated to every strip whether it is below or above the total computed flow depth (*fill depth*).

Many existing surface water routing models assume that the hydraulic radius is equal to the water depth. In a model that allows for variation in water depths across the slope and the development of dynamic flow concentrations, this assumption is unrealistic and the hydraulic radius of groups of deeper strips in which flow concentrates needs to be considered. In the model this problem is dealt with in two ways:

(i) Water above the fill depth is assigned a hydraulic radius equivalent to the flow depth on the assumption that flow is very shallow so that hydraulic radius approaches the flow depth. The flux from each strip is then calculated using the uniform flow equation.

(ii) Where flow concentrates, such as in the micro-channels denoted by the letters A–E in Figure 14.6b, the hydraulic radius for the *group* of strips forming the flow concentration is calculated. The flux from each flow concentration or micro-channel is then calculated.

To avoid defining a relationship between flow velocity and roughness, Chezy's C is held constant for each cell on the assumption that it represents the skin friction. Variations in the efficiency of overland flow are controlled by the hydraulic radius and slope factors in Equation 2 above.

Two clear advantages of the new model over traditional approaches can be identified. The first is that the model allows flow concentrations to change in shape, size and number according to the discharge, as indicated in Figure 14.6b. The second advantage is that it gives the depth and velocity of the flow for each strip. By simulating varying depths and

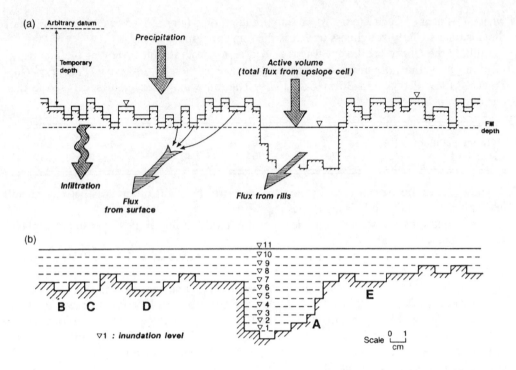

FIGURE 14.6   (a) Example relationship between depth and active volume. (b) Hypothetical slope
cross section consisting of a series of depths of unit width below an arbitrary datum

velocities of water across the slope the model can more easily approximate real-world
spatial variations of erosion on rilled and gullied slopes during a storm event. Moreover,
these variations affect the runoff hydrograph itself, so that the model is better able to
simulate the impact of erosion on hillslope hydrology.

Simulations of rilled and unrilled surfaces have indicated that the model produces
rational and valid results. For example, consider a surface with a single rill and a surface
with two rills. The first rill on the two-rilled surface is the same size as the single rill on the
one-rilled surface. However, the second rill has a mean depth half that of the first rill. The
slope length is 20 m and the larger rill increases in mean depth downslope from 2 cm at the
divide to 8.3 cm at the slope base. Rain is added to the surface (which is assumed to be
impermeable) at a rate of 72 mm h^{-1} for 5 minutes at the start of both simulations. The
hydrographs of slope base discharge for each surface are shown in Figure 14.7. Both
hydrographs show the same rise until about 145 seconds, and the two-rilled surface then
shows a more delayed response than the one-rilled surface. This occurs because some of
the water arriving as active volume then starts to fill the second rill, resulting in a reduced
hydraulic efficiency. Although very easy to simulate, this type of behaviour is not
generated in conventional flow routing models of rilled surfaces, for which the proportion
of flow reaching each rill is fixed throughout the simulation run.

In common with many hillslope models, overland flow is routed downslope in the
MEDALUS hillslope model as a one-dimensional kinematic wave or cascade. Although

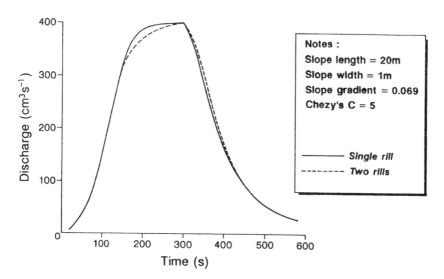

FIGURE 14.7   Example model output for one- and two-rilled surfaces. Discharges at the base of a
uniform 20 m slope, for a period beginning with 5 minutes of rainfall

one-dimensional, each store of the cascade can have different widths so that increasing
store widths downstream represent divergent flow on a nose or spur, and decreasing
widths represent a hollow. Existing approaches tend to assume that the water fills each
store in the kinematic cascade to a uniform depth, and describe the effects of
microtopography using a functional relationship between roughness and flow depth or
velocity (see Abrahams et al., 1989). In the overland flow part of the MEDALUS hillslope
model we adopt a different approach in which the microtopography is described as a series
of depths or strips. Each store in the downslope cascade is made up of a number of strips
of different depths which, like the store, can also be variable in width. These can be
generated using a number of generators within the model. Alternatively field data can be
entered into the model. Preliminary results from the field surveys suggest that the elevation
of microtopographic elements can be described by a normal distribution.

Two clear advantages of this method of simulating overland flow can be identified. The
first is that the model allows flow concentrations to change in shape, size and number
according to discharge. The second is that the model describes variations in water depth
and velocity within a store in the cascade. By simulating varying depths and velocities
across the slope, the model can more easily approximate observed spatial variations of
erosion on complex hillslope surfaces during a storm event.

14.5.3 The sediment transport model

Down the length of the slope catena, a sedimentation balance is calculated for each
successive store used in the kinematic cascade, which represents a small fraction of the
distance between the three main sites. Net erosion is allocated according to water depth in
the topographic lows representing 'rills', and net deposition is allocated inversely with flow
depth in the 'rills'. Transport for each size class is calculated from the process which is

defined by a detachment rate and a travel distance. The use of a travel distance model allows effective separate forecasting of rainsplash, rainflow and rillwash. For this purpose, rainflow is defined as detachment by raindrops and transport in a water flow. Rillwash is defined as detachment by fluid traction, and transport in the flow. Thus we need to define two modes of detachment, by raindrops and by fluid traction; and two modes of travel, splashing through the air and moving as bed or suspended load in the water flow.

Sediment movement and changes in surface elevation (z) under wash and rainsplash is described by the sedimentation balance equation proposed by Kirkby (1980) in which sediment transport is assumed to be 'erosion limited':

$$-\frac{\partial z}{\partial t} = \frac{\partial S}{\partial x} = D - \frac{S}{h} \tag{3}$$

where D is the detachment rate, S is the sediment transport rate, h is the mean travel distance of a sediment grain and x is horizontal distance from the slope divide.

Bennett (1974) notes that Equation 3 assumes either that the change in sediment concentration over time (dc/dt) is negligible when compared to the second term in the equation, or that the flow is quasi-steady state. In the current application the former would be assumed.

The finite difference form of Equation 3 is:

$$\Delta S = \left(D - \frac{S_{in}}{h}\right) \bigg/ \left(\frac{1}{\Delta x} + \frac{1}{2h}\right) \tag{4}$$

where $\Delta S = S_{out} - S_{in}$.

Detachment rates under rainfall are approximately generalized from Foster and Meyer 1975) as:

$$D \propto r^2 \tag{5}$$

where r is the net storm rainfall amount.

Raindrop detachment rates have been noted to show only a weak dependence on gradient, but it should also be noted that detachment is constrained by the depth of flow. Using Manning's equation as a first approximation, then flow depth y corresponding to a discharge per unit width q is given by:

$$y \propto \left(\frac{q}{\Lambda^{0.5}}\right)^{0.6} \tag{6}$$

where Λ is the tangent slope gradient.

Forecast detachment is then attenuated in the ratio:

$$\left(1 + \frac{y}{\bar{y}}\right) \exp\left(-\frac{y}{\bar{y}}\right) \tag{7}$$

where $\bar{y}$ is a critical flow depth.

The critical depth is usually taken (Palmer, 1963) to be 5–6 mm. This effect is important for limiting the effect of rainflow on long, very gentle slopes.

Rainsplash travel distance is strongly, and to a first approximation linearly, dependent on gradient, though largely independent of distance from the divide (representing overland flow collecting area). Empirical exponents of gradient vary, with values as low as 0.5 and occasionally greater than 1.0, but the value of 1.0 is adopted here for simplicity.

On steep slopes and for relatively large grains, the normal Shields/Andrews analysis for detachment by flow forces requires considerable revision. Firstly it is not fair to assume that the depth of flow is large relative to the grain diameter, and secondly the gradients are large enough that the gravity term in grain traction cannot be ignored.

The forces acting are then:

(i) a downslope component of the submerged grain weight;
(ii) an upslope frictional resistance, related to submerged grain weight and angle of friction; and
(iii) a fluid traction stress integrated over the range of depths of grain submergence.

These can be combined to give:

$$\frac{y}{d_c} = \Psi + 0.5 \quad \text{for } \Psi \geqslant 0.5$$

$$= 2\Psi \quad \text{for } \Psi \leqslant 0.5 \tag{8}$$

$$\Psi = 0.06\Delta(\cot \alpha - \cot)\phi$$

where Δ is the ratio of submerged grain to water densities, y is flow depth, d_c is grain diameter, ϕ is angle of friction and α is slope angle ($\Lambda = \tan \alpha$).

This relationship gives a roughly linear increase in critical grain diameter with gradient up to angles of 20–25° and then a very rapid rise as the angle of friction is approached. It is assumed here that the relevant grain size is the mean, and that there is equal mobility during grain detachment. Thus, for a surface layer containing a mixture of grain sizes, the detachment rate for each size class is proportional to its concentration in the surface layer. In this analysis it is assumed that the traction threshold is related solely to grain size, though it is recognized that a substantial part of the total flow resistance is attributable to form drag associated with the microtopography in the downslope direction. Implicitly we assume here that form drag combines with turbulent fluctuations to modify the constant 0.06 in the above equations, so that it should generally be evaluated empirically.

Over the threshold, flow detachment is modelled as proportional to the excess flow power. The expression uses a modified gradient term which provides a significant correction on steep slopes:

$$D = A(q\Lambda' - u\theta) \tag{9}$$

where D is the detachment rate, A is a constant, taken as $0.1 \text{ g s}^{-1} \text{ cm}^{-2}$, q is the mean discharge per unit width, Λ' is the modified gradient, $\Lambda/(1-\Lambda/\tan \phi)$, θ is the traction threshold, and u is the flow velocity.

In this expression the flow velocity, u, is taken as constant at 0.01 m s^{-1}, and the traction threshold is derived from Equation 9 above as:

$$\theta = 0.06\Delta d_c \tag{10}$$

We propose that the balance of force on a particle just before it starts to move on a slope with gradient S and flow depth R may be expressed in the form:

$$\Delta R D^3 \tan\phi + \lambda D^n = R \Lambda D^2 \qquad (11)$$

where λ and n are constants of the cohesive force term.

The threshold value of the depth slope product $R\Lambda$ should therefore be expressed as:

$$\theta = \Delta D \tan\phi + \lambda D^{n-2} \qquad (12)$$

where $\lambda = 2.762 \times 10^{-4}$ mm^{3-n}, $n = -0.5$ and D is measured in mm, giving the final factor of 10 in the equation.

The cohesive constants are set to ensure that the threshold of motion will reach its minimum value when $D = 0.125$ mm, which is the D_{50} value for silt, the most erodible soil. Figure 14.8 shows the traction threshold against soil particle diameter.

For travel distance, there is the potential for strong size selectivity, which appears to be demonstrated in many downstream and downslope fining sequences. If grains are given an equal initial velocity at detachment, then travel distance may be calculated from the net deceleration of the grains. If the angle of friction is kept constant, then deceleration is greatest for large grains, and grains below a critical size (for the flow) will be transported indefinitely. However, the effective angle of friction has been shown to increase as the size of a moving grain is reduced relative to the mean bed size, producing an opposite effect. Another factor is that initial grain velocities are not equal, and that grains are initially displaced upwards into the flow. An alternative view of travel distance is that it is the distance taken to fall to the bed from a position initially high in the flow.

On this view, the travel distance is given by:

$$h = y \frac{u}{w} = \frac{q}{w} \qquad (13)$$

where h is travel distance, y is the flow depth and w is grain settling velocity.

This estimate is more realistic in forecasting much greater travel distances for smaller grains. For present purposes, it is proposed to use a travel distance which broadly follows Equation 13, with distance inversely proportional to grain size, and proportional to discharge. The flow velocity is calculated using the Chezy equation (Equation 2 above) and depth of flow is worked out within the concentrated flow module. The fall velocity of a sediment particle is then obtained from the balance of gravity and hydrodynamic forces acting on the settling sediment particle:

$$\frac{\pi}{6} \Delta g D^3 = \frac{1}{2} \frac{\pi}{4} C_d u^2 \qquad (14)$$

where C_d is the dimensionless resistance coefficient.

For a sphere, the aerodynamic resistance coefficient is:

$$C_d = 24/Re \text{ for Particle Reynolds number } Re < 0.4$$
$$= 0.45 \text{ for } Re > 1000 \qquad (15)$$

where $Re = vD/v$ and v is the kinematic viscosity of water ($= 1$ mm^2 s^{-1}).

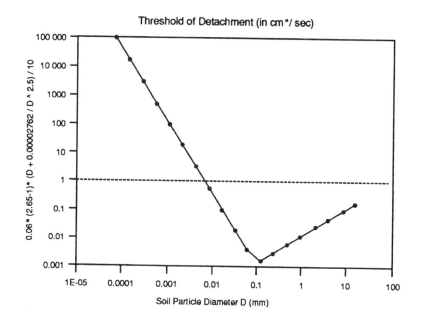

Threshold of Detachment (in cm*/sec)

0.06 * (2.65-1) * (D + 0.00002762 / D ^ 2.5) / 10

Soil Particle Diameter D (mm)

FIGURE 14.8   Modelled relationship between traction threshold and grain size

where $Re = vD/v$ and v is the kinematic viscosity of water ($=1\,mm^2\,s^{-1}$).

Substituting for the Reynolds number, the end members of Equation 15 may be combined in the log–log relationship:

$$C_d = 1.7928D^{-1.3648} \tag{16}$$

The fall velocity (w) in mm s^{-1} then becomes:

$$
\begin{array}{ll}
w = 898.333\ D^2 & (D < 0.076\,mm) \\
w = 109.662\ D^{1.1824} & (0.076 < 2.755\,mm) \\
w = 218.717\ D^{0.5} & (D > 2.755\,mm)
\end{array}
$$

The relationship of travel distance with velocity of flow is shown in Figure 14.9, for a flow depth of 5 mm.

The overall effect of these wash processes is to assume that all sizes are detached in direct proportion to their frequencies in the surface material (armour layer if present or underlying soil if not). This is seen as equivalent to the assumption of equal mobility in the fluvial literature. Travel distance is, however, strongly size selective, so that effectively all the detached fines are removed far downslope whereas coarse material is moved only slightly. In effect there is a continuous gradation—source or detachment-limited removal of fines, and transport-limited removal of the coarse gravels. Where there is net erosion, the detached layer is assumed to be remixed, generally providing a coarsening of the surface armour layer. Where there is net deposition, then downslope fining occurs through selective transportation (Kirkby, 1992).

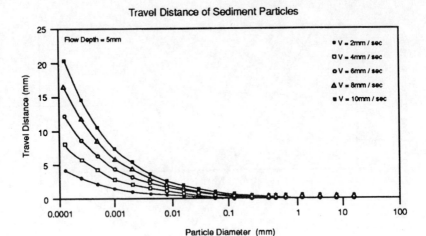

FIGURE 14.9   Modelled relationship between grain travel distance and grain diameter at a range of flow velocities, with a 5 mm flow depth

14.6 SOIL MOISTURE AND SOIL PROPERTIES

14.6.1 Conceptual basis

Beneath the soil surface the most important processes to take into account are water movement, organic matter cycling and changes in composition due to erosion and deposition at the surface. Cation mobilization and nitrogen release may also be significant, but have not been incorporated into the present model. These processes influence one another, but also modify the water reservoir for plant growth and evapotranspiration between rainfall events. There is also a strong interaction with surface processes: soil conditions control infiltration rates, and overland flow erodes the soil, removing organic matter and the more fertile soil layers.

Infiltration and downward movement of soil water is computed by application of the Darcy/Richards equation for unsaturated flow in a vertical column. In line with the measurement programme set out in the field manual, the soil moisture retention characteristic will be described by the van Genuchten (1980) equations for hydraulic potential and conductivity. A current summary of relevant methods may be found in Anderson and Burt (1990). Conditions at depth will normally be related to constant (dry) conditions or, where appropriate to a water-table. Computations of flow will be made over vertical intervals which increase in depth geometrically from the surface down. Soil hydraulic properties are allowed to vary with depth in the soil, dynamically with soil organic matter and with soil dispersion, particularly in the context of the lower infiltration rates associated with surface sealing. If suitable data are available, it may also be possible to take account of hydrophobic soil properties and their breakdown.

It is not practicable to go to the extreme of simulating every individual point on a closely spaced grid on the slope surface, and this level of detail will not in any case be supported by an adequate data-base for the variability in soil properties. Instead we seek a compromise which represents average soil conditions and soil layering, but uses available information about microtopography, so that rills etc. intersect the upper soil layers. Similarly erosion, deposition and/or changes in organic soil content must be able to change horizon depths. We therefore introduce the concept of the mean soil surface as follows.

The mean soil surface (MSS) acts as the reference level for soil properties and erosion or deposition of sediment and/or litter. It is dynamically defined as a moving average of local elevations over a width of 1–2 m or as a locally fitted linear trend surface. This allows the MSS to follow (and therefore ignore) gentle undulations in the surface, and to respond to the locally different soil characteristics at the base of rills and small gullies. For any particular elevation relative to the MSS, the pattern and history of surface water depths is expected to show strong similarities, so that relative elevation is treated as a basis for stratifying the population points on the ground surface. Soil moisture models are run in parallel for a series of elevations relative to the MSS, to provide local profiles of soil moisture, infiltration and bare soil evaporation. The estimated local values of infiltration rate are then used to distribute rates across the microtopography on the basis of the height distribution.

For percolation to groundwater (if relevant) and downslope flow, the local values of soil moisture are combined into weighted averages, on the assumption that these processes span larger areas than the very local scale of surface/subsurface interactions. Gains or losses of soil moisture through these broader-scale processes are allocated in proportion to existing soil moisture contents, but these processes are not generally significant under Mediterranean climates.

Transpiration is closely linked to the distribution of vegetation, which is itself associated with local microtopography. Large shrubs tend to grow on, and to some extent form, higher mounds on the surface, while eroded areas are mainly covered with annual herbs. The (normal) distribution of microtopography is combined with an overlapping vertical layering of vegetation types to associate micro-elevations with proportions of plant types, and infiltration and evapotranspiration rates are separately calculated for each micro-elevation class. Micro-elevation is also associated with different stone covers and different degrees of armouring and/or surface sealing. This partition, although oversimplified, allows some differentiation to be made on the basis of different vegetation covers, and different durations of ponding after rainfall, and yields significant variations in near-surface soil water within a plot, which have been well documented for Kenya (Dunne and Aubry, 1986).

As microtopography evolves through surface erosion, there is a need to maintain a realistic distribution of plant types with respect to the detailed surface form in the model. We propose to distribute plant types in a fixed way, established initially by measurement, across the elevation range as it is at any time. We therefore propose to relate the vegetation distribution dynamically to the mean and standard deviation of this distribution. For example, if the surface is eroded, there is usually some increase in microrelief, with the mound tops relatively unaffected. For this scenario, large vegetation is implicitly assumed to migrate outwards from the mound tops, without destruction of its roots, while grasses and annuals re-colonize the eroded lower areas. Figure 14.10 shows illustrative values for a site with four vegetation components present.

Surface erosion and deposition rates are also strongly stratified by relative elevation classes, and show up in extreme cases through the concentration of rill activity in topographic lows. Local modifications of relative elevation by sediment transport over a period lead to gradual changes in both MSS and in the distribution of relative erosion. As the surface is lowered, material is released to the surface in proportion to its grain size composition in the soil, taking account where appropriate of stones which can armour the surface and influence its infiltration rates.

The accumulation of organic soil from surface litter is also thought to be concentrated preferentially on topographic highs such as vegetation mounds, and to be less along rill courses and in the generally lower bare areas between plant stems. Surface organic matter is therefore distributed preferentially, and is modelled as concentrating directly beneath its parent plants. Organic matter is also introduced more uniformly over the area, from root decay within the soil. This organic matter is mixed through the profile by meso-organisms such as ants, and decomposes progressively over time, releasing CO_2 which diffuses back to the atmosphere. The organic matter greatly enhances soil moisture retention and adds to the soil bulk. There are therefore important implications for the hydraulic properties of the soil, which may change significantly as soil degrades.

14.6.2 Soil water model

The soil moisture model used is based on a series of solutions to the Richards' equation in one dimension, using the van Genuchten (1980) equations to give the interdependence of soil moisture content, hydraulic potential and hydraulic conductivity. Use of the van Genuchten forms for the soil moisture retention curves has been agreed as a standard within the project. The parameters of the equations are being modified from standard values corresponding to field measurements (e.g. Hendrickx, 1990) and soil textural classes to take account of the organic matter profile within the soil. Stones in the soil are also allowed to influence the parameters, by reducing the total water-filled porosity of the soil, and by restricting the volume within which the organic matter can mix. The detailed soil water model has now been expanded to take account of the combination of evaporation and transpiration within the soil profile.

$$\frac{\partial \theta}{\partial t} = -\frac{\partial q}{\partial z} = -\frac{\partial}{\partial z}\left[K\left(1 + \frac{\partial h}{\partial z}\right)\right] \tag{17}$$

where θ is volumetric moisture content, q is discharge measured downwards, z is depth below the mean soil surface, K is hydraulic conductivity at θ and h is soil moisture tension (cm water).

$$S = \frac{\theta - \theta_r}{\theta_s - \theta_r} = \left[1 + \left(\frac{h}{h_0}\right)^n\right]^{-m} \tag{18}$$

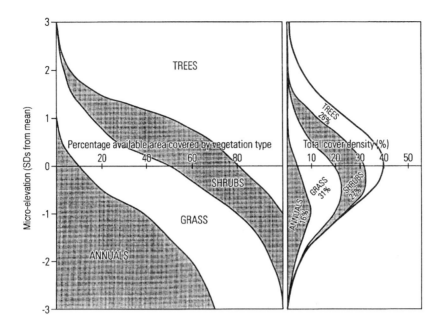

FIGURE 14.10   Illustration of how vegetation classes are distributed over microtopography in the MEDALUS model. Fixed vertical distributions, shown at left, are distributed across the normal distribution of micro-relief after normalization for mean and standard deviation, giving distributions shown at right. If micro-relief grows through erosion, trees and shrubs spread downhill and annuals invade newly eroded rill areas

where θ_r and θ_s are residual and saturated moisture contents, S is proportional saturation, h_0 and n are parameters of the equation and $m = 1 - 1/n$.

$$K = K_s S^{\lambda}[1 - (1 - S^{1/m})^m]^2 \tag{19}$$

where K_s is the saturated hydraulic conductivity, λ is a parameter of the equation and other parameters and variables are as before.

These equations have been solved numerically using a finite difference scheme. Although there are many algorithms available which follow this approach, there has been some difficulty in selecting an optimal algorithm which combines reasonable computing speed with adequate accuracy, in particular in maintaining a correct water balance. The difficulties are greatest when the soil is strongly layered.

Two alternative approaches have been tested: one an explicit scheme and one fully implicit. There is a trade off between the larger number of calculations involved in inverting the tri-diagonal matrix for the implicit solution, and the shorter time steps required for stability of the explicit solution. Computational stability is maintained by using a dynamically variable time step which is chosen so that the gradient in soil moisture with depth changes by a maximum of 20% of its current value in a single time step. This lies well within stability limits, but must be relaxed at very low gradients to allow reversals in gradient over time. For example, with a layered (loam over silt) soil divided into 20 layers of 5 cm each, 213 iterations were needed over a 30 hour simulated period. Soil moisture and tension values are associated with the mid-point of each soil layer modelled, with any sharp interfaces at the boundaries between cells, where there will be continuity of tension, but not generally of soil moisture where properties change. At each time step, the value of pore water tension at the boundaries is iterated to provide for flow continuity. This method was found to give the most consistent results and minimized water balance difficulties which can be severe for infiltration models.

14.6.3 Model dimensionality

Soil infiltration modelling is usually one dimensional, which is an adequate approximation provided that the horizontal dimensions of the soil which are dealt with as one 'column' are substantially greater than the significant depth of the soil, so that we can assume that horizontal water movement is not a significant factor. However, in the case of Mediterranean soils, where surface sealing and rill formation are important, this assumption breaks down. There will be significant horizontal diffusion from and immediately beneath rill beds and, while surface sealing is effective, between vegetated and bare areas, due to the differences between sealed and protected areas.

On the other hand we are likely to lack both the data and the computing power to go to a fully three-dimensional system. What we therefore propose is to divide the soil column into five representative domains, essentially to be regarded as a statistical rather than a spatial division. Each domain is subject to the usual vertical infiltration but the flow into the top layer is different for each. In addition there is a horizontal diffusion between domains based on an average distance to be computed as the area of the domains divided by their shared perimeter.

14.6.4 Infiltration domains

Rills are a temporary phenomena and therefore the proportion of the soil surface which is covered with water varies with time and is usually zero. The weighting of this wet domain in the model therefore varies to reflect this. As domain weights vary the model assumes that boundaries are moved within the soil and the water contents of each domain are adjusted accordingly. When, as is usually the case, no surface water is present, no processing of the wet domain will occur. When surface water appears, the initial water content of the newly activated domain will be that of the annual plant domain.

The infiltration rate for the wet domain is simply equal to the infiltration capacity of the soil. It is expected that the erosion and sediment transport sections of the model will provide figures for the composition and depth of the armour layer, from which its hydraulic characteristics may be estimated.

14.6.5 Integration method

The integration is performed by the Runge–Kutta central difference scheme. The initial guess of the new value is a linear extrapolation of the rate of change in the previous step. Time-step duration is controlled by performing the integration in two short steps and one double-length step and comparing the results. This gives an estimate of the error term and checks the dependency of the solution on the time step chosen. The recommended order for the Runge–Kutta calculations to produce the maximum model speed is four. However, in stable conditions the step length calculated for the fourth-order calculation is almost never completed without some external interruption so a third-order calculation is currently being used although the order is a simple program constant which can readily be changed.

The hydraulic conductivity used both for transport between layers and lateral transport between sub-columns is a geometric mean suitably weighted where the thickness of the layers varies. The matric potential is the variable which is assumed to be continuous at layer boundaries. Water movement is calculated from the centre of one layer to the centre of the next.

14.6.6 Soil organic matter profile

Within each vegetation type, dead leaves and stems eventually fall to the surface, while roots die *in situ*. Decomposition of organic matter is usually considered to be at a fixed proportional rate (e.g. 5% per month), with seasonal variations and partitioning into more and less stable components. The more stable component can be caricatured as lignin, the more labile as carbohydrate, though there are in fact many components and a range of breakdown rates. Decomposition rates correspond primarily to temperature, provided that soils are well aerated, as in most Mediterranean examples, and is accomplished mainly by soil organisms. Mesofauna such as isopods, worms and termites also play a substantial role in vertical mixing of the soil, which can be modelled as a diffusion process. Vertical mixing affects all soil fractions, but has greatest significance for mixing decomposed leaf-fall material down into the soil. The rate of diffusive mixing is related to the soil as a life support medium for mesofauna, both in terms of nutrient status and pore space, and therefore tends to decline sharply below the root zone. For a single decomposing fraction, we can solve for the steady-state organic profile as follows:

$$
\left.
\begin{aligned}
c &= \frac{1}{k}\left[\left(\frac{L}{z_M^2} - \frac{R}{z_R^2 - z_M^2}\right)z_M\exp\left(-\frac{z}{z_M}\right) + \frac{R}{z_R^2 - z_M^2}\exp\left(-\frac{z}{z_R}\right)\right] \\
z_M &= \left(\frac{D}{k}\right)^{0.5} \\
R(z) &= \frac{R}{z_R}\exp\left(-\frac{z}{z_R}\right)
\end{aligned}
\right\} \quad (20)
$$

where z_R, z_M are respectively the rooting and mixing depths, D is the diffusion rate for vertical mixing, k is the proportional rate of decomposition, L and R are the total rates of leaf and root fall and $R(z)$ is the distribution of root fall, following the roots.

This solution is only meaningful where the mixing depth is less than the root depth. An example of this distribution is shown in Figure 14.11 for plausible parameter values. There is a surface zone dominated by decomposed leaves, and below this a zone where decomposed root material is dominant.

These additions of organic material are expected to modify the hydraulic properties of the soil, though some measurements (Imeson, personal communication) suggest that the

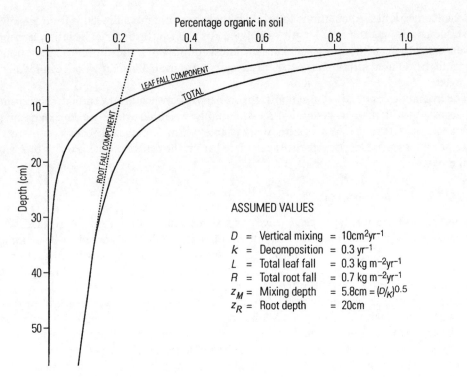

Percentage organic in soil

FIGURE 14.11   Example of equilibrium organic matter profiles in the soil. Near the surface, OM is dominated by leaf fall, and at depth by root fall component

differences may be rather small. During active desertification it is important to monitor and model changes in organic status, which will no longer follow the equilibrium distributions, although response times are expected to be only a few years. At present we are assuming that organic matter modifies the saturated moisture content of the soil without changing other parameters of the van Genuchten moisture retention curve, but we seek further enlightenment!

As vegetation patches have the scope to change over time, there is the scope for some areal averaging, but the results from the dynamic modelling group and field observations suggest positive reinforcements of existing patterns. We are therefore allowing differences between vegetation types to accumulate over time, on the assumption that the vegetation types, and the soils beneath them, are stable features for periods which are longer than our time spans of interest.

14.6.7 Soil stoniness

The work of J. Poesen's group (Chapter 12) has shown the importance of both surface and soil stoniness for physical processes and plant growth. At the surface, the armouring through selective transportation reproduces the general form of experimental results. As stone cover is progressively increased, for a given discharge, the increase in roughness produces an initial increase in sediment yield, and then a subsequent decrease as less of the surface becomes mobile.

Within the soil, the effects of soil stoniness are to some extent taken into account through the field measurements of soil hydraulic properties, which are used as a model input for direct comparisons with field sites. Where these data are not available, it is important to understand the relationship between components of the soil composition. For an initial analysis, three components are considered: fine earth, stones and organic matter. As the stone fraction increases, the organic matter remains mixed within the fine earth fraction, which benefits from its structural properties. The water-retaining properties of the organic fraction are thus able to increase the water available to plants in stony soils, so that, in semi-arid conditions, productivity increases with stoniness over an appreciable range. Eventually, of course, the trend is reversed, as the total volume of fine earth is severely reduced by high stone contents.

To a first approximation, the water content available to plants in a fine earth and organic mixture with a proportion p^* of organics, can be expressed as:

$$w = w^*p^* - w_0 \text{ for small } w$$
$$= r_0 \text{ for large } w \tag{21}$$

where p^* is the effective proportion of organic matter in the soil, w^* is the extra water-holding capacity of organic matter, w_0 is the threshold below which water is unavailable and r_0 is the mean rain per rain day.

Since the organic matter is mixed only in the fine earth, we also have:

$$p^* = \frac{p}{1 - p_s} \tag{22}$$

where p is the proportion of organic matter in the soil as a whole and p_s is the proportion of stones.

Figure 14.12 illustrates the variation in effective water-holding capacity in terms of soil organic matter and stoniness. It may be seen that productivity is greatest at moderately high stone content.

Similarly the bulk density of the non-stony fraction may be expressed as:

$$\rho = \rho^* \frac{1 - p_s + p}{(1 - p_s)/\alpha + p} \tag{23}$$

where ρ^* is the density of the organic fraction
and α is the density ratio of fine:organic fractions.

14.7 PROGRAMMING FRAMEWORK

The model has been written for MSDOS machines, using TURBO Pascal for Microsoft Windows 3.1. This will provide executable programs suitable for a 386/486 system with VGA graphics, and is available to all project groups. Updated versions are available by EMAIL links, on 3½ inch disk from the School of Geography, University of Leeds, and a

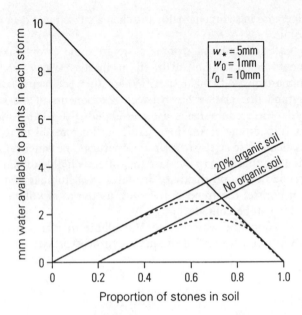

FIGURE 14.12   Illustration of conceptual dependence of soil water for plants and soil stoniness and organic content. Solid lines indicate asymptotic response. Broken lines sketch the actual relationship

copy is also available on the MEDALUS CD-ROM. Development of the model is continuing, particularly with respect to vegetation components, in MEDALUS II.

Windows allows user-friendly and self-documenting control. It also facilitates flexible graphical output, both in real time while the model is running and output on a wide variety of ordinary printers. Whereas Windows does impose some overheads on minimum hardware configuration, it has been found that little or no overhead is experienced on actual CPU time once the model is running (except the inevitable overheads associated with graphical output which would exist however such output was produced). Some time overheads may however be experienced on machines with limited memory. It is believed that the extra functionality more than justifies the time costs.

We have provided a common core for the model which is intended to coordinate and provide facilities for the various sections envisaged. The core allows the various sections to cooperate with minimal knowledge of the internal workings of the others. This reduces the inevitable misunderstandings about interfacing and allows totally different algorithms to be tried for one section without much, if any, modification to the rest of the program.

The overall program sequence is roughly as follows:

(i) The core initializes the Windows facilities. It then calls a set up procedure in each of the sections of the model and this procedure sets up any defaults for its input values and registers menu entries which allow the user to interact with that particular model section.

(ii) The user selects menu items contributed by the various sections which then allow him or her to enter settings and specify data files for the run. This means that the user can enter the data in any order that suites.

(iii) The user selects the 'Run Model' menu entry. Each section is given the opportunity to check that it has all the information necessary for the run and, if not, either give a message and abort the start up process or request additional input.

(iv) Each section of the model initiates a set of processes that actually perform the model activities.

(v)   The system goes into run mode. Each process is run at intervals in simulated time which can be different for different processes and may vary dynamically as conditions require. The menu entries and other user controls continue to function while the model is running allowing the user to dynamically change run parameters if the model sections permit. He or she will also be able to request and cancel dynamic displays and pause the model to study some particular state.

(vi) The model terminates at the request of the user or one of the sections (e.g. because the available weather data is exhausted or after a set run time). The system now returns to step (ii) which allows new runs with different data or parameters.

In addition to the process management functions the core provides facilities for state variables which can be instructed to vary in a time linear fashion between steps of the controlling process or processes giving a reasonable estimate of the current value in the situation where the using and controlling processes are not synchronized. There is also a set of generally useful facilities for such activities as interpreting data files and plotting graphs which help protect the section writers from the intricacies of Windows.

The common core for the model coordinates and provides facilities for the various sections envisaged. It allows the various sections to cooperate with minimal knowledge of one another's internals. This reduces the inevitable misunderstandings about interfacing and allows totally different algorithms to be tried for one section without much if any modification to the rest of the program.

14.7.1 Using the computer model

The catena data for the model is input through a structured text file which contains data for all parts of the model laid out in a simple, hierarchical format. A separate text file contains the numerical data for the different soils which are referred to in the main catena file by name. A third text or spreadsheet file supplies the AWS data.

User interaction occurs through a fairly standard Windows interface. Data displays are keyed through a simple cross-sectional view of the catena. Clicking the right mouse button on any part of the slope brings up a short menu of possible displays. The number of displays viewed concurrently is limited only by the limitations of screen territory. Data displays appear as independent windows which can be positioned on the screen or re-sized at will. Obviously there is a processor overhead involved in maintaining multiple displays. It is up to the user to choose how much detail he or she wants to see at any stage of the model's progress, and what to include in a log file. Figure 14.13 shows an example screen layout.

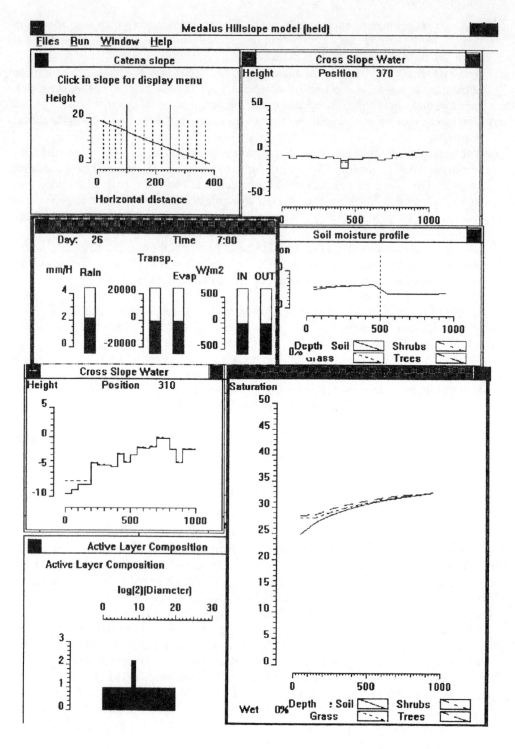

The detail of microtopography form a considerable amount of data which is felt to be too unwieldy to ask the user to generate on speculative simulations or where it is not available from the field. A random cross-slope profile generator is therefore supplied within the program. The intention is that a selection of different generators with different characteristics will be made available.

14.8 MODEL TESTING

14.8.1 General considerations

The integrated model responds to inputs with a wide range of characteristic time scales, which are seen within the model as the frequency of calls to the central sequencer. During rainfall events, the model runs slowly. All aspects of the atmospheric and surface hydrology are changing very rapidly. Interception and throughfall take place through the plant canopy. At the surface, there are two domains, dominated respectively by ponded infiltration and by infiltration at the rainfall rate. The pattern of infiltration establishes the total amount of overland flow generated. As this total is distributed over the microtopography, it modifies the subsequent pattern of ponding and generates overland flow discharges which are routed downslope along the hillslope catena. This discharge generates sediment detachment, transport and deposition or erosion, which progressively modifies the microtopography. Within the soil, there are sharp moisture tension gradients which drive percolation. All of these process are rapid, requiring time steps of fractions of a minute. For erosion and soil water storage, the most significant events are short-lived, high intensity storms, so that it is critical for the model to follow their progress in detail. During dry periods, the surface dries out and evapotranspiration and percolation slow down, but remain as background processes which control the rate at which the model runs.

Although vegetation growth responds to the daily march of radiation and available water, the relevant rates can readily be integrated over the day to give adequate estimates of growth rates over the year, but the most important fluctuations in the plants are on the seasonal cycle, with significant trends generally developing only over several years.

Many of the other changes which are contained in the model are only effective over a period of many years. Changes in the vegetation cover can take place within a year or two, or more rapidly if due to human intervention. Over a period of 5–50 years, changes in the vegetation lead to changes in the organic content of the soil, with its impact on the water economy of the soil. Erosion gradually builds up an armour layer over a period of decades, and takes decades to centuries to strip the soil to an extent which changes its physical properties (Kirkby, 1989).

It has only been possible to begin making proper comparisons with field site data. Our work is still not complete, especially with respect to integration of a sufficient range of vegetation model types. Further work on model development is continuing within MEDALUS II and extensive tests of the model have been published (Thornes et al., 1996).

FIGURE 14.13 Example of a possible screen layout during running of the integrated MEDALUS model

14.8.2 Summary of the MEDALUS model test runs

A programme of runs with the MEDALUS model have been set up in order to debug the program. The general philosophy has been to start with simple realizable conditions in order to detect major problems as quickly as possible. As new problems arise we have been forced to review earlier runs to check for consistency with the corrected versions.

Tests started with bare soils and tested hydrograph generations against simple 1-hour storms to test the runoff generation component. Test characteristics are the ratio of rainfall to runoff, the hydrograph shape, peak and duration, the sediment yield and the sediment concentration. These characteristics have been carried into the analysis of plots with vegetation of different cover densities and evaluated as a function of intensity and cover.

In early runs, there has been a problem of excessive soil moisture during droughts. When these problems have been overcome, investigations will move on to testing different combinations of vegetation cover and different slope distributions. Meanwhile attention is being turned to the effect of slope shape on runoff for bare slopes, of different materials and of further experiments on slope armouring.

A series of test runs have been designed to check the sensitivity of model results with regard to the vegetation cover on the slope. The basic slope conditions, storm events and condition of plant cover are listed in Table 14.2 below.

Plant cover on the slope made a distinct difference to the runoff expressed as a percentage of rainfall. A 100 mm h^{-1} rainfall over 1 hour resulted in a 80 mm runoff on bare slope, while the same rainfall intensity only created a 7 mm runoff when 99% of the slope area was covered by plants (*Arbutus*). Figure 14.14 shows the runoff results on a silt soil slope against different rainfall intensity and plant cover. The shapes of the curves are very similar to those reported in earlier work, though absolute values still have to be properly calibrated.

The hydrographs on bare slope have a typical familiar form, but as the plant-covered area increases, the shape of the hydrograph becomes more irregular. Irregular runoff hydrographs are the norm on irregular, rilled and bush-vegetated plots (see Francis and Thornes, 1990)

Erosion on the slope is reduced by a considerable amount as vegetation cover increases. Figure 14.15 is a typical example of such a reduction in erosion by vegetation cover under *Arbutus*. Its form is quite similar to the relationship first described empirically by Elwell and Stocking (1976). Figure 14.16 shows total erosion (i.e. the total amount of sediment transport at the slope base) against various plant cover and rainfall intensities. At present there is an inexplicable increase in erosion as plant cover grows from 80 to 99% during 80 mm h^{-1} and 100 mm h^{-1} storms, and further work is needed to find out the cause of this. The curves show that on bare slopes or with low cover the effect of intensity is significantly greater than on vegetated slopes, in good agreement with field observations.

The size distribution of the top soil layer changes when erosion or deposition occurs. Model runs for silt soils gave modest increases in armouring, which were much greater for bare than for vegetated surfaces. From an initial D_{50} of 0.488 mm, there was an increase to 0.489 mm after a single 30 mm storm on a bare surface. In a single 100 mm storm, the increase was to 0.500 mm for a bare surface, and to 0.492 mm for a fully (99%) vegetated surface.

TABLE 14.2   Basic model conditions for test runs

SITE GEOMETRY

Slope area:	2.5 × 10m
Slope gradient:	5% (2.86°)
Grid:	longitudinally 15 nodes; laterally 20 strips
General description:	planar, constant initial topographical roughness for each storm and soil type
Soil:	silt
Initial soil moistures:	40, 80, 99% saturation

STORM EVENTS

Rainfall Intensity:	10, 30, 50, 80, 100 mm h^{-1}
Duration of storms:	1 hour for most runs; 2 hours for a few runs
Modelled time in run:	24 hours for most runs; 15 days for one run (in order to check soil moisture changes)

PLANT COVERS

Plant type:	*Arbutus* (shrub model)
Leaf area index (LAI)	4.0 for each bush
Initial height:	300 mm
Root depth:	300 mm
Area covered by plants:	0, 20, 40, 80, 99%

SOIL TYPES

The current model allows the use of various soil types with different size distribution (as different combinations of three basic types: clay, silt and sand), different porosity and coefficients in the van Genuchten which determines the hydraulic conductivity. Parameters used are as follows.

	Saturated moisture (volumetric)	Saturated conductivity (cm day^{-1})	a	Residual theta	Power	Soil proportion	
						Clay	Silt
Clay	0.38	4.8	0.008	0.068	1.09	100	0
Clay loam	0.41	6.2	0.019	0.095	1.31	70	30
Loam	0.43	25	0.036	0.078	1.56	30	70
Loamy sand	0.43	350.2	0.124	0.057	2.28	0	50
Silt	0.46	6	0.016	0.034	1.37	0	100
Silty loam	0.45	10.8	0.02	0.067	1.41	0	80
Silty clay	0.36	0.5	0.005	0.07	1.09	85	15
Silty clay loam	0.43	1.7	0.01	0.089	1.23	80	15
Sand	0.43	712.8	0.145	0.045	2.68	0	0
Sandy clay	0.38	2.9	0.027	0.100	1.23	40	0
Sandy clay loam	0.39	31.4	0.059	0.100	1.48	20	10
Sandy loam	0.41	106.1	0.075	0.065	1.89	0	10

The total amount of erosion from the simulated 2.5 × 10 m plot is quite small (1667 g on bare slope after the 100 mm h^{-1} storm, 531 g when there is 99% plant cover); therefore the armouring is almost negligible. However, the absolute amount of erosion needs to be calibrated according to field observations, and the values of constant 1 in Equation 1 should be adjusted accordingly.

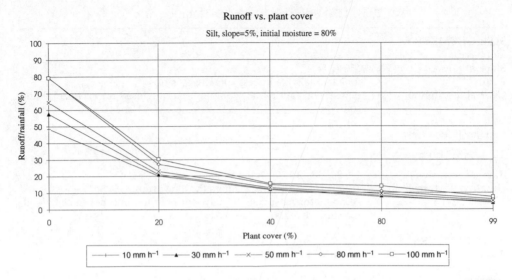

FIGURE 14.14   Example model output showing total runoff as a percentage of storm rainfall as a function of plant cover, in 60-minute storms of 10 to 100 mm

The modelled soil moisture evolution through time for initial moisture conditions without rain suggest a simple first-order exponential decay with a time constant of 26 days for bare soil and 22 days for shrubs. The form of this curve is very close to that observed for field conditions in drying soils under *Anthyllis cytisoides* by Francis and Thornes (1990) though there the decay times were between 25 and 33 days, presumably reflecting a different climatic regime as well as different species. The model performance seems to be quite satisfactory including the initially higher value of moisture under the bushes.

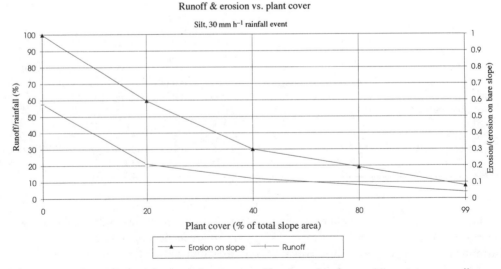

FIGURE 14.15   Example model output showing runoff and erosion from a 30-mm storm on silt, as a function of plant cover

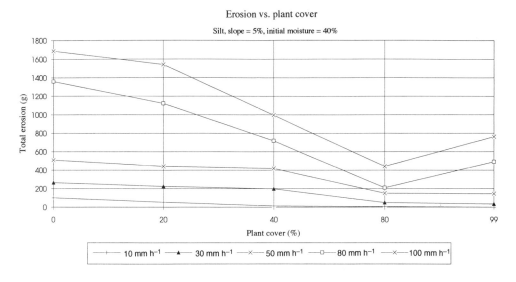

FIGURE 14.16   Example model output showing total soil loss from a 2.5×10 m plot in storms of different sizes as a function of plant cover

The changes in soil moisture with depth in the model suggests a slight but probably not very significant difference between shrubs and bare soil. Overall the absolute values seem on the high side for Mediterranean summers, perhaps reflecting difficulties in water extraction at very low soil moistures in the model. This may require some revision of the forms used for the moisture retention curves.

14.9 CONCLUSIONS

Our provisional results from the modelling suggest that perhaps the most significant physical processes of desertification may be summarized in Figure 14.17. In this flow diagram the arrows indicate both the direction of expected change and indicative time periods required for change. Where the direction of change is indicated as $=/+$, for example, it shows that there is little change at low values of the variable, and positive change at high values.

External controls on the system are not shown. Both land-use change and climatic change impact most significantly on vegetation cover, though climate also has some direct effect on runoff and erosion rates, and on soil water capacity.

Within the diagram, the closed cycles provide the feedback loops which control the evolution of the system in response to external change. The response time of a complete cycle is roughly given by the longest response time within the loop. Negative feedback loops provide a homeostatic response which limits change, and generally indicate a system which responds only slightly to external stress. Positive feedback loops provide a chain reaction, which is only corrected when the resource is exhausted, or the direction of change is reversed. In the diagram there are three main feedback cycles identified. The first (the

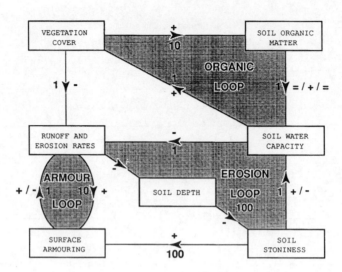

FIGURE 14.17   Dominant physical interactions involved in desertification. Arrows show direction of change, and illustrative response times. $+/-$ etc. indicate positive change at low values, negative at high values. Main feedback loops are highlighted

organic loop) is from vegetation cover to soil organic matter to soil water capacity, and back to vegetation cover. The second (the erosion loop) is from runoff and erosion to soil depth to soil stoniness to soil water capacity, and back to runoff and erosion. The third (armour) loop is from runoff and erosion to surface armouring and back to runoff and erosion rates.

The organic loop provides the main control on vegetation cover. Let us suppose that there is a reduction in vegetation cover, either through a drier sequence of years or through an increase in grazing pressure, for example. Perhaps after an initial brief period during which the plant dies back, less organic litter goes into the soil, and, over a decade or so, the organic matter in the soil falls. It may be further depleted directly by erosion. At low levels, water is mainly stored by the inorganic fine earth, and at high levels storage is constrained by available rainfall and/or by the transpiration needs of the plant. At modest organic matter levels of 1–10%, the loss of organic matter seriously decreases water availability for plant growth, particularly close to the surface. It is argued that vegetation cover will then decline in response, until a platform is reached where soil storage is no longer dependent on organic matter content. A sparse vegetation cover then persists, depending on the inorganic fine earth, and generally consisting of perennial shrubs. This positive feedback loop provides one important mechanism of desertification, but appears to be largely reversible if it is not accompanied by substantial erosion. Active badland areas appear to be among those which suffer from low levels of available water storage, and minimal soil organic matter.

Under increasingly favourable conditions, a higher plateau is reached, at which organic soil is sufficient to hold almost all of the available rainfall and a dense vegetation cover can be supported. It is argued that there are two neutrally stable states, and that the vegetation and organic matter tend to flip between them over a period of a decade or so. The erosion characteristics of the two states differ markedly: sparse vegetation and low water storage capacities lead to high rates of runoff and therefore sediment transport.

The erosion and armour loops are effective only for stony soils. On shales, clays and marls, where there are very few stones in the soil or at the surface, erosion is constrained only by available relief. Thus active badlands continue to erode until deposition lowers basal gradients, or divide relief is reduced by headward cutting. Only then is it possible for the organic loop to re-establish a soil and vegetation cover.

For stony soils, the erosion loop is the crucial step in irreversible desertification. Even very modest rates of erosion (0.01 mm yr^{-1} or 0.2 t ha^{-1}) exceed geological rates of renewal of fines through weathering, so that soils gradually become more stony. Initially, the increase in stoniness due to increased erosion leads to an increase in effective water storage capacity, through concentration of roots and organic matter in the fine earth. This provides a negative feedback loop, so that a modest increase in erosion leads to a stonier soil with increased vegetation cover and a new equilibrium is established. The stonier soil requires less weathering to produce it, so that geological rates may also come into balance.

Once a critical stoniness has been reached, however, the direction of feedback is reversed. We have seen that soils with more that 50–70% stones begin to show a reduction in effective water storage due to the lack of fine earth and organic volume in which to store water. Beyond this point, increases in erosion lead to progressive loss of water storage, and further enhance the erosion rates. This cycle can only end in complete stripping of the soil to an almost bedrock slope which represents the end point of irreversible desertification. Since stony soils are generally thin, this stage is commonly reached before erosion is controlled through loss of available relief. Because significant soil stripping takes an appreciable time period, and because the initial stages of stripping may be buffered by negative feedbacks, it may be difficult to recognize the early stages of this sequence which ends in irreversible soil loss.

The arguments above apply to areas of net erosion. The corollary of erosion is deposition elsewhere. Where the products of erosion can be contained within terrace system downslope or downstream, then some of the loss of productivity can be recovered. In some areas of very high erosion from steep slopes, it is even possible to increase the areas of cultivable land in terraces, replacing steep slopes of marginal use for cultivation.

The armour loop contains a positive feedback at low levels of armour cover, since the armour increases overland flow roughness and so increases erosion. The armour layer therefore builds up to a density (30–50% cover) beyond which this trend is reversed. Thereafter, increases in armour reduce erosion rates. The effect of this feedback is to slow down, but not eliminate, the effect of the erosion loop.

Taking these processes together, two main physical routes towards desertification are identified. The organic cycle, which leads to a reduction in soil organic matter and a sparser vegetation cover, is not, on its own, irreversible, but may be the precursor of the erosional cycle. On rocky hillslopes, moderate erosion leads to a shifting dynamic equilibrium and thinner, stonier soils, but, as the soil becomes more and more stony, a threshold is reached beyond which erosion and increased stone content reinforce one another, leading to irreversible loss of the remaining soil. On shale or marl bedrocks, and in depositional areas, erosion, once started by destabilizing the vegetation cover and/or stream incision, in some areas associated with neotectonics, is limited only by available relief.

14.10 ACKNOWLEDGEMENTS

The work reported here has been fully funded as part of EC contract EPOC-CT90-0014(SMA) for the period January 1991 to December 1992. Our thanks are also due to our very many scientific partners in Spain, Italy, Portugal, France and Greece, whose field sites and carefully collected data have been critical in making our modelling efforts meaningful. Models may be able to exist in isolation but their credibility can only be built on exacting field studies, for which we are very grateful. Good models have the power to generalize from a small number of field studies, but good fieldwork can destroy a model overnight, and send us all back to the drawing board!

14.11 REFERENCES

Abrahams, A.D., Parsons, A.J. and Luk, S-H. (1989). Distribution of depth of overland flow on desert hillslopes and its implications for modelling soil erosion. *Journal of Hydrology*, **106**, 177–184.

Anderson, M.G. and Burt, T.P. (eds) (1990). *Process Studies in Hillslope Hydrology*. Wiley, Chichester, 539 pp.

Arianoutsou, M. (1989). Timing of litter production in a maquis ecosystem of north-eastern Greece. *Acta Oecologia/Oecologia Plantarum*, **10**, 371–378.

Arianoutsou-Faraggitaki, M., Psaras, G. and Christodoulakis, N. (1984). The annual rhythm of cambial activity in two woody species of the Greek "maquis". *Flora*, **175**, 221–229.

Baird, A.J., Thornes, J.B. and Watts, G.P. (1992). Extending overland flow models to problems of slope evolution and the representation of complex slope-surface topographies. In Parsons, A.J. and Abrahams, A.D. (eds) *Overland Flow: Hydraulics and Erosion Mechanics*. UCL Press, London, 199–224.

Bennett, J.P. (1974). Concepts of mathematical modelling of sediment yield. *Water Resources Research*, **10** (3), 485–492

Diamantoglou, S. and Mitrakos, K. (1981). Leaf longevity in Mediterranean evergreen sclerophylls. In Margaris, N.S. and Mooney, H.A. (eds) *Components of Productivity of Mediterranean-Climate Regions*. Junk, The Hague, 17–19.

Dunne, T. and Aubry, B.F. (1986). Evaluation of Horton's theory of sheetwash and rill erosion on the basis of field experiments. In Abrahams, A.D. (ed.). *Hillslope Processes*. Allen & Unwin, Boston, 31–53.

Elwell, H.A. and Stocking, M.A. (1976). Vegetative cover to estimate soil erosion hazard in Rhodesia. *Geoderma*, **15**, 61–70.

Foster G.R. and Meyer, L.D. (1975). Mathematical simulation of upland erosion by fundamental erosion mechanics. In *Present and Prospective Technology for Predicting Sediment Yields and Sources*. Proceedings Sediment Yield Workshop, US Dept Agriculture Sedimentation Laboratory, Oxford, MS, Nov. 1972 (ARS Report ARS-S-40), US Dept Agriculture, Washington, DC, 196–206.

Francis, C.F. and Thornes, J.B. (1990). Runoff hydraulics for three Mediterranean vegetation types. In Thornes, J.B. (ed.) *Vegetation and Erosion*. Wiley, Chichester, 363–384.

van Genuchten, M.Th. (1980). A closed form equation for predicting the hydraulic conductivity of unsaturated soils. *Soil Science Society of America Journal*, **44**, 892–898.

Gerwitz, A. and Page, E.R. (1974). An empirical mathematical model to describe plant root systems. *Applied Ecology*, **11**, 773–781.

Gregson, K., Hector, D.J. and McGowan, M. (1987). A one-parameter model for the soil water characteristic. *Soil Science*, **38**, 483–486.

Hendrickx, J.M. (1990). Determination of hydraulic soil properties. In Anderson, M.G. and Burt, T.P. (Eds) *Process Studies in Hillslope Hydrology*. Wiley, Chichester, 43–92.

Hsiao, T.C., Silk, W.K. and Jing, J. (1985). Leaf growth and water deficits: biophysical effects. In Baker, N.R., Davies, W.J. and Ong, C.K. (eds) *Control of Leaf Growth*. (Cambridge University Press; SEB Seminar Series No. 27, 239–266.

Jones, H.G. (1983). *Plants and Microclimate*. Cambridge University Press, Cambridge.

Kibler, D.F. and Woolhiser, D.A. (1970). The kinematic cascade as a hydrologic model. *Colorado State University Hydrology Paper*, No. 39.

Kirkby, M.J. (1980). Modelling water erosion processes. In Kirkby, M.J. and Morgan, R.P.C. (eds) *Soil Erosion*. Wiley, Chichester.

Kirkby, M.J. (1988). Hillslope runoff processes and models. *Journal of Hydrology*, **100**, 315–339.

Kirkby, M.J. (1989). A model to estimate the impact of climatic change on hillslope and regolith form. *Catena*, **16**, 321–341.

Kirkby, M.J. (1990). A one-dimensional model for rill inter-rill–interactions. In Bryan, R.B. (ed.) Soil erosion—Experiments and Models. *Catena Supplement*, **17**.

Kirkby, M.J. (1992). An erosion-limited hillslope evolution model. *Catena Supplement*, **23**, 157–187.

Kirkby, M.J. and Neale, R.H. (1987). A soil erosion model incorporating seasonal factors. In Gardiner, V. (ed.) *International Geomorphology, Volume 2*. Wiley, Chichester, 189–210.

Kummerow, J. (1981). Carbon allocation to root systems in Mediterranean evergreen sclerophylls. In Margaris, N.S. and Mooney, H.A. (eds) *Components of Productivity of Mediterranean-Climate Regions*. Junk, The Hague, 115–120.

Landsberg, J.J. (1986). *Physiological Ecology of Forest Production*. Academic Press, London.

Lockwood, J.G. (1989). Modelling evaporative loss from vegetation under warm conditions. Paper presented at symposium on hydrological and desertification processes at land surfaces, European Geophys. Soc., Barcelona, 13–17 March 1989.

Lockwood, J.G. (1992). Sensitivity study of the influence of changes in canopy characteristics on evaporation loss and soil moisture using a sparse vegetation model. *Climate Research*, **2**, 151–165.

Lockwood, J.G. (1993). Impact of global warming on evapotranspiration. *Weather*, **48**, 291–299.

Lockwood, J.G., Jones, C.A. and Smith, R.T. (1989). The estimation of soil moisture deficits using meteorological models at an upland moorland site in northern England. *Agricultural and Forest Meteorology*, **46**, 41–63.

Lo Gullo, M.A. and Salleo, S. (1988). Different strategies of drought resistance in three Mediterranean sclerophyllous trees growing in the same environmental conditions. *New Phytologist*, **108** 267–276.

Marshall, J.D. and Waring, R.H. (1985). Predicting fine root production and turnover by monitoring root starch and soil temperature. *Canadian Journal of Forest Research*, **15**, 791–800.

Moore, R.J. and Clarke, R.T. (1978). A distribution function approach to runoff modelling. *Water Resources Research*, **17** (5), 1367–1383.

Palmer, R.S. (1946). Waterdrop impactometer. *Agricultural Engineering*, **44**, 198–199.

Parsons, D.A. (1949). Depths of overland flow. *Soil Conservation Service, Technical Paper 82*, 33 pp.

Petryk, A. and Bosmajian, G. (1975). Analysis of flow through vegetation. *Journal Hydraulics Division, American Society of Civil Engineers*, **101** (HY7).

Romero-Diaz, M.A., Lopez-Bermudez, F., Francis, C.F., Fisher, G.C. and Thornes, J.B. (1988). Variability of overland flow erosion rates in a semi-arid environment under mattoral cover, Murcia, Spain. *Catena Supplement*, **13**, 1–11.

Ross P.J. and Bristow K.L. (1990). Simulating water movement in layered and gradational soils using the Kirchoff Transform. *Soil Science Society of American Journal*, **54** (6), 1519–1524.

Rundel, P.W. and Nobel, P.S. (1991). Structure and function in desert root systems. In Atkinson, D. (ed.) *Plant Root Growth: An Ecological Perspective*. Blackwell, Oxford, 349–378.

Rutter, A.J., Kershaw, K.A., Robins, P.C. and Morton, A.J. (1971). A predictive model of rainfall interception in forests. 1. Derivation of the model from observations in a plantation of Corsican pine. *Agricultural Meteorology*, **9**, 367–384.

Salleo, S. and Lo Gullo, M.A. (1990). Sclerophylly and plant water relations in three Mediterranean *Quercus* species. *Annals of Botany*, **65**, 259–270.

Sheehy, J.E. and Johnson, I.R. (1988). Physiological models of grass growth. In Jones, M.B. and Lazenby, A. (eds) *The Grass Crop*. Chapman & Hall, London, 243–275.

Sheehy, J.E., Cobby, J.M. and Ryle, G.J.A. (1979). The growth of perennial ryegrass: a model. *Annals of Botany*, **43**, 335–354.

Sheehy, J.E., Cobby, J.M. and Ryle, G.J.A. (1980). The use of a model to investigate the influence of some environmental factors on the growth of perennial ryegrass. *Annals of Botany*, **46**, 343–365.

Shuttleworth, J.W. and Wallace, J.S. (1985). Evaporation from sparse crops—an energy combination. *Quarterly Journal of the Royal Meteorological Society*, **111**, 839-8-55.

Szeicz, G. (1974). Solar radiation for plant growth. *Journal of Applied Ecology*, **11**, 617–636.

Terry, N., Waldron, L.J. and Taylor, S.E. (1983). Environmental influences on leaf expansion. In Dale, J.E. and Milthorpe, F.L. (Eds) *The Growth and Functioning of Leaves*. Cambridge, 179–205.

Thompson, N., Barrie, I.A. and Ayles, M. (1981). Meteorological office rainfall and evaporation calculation scheme: MORECS (July 1981). *Hydrological Memorandum 45*, Meteorological Office, Bracknell, UK.

Thornes, J.B. (1988). Erosion and equilibria under grazing. In Bintliff, J., Davidson, D. and Grant, E. (eds), *Conceptual Issues in Environmental Archaeology*. Edinburgh University Press.

Thornes, J.B. (1989). Geomorphology and grass-roots models. In MacMillan, W. (ed.), *Remodelling Geography*. Blackwell, Oxford.

Thornes, J.B., Shao, J., Diamond, S., McMahon, M. and Hawkes, J.C. (1996). Testing the MEDALUS model. *Catena*, **26**, 106–156.

Thornley, J.H.M. and Johnson, I.R. (1990). *Plant and Crop Modelling*. Oxford University Press, Oxford.

Thornley, J.H.M. and Verberne, E.L.J. (1989). A model of nitrogen flows in grassland. *Plant, Cell and Environment*, **12**, 863–886.

Thornley, J.H.M., Fowler, D. and Cannell, M.G.R. (1991). Terrestrial carbon storage resulting from CO_2 and nitrogen fertilization in temperate grasslands. *Plant, Cell and Environment*, **14**, 1007–1011.

Troughton, A. (1981). Length of life of grass roots. *Grass and Forage Science*, **36**, 117–120.

Wallace, J.S., Roberts, J.M. and Sivakumar, M.V.K. (1990). The estimation of transpiration from sparse dryland millet using stomatal conductance and vegetation area indices. *Agricultural and Forest Meteorology*, **51**, 35–49.

Woodward, F.I. and Bazzaz, F. (1988). The responses of stomatal density to CO_2 partial pressure. *Journal of Experimental Botany*, **39**, 1771–1781.

15

Modelling the Impacts of Climate and Land-Use Change on Basin Hydrology and Soil Erosion in Mediterranean Europe

J. C. BATHURST[1], C. KILSBY[1] and S. WHITE[2]

[1]*Water Resource Systems Research Unit, Department of Civil Engineering, University of Newcastle upon Tyne, UK*
[2]*Instituto Pirenaico de Ecología, CSIC, Campus de Aula Dei, Zargoza, Spain*

15.1 INTRODUCTION

An important element of MEDALUS is the development of mathematical models for predicting the impact of climate and land-use change on river basin response and the process of desertification. Such models can then form a basis for developing guidelines for sound land management, aimed at limiting the spread of desertification. As its contribution to this effort, the Water Resource Systems Research Unit (WRSRU) at the University of Newcastle upon Tyne tested its SHETRAN hydrology and soil erosion modelling system as a means of representing river basins in areas undergoing desertification and of predicting the impacts of changes in climate and land use on basin hydrology and soil erosion. The principal components of the work were as follows.

(i) SHETRAN was used to examine the degree to which data and process understanding obtained at the small spatial scale of the MEDALUS field studies can be extrapolated for use in simulations at the basin or regional scale at which planning decisions are made.
(ii) Following calibration of SHETRAN for focus basins in Portugal and Spain, applications were carried out to examine the response of the basins for specified scenarios for future climate and land use.

The results of the study are discussed in terms of scale effects in model parameter evaluation and the impacts of environmental change on basin response.

Mediterranean Desertification and Land Use. Edited by C. Jane Brandt and John B. Thornes.
© 1996 by John Wiley & Sons, Ltd.

15.2 SHETRAN

SHETRAN is a physically based, spatially distributed modelling system for water flow, sediment transport and contaminant migration, applicable at the scale of the river basin, which has been developed by the WRSRU. It is based on an enhanced version of the Système Hydrologique Européen (SHE) hydrological modelling system (Abbott et al., 1986a,b), which provides an integrated surface and subsurface representation of water movement through a river basin, incorporating the major elements of the land phase of the hydrological cycle (interception, evapotranspiration, snowmelt, overland and channel flow, unsaturated and saturated zone flow). SHETRAN also incorporates an enhanced version of the SHE sediment transport component SHESED, which accounts for soil erosion by raindrop impact and overland flow, and transport by overland flow and channel flow (Wicks, 1988; Wicks et al., 1988, 1992; Wicks and Bathurst, 1996). Each of the processes is modelled either by finite difference representations of the partial differential equations of mass and energy conservation or by empirical equations derived from independent experimental research. The spatial distribution of catchment properties, rainfall input and hydrological response is achieved in the horizontal direction through the representation of the catchment by an orthogonal grid network and in the vertical direction by a column of horizontal layers at each grid square.

Within each model grid square, each physical characteristic is represented by one parameter value. As long as the grid square is small compared with the distances over which there is significant spatial variability in basin properties and hydrological response, this does not compromise the model's ability to represent local variations in response. However, as grid scales increase, the local spatial variability in properties and response becomes subgrid. There are then difficulties in applying the equations of small-scale physics which make up SHETRAN and evaluating their parameters, at the grid scale (e.g. Beven, 1989). In particular, the field measurements which form the basis of parameter evaluation are most easily carried out at the point or plot scale, which may not be representative of the large grid scales (typically 1 km) used in modelling river basins. The solution has been to use 'effective' parameter values, which represent the subgrid spatial variability, to give a grid-scale response. However, this is a pragmatic approach and it is recognized that the concept may not allow an accurate reproduction of the observed response in all circumstances (as shown for example by Binley et al., 1989). It was an important aim of this project to establish the relationship between the effective model parameter values and the relevant field measurements and to study scale effects in this relationship.

The principal soil parameters and functions in SHETRAN are the soil depth, the saturated zone conductivity, the saturated values of conductivity and moisture content for the unsaturated zone and the water retention (moisture content/tension) and moisture content/conductivity relationships for the unsaturated zone. These characteristics do not vary through a simulation. The proportion of ground covered by vegetation at the grid scale (i.e. the proportion of the grid square which is not bare soil) is accounted for by a proportional index on a scale from 0 to 1. This represents the integrated cover provided by the full range of vegetation present. It is determined from field surveys and aerial photographs and can be varied in a predetermined manner through the simulation. The vegetation parameters are interception drainage and storage terms, properties affecting evapotranspiration, and root distribution, and are mostly time invariant. Overland flow resistance is quantified by the Strickler resistance coefficient (the reciprocal of the Manning coefficient). The coefficient is specified by the modeller, usually according to land use, and does not vary through the simulation. The ease with which the soil can be eroded is quantified by two coefficients, representing raindrop impact erodibility and overland flow erodibility respectively. These coefficients cannot yet be determined directly from measurable soil properties and therefore require calibration. Topographic evaluations are determined from appropriate maps or digital elevation models. Channel characteristics are quantified in terms of the channel cross-sectional shape, elevation and Strickler resistance coefficient. All the above parameters and functions are spatially variable between grid squares (or channel links as appropriate), as are also the specified time varying rainfall and meteorological variables determining potential evapotranspiration, and the simulated time varying hydrological responses.

15.3 SELECTION OF FOCUS BASINS

The focus basins (the Cobres in Portugal and the Mula in Spain) were chosen to be in areas suffering from desertification, to be near MEDALUS field study areas and to provide simulation opportunities at several spatial scales.

The Cobres basin lies in the Alentejo region of southern Portugal. Data were available at three scales:

(i) 167-m^2 soil erosion plots at the Centro Experimental de Erosão de Vale Formoso, just to the east of the Cobres basin (also used by the New University of Lisbon MEDALUS team).
(ii) A 32-ha subcatchment at Santa Clara do Louredo near Beja at the northern edge of the Cobres basin.
(iii) The 701-km^2 Cobres basin above the Monte da Ponte gauging station, incorporating two gauging stations within the basin (Albernoa, basin area 172 km^2, and Entradas, basin area 51 km^2).

The Mula basin lies near Murcia in southeast Spain (Figure 8.1). Data were available at two scales:

(i) 20-m^2 experimental plots at the El Ardal MEDALUS site of the University of Murcia team.
(ii) The 159-km^2 upper Mula basin above La Cierva reservoir.

In general the data availability and quality were poorer for the Mula than for the Cobres. In particular, plot data were available only from January 1991 to May 1992 (a period of low rainfall and very little runoff), autographic rainfall data were available for only one site (El Ardal) and only from January 1990 to May 1992, and the time series of basin discharge to La Cierva reservoir had to be constructed from a mass balance procedure for the reservoir. Application of SHETRAN within the Mula basin therefore constituted only an approximate validation exercise. However, it provided valuable experience in the use of SHETRAN within an area of active desertification processes and, with the Cobres application, enables comparison to be drawn between applications under two different rainfall regimes.

15.4 SIMULATION APPROACH

Simulations at the plot scale were carried out first, primarily to calibrate those model parameters which were not well defined by the available data. These were principally the soil saturated zone conductivity (SZK) and the saturated value of conductivity for the unsaturated zone (UZK) (for which the available measurements covered a relatively wide range), the Strickler resistance coefficient for overland flow (KSTR) (evaluated in the first instance from data in the literature, e.g. Engman, 1986) and the soil erodibility coefficients k_r and k_f representing erodibility by raindrop impact and overland flow respectively (which cannot yet be evaluated directly from measurable soil properties). Simulations were then carried out at the larger scales to investigate the scale dependency of model parameters and to assess the degree to which parameter values calibrated at the plot scale can be applied at larger grid scales. Calibration at the basin scale was carried out largely by adjusting the above-mentioned parameters and soil depth (SDEP), although keeping them within physically realistic bounds.

Where there were sufficient meteorological data, potential evapotranspiration (PE) was calculated using the Penman combination method; otherwise the Blaney–Criddle method (based on temperature) was used. A comparison of the two methods using data from the Vale Formoso site in Portugal showed that both give similar annual totals. However, the Blaney–Criddle method produces much less seasonal variation, with higher winter PE and lower summer PE than the Penman method.

15.5 APPLICATION OF SHETRAN TO COBRES BASIN

15.5.1 Vale Formoso plots

The soil erosion plots are of the Wischmeier type, of area 1/60 ha (20 m×8.3 m). Runoff and sediment yield data are available as bulk values for periods containing one or more rainfall events.

Data

Two plots were selected for modelling, lying side by side on an 11° slope and subject to the 2-year winter-wheat/fallow cycle employed in the region. These plots are kept one year out of phase so that one is under wheat while the other lies fallow. The available data include times of ploughing and seeding and a description of the vegetation cover, crop height and soil surface condition after each event. These provided the basis for the time varying vegetation cover index (which changed though the cycle from nearly zero to 1) and were used as supplied, i.e. the vegetation parameters were not used for calibration purposes.

Rainfall chart recordings were digitized at 12-minute resolution for three 2-year periods (each beginning in October), 1977–79 (wetter than average), 1980–82 (drier than average) and 1983–85 (average), referred to henceforth as Wet, Dry and Mean respectively. (Rainfall data for the three periods are compared in Tables 15.1 and 15.2.) Daily potential evapotranspiration (PE) was calculated from the Vale Formoso meteorological record using the Penman combination formula. Corresponding records of bulk runoff and sediment yield were obtained for both plots.

TABLE 15.1   Comparison of measured and simulated runoff depths and sediment yields for the Vale Formoso erosion plots for the Wet (calibration) and Dry and Mean (validation) periods

			Plot 1				Plot 2			
			Runoff (mm)		Sediment yield (kg)		Runoff (mm)		Sediment yield (kg)	
Event no.	Date of event (from–to)	Rainfall (mm)	obs.	SHE	obs.	SHE	obs.	SHE	obs.	SHE
	Wet years									
1	14–21 Oct. 77	107	4	1	0.1	0.1	3	1	0.3	0.2
2	19–21 Nov. 77	21	1	0	0.0	0.0	1	0	0.0	0.0
3	3–12 Dec. 77	82	42	17	0.6	1.1	12	18	0.8	1.0
4	18–22 Dec. 77	98	86	83	8.8	(57.9)	92	84	60.8	(60.0)
5	1–4 Jan. 78	30	11	13	0.2	0.2	13	15	1.1	0.2
6	28 Feb.–3 Mar. 78	44	29	9	0.9	0.5	29	9	0.9	0.2
7	28 Apr.–5 May 78	57	1	0	0.1	0.0	1	0	0.0	0.0
8	27 May 78	8	0	0	0.1	0.0	0	0	0.0	0.0
9	2 Sep. 78	29	1	0	1.2	0.0	1	0	0.0	0.0
10	9–12 Oct. 78	61	2	1	0.4	0.1	2	0	0.1	0.0
11	6–10 Nov. 78	61	3	2	0.2	0.7	2	3	0.0	0.7
12	27 Nov.–3 Dec. 78	18	0	0	0.0	0.0	1	0	0.0	0.0
13	5–15 Dec. 78	44	3	2	0.3	0.5	3	2	0.1	0.2
14	24–27 Dec. 78	16	2	0	0.0	0.0	2	0	0.1	0.0
15	30 Dec. 78– 7 Jan. 79	47	20	13	0.6	1.2	13	13	0.3	2.1
16	14 Jan.–14 Feb. 79	225	62	99	0.8	1.5	90	119	1.3	2.1
17	17–21 Feb. 79	15	2	0	0.0	0.0	2	1	0.0	0.0
18	13–14 Apr. 79	18	1	0	0.0	0.0	0	1	0.0	0.1
19	2 Jul. 79	14	0	0	0.0	0.0	0	0	0.1	0.0
20	11 Jul. 79	14	0	0	0.0	0.0	0	0	0.0	0.0
21	14–16 Sep. 79	9	0	0	0.0	0.0	0	0	0.1	0.0
	Totals for wet years		271	240	14	5.9	266	266	66	4.9
	Dry years									
22	15–16 Oct. 80	22	0	0	0.0	0.0	1	0	0.0	0.0
23	23 Oct. 80	25	1	0	0.1	0.0	1	0	0.0	0.0
24	1–12 Nov. 80	91	3	8	1.1	0.7	3	8	0.1	0.9
25	1–2 Apr. 81	16	0	0	0.0	0.0	0	0	0.0	0.0
26	8–13 Apr. 81	14	0	0	0.0	0.0	0	0	0.0	0.0
27	16–29 Apr. 81	9	0	0	0.0	0.0	0	0	0.0	0.0
28	6–10 May 81	8	0	0	0.0	0.0	0	0	0.0	0.0
29	21 Sep. 81	16	0	0	0.0	0.0	0	0	0.1	0.0
30	25 Sep. 81	11	0	0	0.0	0.0	0	0	0.1	0.0
31	1–5 Oct. 81	10	0	0	0.0	0.0	0	0	0.2	0.0
32	9–13 Dec. 81	10	0	0	0.0	0.0	0	0	0.0	0.0
33	15 Dec. 81– 5 Jan. 82	148	45	27	32.7	(26)	31	27	39.0	(17.0)
34	10–12 Jan. 82	44	14	3	2.1	0.1	12	4	5.2	0.0
35	15–19 Jan. 82	14	0	0	0.0	0.0	0	0	0.0	0.0
36	3–17 Feb. 82	23	1	0	0.0	0.0	1	0	0.0	0.0
37	28 Mar.–1 Apr. 82	67	7	0	0.3	0.0	2	0	0.0	0.0
38	10–14 Apr. 82	12	0	0	0.0	0.0	0	0	0.0	0.0

(continued)

Table 15.1 *(continued)*

Event No.	Date of event (from–to)	Rainfall (mm)	Plot 1 Runoff (mm) obs.	SHE	Sediment yield (kg) obs.	SHE	Plot 2 Runoff (mm) obs.	SHE	Sediment yield (kg) obs.	SHE
39	18–26 Apr. 82	15	0	0	0.1	0.0	0	0	0.0	0.0
40	2 Jun. 82	8	0	0	0.0	0.0	0	0	0.0	0.0
41	3–4 Jul. 82	12	0	0	0.0	0.0	0	0	0.0	0.0
42	26 Aug. 82	9	0	0	0.1	0.0	0	0	0.0	0.0
43	18–19 Sep. 82	62	2	0	0.2	0.0	2	0	0.0	0.0
	Total for dry years		76	38	37	0.8	54	39	45	1.0
	Mean years									
44	27 Oct.–5 Nov. 83	103	10	18	7.3	0.7	3	6	0.4	1.0
45	7–22 Nov. 83	168	57	58	28.0	(32)	49	83	7.4	(41.0)
46	11–20 Dec. 83	53	11	7	0.2	0.0	6	7	0.3	0.2
47	12–21 Mar. 84	46	1	0	0.0	0.0	1	0	0.0	0.0
48	27 Apr.–2 May 84	20	0	0	0.0	0.0	0	0	0.0	0.0
49	16–21 May 84	18	0	0	0.0	0.0	0	0	0.0	0.0
50	19 Oct. 84	25	1	0	0.0	0.0	1	0	0.0	0.0
51	5–18 Nov. 84	56	1	0	0.0	0.0	1	0	0.0	0.0
52	13–22 Jan. 85	76	14	8	0.1	1.3	18	7	0.0	0.2
53	16–28 Feb. 85	32	4	0	0.0	0.0	7	0	0.0	0.0
54	30 May–2 Jun. 85	16	0	0	0.1	0.0	0	0	0.0	0.0
	Totals for mean years		100	91	35.8	2.0	87	103	8.2	1.4

Note: The sediment yields in parentheses were determined using the extreme values of the erodibility coefficients. The other yields were determined using the normal values.
obs. = measured value; SHE = SHETRAN simulated value.

The soil is a thin, poor quality, red Mediterranean soil overlying schists. A survey by the University of Amsterdam MEDALUS team of a similar soil at Santa Clara do Louredo gave saturated hydraulic conductivity values between 0.03 and 0.4 m day^{-1}. A water retention curve was derived from a mean of several curves taken during this survey. Another soil survey, by the University of Cagliari team, at Vale Formoso, found 50% stoniness at the surface, with A and B horizons between 13 and 33 cm thick.

SHETRAN application

The SHETRAN model consisted of a single grid square representing the erosion plot and two grid squares, one above and one below, to provide a slope and boundary conditions. This configuration allows surface processes (i.e. runoff, erosion and sediment transport) to be simulated, but subsurface process simulation must be considered less reliable owing to the effect of poorly defined boundary conditions on water balance. Rainfall was applied at 12-minute resolution, PE at daily resolution and actual evapotranspiration (AE) was calculated through a dependency of the ratio AE/PE on the simulated soil moisture content based on that found by Denmead and Shaw (1962).

Calibration was carried out for 21 events on each plot for the Wet period. Validation was subsequently carried out for the Mean and Dry periods, consisting of 22 and 11 events respectively. The simulations were performed on a continuous basis for each period, holding the model parameters (except the time varying vegetation cover index) constant through time regardless of land treatment and condition.

Goodness of fit was assessed in terms of total runoff volume and sediment yield and the residual sum of squares of the events treated as a single series for each period.

Results

The results of the hydrological and sediment yield calibrations (Table 15.1) are considered very satisfactory on an overall basis. At the level of individual events there are differences between the simulated and measured results but, apart from one extreme event (No. 4, 18–22 December 1977), these are not large and tend to average out over the 2-year period. A good fit was obtained for the runoff calibration for both plots, with a best parameter set of SDEP = 0.45 m, UZK = 0.035 m day^{-1} (both values within the measured ranges) and KSTR = 5 (within the range of values in the literature, e.g. Engman (1986)). In addition, SZK was set at 0.05 m day^{-1}. The erosion calibration was more difficult, as event 4 accounts for the major part of the sediment yield from both plots. It was found impossible to simulate this event adequately using the same erodibility coefficients as for the remainder. Two sets of coefficients were therefore used—a 'normal' set ($k_r = 0.13$ J^{-1}, $k_f = 1.3$ mg m^{-2} s^{-1}) calibrated on all events other than event 4; and an 'extreme' set ($k_r = 2.0$ J^{-1}, $k_f = 20.0$ mg m^{-2} s^{-1}) calibrated on event 4 at plot 2, the largest erosion event in the calibration period.

The extreme sediment yield of event 4 remains largely unexplained. The rainfall for the event, although heavy and prolonged, was no more severe than that of events 1 and 16 which were adequately modelled and produced relatively small sediment yields. The runoff ratio for the event (about 90%) was much higher than for the others but was adequately simulated. However, the soil was recently ploughed and planted and this may have enabled severe soil loss by rill erosion, a non-linearity which is not currently represented in SHETRAN, which represents erosion by a linear process based on sheet flow.

The runoff validations for the Dry and Mean periods are generally good, although runoff in the Dry period is somewhat underestimated (Table 15.1). The erosion validations were less satisfactory and again an extreme event was found in each of the periods—event numbers 33 and 45. If the extreme values of the erodibility coefficients are used for these events, then reasonable agreement is obtained on average for both the normal and extreme events.

A further simulation using rainfall aggregated to 1-hourly resolution was performed to investigate the effect of coarser time resolution on the simulated runoff and sediment yield. The effect on runoff was found to be negligible while sediment yield was found to be reduced overall by an average of some 16%. This suggests that it is not essential, at least for the rainfall regime encountered in the Alentejo, to resolve rainfall at finer than hourly resolution. This result allows the use of 1-hourly resolution rainfall records for the climate change impact studies for the Cobres basin.

15.5.2 Santa Clara do Louredo subcatchment

The 32-ha subcatchment was run for some years by the Escola Superior de Agraria at Beja as part of a soil conservation study. It has a mean slope of 9.4° and is subject to the normal

2-year wheat cropping cycle. An ephemeral channel system develops during wet periods but is subsequently destroyed by ploughing.

Data

Discharge was measured at a weir at the outlet, and digitized values of stage are available for the period January 1982 to February 1987 with a number of periods of missing data. Sediment discharge was also measured but the data could not be located for this study. Autographic rainfall records for Beja were obtained from the Instituto Nacional de Meteorologia e Geofisica MEDALUS team and digitized with 12-minute resolution for the Wet, Dry and Mean periods as for Vale Formoso. Daily PE was calculated from the Beja meteorological record using the Penman combination formula. Soil hydraulic properties were measured by the University of Amsterdam. Elevations were taken from a 1:5000 scale topographic map.

SHETRAN application

Simulations were carried out for a subset of the Mean period (November 1983–September 1985) only, as the length of the discharge record did not allow comparison with the Wet and Dry periods. The subcatchment was represented by 129 grid squares of size 50 m × 50 m and a channel system (Figure 15.1). The model parameters derived from the Vale Formoso calibration were used unchanged, except that the soil depths were set to 0.3 m and 0.5 m at

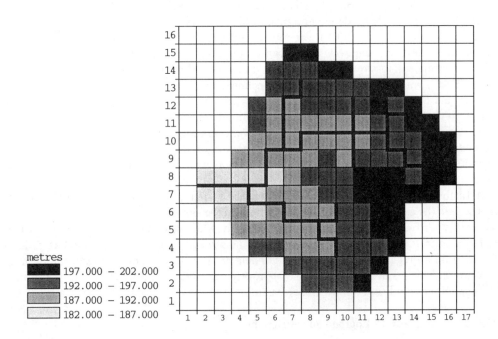

metres
- ■ 197.000 – 202.000
- ▨ 192.000 – 197.000
- ▧ 187.000 – 192.000
- □ 182.000 – 187.000

FIGURE 15.1   SHETRAN grid network, channel system and elevation distribution for the Santa Clara do Louredo subcatchment. The grid squares have dimensions 50 m × 50 m

the top and base of slopes respectively as measured in the soil survey. The land cover was specified by applying the wheat/fallow cycles of the two Vale Formoso plots (a year out of phase) to alternate grid squares. The model was run once with the normal values of the soil erodibility coefficients and once with the extreme values.

Results

Reasonable agreement was obtained between the observed and simulated hydrographs (Figure 15.2). Peak discharges are simulated to a good order of magnitude and the total simulated runoff depth (152 mm) is close to that measured (168 mm) considering the total rainfall of 947 mm for the simulation period. A large proportion of the apparent model

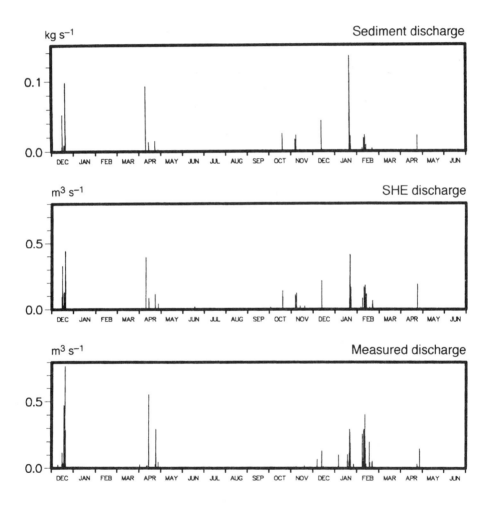

FIGURE 15.2 Time series of measured and simulated discharge and simulated sediment discharge at the outlet of the Santa Clara do Louredo subcatchment for the period December 1983–June 1985

error is thought to be a result of the missing periods of data in the discharge record and the use of the rainfall record from Beja (some 25 km distant) as model input.

There are no measured data on soil moisture content for comparison but the simulated pattern in Figure 15.3 is considered realistic. Moisture content is relatively high in the winter period and decreases under the influence of evapotranspiration during the summer periods. Comparison with Figure 15.2 shows runoff to be simulated both by upward saturation of the soil and by excess of rainfall over infiltration. The latter process, involving also surface sealing by swelling clay, was highlighted by the University of Amsterdam team's rainfall simulator experiments.

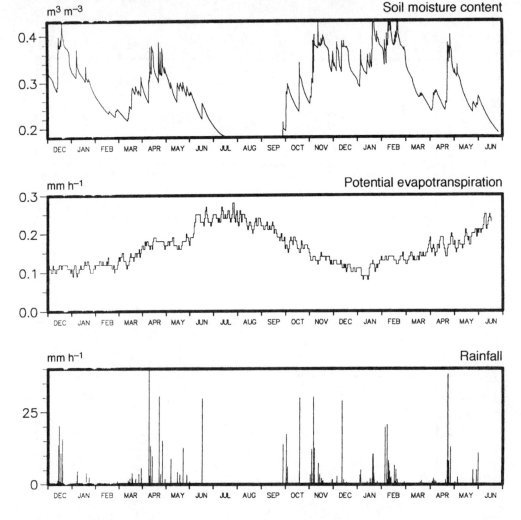

FIGURE 15.3   Time series of rainfall and calculated potential evapotranspiration applied in the Santa Clara do Louredo simulations, and the simulated unsaturated zone moisture content of the surface node in the soil column at grid square (14,10) for the period December 1983–June 1985

The sediment yield simulations (Figure 15.2) cannot be tested directly but the overall sediment yields of 0.08 and 0.25 t ha^{-1} year^{-1} for the 2-year simulation period (obtained with the normal and extreme values of the erodibility coefficients respectively) compare reasonably with the range 0.2–20 t ha^{-1} year^{-1} observed for small catchments (e.g. Walling, 1983).

15.5.3 Cobres basin

The Cobres basin is an area of relatively low relief in the Baixo Alentejo, with some significant local dissection. It has been subjected since the 1930s to a large-scale wheat-growing campaign, which reached its peak in the 1960s and is now in decline, with increasing land abandonment and change of land use to pasture and eucalyptus plantations. The region has suffered long-term soil erosion with a corresponding loss of fertility and drop in agricultural yield.

Data

The basin topography was digitized from a 1:50 000 scale chart, from which the major river network structure and bed elevations were also taken by eye. On the basis of a 1:50 000 scale soil map and a field visit in 1991, the soil properties were assumed to be relatively uniform throughout the basin and to be characterized by the measurements made at Santa Clara. Soil depths were estimated from thicknesses observed at road cuttings and are also relatively uniform. Combination of a 1957 land-use map, 1985 aerial photography and the 1991 field visit showed that wheat has been grown on at least 90% of the basin, although this proportion is now in steady decline.

The Wet, Dry and Mean periods were used for the Cobres study in the same way as for the Vale Formoso study. Autographic rainfall data at 12-minute resolution were available for Vale Formoso and Beja and these were used in conjunction with daily rainfall data for four other sites to provide coverage for the whole basin. SHETRAN requires rainfall input at a short time resolution in order to simulate the non-linear dynamic processes responsible for the runoff and sediment yield responses. Disaggregation of the daily gauge totals using 12-minute data from the nearest autographic gauge was therefore carried out, analysis showing that there was sufficient temporal correlation between the gauges at the daily level for the method to be acceptable. Daily PE totals obtained from Vale Formoso meteorological data were taken to be representive of the whole basin.

Stage records for the three gauging stations were supplied in chart form by the Direcção Geral dos Recursos e Aproveitamentos Hidraulicos (DGRAH) and were subsequently digitized and converted to discharge. A few spot values of suspended sediment discharge were provided by DGRAH for the Monte da Ponte and Entradas gauging stations.

SHETRAN application

The basin was represented by 175 grid squares of size 2 km×2 km and 132 channel links (Figure 15.4). The grid square elevations were taken from a 2-km digital elevation model (DEM) derived from the digitized topographic map. The 2-year wheat-cropping cycle was represented by applying the cycles of the two Vale Formoso plots to alternate grid squares. Rainfall was applied at 12-minute resolution in five areas based on Thiessen polygons

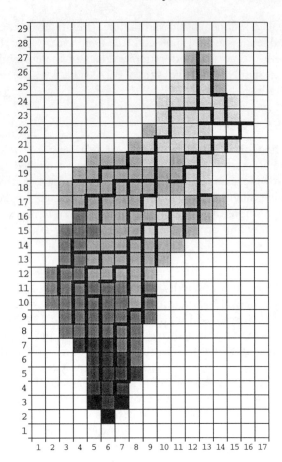

FIGURE 15.4   SHETRAN grid network, channel system and elevation distribution for the Cobres basin. The grid squares have dimensions 2 km×2 km

centred on the daily gauges. Daily PE calculated using the Penman combination method was applied uniformly to the whole basin. The simulations were performed on a 2-hour basic timestep, reducing to several minutes during storm events.

The model was calibrated for the Wet period and the calibrated parameter set was then validated for the Dry and Mean periods. Goodness of fit was assessed for the outlet hydrograph at Monte da Ponte in terms of model efficiency (r^2), standard error of estimate, volumes and peak discharges for the 10 largest events, total runoff volume and graphical comparisons. The rainfall, runoff and PE totals and r^2 values for the calibration and validation simulations are shown in Tables 15.2 and 15.3.

Results

The calibration produced a best set of parameters as follows: SDEP = 0.4 m, UZK = 0.035 m day^{-1} and KSTR = 6. In addition SZK was set at 0.05 m day^{-1}. These

TABLE 15.2 Statistics for the Cobres (Monte da Ponte gauging station) calibration and validation simulations

Simulation	Year[a]	Rainfall[b] (mm)	Potential evapotranspiration[c] (mm)	Runoff (mm) Measured	Simulated	Simulated runoff r^2	Simulated sediment yield[d] ($t\,ha^{-1}\,year^{-1}$) Normal	Extreme
Wet	1977–78	657	1362	199	213	—	—	—
(calibration)	1978–79	693	1419	369	370	—	—	—
	1977–79	1350	2781	568	583	0.83	0.28	0.62
Dry	1980–81	250	1525	0	8	—	—	—
(validation)	1981–82	483	1489	86	100	—	—	—
	1980–82	733	3014	86	108	0.81	0.07	0.15
Mean	1983–84	509	1505	124	134	—	—	—
(validation)	1984–85	541	1583	108	120	—	—	—
	1983–85	1050	3088	232	254	0.61	0.09	0.17

[a]Years are defined October to September.
[b]Basin mean rainfall based on five rain gauges.
[c]Potential evapotranspiration calculated by Penman combination method from Vale Formosa meteorological data.
[d]Sediment yield simulated using the normal and extreme values of the erodibility coefficients; yields are the mean annual values for the relevant 2-year period.

TABLE 15.3 Comparison of measured and simulated runoff depths for the Albernoa and Entradas
gauging stations based on the Cobres calibration and validation simulations

| Simulation | Year[a] | Albernoa runoff (mm) | | Entradas runoff (mm) | |
		Measured	Simulated	Measured	Simulated
Wet	1977–78	154	206	187	220
(calibration)	1978–79	374	402	414	417
	1977–79	528	608	601	637
Dry	1980–81	1	5	0	6
(validation)	1981–82	96	102	98	119
	1980–82	97	107	98	125
Mean	1983–84	142	166	102	189
(validation)	1984–85	123	122	131	123
	1983–85	265	288	233	312

[a]Years are defined October to September.

values are within the measured range, or the range given in the literature, and are very
similar to the values derived from the Vale Formoso plot calibrations. A model efficiency
of 0.83 was obtained for the 2-year period. The simulated and observed hydrographs are
shown in Figure 15.5 and an example of an individual storm hydrograph is given in Figure
15.6.

The simulation is generally very satisfactory, considering the large grid size employed
and the relatively crude spatial and temporal disaggregation of the input rainfall.
Runoff production is simulated by a combination of saturation excess (because of the
shallow soils) and infiltration excess (because of the low UZK and intense rainfall), a
combination which seems physically reasonable since both mechanisms appear to be
present in the observations. The rising limbs and peak discharges of hydrographs are
well reproduced but the simulated recessions are rather slower than those observed,
probably because of the effect of large grid squares. The annual instantaneous
maximum discharges and total runoff are reproduced to within about 5%. The
simulated runoff depths at the internal gauging stations (Table 15.3) are less accurate,
although still reasonable. The lower accuracy probably reflects the relatively coarse
level at which the internal catchments are represented within the SHETRAN grid
network for the Cobres, with a relatively small number of grid squares. However, the
simulated hydrographs (not shown here) reproduce, as observed, the more flashy
responses at the internal gauging stations relative to conditions at the Monte da Ponte
gauging station.

The validations for the Dry and Mean periods constitute a rigorous test of
the model, since the rainfall totals in each were substantially less than in the Wet
period and the soil conditions antecedent to rainfall might therefore be expected to
differ considerably. The observed and simulated hydrographs for the Dry period are
shown in Figure 15.7. The fits are generally satisfactory, with r^2 values of 0.81 and 0.61
for the Dry and Mean periods respectively. A large proportion of the error can
probably be attributed to the problems associated with disaggregating the rainfall
input.

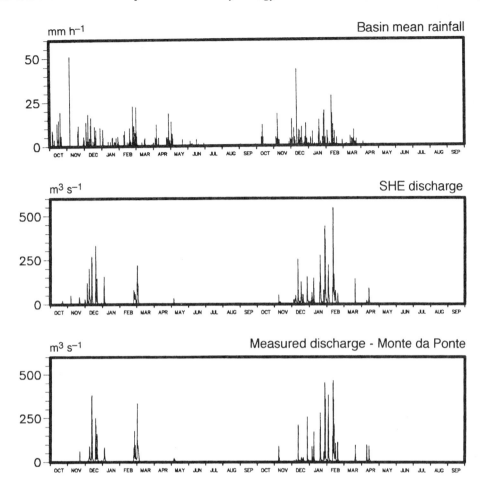

FIGURE 15.5 Time series of mean basin rainfall (area weighted) and measured and simulated discharge at the Monte da Ponte gauging station in the Cobres basin for the Wet (calibration) period (October 1977–September 1979)

Sediment yields were simulated for each period using the normal and extreme values of the erodibility coefficients (Table 15.2). The 'extreme' yields are approximately twice the 'normal' yields but both sets are within the range $0.1-10\,t\,ha^{-1}\,year^{-1}$ observed for catchments of around $1000\,km^2$ (e.g. Walling, 1983). There is little difference between the respective yields for the Dry and Mean periods. However, the yields are significantly increased for the Wet period.

Only a small number of suspended sediment load measurements are available, during the period 1985–88. These lie in the range $0-0.4\,kg\,m^{-3}$ and compare reasonably with those simulated, in the range $0-0.2\,kg\,m^{-3}$.

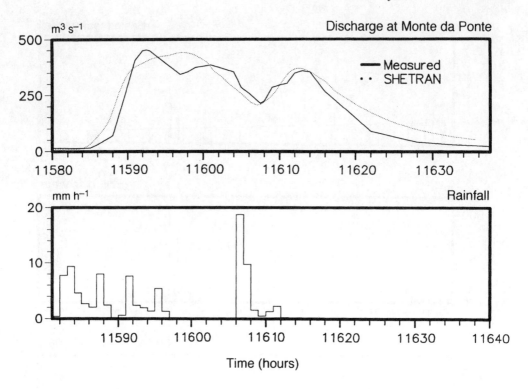

FIGURE 15.6   Comparison of the measured and simulated discharge hydrographs for the event of 27
January 1979 at the Monte da Ponte gauging station in the Cobres basin. The rainfall was measured
at the gauge representing the largest Thiessen subdivision in the basin

15.6 APPLICATION OF SHETRAN TO MULA BASIN

15.6.1 El Ardal plots

The experimental plots have the dimensions $2\,m \times 10\,m$ and are located at various positions
and slopes on the hillside, under a range of land uses. Runoff and sediment yield are
available as bulk values for periods containing one or more rainfall events (López
Bermúdez et al., 1991).

Data

Two plots were selected for modelling, representing abandoned agricultural land (No. 13)
and matorral (No. 16). Details of plant types, vegetation cover, vegetation height and root
profiles were available on a monthly basis for plot 13, while vegetation data for plot 16
were derived from a combination of measurements at El Ardal and data from other
matorral sites. Rainfall data obtained from the El Ardal automatic weather station with a
resolution of 10 minutes were used directly as model input, while temperature data were
used via the Blaney–Criddle model to calculate daily potential evapotranspiration.

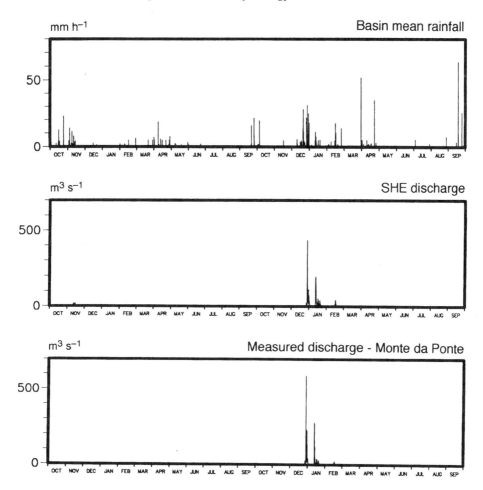

FIGURE 15.7 Time series of mean basin rainfall (area weighted) and measured and simulated discharge at the Monte da Ponte gauging station in the Cobres basin for the Dry (validation) period (October 1980–September 1982)

The soils of the area are thin (20 cm) gravelly silt loams overlying an intermittent rock-hard calcium carbonate layer and limestone. Measurements of the soil saturated hydraulic conductivity by the University of Amsterdam and University of Murcia teams vary widely, from 0.2 to 40 m day^{-1}, so the conductivity was used in model calibration. Soil moisture retention curves showed less variability, so an average of the measured curves was used.

SHETRAN application

Two simulation periods covering respectively three and two events at plot 13 were chosen for model application. The first (19 February to 30 June 1991) was used for model calibration and the derived parameter values were then applied in the second (1 November to 11 December 1991), which included the largest recorded runoff of 10 litres. The same calibrated parameter values, but with changed vegetation parameter values, were used to run the plot 16 simulation period. Each period was simulated on a continuous basis, holding the model parameters (except the time varying vegetation cover index) constant through time regardless of land treatment and condition. Calibration was achieved by matching the simulated and measured bulk runoff and sediment yields for the events, treating the series as a whole.

Results

Considerable difficulty was experienced in achieving calibration because of the very low and intermittent amounts of runoff observed (mostly less than 1% of rainfall). Identification of a parameter set which struck the fine line between giving either zero runoff or an overestimate with continuous runoff was almost impossible, especially within the measured range of values. Thus the optimum value of UZK for the topsoil was found to be 0.11 m day^{-1}, lower than any values measured at the plots. SZK was set at the similar value of 0.1 m day^{-1} and KSTR was set at the physically reasonable value of 7. Calibrated runoff for the first simulation period for plot 13 was then 5.1 litres, compared with a measured runoff of 8.5 litres, while validation for the second period gave a simulated runoff of 25.9 litres compared with a measurement of 11 litres. The simulation for plot 16, on the other hand was a considerable underestimate. Also, the representation of individual events within each period was uneven. Because of the errors in the simulated runoff, sediment yield could not be determined with much confidence. However, results of the correct order of magnitude were achieved using values for the soil erodibility coefficients of $k_r = 0.25 \, J^{-1}$ and $k_f = 2.5 \, mg \, m^{-2} s^{-1}$, obtained from an early evaluation of the coefficients for the Vale Formoso plots near the Cobres basin.

Given the difficulty of achieving consistently realistic results for the plots, the calibrated parameter values must be considered as very approximate. A more extensive data-set is required for a methodical examination of the plot response mechanisms and means of evaluating the model parameters. The problems with the plot simulations also meant that it was not possible to investigate the effect of a coarser time resolution in the rainfall input on the simulation results. In contrast with the case for the Cobres basin, therefore, it is not clear if the subsequent use of 1-hourly resolution rainfall records for the Mula basin scale and climate change impact studies is fully justified.

15.6.2 Mula basin

The Mula basin varies in elevation from 400 m at La Cierva reservoir to 1200 m in the southwest corner (Figure 8.1). The topography is generally steep in the west and south of the catchment with flatter areas in the centre and east. There are numerous ephemeral channels, or ramblas. Climatic conditions are semi-arid, with around 300 mm of rainfall and 1100 to 1200 mm of potential evaporation annually. Rainfall is very variable with prolonged dry periods and intense rainfall events, particularly in the autumn. Rainfall intensities can exceed 50 mm h^{-1} for short time intervals.

Data

A period from 1 September 1988 to 30 June 1991 was chosen for validating SHETRAN, in order to include some large runoff events and thence avoid the problems encountered with the plot applications. For this period data were available from five daily rain gauges around the catchment and from a Piche evaporimeter at La Cierva reservoir. Data from the evaporimeter were used as direct input to SHETRAN.

No autographic rain gauges were operational for the full length of this period, so daily rainfall totals were disaggregated using a statistical distribution of recorded rainfall from El Ardal. Ten-minute intensities were sampled from the El Ardal intensity distribution and, using these, the daily totals were distributed symmetrically around midday in an appropriate number of 10-minute intervals. Rainfall from the six wettest days at each site was assigned the highest intensities from the distribution, thus ensuring some reasonably long periods of intense rainfall.

Combination of a 1:50 000 scale land-use map and a field visit in 1992 showed that land cover in the catchment varies from sparse conifer forest on the hills at the western edge of the catchment to matorral on the valley sides and agricultural land in the valleys and on terraces on the gentler slopes. Some barley is grown but most of the cultivated area is used for vines, almonds, olives, apricots and citrus fruits. Where vine and tree crops are grown, land around the plants is ploughed so that the soil is open to erosion during most of the year. Parameters describing the water use and interception properties of the plants in the catchment were derived from data in the literature and measurements carried out at the El Ardal plots by the University of Murcia team. Seasonal variations in vegetation cover were represented by variations in the proportional cover index of about 0.7 to 1 for the matorral and about 0.1 to 0.2 for the vines and tree crops. The sparse conifer cover was represented by a constant index of 0.85.

The soils of the Mula catchment are generally shallow (between 20 cm and 1 m deep) and overlie limestones and sandstones. Details of soil texture and water retention properties were obtained from a detailed set of soil maps at 1:100 000 scale produced as part of the Proyecto Lucdeme (ICONA and Universidad de Murcia, 1986). An average of the values for each soil type was then used in conjunction with work by Ritjema (1969) and Martinez Fernandez (1992) to derive the soil hydraulic properties required by SHETRAN. Topographic elevations and details of the channel network were taken from a 1:50 000 scale map.

Daily discharge from the catchment to La Cierva reservoir was calculated by means of a mass balance procedure involving the reservoir outflow discharge record, the surface elevation variation of the reservoir and the volume/elevation relationship. This gave no information on flood peaks which, from the rainfall pattern observed at El Ardal, could be expected to exceed by far the daily average flows. Validation of SHETRAN was therefore carried out at the daily level. A bathymetric survey of the reservoir sediment deposit gave a long-term (1929–85) average sediment yield of $2.42\,t\,ha^{-1}\,year^{-1}$ (Romero Diaz et al., 1992) which formed an approximate basis for testing the sediment yield model.

SHETRAN application

The basin was represented by 159 grid squares of size 1 km × 1 km and 94 river links (Figure 15.8). KSTR was set at 10 and UZK and SZK for the topsoil were varied between 0.13 and

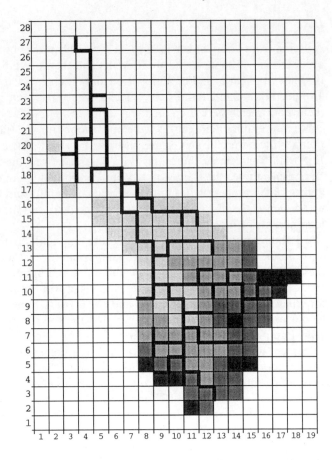

FIGURE 15.8   SHETRAN grid network, channel system and elevation distribution for the Mula basin. The grid squares have dimensions 1 km × 1 km

0.3 m day^{-1} and between 0.2 and 0.5 m day^{-1} respectively according to soil type and the available measurements. The soil profile was represented by soils of depths between 0.2 and 0.5 m, overlying the rock-hard calcium carbonate layer which was represented by a low vertical hydraulic conductivity (UZK = 0.05 m day^{-1}). The vegetation parameters were used as derived from the available data. For the sediment yield simulations the soil erodibility coefficients were set at the values $k_r = 0.25$ J^{-1} and $k_f = 2.5$ mg m^{-2} s^{-1} used for the plots. Rainfall input was supplied with a resolution of 1 hour; disaggregation of the measured daily rainfall to a finer resolution was not felt to be justified on the basis purely of the limited El Ardal autographic data. The use of hourly rainfall is also compatible with the provision of rainfall data for the climate change impact studies.

Results

An excellent agreement between the observed and simulated cumulative discharges into La Cierva reservoir (140 mm and 141 mm respectively) was achieved for the simulation period (Figure 15.9). Furthermore, the daily flows simulated by SHETRAN, although not always on exactly the right days because of the rainfall disaggregation technique employed, are of the same order of magnitude as those observed (Figure 15.10). The simulated instantaneous discharges (Figure 15.10) cannot be checked but, as expected for a region with high intensity rainfall, are much larger than the daily means.

The simulated sediment yield was $1.1\,t\,ha^{-1}\,year^{-1}$. Given the difficulties of measuring reservoir sedimentation accurately, the unknown degree to which the sediment yield during the simulation period represents the long-term sediment yield, and the uncertainties associated with the hydrological simulations, the close agreement

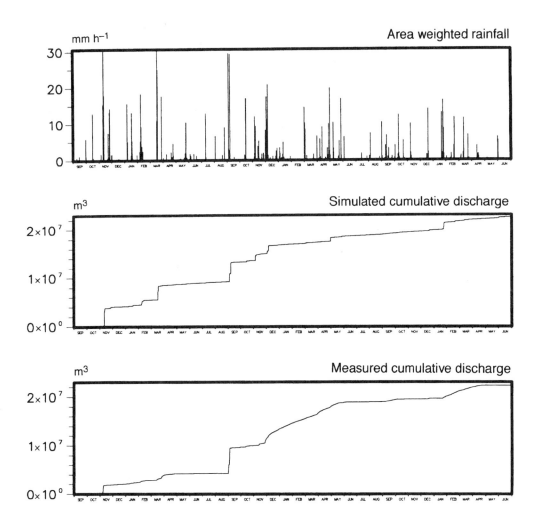

FIGURE 15.9 Time series of mean basin rainfall (area weighted) and measured and simulated cumulative discharges for the Mula basin above La Cierva reservoir for the period September 1988 to June 1991

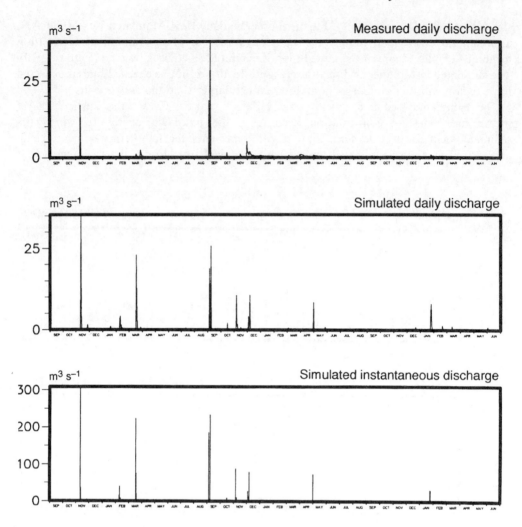

FIGURE 15.10   Time series of measured and simulated daily discharges and simulated instantaneous discharges for the Mula basin above La Cierva reservoir for the period September 1988 to June 1991

of this figure with the long-term measured yield of $2.42\,t\,ha^{-1}\,year^{-1}$ should be treated with caution. However, the result is encouraging for the use of plot calibrated erodibility coefficients at the basin scale.

Finally it should be noted that the available rainfall data for El Ardal are for a period of relatively low intensity rainfall, not fully concurrent with the simulation period, and it is likely that the hourly rainfall intensities for the latter have been underestimated in the disaggregation process. The simulated instantaneous discharges, total runoff volume and sediment yield should therefore be viewed in this light.

15.7 SCALE EFFECTS

The results of the SHETRAN applications were reviewed to find out if model parameter values calibrated at the plot scale were representative, or 'effective', at the large grid scales used in simulating the full catchments.

Table 15.4 compares the values of the saturated zone conductivity (SZK), the saturated conductivity for the unsaturated zone (UZK) and the Strickler resistance coefficient for overland flow (KSTR) for the three scales simulated in the Cobres basin. It shows that SZK was kept at 0.05 m day^{-1} and UZK was kept at 0.035 m day^{-1} at all scales. This is probably a fortuitous result, reflecting the general homogeneity of the Cobres catchment and surrounding region—a relatively uniform topography, soil, vegetation cover and hydrological response. The constancy of UZK may also reflect the use of a plot, rather than point measurements, to obtain the basic calibrated value: the plot area may already be large enough to incorporate the effects which determine the effective UZK for large scales. The UZK and SZK values are close enough to form a physically consistent representation of the soil properties. They are also both within the permeability range 0.03–0.4 m day^{-1} measured by the University of Amsterdam team.

A decrease in the overland flow resistance might be expected, to account for the effect of a subgrid channel network on intergrid water transfers at large grid scales. The increase in KSTR from 5 to 6 at the full basin scale is therefore in the expected direction. However, it is too small compared with general uncertainties in the value of KSTR to provide firm evidence of a scale dependency.

The Mula data-base is not of sufficient quality to enable meaningful deductions on scale effects to be made. However, it may be noted that similar UZK values are applied at the plot and catchment scales and that KSTR undergoes a small increase (7 to 10) from the plot to the catchment scale.

The measured sediment yield record at the subcatchment and catchment scales is not sufficient to enable scale dependencies in the soil erodibility coefficients to be investigated. However, use of the plot calibrated values produced physically plausible annual sediment yields at the larger scales, which is consistent with the finding of Wicks et al. (1988) that plot calibrated values could be successfully applied at the 1-ha scale. For the Cobres the decrease in yields from the (measured) plot values (0.44, 1.10 and 1.10 t ha^{-1} year^{-1} for the

TABLE 15.4   SHETRAN parameter values for three scales in the Cobres basin

Parameter	Scale		
	Vale Formoso plots	Santa Clara do Louredo subcatchment	Cobres basin
Saturated zone conductivity (SZK) (m day^{-1})	0.05	0.05	0.05
Saturated conductivity for unsaturated zone (UZK) (m day^{-1})	0.035	0.035	0.035
Strickler overland flow resistance coefficient (KSTR)	5	5	6

Wet, Dry and Mean periods for plot 1 and 2.0, 1.34 and 0.24 t ha^{-1} year^{-1} for the same periods for plot 2) to the (simulated) basin values (0.28, 0.07 and 0.09 t ha^{-1} year^{-1} for the three periods using the normal values of the erodibility coefficients and 0.62, 0.15 and 0.17 t ha^{-1} year^{-1} using the extreme values) is also in line with observations at a wide range of scales (e.g. Walling, 1983).

Model vegetation parameters were not used for calibration and were not investigated for scale dependencies. It may be noted, though, that the use of plot-derived parameter values at the two larger scales in the Cobres basin was not inconsistent with successful simulations.

Plot calibration therefore has the potential to provide parameter values which are applicable at large grid scales (up to 2 km×2 km), at least under the following conditions:

(i) As the boundary conditions for lateral saturated zone flow are poorly defined, plots should in general be used for the calibration of model overland flow and soil erosion parameters only.

(ii) Plot calibration is most appropriate where overland flow generation is unaffected by lateral saturated zone flow, i.e. where overland flow is generated by rainfall exceeding infiltration capacity. In such circumstances plots can be used to calibrate UZK.

(iii) Plot calibration should be applied with caution where overland flow is generated by upward saturation.

(iv) Best results are achieved where the plot calibrated values are applied in catchments with uniform geomorphologies and hydrological responses. The apparent lack of scale dependency for the Cobres catchment cannot be extrapolated to areas with less uniform land use, geomorphology and hydrological response.

(v) Plot calibration is difficult if the runoff is only a small percentage of the rainfall.

The significance of the successful use of plot calibrated parameters is that it is currently very much easier and cheaper to take measurements at the plot scale than at larger scales.

15.8 CLIMATE AND LAND-USE CHANGE IMPACTS

Simulations of the impacts of climate and land-use changes on hydrological response and sediment yield were carried out using the calibrated models for the Cobres and Mula catchments. Scenarios were generated for possible future conditions and were applied to the models by altering either the model input data or the model parameters from the calibrated values. The effects of the changes on model output were examined by comparison with the output for the calibrated models, representing current conditions.

15.8.1 Future climate scenarios

In consultation with the University of East Anglia team, output from the Canadian Climate Centre Model (CCCM) was used to derive future climate scenarios for input to SHETRAN. The CCCM is a general circulation model (GCM) which has been used to simulate the current climate (with current levels of atmospheric CO_2, denoted $1 \times CO_2$) and a climate represented by doubled CO_2 levels (denoted $2 \times CO_2$). The model output for the latter state was chosen to provide the basis of the future climate conditions to be used in the impact studies. The CCCM data consist of 10-year series for $1 \times CO_2$ and $2 \times CO_2$ conditions of the 12-hourly mean values of a range of meteorological variables. For this study the rainfall and temperature series were extracted for the CCCM grid points nearest the Cobres and Mula catchments and were used to derive hourly time series of rainfall and daily time series of potential evapotranspiration (PE) for input to SHETRAN for each catchment.

As GCM grid-point variables (particularly rainfall) are unreliable (e.g. Grotch and MacCraken, 1991) and available only at coarse temporal resolution, it was necessary first to assess the CCCM data for the current climate by comparison with observations and second to disaggregate the rainfall data temporally to the hourly level.

15.8.2 Assessment of CCCM output

The frequency distribution of daily totals of the $1 \times CO_2$ CCCM rainfall was compared with those of observed series for the Cobres and Mula. Good agreement is apparent in the monthly and annual means (Table 15.5) but the observed distribution of totals above 15 mm was entirely absent from the CCCM data. This high-intensity rainfall is crucial for hydrological and erosion responses, so it was decided to use, from the CCCM data, only the mean monthly rainfall totals for scenario production. Since it is not possible at present to predict the variability of future rainfall with any certainty, the approach adopted here was to use present-day daily and sub-daily rainfall statistics with altered monthly means as the basis for future scenarios.

The CCCM daily mean surface temperatures were used with the Blaney–Criddle method to calculate monthly PE. The monthly and annual totals again compare well with observations (Table 15.5). The CCCM $2 \times CO_2$ output shows a considerable change in climate from current conditions, consisting of a decrease in annual rainfall of 20–30%, an increase in temperature of some 3.5°C and an associated increase in annual PE of some 12% (Table 15.5).

TABLE 15.5 Comparison of measured annual rainfall and potential evapotranspiration with values modelled by the Canadian Climate Centre Model (CCCM) for the Cobres and Mula basins

	Cobres		Mula	
Data source	Annual rainfall (mm)	Annual potential evapotranspiration (mm)	Annual rainfall (mm)	Annual potential evapotranspiration (mm)
Measured	480[a]	1588[b]	300[c]	1625[d]
CCCM $1 \times CO_2$	518[e]	1543[f]	331[e]	1660[f]
CCCM $2 \times CO_2$	429[e]	1749[f]	235[e]	1825[f]

[a]25-year mean for 10 stations.
[b]Mean of 10 years at two sites.
[c]Long-term mean.
[d]1988–90 at La Cierva reservoir.
[e]Obtained directly from CCCM.
[f]Calculated from CCCM daily mean temperature using Blaney–Criddle formula.

15.8.3 Neyman–Scott rainfall modelling

Temporal disaggregation of the CCCM rainfall data was performed using the Neyman–Scott model (Rodriguez-Iturbe et al., 1987; Cowpertwait, 1991a,b). This is a stochastic scheme which may be used to generate continuous time series of rainfall at time intervals of one hour or greater. Statistics from a long-duration observed rainfall time series must be supplied for model calibration. This series may also be at any resolution greater than hourly but a synthetic series of finer resolution than the observed series may not accurately represent the true rainfall statistics at the finer resolution. For this study three 100-year synthetic series of hourly rainfall data were generated for the Cobres and Mula basins, denoted 'historic', '$1 \times CO_2$' and '$2 \times CO_2$'. The historic series used calibration statistics entirely from measured daily data for the period 1977–87, while the $1 \times CO_2$ and $2 \times CO_2$ series used measured statistics with the exception of monthly-mean daily rainfall totals which were taken from the CCCM output. The mean annual rainfalls obtained for the 100-year periods for the $1 \times CO_2$ and $2 \times CO_2$ series therefore matched those obtained from the CCCM 10-year data-sets. It was assumed that the mean daily variance of the rainfall remains constant under climate change, which implies that, as the mean rainfall is reduced, the relative variability increases.

The historic series provided a means of investigating whether the Neyman–Scott model generated unrealistic effects by comparison with observations. For both basins, annual maxima of daily and sub-daily rainfall totals were compared satisfactorily with observations using Gumbel plots but a detailed analysis of extreme rainfall intensities revealed a general under-representation of short-period, high-intensity rainfall. This effect was more marked for the Mula basin, where the observed rainfall is very sporadic and intensities frequently exceed $20\,mm\,h^{-1}$ (e.g. Figure 15.9).

This shortcoming in the hourly rainfall was due to the model calibration being performed on daily rainfall statistics. The distribution of intensities sampled by the model to generate hourly rainfall contained insufficient extreme values to represent Mediterranean-type rainfall at the hourly level. For the Cobres basin (but not the Mula) sufficient hourly observed rainfall data were available to fit a more representative model. This was achieved by a two-stage process. First, an extra set of statistics was calculated from regression of rainfall totals at sub-daily time intervals (1, 3, 6 and 12 hours) against daily totals and used as additional fitting parameters in the Neyman–Scott calibration. Second, the generated series were subjected to an 'intensity-stretching' technique. This technique re-samples the hourly rainfall on a storm-by-storm basis so that an exponential distribution is obtained, containing higher peak intensities. The storm duration and rainfall total are conserved, and the overall series mean and structure are unaffected. The frequency distribution of the hourly rainfall totals for the modified Cobres historic series was found to match closely that observed.

15.8.4 Potential evapotranspiration

Because the synthetic rainfall series are stochastically generated, there is no deterministic means of associating PE with rainfall at the hourly or daily level. However, a detailed specification of PE is of secondary importance, behind rainfall, for modelling semi-arid hydrology. For the Cobres, therefore, the PE input was specified using a repeated, annual sine-curve, best-fit approximation to the Penman daily series used for the Wet, Dry and

Mean years. Daily variation is smoothed and the annual totals are fixed at the mean scenario total but the important seasonal variation is well reproduced. A sensitivity study was performed to assess the effects of each stage of this approximation and an overall change of less than 10% in annual runoff may be assumed relative to the case where daily and annual variabilities are taken into account.

For the Mula, 3 years of daily temperature record were selected from the CCCM output and used to calculate PE using the Blaney–Criddle formula. The years were selected such that the CCCM annual rainfall totals matched the 3 years of synthetic rainfall data used for the Mula climate scenarios.

15.8.5 Cobres climate change impact study

Using the calibrated Cobres SHETRAN parameter set, with current land use (i.e. the 2-year wheat cropping cycle), the historic, $1 \times CO_2$ and $2 \times CO_2$ simulations were each run for a period of 40 years. The historic simulation was designed as a control to validate the Neyman–Scott rainfall generation method. The $1 \times CO_2$ simulation was designed as a control to validate the CCCM present-day climate, and the $2 \times CO_2$ simulation was designed to assess the impacts of the predicted climate change.

The results of the three simulations, shown as a plot of runoff against rainfall for each hydrological year in Figure 15.11, indicate that the simulated basin hydrological response is similar to the observed for each of the scenarios. The annual rainfall/runoff data points fall in overlapping regions of the same general response, with the centre of gravity of the points dependent on the mean annual rainfall in an approximately linear manner, giving confidence in the model as a predictive tool outside the climate regime for which it was first calibrated and validated. Analysis of the individual storm peak discharges in the historic case also showed a similar response to that observed.

The mean annual rainfall, runoff depth and sediment yield for the simulations are shown in Table 15.6, with the mean of measured values for comparison. The predicted reduction in mean annual rainfall from the historic (495 mm) to the $2 \times CO_2$ (414 mm) condition is considerable but Figure 15.11 shows that this is small compared with the natural variability. A range of 230 to 880 mm is seen in the observed annual rainfall for the period 1977–1987 alone, and the standard deviations of the annual rainfall totals for the synthetic series are between 80 and 100 mm. This implies that, on an annual basis, more significant hydrological impacts will arise from inter-annual variability than from any shift in climate. (The difference between the rainfall and evaporation figures for the $1 \times CO_2$ and $2 \times CO_2$ series in Tables 15.5 and 15.6 arise because the values in Table 15.6 are based on only 40 of the 100 years of synthetic data generated from the CCCM data and do not therefore retain the mean values of the CCCM data exactly.)

Table 15.6 and Figure 15.11 show an overall reduction in runoff between present and future ($2 \times CO_2$) conditions but with individual years out of line with this trend. Sediment yields also show a decrease to the $2 \times CO_2$ condition. However, there is greater variability in yield between the Wet, Dry and Mean periods (Table 15.2) and it may be concluded that (a) annual sediment yield is critically dependent on only one or two events in each year, and (b) the effect of inter-annual variability on sediment yield is likely to be more significant than the effect of climate change.

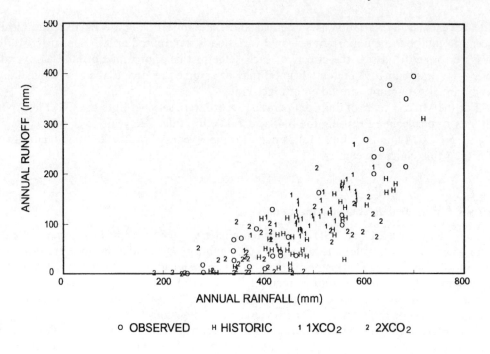

FIGURE 15.11   Relationship between annual runoff and annual rainfall for the Cobres basin. Twenty-five years of measured data are compared with 40 years of data for each of the climate scenarios

TABLE 15.6   Simulation results for Cobres climate and land-use change impact studies

Simulation	Annual rainfall (mm)	Annual potential evapotranspiration (mm)	Annual runoff (mm)	Runoff/ rainfall ratio (%)	Annual sediment yield (t ha⁻¹) Normal	Annual sediment yield (t ha⁻¹) Extreme
Measured[a]	479	1458	129	27	—	—
Historic[b]	495	1458	92	19	0.083	0.100
$1\times CO_2$[b]	501	1458	125	25	0.091	0.140
$2\times CO_2$[b]	414	1720	56	14	0.066	0.084
Abandoned land[c]	675	1390	279	43	0.260	0.590
Natural vegetation[c]	675	1390	280	43	0.220	0.510

[a]Measured rainfall (basin mean) and runoff figures are based on 25 years of data (1962–87); evapotranspiration is the mean value for the three 2-year test periods.
[b]Climate change impacts are simulated for 40 years from the 100-year synthetic series.
[c]Land-use change impacts are simulated for the 2-year Wet period (1977–79); the results may be compared with the relevant figures in Table 15.2.

15.8.6 Cobres land-use change impact study

Two land-use change simulations were performed for the Cobres basin, for the Wet period. The first, denoted 'abandoned', represents the period following abandonment of the existing practice of wheat cropping and was simulated using a sparse cover of matorral (time varying vegetation cover index of 0.2–0.6). The second, denoted 'natural', represents the period when the vegetation cover has returned to a natural equilibrium condition, consisting of a denser cover of matorral (vegetation cover index of 0.9). In each case the land-use change was applied to the entire catchment.

As simulated, the existing practice of wheat cropping leaves half the basin ploughed and poorly vegetated in any one year. The more continuous vegetation cover of the two impact scenarios therefore increased rainfall interception, transpiration and protection against soil erosion, so that simulated runoff and sediment yield for the basin (Table 15.6) are reduced relative to the calibrated Wet period results (Table 15.2). However, the reduction in runoff is relatively minor, probably because most runoff takes place during major rainfall events when vegetation cover has relatively little impact on response. For these particular simulations this effect is accentuated by the use of the higher than average rainfall of the Wet period. It should also be remembered that the effects of land-use change are likely to be most noticeable at the small (plot or field) scale. The catchment scale response to change is likely to be attenuated as a result of the checks and storages which delay the passage of water to the catchment outlet.

The impact studies are to some extent simplistic and do not take into account such factors as the collapse of current soil conservation measures in any land abandonment, which might be expected to increase erosion locally. Nevertheless, they suggest that wheat cropping with a period of unprotected ploughed soil significantly increases soil erosion relative to conditions with a natural vegetation cover.

15.8.7 Mula climate change impact study

Using the calibrated Mula SHETRAN parameter set, with current land use, the historic, $1 \times CO_2$ and $2 \times CO_2$ simulations were each run for a 3-year period selected from the respective 100-year synthetic series of rainfall data. It is recognized that a 3-year sample is not statistically representative. It was also apparent that there were insufficient observed autographic rainfall data to condition the Neyman–Scott model to the Mula catchment as closely as had been done for the Cobres. Furthermore, the autographic data which were available and which were used in testing the historic series were from a period of low rainfall characterized by an under-representation of high-intensity rainfall. Consequently the rainfall time series generated by the Neyman–Scott model all contain too few high intensity rainfalls and the generated rainfall for the historic time series is low by comparison with the observed rainfall for the period 1988–91. The simulated runoff amounts and sediment yields for the three cases, shown in Table 15.7, are therefore low. However, the relative decrease in runoff and sediment yield from present-day ($1 \times CO_2$) conditions to $2 \times CO_2$ conditions suggests that, unless a major increase in intensity accompanies the decrease in mean rainfall, annual sediment yield will decrease with the expected change in climate. As with the Cobres, though, the effects of current inter-annual variability are likely to be comparable with those of climate change at the annual scale. (The differences between the rainfall and evaporation figures for the $1 \times CO_2$ and $2 \times CO_2$

TABLE 15.7   Simulation results for Mula climate and land-use change impact studies

Simulation	Annual rainfall (mm)	Annual potential evapotranspiration (mm)	Annual runoff (mm)	Runoff/ rainfall ratio (%)	Annual sediment yield (t ha^{-1})
Measured	300[a]	—	—	—	2.42
1988–91 calibration	476	1625	47.2	9.9	1.11
Historic[b]	236	1625	12.6	5.3	0.05
1×CO$_2$[b]	328	1656	23.4	7.1	0.44
2×CO$_2$[b]	298	1827	20.0	6.7	0.21
Bare soil[c]	476	1625	53.4	11.2	1.08
Annual weeds[c]	476	1625	51.2	10.8	1.04
Matorral[c]	476	1625	46.6	9.8	0.96
Matorral and 2×CO$_2$[b]	298	1827	19.5	6.6	0.19

[a]Long-term mean.
[b]Climate change impacts are simulated for 3 years from the 100-year synthetic series.
[c]Land-use change impacts are simulated for the 1988–91 calibration period.

series in Tables 15.5 and 15.7 arise because the values in Table 15.7 are based on only 3 of the 100 years of synthetic data generated from the CCCM data and do not therefore retain the mean values of the CCCM data exactly.)

15.8.8 Mula land-use change impact study

A series of land uses, representing different stages of abandonment of agricultural land in the basin, were simulated for the calibration period (1988–91). These are respectively bare soil, annual weeds and a natural mattoral cover (the latter two covers being given time varying cover indices of about 0.03 to 0.3 and 0.7 to 1 respectively). A combination of mattoral and a 2×CO$_2$ climate was also simulated.

The results (Table 15.7) suggest that the current agricultural land use, with its high percentage of bare (ploughed) soil, supports higher runoff and soil erosion than that associated with a complete cover of matorral. However, during the process of abandonment and the recolonization of matorral species, via bare soil and a cover of annual weeds, it is possible that annual runoff may increase temporarily, probably because of reduced interception loss and plant transpiration. The simulated changes are not large, probably because, even more than in the Cobres basin, most runoff takes place during a few intense rainfall events and the vegetation therefore has relatively little impact on response, at least at the basin scale. Also, the overall effect is limited because only 42% of the basin is at present under agricultural cultivation and because the proportion of bare soil in the cultivated area is already comparable with that of the bare soil and annual weed land uses.

A higher sediment yield is simulated for the current land use than for bare soil. This is probably a result, within the simulation, of the additional erosion caused by leaf drip from the cultivated trees relative to that caused simply by direct raindrop impact (which is all that would affect the bare soil). The combination of the matorral land cover and 2×CO$_2$

climate produces a large decrease in runoff and sediment yield relative to present-day conditions.

Given the uncertainties attached to the initial Mula calibration and the derivation of the climate scenarios the figures in Table 15.7 should be treated with caution. Nevertheless they suggest that any probable combination of climate and land-use change is unlikely to cause significant increases in basin annual sediment yield. However, this does not preclude the possibility of locally severe erosion occurring when either existing, terraced cultivation is abandoned and collapses, or existing natural cover is converted to irrigated viticulture or aboriculture with extensive areas of bare soil. Also, the response to land-use change is likely to be more marked at the plot or field scale than is simulated at the basin scale.

15.9 CONCLUSIONS

The study has demonstrated the ability of SHETRAN to simulate the hydrological and sediment yield regimes of Mediterranean river basins for current land use and climate conditions and thence to form a sound basis for predicting the impacts of possible future changes in those conditions. In particular the flexibility of SHETRAN apparent in its applications at different spatial scales, in different geographical areas and for different land use and climate conditions demonstrates the advantages of physically based modelling systems for such studies.

The applications at plot, subcatchment and full basin scale for the Cobres and Mula basins show that, at least for certain conditions, it is possible to calibrate model parameters at the plot scale and apply them unchanged at model grid scales of up to 2 km. In this way basins of at least 700 km² in area can be successfully modelled. In the long term it may become possible to achieve the desirable aim of measuring parameter values directly representative of the model grid scale, for example by remote sensing. However, given that it is currently much easier to measure parameters at the point or plot scale, the above finding holds considerable practical significance for basin modelling.

It is also clear that successful SHETRAN simulations require large and detailed data-sets. The applications demonstrate the ability of SHETRAN to exploit the available data to the full (especially, via disaggregation, the rainfall data). However, where the full range of required data and parameters is not available from measurements and has to be generated or approximated in some way, the simulation results must be interpreted with an understanding of the corresponding uncertainty in the inputs. Thus the Mula simulations were compromised by the lack of rainfall events for the plot study and by the poor quality of discharge data and the absence of within-basin autographic rain gauges for the basin study. The latter problem is particularly severe for an area with such a spatially variable and short-period, high-intensity rainfall regime. Disaggregation of daily data can play a helpful role but is no substitute for measured data. Given the typically wide variability in rainfall regime for the Mediterranean area even the Cobres simulations, based on only 2-year periods, should be treated with some caution. Clearly certain aspects of extreme events, for example the erosion in event 4 for the Wet period for the Vale Formoso plots, are not properly integrated into the simulations. Longer calibration periods would be preferable, although the validation of the Cobres Wet period calibration for the Dry and Mean periods shows that short-period calibrations can be used with reasonable confidence, a capability which is likely to be beyond non-physically based models.

Combination of GCM predictions of mean rainfall with present-day daily statistics to produce hourly time series using the Neyman–Scott stochastic rainfall model is seen as a very powerful tool for generating future climate scenarios. It is essential, however, to fit the Neyman–Scott model to a basin using long records of fine time resolution rainfall data if the extreme intensities are to be reproduced accurately. Careful assessment of the model is required, ensuring that it represents the present-day rainfall regime faithfully, before it can be used as a basis for generating the future regime.

In this study, doubled levels of atmospheric CO_2 formed the basis of the future climate. Reductions in annual rainfall are apparent for this condition and SHETRAN simulated corresponding and plausible decreases in runoff and sediment yield for the Cobres and Mula. However, in both cases it is likely that these reductions are within the range of current inter-annual variability. In other words, the existing climate variability is sufficient to provide a good basis for examining basin response to reduced rainfall in the Mediterranean region.

Scenarios for future land use were based on simple assumptions of the abandonment of cultivated land to natural conditions. For both basins a reduction in erosion was predicted, owing to the elimination of unprotected bare soil associated with wheat growing (Cobres) and vines and tree crops (Mula). However, the simulations do not allow for local increases in erosion which might occur following the collapse of soil conservation measures such as terracing or the development of uncontrolled gully erosion.

A number of recommendations for future studies can be made. In particular, considerable attention should be paid to data provision, since the level to which measured data and parameter values are available determines the quality of the SHETRAN output. It is especially important to obtain representative autographic rainfall data. Parallel modelling and fieldwork can also be helpful by enabling the field measurements to be concentrated most efficiently in those areas and on those parameters indicated by the model to have the greatest effect on simulation output. Further testing of the methodologies developed in this study is also required, examining applications at larger spatial scales, defining the limits to the use of plot calibrated parameters and refining the future scenarios for climate and land use. Enhancement of the SHETRAN erosion and sediment yield component to account for landslide and (in the longer term) gully erosion is under way, widening its applicability to the erosion problems encountered in the Mediterranean region.

15.10 ACKNOWLEDGEMENTS

Selection of the Cobres and Mula basins and implementation of the data assembly programmes was possible only because of the great help provided by the local MEDALUS groups and other data collection agencies. The authors therefore thank in particular the MEDALUS groups of Dr M. J. Roxo (New University of Lisbon, Portugal) and Professor F. López-Bermúdez (University of Murcia, Spain). Other data were provided by: (in Portugal) the Escola Superior de Agraria at Beja, the Instituto Nacional de Meteorologia e Geofisica and the Direcção Geral dos Recursos e Aproveitamentos Hidraulicos; and (in Spain) the Instituto Nacional de Meteorologia and the Confederación Hidrografica. The provision of soil property data by the University of Amsterdam team and the CCCM data by the University of East Anglia team is also acknowledged. Finally the authors thank the following who contributed to the MEDALUS work at Newcastle: Mr A. Burton,

Mr G. O'Donnell, Mr O. Hamad and Professor P. E. O'Connell. The manuscript was typed by Ms S. McLean and Ms S. McElhatton.

15.11 REFERENCES

Abbott, M.B., Bathurst, J.C., Cunge, J.A., O'Connell, P.E. and Rasmussen, J. (1986a) An introduction to the European Hydrological System—Système Hydrologique Européen, 'SHE'. 1: History and philosophy of a physically-based, distributed modelling system. *J. Hydrol.*, **87**, 45–59.

Abbott, M.B., Bathurst, J.C., Cunge, J.A., O'Connell, P.E. and Rasmussen, J. (1986b) An introduction to the European Hydrological System–Système Hydrologique Européen, 'SHE'. 2: Structure of a physically-based, distributed modelling system. *J. Hydrol.*, **87**, 61–77.

Beven, K. (1989) Changing ideas in hydrology—the case of physically-based models. *J. Hydrol.*, **105**, 157–172.

Binley, A., Beven, K. and Elgy, J. (1989) A physically based model of heterogeneous hillslopes 2. Effective hydraulic conductivities. *Wat. Resour. Res.*, **25**, 1227–1233.

Cowpertwait, P.S.P. (1991a) The stochastic generation of rainfall time series. PhD Thesis, University of Newcastle upon Tyne, UK.

Cowpertwait, P.S.P. (1991b) Further developments of the Neyman–Scott clustered point process for modeling precipitation. *Wat. Resour. Res.*, **27**, 1431–1438.

Denmead, O.T. and Shaw, R.H. (1962) Availability of soil water to plants as affected by soil moisture content and meteorological conditions. *Agronomy J.*, **54**, 385–390.

Engman, E.T. (1986) Roughness coefficients for routing surface runoff. *Proc. Am. Soc. Civ. Engrs., J. Irrig. Drain. Engrg.*, **112**, 39–53.

Grotch, S.L. and MacCracken, M.C. (1991) The use of General Circulation Models to predict regional climatic change. *J. Climate*, **4**, 286–303.

ICONA and Universidad de Murcia (1986) Proyecto Lucdeme, Mapas de Suelos. Instituto Nacional para la Conservación de la Naturaleza (ICONA), Ministerio de Agricultura, Pesca y Alimentación, Madrid, Spain.

López Bermúdez, F., Romero Diaz, M.A. and Martinez Fernandez, J. (1991) Soil erosion in a semi-arid Mediterranean environment. El Ardal experimental field (Murcia, Spain). In, Sala, M., Rubio, J.L. and García Ruiz, J.M. (eds) *Soil Erosion Studies in Spain*. Geoforma Ediciones, Logroño, Spain, 137–152.

Martínez Fernández, J. (1992) Variabilidad espacial de las propiedades físicas e hídricas de los suelos en medio semiarido Mediterraneo. Cuenca de la Rambla de Perea, Murcia. PhD Thesis, Universidad de Murcia, Murcia, Spain.

Ritjema, P.E. (1969) *Soil Moisture Forecasting*. Note 513, Instituut voor Cultuurtechniek en Waterhuishouding, Wageningen, The Netherlands.

Rodriguez-Iturbe, I., Cox, D.R. and Isham, V. (1987) Some models for rainfall based on stochastic point processes. *Proc. Roy. Soc., London*, **410**, 269–288.

Romero Diaz, M.A., Cabezas, F. and López Bermúdez, F. (1992) Erosion and fluvial sedimentation in the River Segura basin (Spain). *Catena*, **19**, 379–392.

Walling, D.E. (1983) The sediment delivery problem. *J. Hydrol.*, **65**, 209–237.

Wicks, J.M. (1988) Physically-based mathematical modelling of catchment sediment yield. PhD Thesis, University of Newcastle upon Tyne, UK.

Wicks, J.M., Bathurst, J.C., Johnson, C.W. and Ward, T.J. (1988) Application of two physically-based sediment yield models at plot and field scales. In *Sediment Budgets, IAHS Publ. No. 174*. Intl. Ass. Hydrol. Sci., Wallingford, UK, pp. 583–591.

Wicks, J.M., Bathurst, J.C. and Johnson, C.W. (1992) Calibrating SHE soil-erosion model for different land covers. *Proc. Am. Soc. Civ. Engrs, J. Irrig. Drain. Engrg*, **118**, 708–723.

Wicks, J.M. and Bathurst, J.C. (1996) SHESED: a physically based, distributed erosion and sediment yield component for the SHE hydrological modelling system. *J. Hydrol.*, **175**, 213–238.

16

Modelling Short-Term Water Resource Trends in the Context of a Possible 'Desertification' of Southern Europe

M. CHABART, J. J. COLLIN and J. P. MARCHAL

Bureau de Recherches Géologiques et Minières (BRGM), Montpellier, France

16.1 INTRODUCTION

The groundwater of Mediterranean countries forms a precious resource in an unfavourable situation, threatened by the combined effects of possible climatic change and, above all, human impact. Our objective here is to make a 30-year projection of what may become of this resource and to suggest guidelines for its continued (or expanded) use.

16.2 TYPES OF PROBLEMS FOUND IN THE MEDITERRANEAN AREA

16.2.1 Physical aspects of groundwater in Mediterranean countries

Main geological reservoirs for groundwater

We may start with the axiom that, where a region's water resources are low in absolute terms, any decline in the renewal of such resources by precipitation will mean that a higher proportion of total needs must be met from groundwater resources. The Mediterranean basin, half-way between the 'opulence' of temperate countries and the 'destitution' of arid regions, suffers less from lack of rainfall than from the damaging effects of its (inter) seasonal fluctuation.

In this setting, the long-term regularization of precipitation yield might be achieved by exploiting the natural reservoirs provided by geological formations. However, the geological conditions of the Mediterranean Basin are less than ideal for large-scale natural water storage. Lying as it does between northern Europe, with its abundance of large and adequately fed aquifers, and Saharan Africa, with its vast sedimentary basins containing only non-renewable fossil groundwater, the Mediterranean region is poorly provided for. Briefly summarized, its geological formations fall into four groups:

1. Sand, sandstone and various other porous rocks.
2. Fissured, karstic limestone.

Mediterranean Desertification and Land Use. Edited by C. Jane Brandt and John B. Thornes.
© 1996 by John Wiley & Sons, Ltd.

3.   Fissured crystalline and metamorphic rocks.
4.   Alluvial deposits from various sources.

Group 1 rocks, which provide large reservoirs, are not widespread in the region.

Group 2, karstic limestone, forms the typical aquifer rock around the Mediterranean. This type of reservoir has a very responsive water regime, which is very similar to that of surface water. This regime, with levels dropping markedly in summer, is not conducive to full and easy control of resources. However, there are a few notable exceptions, all connected with mountainous hinterlands, where discharge varies less (i.e. the Fontaine de Vaucluse, Apulia and the springs supplying Rome and Cannes). Another possibility for development lies in karst water below the level of the springs, which opens the way to inter-seasonal management of large volumes of water. Furthermore, such karst systems, which can be as deep as 100 m below present sea-level, discharge large quantities of fresh water directly into the sea. This can be seen at Agios Nikolaos, Crete, and in France, at Port-Miou near Marseille, or the coastal salt lakes of Salses near Perpignan. The discharge of such undersea springs can be up to several $m^3 s^{-1}$.

Fissured Group 3 'basement rock' formations provide moderate-sized aquifers. Their inertial function is generally due to the presence of weathering zones in the upper part of the formation, which provide a porous medium with reasonable capacity. In contrast to temperate oceanic and tropical conditions, however, the rugged topography and geodynamic conditions of the Mediterranean are not conducive to development and maintenance of a thick weathering zone. So, such basement rocks can make only a modest contribution to regulating flow, and are suitable only for domestic use (isolated dwellings).

The processes that wore down the relief, beginning in the Pliocene and much accentuated at the end of the last Ice Age, laid down considerable quantities of alluvium that can be several hundreds of metres thick. Such deposits characterize only a few per cent of the territory, but they are highly porous and permeable (Group 4 formations). They are mainly concentrated near coasts, and can include a considerable thickness of aquifer medium below sea-level, which makes them very vulnerable to sea-water intrusions. The important exception here is the alluvium of the Po valley, although this is more Alpine than Mediterranean.

Some of the streams, ending in the alluvial plains and deltas of the Mediterranean, flow from distant mountains or are hydraulically linked to karst formations. Plains and deltas thus benefit from some lateral inflow, and can receive more water in this way than from the percolation of rain falling directly on their limited surface area. Examples are the Var plain at Nice, the Llobregat plain in Spain, the area around Navplion, Argolis, Greece, the Naples plain, Venice and Albania.

Groundwater and the overall water balance

Water resources of the Mediterranean parts of some countries (i.e. those areas lying within the catchment basin of the Mediterranean) are already very irregular (Margat, 1992). It should be borne in mind that regularity of supply depends on a relatively long passage through the earth: regular supply is therefore synonymous with a groundwater resource.

Table 16.1 shows that for the most typically Mediterranean regions (exceptions being the Rhône basin in France and Venice with the Piave basin), underground aquifers can provide only small-scale buffer supplies.

When considering individual small catchment areas, aquifers seem to play an even more limited role, with the result that many streams completely dry up in summer when no longer fed by their springs. The groundwater, if there is any during this season, can be harnessed only by pumping. Other streams are at the wadi stage, with residual underflow in the alluvium.

In conclusion, in the Mediterranean regions groundwater can only partly and locally replace the rainwater deficit. Rigorous management of these scarce resources is thus imperative.

16.2.2 Socio-economic role of groundwater

Necessity for groundwater resources management

Mediterranean man learned very early how to manage water resources, for example the Roman aqueducts, the diversion of a part of the Durance to the Crau plain, the Moorish irrigation works in Spain. These ancient engineering feats, and the big projects of the latter half of the 20th century, show that most water resources so harnessed come from mountainous hinterlands, often far away, which are Mediterranean only in the sense of being part of the catchment basin, but are not a part of the ecosystem we are studying. Although the water reaching these streams comes from the mountains and commonly is meltwater, the discharge is irregular. Furthermore, part of the summer discharge is drawn off in the middle courses. As a result, large dams have been built to ensure secure supplies throughout the year.

At the dam-building stage, the role of the aquifers was systematically ignored. Groundwater reserves, being hidden and hard to tap, were not a target for the big public development schemes that sought efficiency of scale rather than self-sufficiency in development. However, as technical progress brought down drilling costs until it was cheaper to drill than to use piped water, more and more farmers opted for the independent raising of groundwater by pumping. The water abstracted by large numbers of these private wells soon outstripped the production capacity of aquifers, especially on the Spanish coast, Greece, Cyprus and Sardinia. The notion of overexploitation must be handled with care, as it can easily lead to mistaken ideas about the exploitation potential of aquifers (Llamas, 1992); falling water-tables have indeed been used to argue for exclusive use of surface water.

TABLE 16.1   Regular or groundwater resource as a percentage of total water resource

Country (Mediterranean area)	Total water resource $(m^3\,yr^{-1} \times 10^9)$	Regular resource $(m^3\,yr^{-1} \times 10^9)$	Percentage
Spain	31.1	7.5	24
France	62.0	35.0	56
Greece	40.8	7.7	19
Italy	180.0	30.5	17

In the past few years, the idea of coordinated and complementary use of surface and groundwaters has taken hold. This is doubtless due more to current economic conditions, which are not conducive to major public investments, than to any recognition of the specific value of groundwater resources, which do not readily lend themselves to collective management. There has been significant progress in the public management of groundwater: water resources are state property under Spanish law, and under a recent French law they have become part of the national heritage. However, the technical and administrative instruments for managing groundwater are ill suited to the specific features of this resource: any user can tap it at will, with no concern of how the water reached his or her well, and on some aquifers there are thousands of such users. Governments must arbitrate this uncontrolled use of the resource, while having no control over input into the hydrological systems concerned.

As noted above, groundwater can only partly, and locally, compensate for irregularities in precipitation. This is certainly the reason why planners of large-scale water development projects have neglected groundwater supplies, knowing little about them or deliberately ignoring them (Llamas, 1992). However, when prospects are bleak, full use seems all the more necessary, in the spirit of the declaration made by the European environment ministers at the Hague in 1992. The final objective of this declaration is to improve the qualitative management of both surface and groundwater sources (JO—CE, 1992).

Water consumption

Hereafter, we shall consider only bottom line water consumption, i.e. that part abstracted and not returned to the natural environment. This seems the most objective approach, because it puts the irrigated farming sector face to face with its responsibilities, and gives purification and re-use of waste water their full future value as a second-hand resource. Some activities (particularly cooling systems in power plants), while consuming little water in absolute terms, depend for their operation on the regular availability of considerable volumes of water, which can be supplied only by dams or large rivers. Groundwater cannot be used in these cases.

From the statistics of the 1980s (Margat, 1992), current water use patterns are shown in Table 16.2. Here it can be seen that agriculture accounts for most water consumption, especially when considering that these are net figures.

TABLE 16.2   Water consumption and consumption indices for some countries of the Mediterranean Basin, and only for that part of the country that lies within the basin (surface and underground water resources)

Country (Mediterranean area)	Total ($m^3 yr^{-1} \times 10^3$)	Net consumption ($m^3 yr^{-1} \times 10^3$)			Total consumption ($m^3 yr^{-1} \times 10^3$)	Consumption index (%)
		Public distribution	Industry	Agriculture		
Spain	31.1	0.1	0.1	12.0	12.2	39.0
France	74.0	0.2	0.1	1.7	2.0	2.7
Greece	58.6	0.2	—	3.5	3.7	6.3
Italy	180.0	1.1	0.7	13.0	14.8	8.0

Taking each country as a whole, Table 16.3 shows the comparative water consumption between irrigated and total-farmland areas (Merillon and Roux, 1992).

16.2.3 Threats to water resources: the rot has set in

Problems and implications

Both quantitative and qualitative effects of water resource use combine, each amplifying the damage caused to available water resources by the other.

Quantitative impact

Roquero and Luque (1992) indicated that, in Spain, there is a very marked downward trend in discharge rates. This is shown by one comprehensive indicator: the mean annual hydroelectricity production potential has fallen in recent years by 10% compared with the 1940–70 period. Over 70 years the drop has been 25% with a 30% drop for August discharge rates. It must be stressed that these figures are for the whole of Spain, rather than its Mediterranean catchment areas: given the prevailing conditions in eastern Spain, the ratio there could be even worse. This decline in hydroelectric production potential is of the same order of magnitude as the consumption index; on the face of it, this shows human impact rather than climatic change.

Streams that were once permanent, now dry up and seasonal streams today dry up months before the usual date, based on past records. This is the result of water being pumped at many sites scattered around the aquifer area. If the notion of a safe yield is applied, i.e. that rainfall reaching the aquifer is available for use, then all water in streams and water courses must be considered to be waste water. The declaration of the International Association of Hydrogeologists at the meeting of European environment ministers in Dublin (Collectif, 1992) stresses this point. In Spain (Llamas, 1992), 45 of the country's aquifers are being exploited faster than they are recharged, and therefore no longer contribute in the normal way to their basin's runoff. In eight hydrogeological units with a combined area of $12\,500\,km^2$, the deficit reaches $650.106\,m^3$ every year, which means a fictitious mean annual discharge of $22\,m^3\,s^{-1}$; this also amounts to a depth of water of 52 mm abstracted each year from the reserve (Lopez-Camacho y Camacho and Sanchez-Gonzales, 1991).

TABLE 16.3   Average water consumption for all types of farming combined and for the countries as a whole

Country (Mediterranean area)	Surface irrigated $(ha \times 10^6)$	Percentage of total national farmland	Total consumption $(m^3\,yr^{-1} \times 10^9)$	Relative consumption $(m^3\,ha^{-1}\,yr^{-1})$
Spain	3.3	12.0	20.0	6000
France	1.2	4.0	3.6	3000
Greece	1.2	12.0	3.5	2700
Italy	3.6	21.0	13.0	3600

If the land concerned has an effective porosity of 5×10^{-2}, it is not surprising that a fictitious mean drop over the whole area of 1 m a year has been recorded. On low-capacity formations near Alicante, a record cumulated drop of -250 m has been reached, according to Pulido-Bosch (1991). Under such conditions, the cost of pumping water becomes a deterrent, and this can act as an economic regulating process.

Contrary to the ecological notion of maintaining a certain proportion of underground runoff, it has been argued that borrowing from reserves is of vital economic interest when renewable resources have already been used up. The definition of overexploitation provided by Llamas (1992), 'all abstraction of undergroundwater which produces effects (physical, economic, ecological or social) with a negative final balance for humanity, now or in future years', leaves a very wide margin for interpretation. It would be Malthusian to argue from this that groundwater resources should not be tapped (Custodio, 1991). Diagnosis of overexploitation, which in Spain has administrative consequences, must thus be viewed in terms of a spatial reference, i.e. the whole water system consisting of ground and surface waters, and covering a period that is long enough to absorb transient phenomena.

Real disasters have already occurred however: Llamas (1991) considers the 'Tablas de Damiel' national park to be in a state of 'ecological coma'. A very picturesque scenery, where springs once fed the head-waters of a river, has disappeared and the openings from which pure spring water once emerged now emit surface runoff of mediocre quality. More generally speaking, the entire La Mancha aquifer is being tapped at twice its annual recharge rate, to irrigate 130 000 ha and feed a population of 300 000. It can be assumed that the ecological damage is merely a precursor of the coming economic damage. Part of the latter damage will be a drastic reduction in irrigated farming, currently fed by water from 15 000 wells, which certainly will have very severe economic consequences.

Qualitative impact

Another aspect, very specific for the area described above, is groundwater that flows directly into the sea. Such outflow is never seen, except were it is intercepted by boreholes. Today, however, the hydrogeological conditions have been reversed and sea-water is drawn into the aquifer. Navarette et al., in Lumsden (1992), show the considerable economic impact of such salt-water intrusion, but, as is often the case with such matters, the cost of this damage can be inferred only from the cost of alternative or remedial solutions, here about 20 billion pesetas per year. In the Almeria plain, the 'Campo de Dalias' is another example of uncontrolled development schemes (Thauvin, 1986). Here, 11 000 ha of greenhouses are found in a plain of 300 km². Government policy made access to water too cheap and easy (water costs accounted for only 2.3% of operating costs), quadrupling the local per capita income. According to Lopez-Camacho y Camacho and Sanchez-Gonzales (1991), the annual water deficit of this area is 21 million m³. The space created by over-pumping is inexorably filled by salt water, putting an end to the short-lived wealth created by irrigation. Here, again, the notion of a safe yield cannot be invoked: to hold back a salt wedge, a counteracting hydraulic head must be maintained, i.e. some water must be allowed to flow out into the sea, helping to preserve the quality of the remaining water, which can be exploited in line with the hydrodynamic model.

Other types of qualitative deterioration lower the value of drinking-water resources in such areas even further. Agricultural pollution commonly is particularly severe: evapotranspiration rates are high and considerable concentrations of pollutants, in particular nitrates, build up in the small amount of water infiltrating the soil. The areas worst hit generally are coastal plains that are already affected by intruding salt-water wedges as described above; examples are the Llobregat and Gerona plains in Catalonia, Spain, and the Italian Campania.

Qualitative impacts must be seen as harmful not only to drinking water, although this is a benchmark; they also damage the natural environment, causing pollution and induced eutrophication of streams and coastal lagoons by inflow from groundwater.

To conclude, human quantitative and qualitative impact, for some decades, has been so intense as to mask the impact of potential climatic changes, and this state of affairs will persist into at least the near future. In the absence of adequate records (commonly, little is known about natural discharge rates, and not much more about water-table levels), the very strong background noise from well pumping makes it impossible to identify hydraulic trends that are due solely to a suggested reduction in the amount of water reaching the aquifers. With a mean flow rate of 21.8 $m^3 s^{-1}$, the Fontaine de Vaucluse (the outlet of a virtually unused aquifer) shows no significant variation in discharge, from which one might conclude that the regime could not have been affected by climatic variation. However, even if the impact of such variation is barely perceptible on the groundwater regime, this regime can also be significantly influenced by changes in surface conditions or by land development in the valleys.

Future prospects

Human impact

Agriculture is the main factor that will influence future water-resource consumption, with human consumption, not counting tourism, increasing very little.

If we adopt the scenarios from J. Margat's 'Blue Plan' (Margat, 1992), the final consumption index trend, for all uses, would be as shown in Table 16.4.

In 30 years' time, it seems likely that Spain will have consumed more than half her total water resources. France is not faced with a major problem, but Italy and Greece could suffer severe damage. It cannot be repeated often enough that the phenomena, here examined on a macroscopic scale, differ sharply from one region to another. As there will be more and more demand for regular water supplies to compensate for the vagaries of climate, one can predict that pressure will be particularly heavy on groundwater resources,

TABLE 16.4   Water consumption index for the period 1992–2025

Country (Mediterranean area)	1992	Consumption index			
		Year 2000		Year 2025	
		Low scen.	High scen.	Low scen.	High scen.
Spain	39.0	48.0	51.0	52.0	60.0
France	2.7	3.0	4.0	4.0	5.0
Greece	6.3	7.0	9.0	9.0	12.0
Italy	8.0	9.0	13.0	10.0	18.0

scen. = scenario.

especially in sectors where it is not possible to transfer large quantities of surface water.

As regards water quality, the use for irrigation may lead to higher concentrations of pollutants (in the water that has not evaporated). With more farmland under irrigation and an increase in diffuse pollutants (nitrates and pesticides), it must be feared that naturally potable water for the population will be increasingly hard to find in the circum-Mediterranean regions.

Climatic aspects

One cannot dissociate the effects of groundwater abstraction—which masks climatic effects—from an increase in the amount of water needed to compensate for the natural deficit which would arise from a potential increase in evapotranspiration. As the areas where water abstraction is possible are small compared to the areas over which the water is used, the effect on resources may show up rapidly in the abstraction areas.

An increase in evapotranspiration would then have a double impact: on the net rainfall received and on water demand, for farming in particular. The consumption index would rise faster than the climatic factor responsible (22% increase in this factor if the trend for potential evapotranspiration and net rainfall were 10%).

Lastly, the way in which rainfall feeds the aquifer can change under the impact of an increasingly seasonal rainfall pattern with altered precipitation intensity.

Recharge of the aquifers may then be very different from one scenario to another, i.e. scattered rain on finely porous soil or flooding of coarse porosity land or karst. Then the role of the groundwater may change, its regulatory function becoming more and more important relative to surface water, owing to natural shortage and socio-economic demand.

For example, the change in climatic parameters like temperature, evapotranspiration and rainfall has recently been studied for Roussillon, France (Chabart, 1995). The impact expected for a 1°C change in global temperature according to Chabart and members of the MEDALUS climatic change groups (Palutikof (1993, 1994) and Palutikof et al. (1992, 1993)) is:

— an increase in temperature between 1 and 2°C,
— an increase in precipitation during winter, spring and summer of between 0 and 10% and a decrease in precipitation during autumn of between 0 and 10%,
— an increase in evapotranspiration of between 0 and 3 mm per day.

These climatic changes combine to produce a decrease in effective rainfall of about 20% and a decrease in piezometric levels of between 1 and 3 m. Should groundwater abstraction also increase, the decrease in piezometric levels could reach between 5 and 10 m, bringing with it a real risk of salt-water intrusion.

16.3 MODELLING AND MANAGEMENT OF AQUIFER SYSTEMS

16.3.1 Aims, objectives and characteristics of groundwater modelling

There is no specific tool for the management of shortages, but we can adopt specific behaviour and methodological precautions, in particular for the data to be collected and the techniques for validating models.

In operational hydrogeology, the term 'model' has increasingly deviated from its general meaning, of a representation made to be copied, to a specific meaning of 'simulator of groundwater behaviour'.

Models began as physically constructed objects, then became analogue features that exploited the similarity between laws describing water flow through porous media and laws for the propagation of electricity through a conductive medium, and now have passed into the realm of computers. Models now consist of numerical matrix software; the progress in computer and peripheral design has made it very easy to converse with machines with satisfactory results. These tools have become truly animated scale models that describe aquifer behaviour and enable future projections for resource management on the basis of selected scenarios to be made.

Modelling is first of all an excellent tool for understanding groundwater. Provided that the space/time variables, which were observed and then introduced into the model, are in harmony with the assumed structure and parameters of the environment, it can be considered that the overall knowledge of the aquifer is coherent. When the model, which is an imaginary construction, faithfully reproduces the real world, it can be inferred that the knowledge obtained on the latter is satisfactory. This makes it possible to detect any defects or gaps in the knowledge, and the model can also be used for designing data-collection campaigns.

Engineers, operators, and public and private developers have an even more utilitarian vision of models, which they use for future projection of changes in 'natural' conditions, in the type of land-use management, in resource exploitation, etc. The modelling then is totally directed by the simulation objective. The most common use is for predicting hydrodynamic influence. This can cover water-table fluctuations, for example as a result of digging gravel pits or a canal, or it can be used for optimizing the individual sites for a well-field; other objectives might include the wish to remove a maximum quantity of water at a minimum cost, or the determination of transit time and path of pollution in an aquifer.

16.3.2 Types of models

The two main types of models are termed lumped and deterministic.

Lumped models

Lumped, or black box, models reproduce certain phenomena by adjusting the signal/response ratio with observed data, but without the description and use of physical parameters, and without compartmentalizing the spatial domain. A mathematical trial and error procedure, guided by knowledge of the phenomenon to be simulated, enables the reasonable reproduction of a cause-and-effect relationship, such as fluctuations of a water-table under the influence of successive rainfalls. It is true that most such models are not 'black boxes' in *sensu stricto*: an integral black box would be a research instrument that, without any prior orientation, seeks a relationship between a triggering signal (such as rainfall) and the system response (such as water level) using deconvolution and optimum tuning methods. Models like 'GARDENIA' or 'GARDENSOL' (Thiery, 1986) can be called 'grey boxes', as they use rainfall data that have been pre-processed to deduct evapotranspiration. Moreover, both programs describe reservoir behaviour as a succession of interrelated steps between the soil surface and groundwater, which provides the modelling with a dose of naturalistic judgement and experience, thus avoiding the excesses imposed by the use of the computer.

Figure 16.1 shows what can be expected from this type of model. Its simplicity and easy use make it possible to simulate many situations, but such models also have their limits. In addition to the lack of spatialization, the main disadvantage is that a lumped model provides answers to a set of unidentified causes. If the model is asked to simulate the response of the system to a completely different situation from that for which it is calibrated, it is not at all certain that the resulting answer will still be reliable. For instance, a well-adjusted model can correctly reproduce the nitrate levels in spring water flowing from an agricultural catchment; however, it is not all certain that the same reliable answer will be obtained if the type of crop is changed, for example from wheat to maize, all other parameters remaining the same.

It is thus recommended that lumped models are used only for conditions that vary little from those that were used for calibration, which makes it impossible to use such models for long-term projections into an uncertain future. Such models are best used for acquiring knowledge on active mechanisms that are unlikely to be subject to much drift, such as the simulation of seasonal variations in groundwater-tables. It should also be stressed that such models are valid only within the boundaries used for calibration, and that they cannot be used for extrapolation.

Deterministic, or grid-based, models

Grid-based, or deterministic, models are also called mechanistic, as some of them are used for the explicit integration of laws describing water behaviour in the subsurface zone. Each function is analysed and the variables are linked by one or more parameter algorithms; for example, flow-rate and water-level are related through Darcy's law describing aquifer permeability.

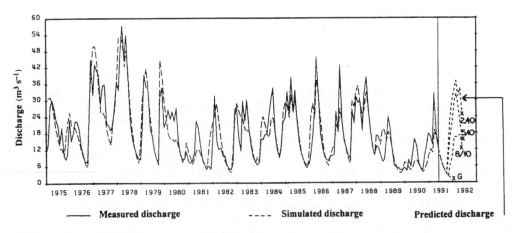

FIGURE 16.1   Measured and simulated discharge for the spring 'Fontaine de Vaucluse' near Avignon (France) (Urgonian limestone). Results from the simulations with the lumped model GARDENIA. After 'Situation hydrologique et prévisions des basses eaux'. Bull. No. 46, Sept. 1991. Ed. BRGM

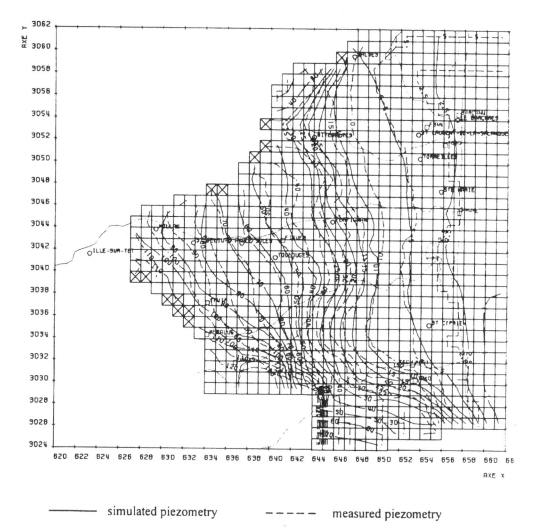

——————— simulated piezometry   − − − − − measured piezometry

FIGURE 16.2   Example of a gridding which divides the aquifer into several meshes showing the simulated and measured piezometry of Aquifer 3 in the case of the model of the Roussillon multilayer aquifer

The gridding serves to divide the area to be studied into a large number of grid elements or meshes (Figure 16.2). This makes it possible to use iterative numerical techniques that model specific elements; the results are automatically transposed from mesh to mesh, leading to an equilibrium solution that is valid up to the boundaries of the modelled area. It is thus clear that boundary conditions of the modelled area must be very well understood. Such conditions can be impermeable, confining the water in the aquifer, or they can be unconfined and leading to exchange with surface waters. There are also lateral boundary conditions which need specifying.

Although such hydrogeological models must be based on a good physical representation of the phenomena and environments that are to be modelled, all models contain a varying number of simplifying assumptions. The most common and schematic of these is the idea that flow is permanent (i.e. invariable), without storage or destorage (i.e. loss, effects over time). Although not entirely realistic, this approach makes it possible to reduce the modelling problem to manageable proportions.

Increasingly, the simulation of unsteady phenomena, such as seasonal recharge, progressive flow reduction, or pumping rates, coupled with techniques of computer graphics, has changed such models into a source of animated virtual images. These are sufficiently realistic to retain the interest of resource managers, who, thanks to attractive presentations, can easily grasp the real issues and decisions for which they are responsible. Models have thus become three-dimensional and multilayered, the latter being several superimposed permeable layers that are separated by semi-permeable layers through which some exchange can take place. Relationships with surface water have been integrated, as are the solute transport, density variations, interaction with the host rock, etc.

The present performance of these calculation tools might give the impression that modelling has shifted into the realm of irrefutable hard science. However, in reality, knowledge particularly of the subsurface, is imperfect and information is available for only a few points in an aquifer reservoir, where there are boreholes. Parameters and coefficients describing such a reservoir are thus extrapolated in space, as fields of values that by necessity are imperfect in their distribution and precision, and this notwithstanding the great progress made in methods of data treatment such as geostatistics. An approach that is commonly guided by the naturalistic understanding of field conditions, leads to the making of assumptions on the geological framework. This framework then has to be divided into meshes, and the time frame must be divided into time steps that are compatible with the phenomena studied.

Validation of the hypotheses is obtained from the fact that, if they are correct, the model will reproduce an image of the aquifer that is very close to the observed one. When the model is thus validated, it can be used for predicting the response to various changes in determining factors of importance. In this manner, one can simulate water abstraction.

When studying desertification, we can apply the model to a reduced rainfall input and examine the repercussion of such reduction on the available resources. However, the model will only be valid for the area and within the time frame for which the representative parameters were collected.

16.3.3 The necessary data

It would be dangerous to persist with the generally accepted, but false, idea of early modelling days, that models can replace data. After an initial period, during which slick engineering salesmen tried to sell miracle projects that were based on non-validated models, a code of good modelling conduct has been established. This has imposed financial problems: the collection, verification and validation of groundwater data require long and costly operations, much more difficult than the modelling itself, which mean that even the most open-minded clients can be easily discouraged.

Data collection, whether through drilling, geophysical surveying, or installing and monitoring recording instruments, is an operation needing great care and scientific rigour.

Two distinct cases must be considered, as the act of modelling, although consistently scientific, is not necessarily connected with research.

1.   In the case of a financially well-endowed R&D project on an experimental site chosen for its wealth of present and past data, it is easy to construct a very elaborate model that is based on a multitude of specifically measured parameters. In this manner, it is possible to develop very powerful tools that can reproduce the most complex mechanisms.
2.   Reality, however, shows that, contrary to the case for such research models constructed under ideal conditions, many simulations of operational projects are faced with inherently poor data. Under such circumstances, it becomes very difficult to calibrate and validate the tool that must be presented to decision-makers. It is then necessary to account for the uncertainty.

Faced with drought and the steady lowering of groundwater-tables through consistent over-pumping that exceeds natural recharge, the various models mentioned above require data that are too complex or expensive for most projects. We will now discuss what type of data are involved and how they should be collected.

The parameters that govern a groundwater resource are 'inflow', 'host rock' and 'outflow'. The inflow is determined by infiltration; host rock parameters are permeability, rock porosity and thickness; outflow parameters depend on whether the water flows out naturally (through springs) or artificially (through pumping). In other words, at a given place and time, the groundwater level is governed by space/time variables, such as climatic phenomena in a (relatively recent) past period, and pumping and/or natural outflow. Groundwater flow is thus conditioned by reservoir behaviour, which can be described by a storage coefficient (characterizing the storage function) and by permeability (characterizing aptitude for transmitting a certain flow).

1.   Modelling is first of all based on the notion of boundary conditions: in deterministic modelling, the functioning of an aquifer can be realistically simulated only if its relationship with surface water, or its independence from such water through geological confinement, is well defined (Figure 16.3). Geo-hydrodynamic analysis makes it possible to define the concept of an aquifer system. The naturalistic approach of geologists, supported by data from drilling, logging and geophysics, enables the definition of effective and functional contours of the three-dimensional space to be modelled. This operation, which is simple with sufficient subsurface data and their proper organization in a databank, can be a major obstacle in countries with underdeveloped monitoring networks. There, major campaigns may be necessary for collecting the required data on aquifer structure and boundaries.
2.   The characteristic parameters of the reservoir (or reservoirs in a composite aquifer) will be known from a few specific points, provided that most existing boreholes have been subjected to pumping tests. It can be said that an aquifer cannot be really well known unless it is already significantly exploited, or even over-pumped. Here, again, the cooperative role of public databanks, such as the French Subsurface Data Bank (operational since the early 1960s and managed by BRGM), is fundamental: the individual data produced by a single borehole are used in a spatial sense for the whole

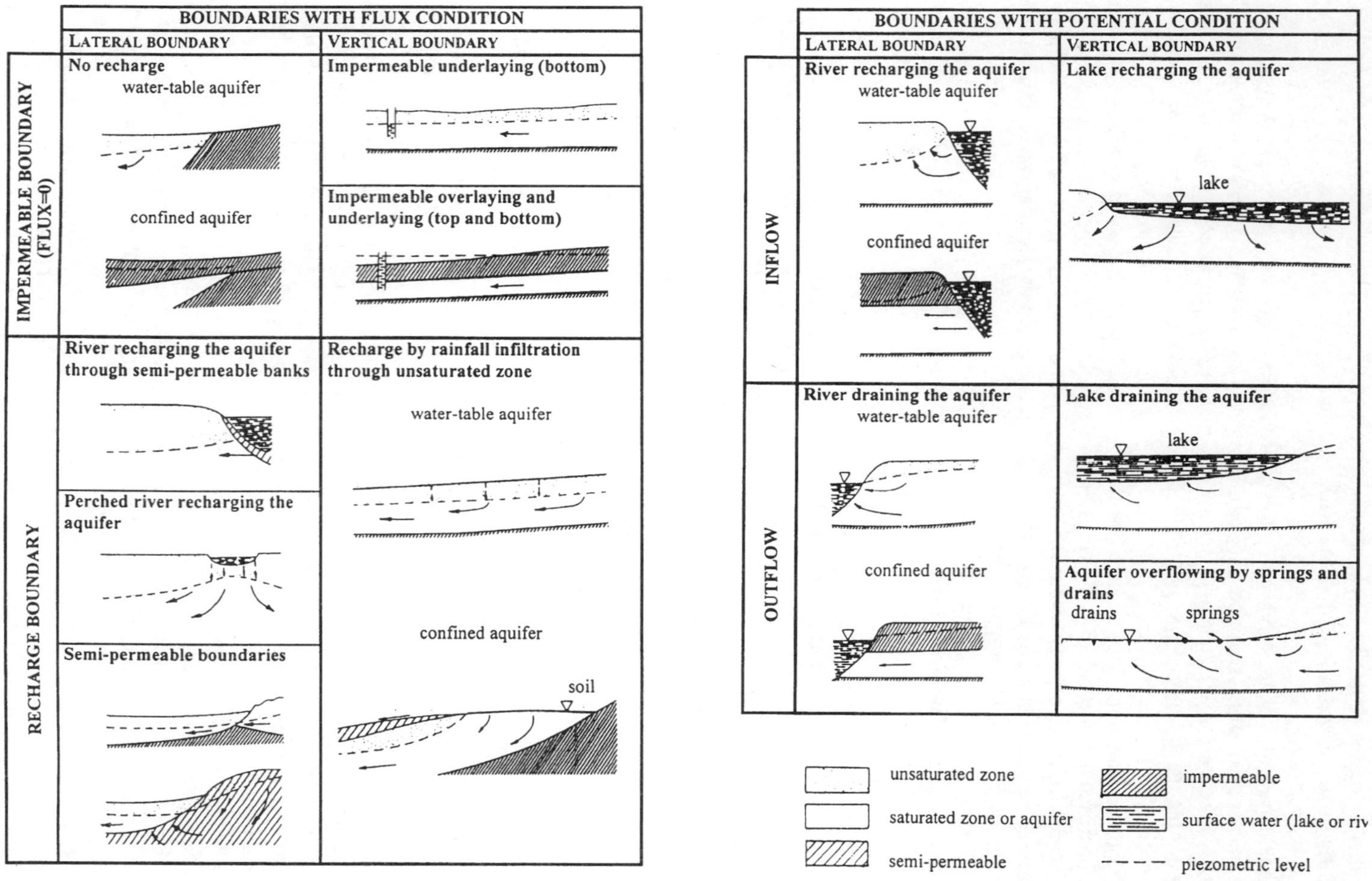

FIGURE 16.3   Some examples of boundary conditions to be simulated in groundwater modelling. Redrawn from Castany and Margat, 1977

aquifer, whereby it is clear that the sum will be greater than its parts. However, it is clear that national, regional, or basin authorities must have the power to impose collection and conservation of all borehole data in the common interest.

3. Space/time variables concern water levels and flow rates; they have to be known simultaneously and, in order for them to be used to validate the model, they should be known for a period that is at least as long as that to be simulated, to avoid unwarranted extrapolation of data. Although the authorities now consider it normal to manage a network of weather stations, it is much more difficult to have them set up a gauging network for measuring flow from aquifers. Surface-flow measurements are mostly used for flood prediction or the calibration of hydraulic development projects and only rarely for understanding the volume of base flow provided by aquifers, which is only really visible after all rainfall event flow has stopped. For groundwater management at the scale of the great aquifer systems found around the Mediterranean, it is thus indispensable to establish a public infrastructure for gauging spring and endogenous-stream outlets.

4. Artificial flow rates (the pumping from all wells) should also be known over time. It might seem that the collection of such data is facilitated by legislation that regulates such pumping through taxation, but in reality the users often defy the law and report only part of what they pump, if any at all.

5. Finally, the best-known variable is probably water level (or hydraulic head), referring to a base level that generally is that of the country. Here, two types of data are required:

    (i) Maps, obtained through interpolation of scattered point data, show the response of an aquifer to water inflow and outflow at a given moment. Such synchronous maps of pressure fields over a predetermined grid will be used for calibrating the model; the latter will be deemed representative and capable of simulating various influences, if the numerical calculations, repeated by iteration for all grid meshes, correctly reproduce observed aquifer conditions.

    (ii) Historical changes, largely influenced by the storage capacity of an aquifer, show the aquifer 'memory', and help in preparing the model for simulation of natural or man-made evolution scenarios.

In reality, these two functions of mapping and historical analysis are intimately related, but the availability of time series aquifer data for a few representative points is the responsibility of public authorities. Only the state can anticipate the request for data that have no precise objective, and which are collected a few decades before they will be used. As an aside, this may be one of the best arguments against privatization of all tasks of regional Water Authorities in Europe, or elsewhere.

Modelling, a modern technique that apparently permits unlimited simulation of the future, is in reality much more constrained by data problems than by the numerical difficulties that are increasingly well mastered by computer specialists. Here, two conclusions are presented as recommendations:

—   It will be useless to develop overly powerful computer tools through R&D projects, if such tools, once operational, will not be provided with all required data, notwithstanding the financial, regulatory and technological difficulties that may have to be faced in obtaining them. R&D in this field should thus bear on improvement of the general representativeness of such models, rather than on the precise simulation of phenomena that in any case cannot be controlled in a realistic manner. However, they can be used to establish methodologies which account for output uncertainty.

— Groundwater management, not counting well or well-field management, is a collective matter that requires a government-level approach. Data-acquisition networks should thus be set up to collect the necessary data for proper aquifer management in the future, in particular of those aquifers that are the most heavily pumped.

16.3.4 Modelling aquifers for groundwater management under critical conditions

The use of the term critical conditions covers shortfall, salt-water intrusion and pollution. The methodological problems mentioned above, which concern the extent to which the model is valid as well as the provision of required data, are much worse when applying such models to an uncertain future and an imperfectly known area.

It is for instance easier to simulate the effect of pumping on the chalk aquifer in the Paris–London basin, or of the canalization of the Rhine on local groundwater conditions, than to predict the condition in 30 years time of the resources in certain karst aquifers of southern Spain that are already overexploited. We must be careful when trying to extrapolate over long periods and modest as to the results. Modelling, which is a professional tool for hydrogeologists, cannot be transformed into an instrument of scientific futurology. Therefore, we must introduce suitable precautions and error ranges, to avoid dangerous use being made of our tools.

Modelling the water resource

We will not discuss here the extraction of fossil and non-renewable groundwater resources, as this type of resource is rare in the European countries along the north side of the Mediterranean. However, certain exploitations of resources that are only slightly renewable and are now in critical condition, are not very far from this concept of groundwater mining. Several cases are known from Spain, where reckless over-pumping of the resource, at the level of several times the annual recharge, has led to almost complete exhaustion of the reserve; in this case, several years without any abstraction will be necessary to restore the aquifer to its original condition, with springs flowing as before.

Evaluation of recharge

When considering the so-called renewable resources, the prediction of their evolution is based on two elements: entry of water into the system, or recharge, and outflow, whether natural or through pumping. Recharge is the element that is most difficult to understand and simulate, as this is a multi-faceted phenomenon that is irregular in space and time, and can be both natural and induced by man. Recharge of an aquifer can occur as overall net infiltration (that part of rainfall that neither runs off nor disappears as evapotranspiration) or it can be through runoff followed by infiltration. Calculation of efficient (i.e. infiltrating) rainfall is based on simultaneous use of the parameters rainfall, temperature and insolation, and must absolutely use small time steps to avoid brief recharge periods being masked by average rainfall values over time steps that are longer than the rainfall events.

Infiltration events are generally short and marked by strong seasonal variations. Even in areas where potential annual evapotranspiration surpasses precipitation such as in the Sahel, infiltration can still take place. Using a time step of 1 month, rainfall of 50 mm associated with evapotranspiration of 150 mm might lead to the conclusion that no recharge can take place. However, if such precipitation were concentrated in a few days, it would probably have provided significant recharge. Therefore, the time step for calculating recharge must not exceed 1 day.

With a few rare exceptions, recharge in the Mediterranean region takes place from November to April (Figure 16.4). Higher winter temperatures and a reduction in rainfall during this period, coupled to stronger sunshine, would strongly reduced recharge from diffuse infiltration. Any increase in summer rains would do little to improve aquifer recharge.

To simulate the possible evolution of diffuse aquifer recharge, one must work on a real rainfall time series from consecutive years, the data for which are modified according to climatological projections, i.e. with a certain reduction in seasonal rainfall and increased temperatures.

In the Mediterranean area with its geological and morphological setting, few areas are sufficiently flat, porous and large to receive diffuse infiltration. The presence of karst that commonly lacks a soil cover can favour infiltration of rainwater that thus escapes evapotranspiration, and that may even support recharge from summer storms, as witnessed by imprudent speleologists engaged in caving. Runoff after precipitation over impermeable formations causes erosion and the creation of talus cones and deep valley fill. It can also lead to temporary puddles and ponds, and even to flooding. However, such water generally disappears after a few hours or days. This is why the evaluation and spatial distribution of such recharge remain excessively imprecise.

When dealing with a flash flood racing down a stream bed with a flow rate of, say, $1000 \text{ m}^3 \text{ s}^{-1}$ (an estimate with an error margin of 100% as few gauging stations can handle this sort of flow), it is difficult to evaluate the small proportion ($1, 10, 50 \text{ m}^3 \text{ s}^{-1}$?) that, for a few hours, might recharge the alluvial aquifer of the main stream bed and, below that, the permeable formations.

Another difficulty is the calculation of the return flow of irrigation-water surplus when using methods that are not optimized. Such flow can reach double or triple the actual crop requirements. This type of recharge, plus that from other water losses and leaks, can be quite substantial. The Crau aquifer in the Provence region of France, for instance, receives 75% of its flow from leakage out of the Durance river. The transfer time of such water was found to be about 300 years.

In any aquifer system, different modes of replenishment co-exist in variable proportions and consequently evaluation of recharge remains one of the most delicate and uncertain tasks in modelling. It is unlikely that traditional data-acquisition networks will help very much to resolve this problem. On the other hand, satellite images are excellent means for establishing which surfaces are subject to runoff or infiltration, and how much land is being irrigated. This at least gives an objective picture of distribution.

Various other methods are available for recharge evaluation. For instance, experimental stations measure water content and pore pressure in the unsaturated

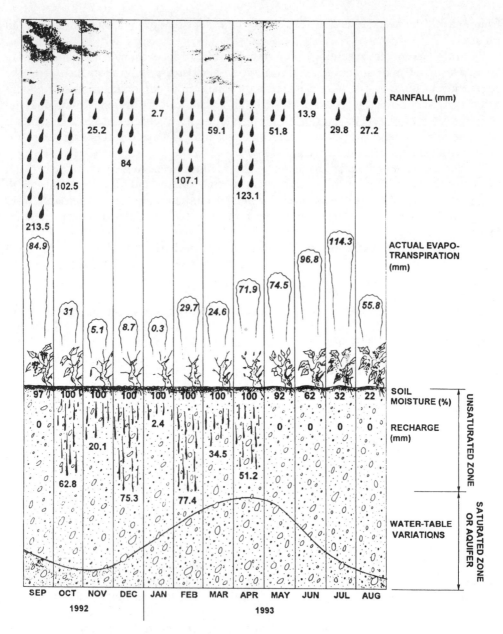

FIGURE 16.4 Water balance under Mediterranean conditions. Results of experiments in the Terrats field site, near Perpignan (Eastern Pyrenees, France)

zone, which gives a figure for vertical flow. This sophisticated and expensive method gives good results, provided the experimental conditions are properly controlled, but these data are representative only for the plot on which they are measured and cannot be extrapolated to the aquifer as a whole. This is because the type, texture and use of soils varies from place to place. At the other end of the spectrum is the generalizing method of measuring the geochemical signatures of groundwater. This detects inflow of water from outside the aquifer, or the proportion of irrigation water finding its way back into the aquifer can be evaluated, especially where the water is of distant origin. The tools available for this work are natural geochemical tracers, stable isotopes that are an effective witness of evaporation phenomena, and radioactive isotopes that mark the time of infiltration; however, such tools are too little used by engineers working on these problems. In any case, such data should be used together with hydrodynamic data and never alone.

No method by itself can provide a proper evaluation of recharge, but satisfactory results can be obtained by combining several types of data, in particular as they will help to eliminate incorrect assumptions.

Evaluation of outflow

The evaluation of natural outflow relies much more on the use of data from monitoring networks than on scientific investigations. However, when pumped volumes are not properly reported or declared, more discreet and indirect enquiries may be necessary. Luckily, the orders of magnitude for agricultural water requirements are quite well known and energy consumption for pumping is an excellent indicator. Combined, this gives an outflow figure with an accuracy of a few tens of percentage points, which is much better than the accuracy of recharge estimations.

Simulation of aquifer behaviour in the case of heavy pumping is relatively easy, as the pumping signal will be much louder than the background noise. Calibration of water-table movements with pumping rates then suffices to dominate any uncertainties on recharge.

Once the basic aquifer-system data for inflow, outflow, internal flow and storage have been acquired, simulation can start. To provide answers to the problems of groundwater resources of the Mediterranean Basin, it is necessary to simulate the climatically induced changes of inflow, based on the conclusions of specialists in this field, *together* with the socio-economically induced changes in outflow. The climatic change, as currently understood, is in principle unfavourable for the resource as, by reaction, it causes a heavier demand for irrigation water. On the other hand, modernization of irrigation methods would lead to drastic reduction in water consumption. However, even though irrigation with such water from far away will contribute to recharge of the aquifer, the effect of such recharge through agricultural soil might be disastrous for the drinking-water quality of this groundwater.

Presentation of results

Groundwater-level maps are the most common presentation of the results of such modelling. Such data are easily translated into pumping potential; beyond a certain depth, pumping-energy costs will render groundwater uneconomic for irrigation of products that commonly have a low market value. The model can also be interrogated for flow rates, providing answers on possible pumping rates for specific exploitation conditions. Even better is that the latest developments in modelling enable the optimization of borehole locations and numbers, to obtain a planned flow rate (El Magnouni et al., 1994).

Situation of water shortage

A fundamental role of groundwater is the feeding of springs or diffuse outflow into streams, and the models can also be interrogated in this sense. In certain dramatic cases in Spain, underground flow has virtually disappeared, and the formerly permanent streams have become channels for occasional flash floods, like North African wadis. Where the aim is to conserve some flow in river beds for ecological or sanitary reasons, models can serve to simulate natural outflow in terms of future inflow and 'permitted' pumping rates.

Based on the economic and environmental ambitions of the various societies, groundwater falls between two poles that are either economic or ecological. In the hands of politicians, a sufficiently user-friendly model can thus be the instrument of a choice of society that has very far-reaching consequences. The experience obtained in dry countries, where many hydrogeologists have earned their spurs, shows that the relative value of groundwater increases as a function of some aridity index or other. Here, water takes a very clear upper hand over the classical ecological values of well-watered countries, and an exploitation model of the resource becomes reasonable that is very close to a safe yield (i.e. pumping rate is close rainwater inflow). In this case all of the resource is used and no permanent surface streams can exist. Application of this hypothesis to developed countries with a relatively dense population is met by double opposition: from ecologists, who defend wet-lands, and on the scientific side from purification specialists, who need surface flow for the dilution of urban effluents, as well as industry that needs water.

Specific case of salt-water intrusion

Salt-water intrusion is one of the worst consequences of over-pumping coastal aquifers. A freshwater flow (gradient) must be maintained to keep the wedge of sea water at bay. In this case even safe yield pumping is a potentially damaging action, as it suppresses most if not all of the flow that opposes salt-water intrusion (Thauvin, 1986). As flow transfer is much slower than pressure changes, the arrival of the first traces of salt water occurs well after the water-table is lowered. However, this means that an enormous volume of salt water has been set in motion, to fill the areas emptied by pumping. Because of several physical phenomena, such as dispersion, the many pumping sites, or the heterogeneity of aquifer formations, the salt front is a diffuse transition zone; the precursor information contained in this zone has not always been used to the full, commonly because of imperfect monitoring facilities.

Modelling of the propagation of the saline front is one of the most difficult operations in hydrogeology, requiring highly specific tools. The presence of the salt creates new hydrodynamic conditions in the aquifer, which needs a progressive modification of the flow-rate parameter as the aquifer becomes saltier. Excellent tools are now available and provide reliable predictions, but here, too, the absence of good data (such as initial conditions or undisturbed profiles in boreholes that are not being pumped) commonly hinders full use of the models.

Problems of pollution

Pollution is the most common degradation of groundwater subjected to intensive pumping. We have already mentioned the problem of stream beds transporting waste water, which may be the most obvious result of over-pumping, but this may hide a more insidious and widespread problem that has many facets. The pumping from aquifers of irrigation water that is mixed on the surface with fertilizers and pesticides, in a warm climate can lead to a progressive build-up of such undesirable substances through evaporation. Nitrates are already extremely concentrated below market-gardening areas, but the build-up of naturally occurring salts in water can occur as well. The water becomes undrinkable and can no longer be used for irrigation, salt crusts form and, finally, the soil is totally destroyed for agriculture.

In this field, both lumped and deterministic models are used, but, because of the large number of explanatory factors and the dearth of pertinent data, most experts prefer to use lumped models for predicting evolution, and this notwithstanding the earlier mentioned problem of extrapolating such data beyond the calibration areas. Deterministic grid-based models are still in the R&D stage, in particular where processes of evolution, degradation, or adsorption of certain substances in the unsaturated zone are concerned.

16.4 AN EXAMPLE OF SUCCESSFUL MANAGEMENT: THE ROUSSILLON MULTILAYER AQUIFER (EASTERN PYRENEES, FRANCE)

16.4.1 Structure and boundaries of the aquifer system

The filling up of the Roussillon ria occurred during the Pliocene with a high sedimentation rate (30–35 m per 1000 years). It consists of detrital series, with alternate semi-permeable and permeable layers, at the depth of up to 250 m, which takes place above the Messinian erosional surface. The geodynamic evolution shows the longitudinal tilting of the basin, induced by a loading subsidence (Clauzon et al., 1987, 1989; Clauzon, 1989).

The permeable Pliocene deposits represent an important groundwater reservoir both in terms of quantity and quality. The Pliocene aquifer and the Quaternary alluvial aquifer form the Roussillon multilayer aquifer, which extends over 850 km². The boundaries of this hydrogeological system are (Figure 16.5):

— to the south and to the west of the plain, metamorphic rocks (gneisses of the 'Massif des Albères' and schists of the 'Massif des Aspres'),
— to the west, igneous rocks (granites of the 'Massif de Millas'),
— to the northwest, sedimentary rocks (karstic limestones of the 'Massif des Corbières'),
— to the east, the Mediterranean Sea.

On the northern boundary of this aquifer system, the supply from the Plio–Quaternary karst formations and from Salses lake, takes place at depth (Gadel, 1966). The main supply or recharge of the system is infiltration of efficient rainfall over the outcrops of the aquifer formations.

16.4.2 The necessity for groundwater management

The alluvial aquifer contains a locally important resource, which is readily exploited and which provides mainly for agricultural and industrial needs. In the areas where the alluvial aquifer is missing or vulnerable to pollution, Pliocene groundwater has been increasingly

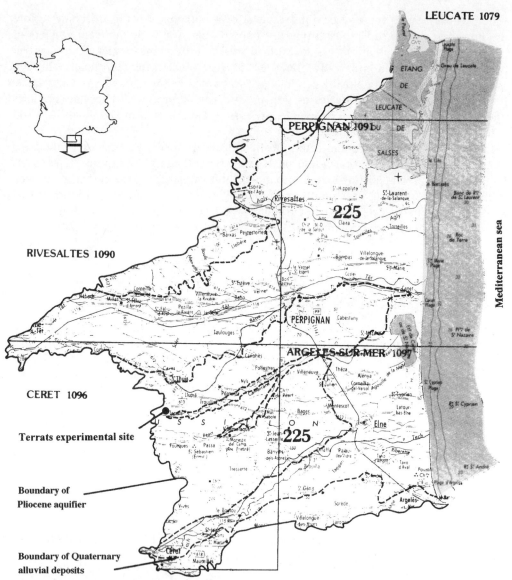

FIGURE 16.5 Map of Roussillon (Pyrénées-Orientales department) and the boundaries of the multilayer aquifer (aquifer system number 225 in the French classification). After Marchal, 1985

exploited for 15 years. The water consumption of the deep Pliocene aquifer, is divided up as follows:

19 million $m^3 yr^{-1}$ Domestic or collective use
5 million $m^3 yr^{-1}$ Agriculture
3 million $m^3 yr^{-1}$ Industry

Between 1975 and 1990, the total withdrawal from this aquifer increased by 60%, mainly to provide public distribution. Indeed, the Pliocene aquifer supplies 300 000 permanent inhabitants with drinking water, representing four-fifths of the population in the Pyrénées-Orientales department and 700 000 persons during the summer period.

Although the deep Pliocene layers are relatively well protected from the surface by thick clayey layers confining the aquifer, there is significant risk of groundwater quality being contaminated by the polluted and interconnected superficial aquifer (which contains nitrates, sulphates, chlorides, pesticides) or by saline intrusion (from over-pumping near the coastal boundary).

Consequently, to deal with all those different risks and to have all the necessary elements for decision-making, a policy of management, controlled by departmental and national authorities (Direction Départementale de l'Agriculture et de la Forêt, Ministère de l'Industrie and Conseil Général des Pyrénées-Orientales), has been undertaken in Roussillon since 1981.

In order to manage the water resources, the local authorities have put the BRGM in charge of:

— collecting geological and hydrogeological data from the Roussillon multilayer aquifer to improve the knowledge of its geometry,
— establishing quantitative and qualitative parameters,
— modelling of the complex aquifer system.

16.4.3 Data necessary for aquifer modelling

Three main aquifers have been identified in the Roussillon multilayer system (Figure 16.6):

Aquifer 1: The superficial aquifer with continuous Quaternary alluvial deposits, intersected by three main rivers, the Agly, the Têt and the Tech.
Aquifer 2: The confined Salanque groundwater body, a coarse detrital deposit (10 to 20 m thick) belonging to Pliocene formations, with limited extension to the northeast of the Roussillon plain.
Aquifer 3: The deep confined Pliocene aquifer (alternate clay, sand and a few gravel layers) which extends over most of the Roussillon plain.

The systematic collection of hydrogeological data has been used to construct several maps depicting the structure and the features of all three aquifers:

— maps of ground surface and bedrock levels,
— topographical map,
— hydraulic conductivity maps,
— storage coefficient maps,
— boundary conditions map,
— piezometric maps.

The other parameters which govern groundwater behaviour are inflow and outflow. Outflow parameters are determined from pumping from all wells and are estimated systematically with volumetric water meters. Volumetric water meters are mainly set up in

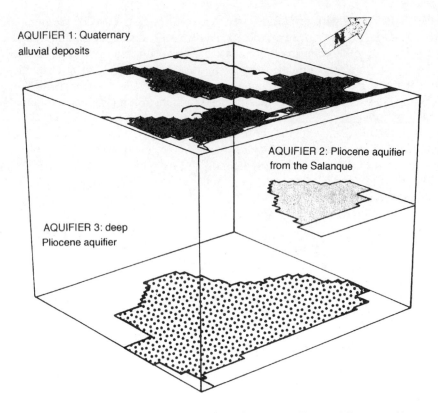

FIGURE 16.6   Schematic representation of the Roussillon multilayer aquifer

pumping for public distribution and rarely in private pumping. Water consumption for private use is evaluated according to owner opinion.

Inflow of water into the hydrogeological system is the most difficult term to estimate. By definition this term is assigned to recharge, i.e. infiltration minus change in the water storage. Infiltration is equal to rainfall and irrigation minus actual evapotranspiration and runoff.

16.4.4 Evaluation of recharge of the Roussillon multilayer aquifer within the context of the MEDALUS project

Study of the unsaturated zone is essential to estimate the deep Pliocene aquifer recharge. To obtain data for modelling purposes, it is necessary to acquire information on water movements in the unsaturated zone; to carry out periodic measurements of soil moisture content and hydraulic head (or water pressure). Construction and monitoring of neutron-tensiometric equipment can help to quantify flows through the unsaturated zone using the variations of the water storage and pressure gradients, and thus characterize the hydrodynamic behaviour of the unsaturated zone in order to model hydraulic transfers.

Flow theory in unsaturated zone

The isothermal flow of water in a homogeneous and isotropic unsaturated zone is given by two equations:

$$\text{The continuity equation: } \frac{\mathrm{d}\theta}{\mathrm{d}t} = \frac{-\mathrm{d}q}{\mathrm{d}z} \text{ (vertical flow)} \tag{1}$$

$$\text{Darcy's law: } q = K(\theta)\frac{\mathrm{d}h}{\mathrm{d}z} \text{ (vertical flow)} \tag{2}$$

where θ = moisture content per volume unit $(\mathrm{m^3\,m^{-3}})$, q = volume flow $(\mathrm{m\,s^{-1}})$, K = hydraulic conductivity $(\mathrm{m\,s^{-1}})$, function of the moisture content per volume unit, H = hydraulic head (negative value) (m), h = water pressure (negative value) (m), function of the moisture content per volume unit, z = depth (m) (assumed positive downwards).

The study of water transfer in unsaturated soil consists of solving a system of five equations, the two main equation (above mentioned) and three additional equations:

$$h = H + z \tag{3}$$

The retention law relating water pressure to moisture content:

$$h = \mathrm{f}(\theta) \tag{4}$$

The permeability law which relates the hydraulic conductivity to moisture content:

$$K = \mathrm{f}(\theta) \text{ or } \mathrm{f}(h) \tag{5}$$

The characteristic relationships of an unsaturated soil (given by the retention law and the permeability law) are determined by simultaneously monitoring the moisture per unit volume of the soil and the water pressure in the soil over 1 or 2 years under natural conditions and also under artificial recharge conditions.

Experimental site

Location

The experimental field site for measuring percolation is located on the Canterrane basin, near Terrats in a vineyard. The hydrological Réart Basin (Pyrénées-Orientales, France) between the Têt and Tech valleys at the foot of the 'Massif du Canigou' extends over a limited area ($137\,\mathrm{km^2}$). In the high valleys of the Réart river and of the Canterrane (its main tributary) there are outcrops of Jujols schists. These valleys are surrounded by hills and the topography results in a large amount of runoff. Flowing out of the primary basement, the Réart and the Canterrane rivers enter the Roussillon plain and flow directly on the Pliocene sands which constitute a water-table aquifer. Many rivers at the western limit of the Plain have the same characteristics (such as the Agly, Têt and Tech).

Water movement in the unsaturated zone of the Terrats experimental vineyard is controlled by special equipment (shown in Figure 16.7) and which consists of:

— 10 tensiometers (porous ceramic cups) with a mercury manometer, 0.2 to 2.5 m deep to measure the hydraulic head,
— two aluminium tubes, 2.5 to 3.2 m deep, for neutron probe access for the measurement of soil moisture,
— an observation well, 60 m deep, for the measurement of piezometric variations,
— a rain gauge.

Measuring hydraulic head in the unsaturated zone
The principle of the tensiometric measurement depends on the pressure created in an airtight system, by the transfer of water through a porous boundary. The hydraulic water pressure in an unsaturated soil is measured on site using tensiometers which consist of:

— a porous ceramic unit (or porous cup, length = 60 mm, diameter = 20 mm) of a very fine porosity and placed at the measurement depth,
— a tensiometric pipe containing the connecting circuit (nylon capillary tube) between the candle and the mercury reservoir,

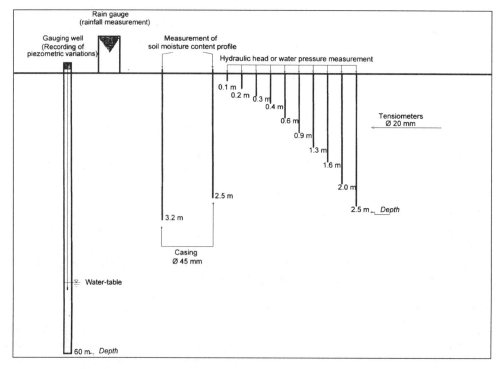

FIGURE 16.7 Terrats experimental site instrumented to study water movements in the unsaturated zone

— a mercury manometer.

The water pressure in the soil is transmitted to the mercury through the water contained in the pipe and the connecting circuit. The mercury manometer is installed so that the value of the hydraulic pressure can be read directly in millibars on a graduated ruler. This relation gives the water pressure in the soil: $h = H + z$, where h = water pressure (m), H = hydraulic head (m), z = depth (m).

The sets of tensiometers (located at various depths between 0.1 and 2.5 m) give the water pressure profile. The gradient indicates the direction of water movement (up or down).

Measuring soil moisture content per unit volume
The soil moisture content may be determined by a neutron moisture meter (or neutron probe) using the neutron elastic diffusion phenomenon. When a high-energy neutron emitted by a radioactive source propagates in the soil, it is decelerated by collisions with the hydrogen atoms in water molecules and slowly loses its energy until it drops to the level corresponding to thermal agitation. It is then said to be 'thermalized'. Soil moisture content is thus determined by measuring the flow of thermalized neutrons close to the high-energy neutron source.

A neutron probe containing a radioactive source composed of a mixture of americium and beryllium and a thermoneutron detector (helium 3 counting tube) is inserted in an aluminium access tube at the depth of the first measurement. Measurements are generally taken at 10-cm vertical intervals and are obtained by taking the average of two or three counts at the same elevation.

Since the source emissivity is random and related to its activity, it is necessary to apply a correction to the measured count, which will be a function of the number of collisions in water, and the individual characteristics of the probe used.

Theory and experience demonstrate that counts in the radioactivity field obey a Poisson probability law which gives the following relative error with a probability of 68%

$$\pm 1/\sqrt{10} \; N \text{ (for a short count time)}$$
$$\pm 1/\sqrt{40} \; N \text{ (for a long count time)}$$

where N is the number of collisions.

A neutron probe is calibrated by defining the relationship between the uncorrected measurement of the number of collisions N_b, given by the apparatus (or N_e after correction) and the soil moisture content per volume unit, HV. For constant dry density and in the case of a SOLO 25 probe, the calibration curve is a straight line: $N = a \times HV + b$, where a and b are soil characteristic parameters.

Two methods have been used to calibrate our neutron probe: calibration by the gravimetric method and calibration by soil neutronic analysis.

The principle of calibration by the gravimetric method is to find a relation between neutron counts made on site using the access tube and the moisture content per unit volume found in samples taken from a pit close to the tube. Undisturbed soil samples taken at several depths in the pit are used to determine the 'moist' density ρh, the 'dry' density ρs, the moisture content by weight W given by the equation $\rho s = \rho h/(1 + W)$.

The method of calibration by neutron analysis was developed at the Centre d'Etude Nucléaire de Cadarache. This method consists of the direct measurement of soil neutron parameters, particularly thermoneutron absorption (Σa) and diffusion (Σb) characteristics of the soils, which determine the response of the neutron probe.

Results of field investigations

Pedological data

Water pressure and hydraulic conductivity (K) vary widely and non-linearly with soil moisture content for different soil textures (Saxton et al., 1986). Experience has shown that the soil texture is the predominant factor in determining the hydraulic characteristics of most agricultural soils. The texture data can be estimated by the simple analysis of soil samples and could readily serve to describe the soil profile and to attribute soil water characteristics to each layer.

The main analysis carried out at the Terrats site allows the establishment of the sedimentological and mineralogical characteristics of the layers, which form the unsaturated zone. The soil profile is characteristic of a Brunisol mesosature (saturation rate between 50 and 80%) with an important coarse fraction (100 to 2 mm: 50%) and divided into three main horizons (Figure 16.8).

— At the surface, a cultural or ploughed L horizon (new pedological reference base, Baize and Girard, 1992) with an important coarse fraction (100 to 2 mm: 60%) down to a depth of 60 cm.
— A structural S horizon, characteristic of a Brunisol, from 60 cm to 1.2 m, slightly more clayey and sandy. Figure 16.8 shows the decrease of the coarse gravel (100 to 20 mm) and fine gravel (20 to 2 mm) percentage from 60 to 40%, an increase of coarse and fine sand from 15 to 25%, and an increase of silt (0.02 to 0.002 mm) and clay (<0.002 mm) from 15 to 25%.
— At the bottom, a deep mineral horizon known as the C horizon, from 1.2 to 3 m, different from the parent rock. Indeed the C horizon is marked by geochemical weathering and does not represent a textural discontinuity. The mean granulometry is: coarse gravel 5%, fine gravel 37%, coarse sand 31%, fine sand 2%, silt 8% and clay 17%.

Hydrologic and hydric data

The following measurements have been made at the Terrats experimental site:

— precipitation,
— soil moisture content,
— water storage in the unsaturated zone,
— hydraulic head.

With 983 mm in Terrats (1009 mm in Perpignan), the year 1992 was one of the most rainy since 1965 because of two main events, in January (from 22 to 24 January: 171 mm) and in September (26 September: 202 mm). In 1993, the precipitation reached 732 mm in Terrats and 670 mm in Perpignan. This is slightly higher than the mean annual

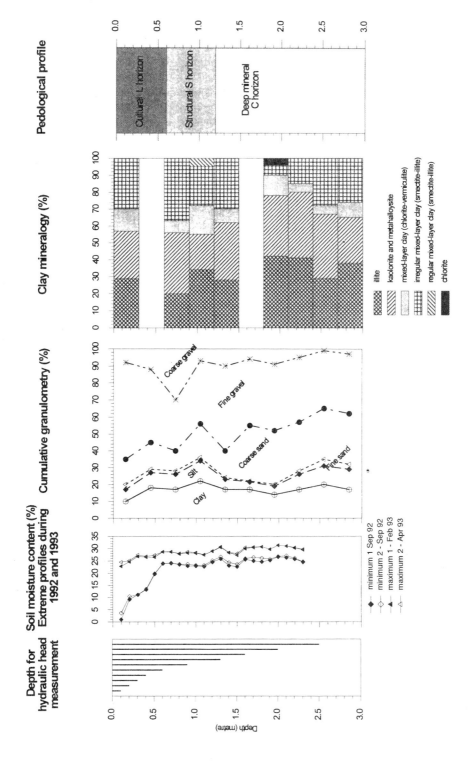

FIGURE 16.8 Synthesis of pedological and hydrological data for the unsaturated zone (Terrats experimental site)

precipitation in Perpignan (578 mm). Generally, the rainy months are February, April and September to November.

For soil moisture content, Figure 16.9 shows a superficial layer where higher variations occur: between 3% (soil moisture content at the surface at the end of summer) and 25% (soil moisture content at the surface in winter). See also the extreme profiles in Figure 16.8. The variations for the lower layers are less important: between 24% (soil moisture content at the end of summer) and 31% (soil moisture content in winter).

The water storage variations in the unsaturated zone are similar for each specific layer and are more important for the superficial horizon. The upper level is settled during winter and spring (October to June) and the lower level at the end of summer. The available storage in the soil is calculated by adding up the available storage in each horizon down to 4 m, the maximum rooting depth for vines. The estimated values are:

RU [L horizon, 0 to 0.6 m] = 80 mm
RU [S and C horizons, 0.6 to 2.3 m] = 85 mm
RU [unsaturated zone, 0 to 4 m] = 280 mm

Figure 16.10 shows the great sensitivity of the hydraulic head to a rainfall event. The time series of the hydraulic head gradient shows the periods of recharge, from October to May. During these periods the hydraulic head gradient is positive, i.e. the water movements are essentially downward. By contrast, during summer (June to September) when the hydraulic head gradient is negative, the water movements are essentially upward (evapotranspiration) between the surface and the depth of 2 m.

Determination of soil water characteristic relationships
The relationship between water pressure (h) and the moisture (θ) fitted to experimental pairs of h and θ is given by Brutsaert (1966) as a homographic law (Figure 16.11).

$$h = h_t \left(\frac{\theta_s - \theta_r}{\theta - \theta_s} \right)^{b_t}$$

where θ = volumetric moisture (without dimension) or (no dimension), θ_s = volumetric moisture at saturation (without dimension), θ_r = residual volumetric moisture (without dimension), h = water pressure (or -suction) (m), h_t = water pressure at half saturation (m), b_t = exponent (without dimension).

The relationship between unsaturated hydraulic conductivity (K) and moisture content (θ) fitted to experimental pairs of K and θ is given by Brooks and Corey (1964) as a polynomial law (Figure 16.11).

$$K = K_s \left(\frac{\theta - \theta_r}{\theta_s - \theta_r} \right)^{b_k}$$

where θ = volumetric moisture (without dimension), θ_s = volumetric moisture at saturation (without dimension), θ_r = residual volumetric moisture (without dimension),

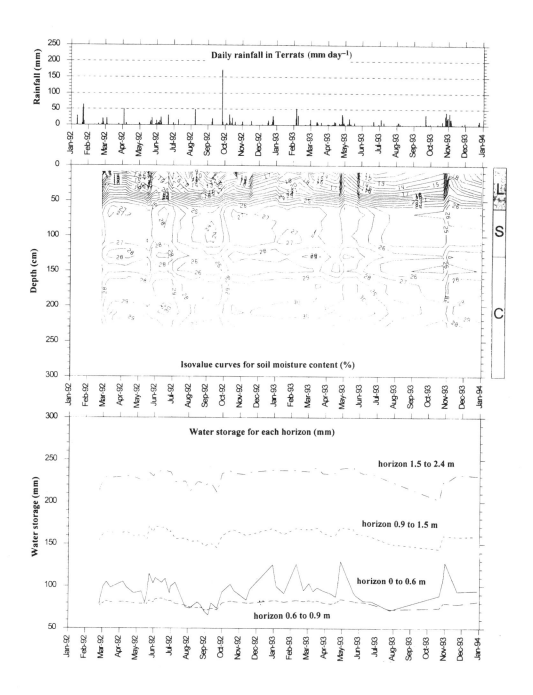

FIGURE 16.9   Time series of precipitation, change in soil moisture content and water storage in the unsaturated zone during 1992 and 1993 at the Terrats experimental site

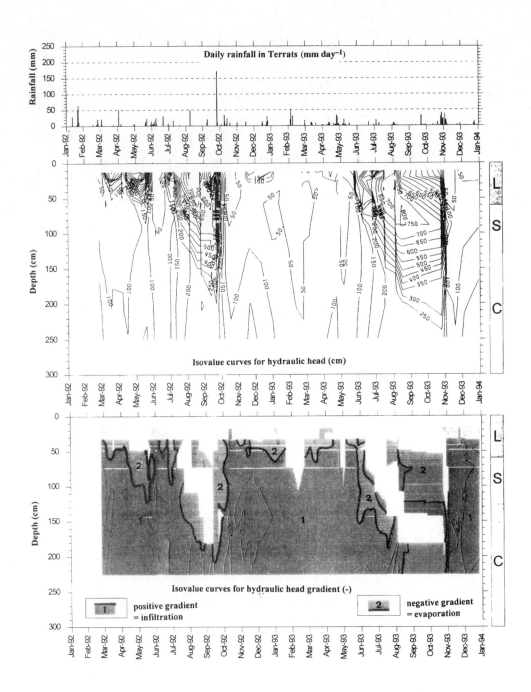

FIGURE 16.10 Time series of precipitation and the change in hydraulic head and hydraulic head gradient in the unsaturated zone during 1992 and 1993 at the Terrats experimental site

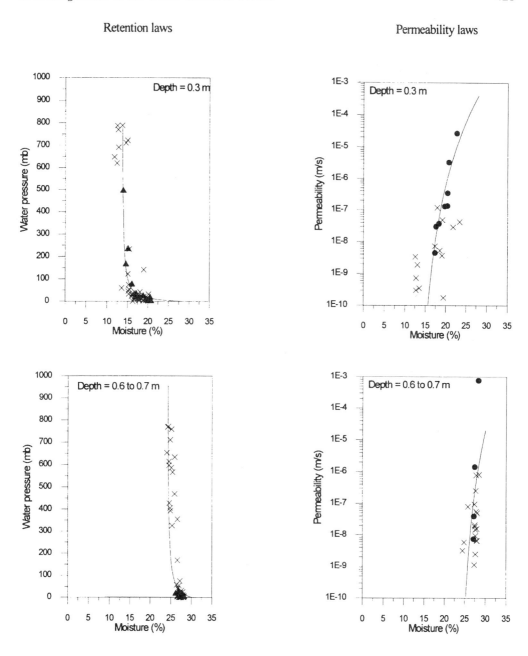

FIGURE 16.11 Some examples of relationships which characterize the different layers of the unsaturated zone (Terrats experimental site)

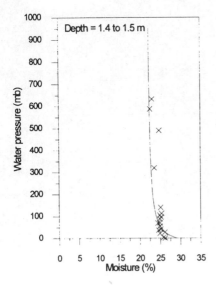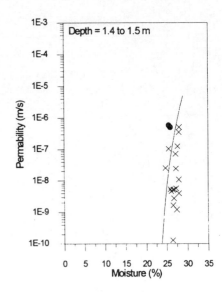

FIGURE 16.11   (*Continued*)

K = unsaturated hydraulic conductivity (m), K_s = unsaturated hydraulic conductivity at saturation (m), b_k = exponent (without dimension).

The characteristic relationships was determined from pairs of experimental data from measurements in natural conditions during the study period 1992-1993 (one measurement every fifteen days) and by measurement during internal drainage experiments (artificial recharge with a water depth about 200 mm).

Approach of hydrological balance by unsaturated zone modelling

Water balance for vineyard (INRA: Riou, 1994)
Water balance terms for the Terrats experimental site (vineyard) are presented for hydrologic years 1992 and 1993 in Table 16.5.

The water storage is never zero and is at a maximum during 4 to 6 months of the year. Recharge represents 33% of the 1992 precipitation and 22.6% of the 1993 precipitation. Recharge occurred from January 1992 to March 1992 and from October 1992 to April 1993 (Figure 16.12).

Lumped model GARDENSOL (Thiery, 1986)
The lumped model, GARDENSOL, simulates the water storage in the unsaturated zone and the piezometric variations. It can be calibrated simultaneously on these two series for a determinate period and can be used by extending data to analyse the resource fluctuations according to climatic variability. It allows us also to calculate hydric balance and especially recharge (Figure 16.12).

Deterministic model MARTHE (Thiery, 1990, 1994)
The deterministic model MARTHE ('Modélisation d'Aquifère par un maillage Rectangulaire en régime Transitoire pour le calcul Hydrodynamique des Ecoulements'),

TABLE 16.5   Water balance for vineyard calculated for the Terrats experimental site

Date	R (mm)	PET (mm)	AET (mm)	RECHARGE (mm)	AS (mm)
Dec-91					280.0
Jan-92	166.2	17.3	17.3	148.9	280.0
Feb-92	41.7	40.3	11.6	30.1	280.0
Mar-92	30.9	95.2	19.0	11.9	280.0
Apr-92	57.8	122.2	69.6	0.0	268.2
May-92	65.0	136.9	89.2	0.0	244.0
Jun-92	116.4	124.8	110.8	0.0	249.6
Jul-92	28.7	162.8	96.5	0.0	181.8
Aug-92	51.5	138.6	90.4	0.0	142.9
Sep-92	213.5	93.7	84.9	0.0	271.5
Oct-92	102.3	47.0	31.0	62.8	280.0
Nov-92	25.2	30.2	5.1	20.1	280.0
Dec-92	84.0	16.1	8.7	75.3	280.0
Jan-93	2.7	19.8	0.3	2.4	280.0
Feb-93	107.1	38.8	29.7	77.4	280.0
Mar-93	59.1	64.6	24.6	34.5	280.0
Apr-93	123.1	82.2	71.9	51.2	280.0
May-93	51.8	124.0	74.5	0.0	257.3
Jun-93	13.9	177.2	96.8	0.0	174.4
Jul-93	29.8	203.2	114.3	0.0	89.9
Aug-93	27.2	170.0	55.8	0.0	61.3
Sep-93	47.1	100.9	32.2	0.0	76.2
Oct-93	147.6	66.7	63.5	0.0	160.3
Nov-93	99.0	31.4	20.7	0.0	238.6
Dec-93	23.2	45.0	6.7	0.0	255.1
TOTAL 1992	983.2	1025.1	634.1	349.1	
TOTAL 1993	731.6	1123.8	591.1	165.5	

R = rainfall.
PET = potential evapotranspiration.
AET = actual evapotranspiration.
$RECHARGE$ = water surplus below 4 m (unsaturated zone depth that we have studied and maximum depth for vine roots).
AS = available storage in the soil (maximum available storage = MAS = 280 mm and readily available storage = RAS = 2/3MAS = 187 mm).

developed by the BRGM, is used for the modelling of a hydrodynamic system of groundwater. In a finite difference three-dimensional model, there are as many parameters to determine for each grid cell in the model (Figure 16.2). The calibration consists of determining the permeability coefficient, storage coefficient and the recharge. The parameters are estimated using a trial-and-error approach to fit the computed head gradients to the observed ones. In the case of the unsaturated zone study, the MARTHE model can also reproduce the vertical water movements. Necessary data are the soil parameters which defined the characteristic relationships. The calibration consists of adjusting these soil parameters to minimize the differences between simulated and observed moisture profiles. The simulation achieved on the basis of the measurement of

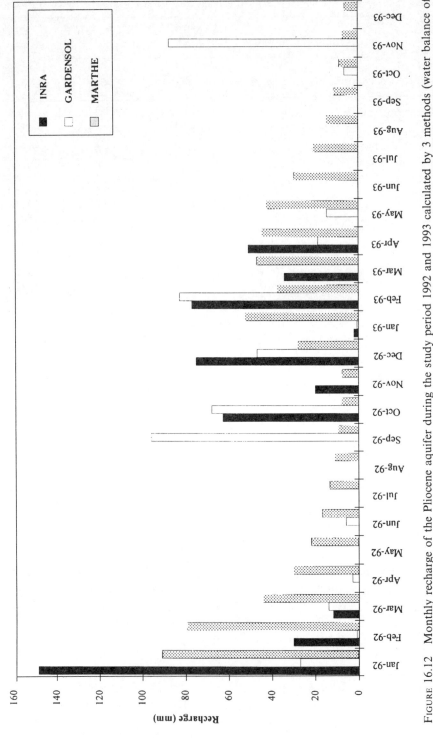

FIGURE 16.12   Monthly recharge of the Pliocene aquifer during the study period 1992 and 1993 calculated by 3 methods (water balance of INRA for a vineyard, simulation with the GARDENSOL model and simulation with the MARTHE model)

soil moisture content and precipitation, under in natural conditions, gives the water balance for the study period 1992 and 1993 (Figure 16.12).

Regionalization of data and production of a recharge map for Roussillon

The unsaturated zone study has provided the necessary data to calculate the recharge in the specific cultural (vineyard), pedological and topographic (slope = 0) conditions of the Terrats experimental site. To extrapolate from the local to the regional scale the solution is to combine the results of modelling (the calculation of runoff and recharge in different condition of slope) with maps of precipitation and agronomic, pedologic and topographic parameters. The final result is a recharge map for the Roussillon multilayer aquifer.

16.4.5 Results of simulations for the Roussillon aquifer model

The present conditions

The behaviour of the Roussillon plain multilayer aquifer (Pyrénées-Orientales) has been analysed and simulated with the hydrodynamic model, MARTHE. Simply, the system consists of three main superimposed aquifers: the alluvial or superficial Aquifer 1, the confined Pliocene Salanque, Aquifer 2, and the deep confined Pliocene Aquifer 3. These aquifers are separated by semi-permeable layers.

The calibration of the model under steady-state conditions (constant flow with potential not changing with time) consists of defining permeability fields and effective rainfall to adjust simulated and observed piezometric levels of the three main aquifers.

— The map of the simulated mean annual piezometry shows negative values in the northeast sector (below sea-level) for Aquifers 2 and 3, created by the high pumping rate in summer.
— The groundwater balance shows that the aquifer system is essentially supplied by rain infiltration and irrigation surplus (66 Mm^3 yr^{-1}). The leakage between aquifers is not negligible at 4 Mm^3 yr^{-1}. The recharge by calcareous rocks bordering the Roussillon plain is estimated to be 3 Mm^3 yr^{-1}. Water supplied for collective, agricultural and industrial use is 35 Mm^3 yr^{-1}. The outflow to the sea is estimated at 20 Mm^3 yr^{-1} and to rivers at 14 Mm^3 yr^{-1}.

The calibration of the model under transient conditions is based on permeability and storage coefficient fields and monthly variations in recharge and pumping. Several consecutive years have been simulated in order to obtain satisfactory piezometric variations compared with measurements at 15 observation wells representative of Aquifers 1, 2 and 3.

The simulated maps show:

— a low water period during summer especially at the coast, when piezometric heads are reduced by the significantly increased pumping,
— a high water level in March which is a period of maximum recharge and minimum pumping.

The case of increased water supply

The increase of water abstraction for collective use from the Pliocene aquifer between 1993 and 2010 has been simulated, for three scenarios (low, high and maximum rates) (Marchal, 1995). The water abstraction increases from 56% (low rate scenario) to 103% (maximum rate scenario).

After 6 years of simulation, the increase in water abstraction gives the following results.

—   For the zones where water abstraction is already low, an increase in abstraction will lower the piezometric level of Aquifer 3 by 4 m (low rate scenario) and by 6 m (maximum rate scenario) in the west sector. The piezometric level of Aquifer 2 decreases by 3 m (low rate scenario) and by 4 m (maximum rate scenario) in the north sector.
—   For the zones where the water abstraction rate is already high, near the coast, the piezometric level of Aquifer 3 decreases by 10 m (low rate scenario) and 20 m (maximum rate scenario) in the south sector (Elne) and by 6 m (low rate scenario) and by 16 m in the north sector (Barcarès). The piezometric level of Aquifer 2 (Barcarès) decreases by 4 m (low rate scenario) and by 7 m (maximum rate scenario).
—   The piezometric head of the coastal aquifers will become negative after 1 year in Aquifer 2 and after 1 to 3 years in Aquifer 3 (according to the scenario used).
—   This depression zone will extend to the west.
—   The water-table of the alluvial aquifer will not vary significantly in comparison with the present level. The decrease is between 0.5 m (low rate scenario) and 1 m (maximum rate scenario).
—   The leakage between aquifers will decrease.

The case of climate change

The impact of a 10 to 20% variation in effective (i.e. infiltrating) rainfall was also simulated and gives the following results after 10 years.

—   A decrease of 10% in the effective rainfall will lower the piezometric level by 3 m in the Quaternary alluvial groundwater (Aquifer 1), less than 1 m in the Salanque groundwater (Aquifer 2) and 2 m in the deep Pliocene groundwater (Aquifer 3).
—   A decrease of 20% in the effective rainfall will cause a piezometric level decrease of 7 m in Aquifer 1, of 2 m in Aquifer 2 and of more than 3 m in Aquifer 3.

All the results of these various simulations demonstrate the sensitivity of coastal groundwaters where the piezometric head will decrease to below sea level. If this were to happen, the coastal aquifer will be more vulnerable to contamination by saline intrusion especially if there is both an increase in pumping and a decrease in rainfall.

16.5 GUIDELINES FOR OPTIMUM FUTURE GROUNDWATER MANAGEMENT AND ARTIFICIAL RECHARGE OF AQUIFERS

Faced with the problems mentioned in the preceding sections, recent and present procedures for public groundwater management have turned out to be poorly effective.

Many such practices are passive, based on limiting water abstraction through taxation levels that are meant to be dissuasive.

The combination of individual access to groundwater, scattered well locations and the independent character of most Mediterranean farmers, does not favour the creation of collective management structures, as already exist for surface-water use. Users of the latter resource manage their field perfectly and know how to take initiatives, in particular for creating storage capacity for water commonly, but erroneously, called a new resource. They are in the habit of launching major projects that are heavily subsidized, making water into a marketable product whose users are captive customers.

At the same time, the difficulty of public groundwater management is easily understood. Public authorities are faced with a bewildering dispersion of economic entities, and developers are not motivated by the expectation of profits. Most groundwater users vaguely hope for God or the state to replenish their reservoirs. However, as public authorities have not yet mastered the technique of rain-making, they try to create equal water shares for all through regulation.

One could imagine that economic regulation would be the result of the damage caused by overexploitation, but the drying-up of aquifers is a slow process, whose unwelcome effects seem too distant to be dissuasive. The usual answer is to deepen the holes and pump a little more.

However, the motivation and might of state support should go beyond the harmonization of agricultural activities. They must also be based on concepts of management and priority protection of drinking-water resources, as well as on the good health of the hydrological system. If, locally in the Mediterranean area, certain groups of irrigation farmers mortgage the future of their primary tool, they effectively put the whole hydrological system in peril.

It is thus necessary to promote the active management of groundwater resources in parallel to that of surface water resources: in order to be able to keep on pumping large amounts of water from aquifers, it will be necessary to modify the hydrological cycle ourselves and help natural recharge with artificial recharge. In addition, we can learn from observations that are easily made of peoples that practise water economy through necessity. Together, recharge and restraint will help us to define a durable water resource development.

Natural recharge of aquifers, the combination of concentration, spreading and infiltration processes, will certainly remain the mode of replenishment least affected by climatic changes. It will thus be necessary to favour this process and develop it for appropriate use. This will help avoid the use of too many artificial-recharge points through boreholes and basins, which require additional expense to counter their clogging up with silt. This means that surfaces with suitable infiltration properties must be identified and set aside for intensified infiltration of water collected in catchment areas for this purpose. Such recharge can be qualified as a 'soft' technology, as it does not ask for a service that greatly surpasses environmental capacity.

As the use of rural space is becoming increasingly specialized, the spaces set aside for recharge can be woodland or pasture ground, provided that good-capacity aquifers lie below such ground. Some slight development work, such as low embankments, dykes and levees, can be patterned on traditional designs, but executed with modern methods. Such structures will serve to combat soil erosion, while helping recharge. It is proposed that this process be called 'soft artificial recharge'.

Both reality as observed today and vestiges from the past indicate that this idea is feasible. However, it will be necessary to adopt an innovative attitude for spreading costs as well as benefits. This will have the effect that all, rather than a happy few, users of the hydrological system will benefit from such recharge. As an example, the losses and leaks from our seasonal storage will, somewhat later, benefit flow in streams and rivers.

Modelling is the best tool for studying the feasibility of such systems. In addition such modelling can be intimately coupled to other computer-based tools for help in decision-making.

To reach an economic dimension that will satisfy future needs for the storage of literally billions of cubic metres, in view of the large area and the many countries involved, the modelling method must be applicable to a great variety of situations. At present, a multi-data-set mapping method is under development, which will serve to identify favourable aquifers and their associated infiltration surfaces.

The management of groundwater is inexorably condemned to an increasing level of artificial intervention; at the same time as this technical evolution takes place a social evolution can be predicted, for the simple reason that such practices can no longer be individual. Both human activities and environmental preservation will have to come under the umbrella of such development schemes.

16.6 REFERENCES

Baize, D. and Girard, M.C. 1992. *Référentiel pédologique, Principaux sols d'Europe*. INRA Techniques et Pratiques, 222 pp.

Brooks, R.H. and Corey, A.T. 1964. *Hydraulic Properties of Porous Media*. Hydrology Paper No. 3, Colorado State University, Fort Collins, 27 pp.

Brutsaert, W. 1966. Probability laws for poresize distributions. *Soil Science*, **101**, 85–92.

Castany, G. and Margat, J. 1977. *Dictionnaire français d'hydrogéologie*. BRGM, Service Géologique National, 249 pp.

Chabart, M. 1995. La recharge de l'aquifere multicouche du Roussillon et les consequences d'un eventuel changement climatique sur la gestion de la ressource en eau (Pyrénées-Orientales). Thèsè de 3ème cycle, Univ. P et M. Cuve, Paris, 300 pp.

Clauzon, G. 1989. Un exemple de régulation accélérée d'une côte à rias: Le littoral Méditerranéen Français au Pliocène inférieur. *Bull. Cent. Géomorphol.* Caen, No. 36, 239–242.

Clauzon, G., Aguilar, J.P. and Michaux, J. 1987. Le bassin pliocène du Roussillon (Pyrénées-Orientales, France): exemple d'évolution géodynamique d'une ria Méditerranéenne succédant à la crise de salinité messinienne. *C. R. Acad. Sci., Paris, II*, **304**, 585–590.

Clauzon G., Aguilar, J.P. and Michaux, J. 1989. Relation temps–sédimentation dans le Néogène méditerranéen français. *Bull. Soc. Géol. Fr.*, **8**, 361–372.

Collectif 1992. Déclaration de l'Association Internationale des Hydrogéologues (AIH) (à la conférence internationale sur l'eau et l'environnement, Dublin, janvier 1992). *Hydrogéologie,* **3**, 187–190.

Custodio, E. 1991. Characterisation of aquifer over-exploitation: comments on hydrogeological and hydrochemical aspects: the situation in Spain. *XXIIIe Congress AIH. Aquifer Overexploitation*. Puerto de la Cruz, Canary Islands, Spain, 3–19.

El Magnouni, S. and Treichel, W. 1994. Multicriterium approach to groundwater management. *Water Res. Res.*, **30** (6), 1881–1895.

Gadel, F. 1966. Contribution à l'étude géologique et à l'hydrogéologie des Corbières Orientales (région Est) et des plaines de Rivesaltes Lapalme-Caves et Sigean. Thèse de 3ème cycle, Univ. Montpellier, 289 pp.

JO—CE 1992. Journal Official de la Communanté Europénne number L247, pp. 10–36.

Llamas, R. 1991. Groundwater exploitation and conservation of aquatic ecosystems. *XXIIIe Congress AIH. Aquifer Overexploitation.* Puerto de la Cruz, Canary Islands, Spain, 115–131.

Llamas, R. 1992. La surexploitation des aquifères: aspects techniques et institutionnels. *Hydrogeologie,* **4**, 139–144.

Lopez-Camacho y Camacho, B. and Sanchez-Gonzalez, A. 1991. Unidades hidrogeologicas con problemas o riesgos de sobreexploitation (territorio peninsular e islas Baleares). *XXIIIe Congress AIH. Aquifer Overexploitation.* Puerto de la Cruz, Canary Islands, Spain, 539–544.

Marchal, J.P. 1985. *Synthèse hydrogéologique de la région Languedoc Roussillon.* Rapport BRGM 85 SGN 349 LRO.

Marchal, J.P. 1995. *Simulations d'exploitation complémentaire de l'aquifère multicouche du Roussillon pour assurer les besoins en eau à l'horizon 2010.* Rapport BRGM N1878, 19 pp.

Margat, J. 1992. L'eau dans la bassin Mediterranéen. *Les fascicules du Plan Bleu. Economica.*

Merillon, Y. and Roux, A. 1992. Evolution des besoins en eau agricole. Maîtrise de la demande suite aux sécheresses de 1989, 90 et 91. *XXIIe Journées de l'hydraulique. Société Hydrotechnique de France,* Paris. 15–17 Sept. 1992. Question No. II.

Navarette, P., Del Castillo Gonzales, L.I., Monserrat i Rebult, F.X. and Lopez-Geta, J.A. 1992. The socio-economic impact of marine intrusion as a result of overpumping aquifers on the Spanish coast. In Lumsden, G. (ed.), *Geology and the Environment,* Clarendon Press, Oxford, 268–271.

Palutikof, J.P. 1993. Mediterranean temperature and precipitation trends and assessment of the field site data; extreme precipitation events; and development of GCM based scenarios. *Final Report, MEDALUS I,* University of East Anglia, Norwich, UK.

Palutikof, J.P. 1994. Implications of greenhouse effect for extreme climatic events and agricultural and water impacts. *First Annual Report 1993, MEDALUS II, Project 2,* University of East Anglia, Norwich, UK, 110–124

Palutikof, J.P., Guo, X., Wigley, T. M. L. and Gregory, J. M. 1992. *Regional Changes in Climate in the Mediterranean Basin due to Global Greenhouse Gas Warming.* UNEP Mediterranean Action Plan Technical Reports Series, No. 66, UNEP, Athens, 172 pp.

Palutikof, J.P., Goodess, C. M. and Guo, X. 1993. Climate change, potential evapotranspiration and moisture availability in the Mediterranean Basin. Climatic Research Unit, School of Environmental Sciences, University of East Anglia, Norwich, U.K. (unpublished).

Pulido-Bosch, A. 1991. The overexploitation of some karstic aquifers in the province of Alicante, Spain. *XXIIIe Congress AIH. Aquifer Overexploitation.* Puerto de la Cruz, Canary Islands, Spain, 3–19.

Riou, C. 1994. *Le déterminisme climatique de la maturation du raisin: Application au zonage de la teneur en sucre dans la communauté européenne.* Centre commun de recherche, Commission Européenne, EUR 15863 FR/EN.

Roquero, E. and Luque, M. 1992. Méthodologie pour l'estimation à moyen et long terme de l'usage de l'eau pour l'irrigation et la production électrique. *XXIIe Journées de l'hydraulique. Société Hydrotechnique de France,* Paris. 15–17 Sept. 1992. Question No. II.

Saxton, K.E., Rawls, W.J., Romberger, J.S. and Papendick, R.I. 1986. Estimating generalized soil-water characteristics from texture. *Soil Sci. Soc. Am J.,* **50**, 1031–1036.

Thauvin, J.P. 1986. Etude hydrogéologique, modélisation et gestion des aquifères du Campo de Dalias, Province d'Alméria, Espagne. Thèse sciences, Nice, France.

Thiery, D. 1986. *Un modèle hydrologique semi-global pluie–zone non saturée–nappe: le modèle GARDENSOL. Trois exemples d'application.* BRGM, note technique 86 22 EAU, 33 pp.

Thiery, D. 1990. Modèle MARTHE: *Modélisation d'Aquifère par un maillage Rectangulaire en régime Transitoire pour le calcul Hydrodynamique des Ecoulements—version 4.3.* Rapport BRGM R32210 EAU 4S 90.

Thiery, D. 1994. *Modélisation 3D des écoulements en Zone Non Saturée avec le logiciel MARTHE, version 5.4.* Rapport BRGM R 38 108 HYT/DR/94, 110 pp.

17

Dynamic Modelling of Complex Systems

F. Pérez-Trejo

UNITAR, Geneva, Switzerland

and

N. Clark

Science Policy Research Unit, University of Sussex, Brighton, UK

17.1 INTRODUCTION

Desertification is the kind of environmental problem which challenges our most advanced scientific know-how. The difficulty in detecting and combating desertification is related to a key characteristic of the phenomenon, namely that it is the result of changes in the structural properties of semi-arid regions which can be irreversible. A major problem in desertification policy analysis is that of modelling the linkages between economic systems and their environmental effects in such a way that the models reflect the underlying dynamics of change. The normal procedure is to concentrate on some limited sector or aspect of the problem, and therefore to specify a number of interrelationships among key variables. Such models have limited usefulness because they fail to address in an explicit way the interconnectedness of the different components of the system that is being affected by desertification.

However, what are less well recognized perhaps are underlying methodological features which have the effect of detracting from the capacity of such models to capture evolutionary behaviour. The main reason for this is that the models are mechanical representations of past average behaviour of physical, ecological or economic agents and do not therefore capture the local-level fluctuations which give systems their creativity. The main purpose of this chapter is to present a complex system modelling scheme which explicitly attempts to capture the interrelationships between processes at the local level and macro-level behaviour in terms of patterns, and in so doing provides a potentially powerful tool for understanding the nature of the changes which can bring about desertification. The modelling example illustrates how a process-based dynamic framework can serve as a means for understanding the linkages between socio-economic dynamics and the process of desertification.

Mediterranean Desertification and Land Use. Edited by C. Jane Brandt and John B. Thornes.
© 1996 by John Wiley & Sons, Ltd.

An important property of the model is its spatial dimension—an element usually not present in conventional system modelling, which tends to concentrate exclusively on the behaviour of aggregate variables over time. However, it is precisely where environmental considerations become important that spatial impacts have to be considered, since what the policy analyst often needs to predict are the future consequences, upon specific regions and zones, resulting from interventions or land-use changes in previous time periods, interventions which need not have taken place in the zones where the greatest impacts will be felt. Given the growing inter-relatedness of modern economic systems, there are, we should argue, very strong grounds for thinking spatially about patterns of change and degradation in the landscape.

The core model has been developed as a pilot study for the island of Crete, which will enable us to explore, within an insular subregion, the socio-economic processes which lead to land degradation and help in determining to what extent socio-economic change has been the culprit in surpassing the ecological limitations of these systems. This chapter is aimed at showing how such a framework could be used to study the interactions between human and climatically induced land degradation so as to develop decision tools which could be applied across the whole of the Mediterranean Basin.

The following section of the chapter outlines the broad framework of how the model has been applied to Crete and the kinds of preliminary results that have been achieved. It then proceeds to set out in some detail the model's logic in terms of its mathematical structure and core assumptions, and it provides a qualitative discussion of the model's properties and its use as a decision tool for policy interventions. Section 17.5 summarizes the chapter's broad conclusions. Reference should be made also to Pérez-Trejo et al. (1993) which presents a more general case for the type of modelling scheme used, in contrast with more traditional approaches to development planning. The main message of this earlier paper was the need to 'endogenize' environmental factors in policy research of this kind and not to rely simply upon 'pricing' the environment as a kind of optional extra to traditional models. The ensuing discussion, hopefully, will explain what we mean by this statement.

17.2 THE CRETE MODEL

The aim of this section is to present a spatial, dynamic modelling framework as a tool for exploring the effects that different development scenarios could have on the environmental degradation that can result from land-use change driven by short-term economic incentives. The modelling framework consists of several sub-models (see Figure 17.1) in which the central element is a spatial model of the socio-economic dynamics of Crete. The model gives spatial dimensionality to the space available for any given economic activity, the choice of residence or migration of the population, and most importantly to the choices that the different sectors of the economy or consumers make in purchasing goods and services locally, from other regions, or from the international markets; depending on location, price and choice of supplier. Essentially the central model consists of two sets of actors (or agents) viz: business firms which invest in different regions in response to economic signals, and households which migrate between regions, again largely for economic reasons. As each set of agents vary their behaviour they achieve impacts on regions which alter the behaviour of the other set in a continuous cycle of activity. For

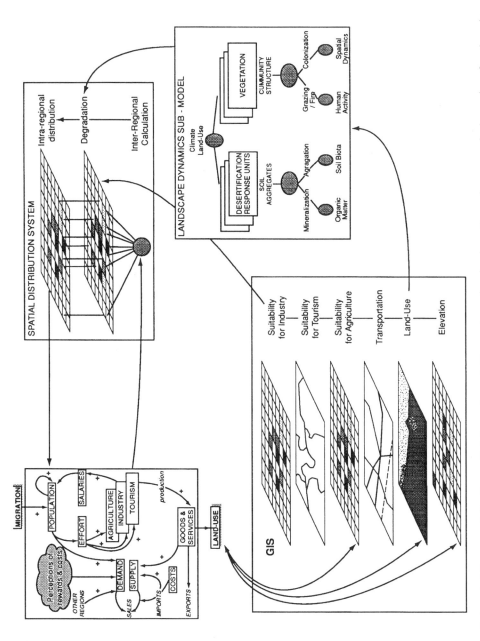

FIGURE 17.1   Integrated framework for the spatial dynamic modelling system: showing how the socio-economic dynamics of each region affect land-use patterns, which are given spatial reference using a set of suitabilities for each sector of the economy. These changes are then evaluated in terms of their effect on landscape dynamics and land degradation, feeding back into the long-term productivities in the socio-economic model

example, the decision of households to migrate from region *i* to region *j* in response, say, to improved income earning opportunities will alter relative economic conditions between the two regions which in turn will alter relative investment behaviour of firms (Figure 17.2). If as a result investment increases in *j*, then there will be further incentives favouring migration from *i* to *j* and (possibly) further investment. Such a positive feedback loop may well continue until some other factor (say environmental stress) turns the balance of advantage in a different direction—and so on.

In this way the socio-economic model is capable of simulating the structural changes which result in fundamental transformations in the way in which people earn their living and the very different economic environments in which the dynamics of economic activity unfold. These, in turn, generate changes in employment opportunities, and land-use changes which are then analysed and displayed in map form by the graphical display component of the modelling framework shown in Figure 17.1.

Land-use changes can have diverse effects on the different landscapes where the changes are taking place. The actual impact of land use depends on many factors, such as the particular land-use history of each landscape, the soil types, the current land-use practices, and their physical, ecological and climatological characteristics. The potential degradation effects are evaluated using the concept of desertification response units (DRUs) (Imeson, Chapter 18) to explore the possible response of the soils and land units to the changes in land use generated by the socio-economic model. The landscape response model is used to explore the possible restructuring of the DRUs to provide representation of what the soil structural properties might be like under these new land-use changes as expressed by new water re-distribution patterns in the landscape, micro-biological changes in the soils, and patterns of vegetation distribution as a source of organic carbon to the soils.

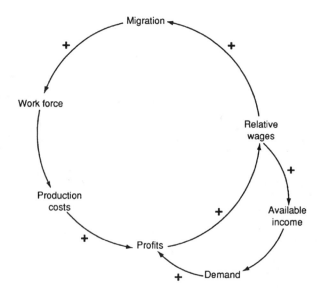

FIGURE 17.2 Causal-loop diagram showing the feedbacks that link population migration, productivity and the relative advantage reflected in profits in each region

These changes in the soil structural characteristics are dynamic indicators of possible degradation which can bring about changes in productivity, in costs of production, and ultimately in the sustainability of the economic activities that these soils support. The spatial, dynamic model for the Crete economy consists of two internal regions: the urban centres, consisting of Heraklion, Chania and Rethymnon; and the rural region of the island. Each of these regions is represented in the model in terms of the amount of area available for any economic activity to take place and grow, the costs that they incur in terms of rent in that region, the distances between them and the costs of transportation. The model also includes the movement of the population from one region to the other, which follows their place of employment. Finally, the model allows for international influences by the inclusion of a 'rest of the world' region whose behaviour is influenced exogenously.

The model is calibrated using census and economic data from Crete to generate the temporal evolution of the system for a period between 1971 and 1981. The results presented in Figure 17.3 show the urban and rural populations generated by the model compared to data from the 1981 census. The model is then used to explore the economic, demographic and environmental dynamics of the system to assess the consequences of future scenarios in the year 2001, and to provide a tool for evaluating development policies in Crete. For example, Figure 17.4 presents the model results for the employment dynamics in agriculture, industry and services for the 1981–2001 period; and Figure 17.5 illustrates a land-use change scenario generated by these economic

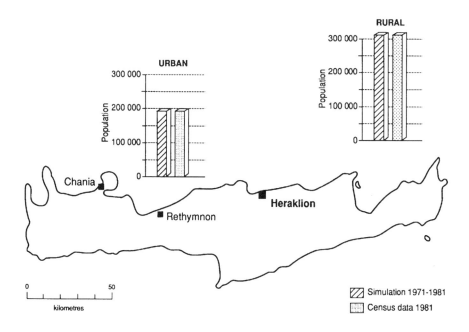

FIGURE 17.3 Comparison of the model results for urban and rural population versus the 1981 census data in Crete

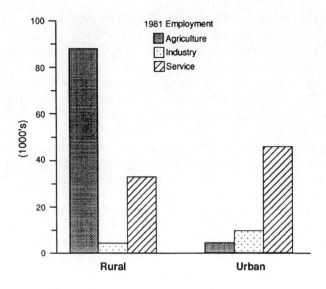

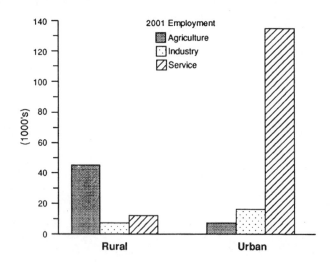

FIGURE 17.4   Results from the simulation of the dynamics of the Crete economy showing the changes in employment in Crete for three economic sectors within the rural and urban areas between 1981 and 2001

dynamics. The interesting outcome of this simulation is the change in agricultural land in the rural areas of Crete which would decrease by almost half, while in the urban region the land in the services sector increases by 2.5 times, mostly due to the development in the tourism sector.

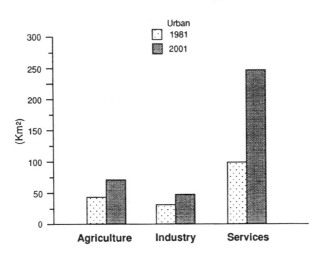

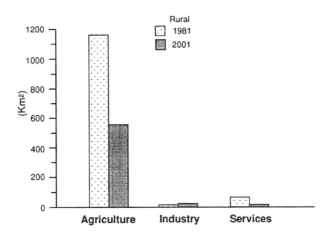

FIGURE 17.5 Results from the simulation of the dynamics of land-use changes in Crete for three economic sectors within the rural and urban areas between 1981 and 2001

17.3 THE CORE MODEL

The Crete model illustrates the way in which the complex systems approach explicitly represents the spatial dynamics of sectorial linkages. The equations presented in the following sections might seem difficult to follow for those without a mathematical background. However these equations should be considered as a formal way of expressing the way in which the firms in each economic sector perceive their 'environment', that is their choice of suppliers, their spatial perception of costs, and how these in turn generate spatial patterns of demand. Similarly, the equations for residential attractivity represent the spatial dynamics of choice of employment and how these generate migration patterns across regions. These equations have been adapted from earlier versions of complex systems models developed for Belgium (Allen and Sanglier, 1979) and for Senegal (IERC, 1991; Allen et al., 1992). A more detailed explanation of the meaning of these mathematical expressions is presented in Allen et al. (1992).

17.3.1 Demand

It is useful to describe the model in detail by starting off with the notion of 'attractivity' which is used to define conditions under which economic agents (consumers, for example) will be induced to alter their behaviour in specific ways. We start by splitting the economic system under consideration into n regions $(1, 2, 3, \ldots, i, j, \ldots, n)$ and m sectors $(1, 2, 3, \ldots, l, \ldots, m)$ and ask the question—what factors will influence the final consumption behaviour of consumers in j with respect to purchases of l from all other regions $i = 1, 2, 3, \ldots, i, \ldots, n$. We then define the attractivity of goods produced in i from the stand-point of consumers in j as:

$$A_{ij}^l = \exp\left(-r*\left(P_i^l + ts^l\, d_{ij}\right)\right) \tag{1}$$

where:

$$-10 < \left(-r*\left(P_i^l + ts^l\, d_{ij}\right)\right) < 20 \tag{2}$$

where 'r' represents a response rationality parameter, $P_i^l =$ price of l in zone i, and $ts^l\, d_{ij} =$ transport costs for a unit of l between i and j.

In order to find the relative attractivity of zone i viewed from zone j we must now sum over all the potential sources of l, to give:

$$SA_i^l = \sum_j A_{ij}^l + \exp\left(-\rho* world\, P_i^l\right) \tag{3}$$

Then, the relative attractivity of sector 'l' in zone i as calculated from j is given by A_{ij}^l/SA_i^l, where *world* $P_i^l =$ world price of sector l and $SA_i^l =$ total economic attractivity of goods produced in sector i.

Total demand falling on sector i will then consist of three components—external demand from overseas, consumption demand and intersectorial demand (which we assume at this stage includes investment demand). External demand is exogenously determined and we call it $extD^l i$. Consumption demand is calculated as the fraction of total disposable incomes falling on zone i and sector l and is given by:

$$\sum_j \left(\beta^l * AIn_j * \frac{A_{ij}^l}{SA_j^l}\right) \tag{4}$$

Where β^l is the fraction of total incomes that is directed to sector l from all points within the system, and AIn_j is the disposable income of consumers in zone j.

Intersectorial demand is given by:

$$Scross_i^l = \sum_{l'} cross^{l'l} * J_i^{l'} * Pry(t)_i^{l'} \tag{5}$$

where $J_i^{l'}$ = number of jobs in sector l' in zone i, $cross^{l'l}$ = the production coefficient for trade between sectors l' and l, and $Pry(t)_i^{l'}$ = productivity of labour in sector l' in zone i.

Labour productivity has been parametrized on the basis of data obtained for each of economic sectors in the model (Chania Statistical Office, 1992).

Total final demand for l falling on region i is then given by:

$$D_i^l = extD_i^l + \sum_j j \left(AIn_j + Scross_i^l * P_i^l\right) * \frac{A_{ij}^l}{SA_j^l} \tag{6}$$

17.3.2 Incomes

Incomes are of two kinds, wages and rents, and are computed as follows. First of all the model calculates an equation which represents the fractional filling of the available space in each region, as a measure of rent costs, including the area used for residential buildings.

$$Rent_i^l = \frac{space_i^l * J_i^l}{area_i^l} \tag{7}$$

where J_i^l = the number of jobs in sector l in zone i, $space_i^l$ = the amount of space required per 1000 jobs in sector l and zone i, and $area_i^l$ = total area available for sector l in zone i.

Next the available income for each of the zones is calculated as a function of total wages earned by the labour force in that region, plus the earnings obtained from renting space to each of the economic sectors. The computational sequence follows by calculating the input costs in each economic sector, in each region. These include the expense of acquiring the goods and services of the intersectorial demand, plus transportation costs for those goods. These payments costs are fractionalized among the regions where the goods could originate from, including the region in consideration. The fractionalization is done according to the 'relative attractivities' calculated above. Total disposable incomes in any zone are thus given by:

$$Ain_i = \sum_l lJ_i^l * \left(wage_i^l + Rent_i^l * space_i^l\right) \tag{8}$$

where wages are given by:

$$\frac{\delta wage_i^l}{\delta t} = 0.1 * wage_i^l * \left(\frac{price_i^l}{cost_i^l} + 0.1 - 1\right) \tag{9}$$

Supply is given by:

$$S_i^l = J_i^l * P_i^l * Pry_i^l \tag{10}$$

17.3.3 Disequilibrium

Having calculated total demand in equation 6 and supply in equation 10, we are now in a position to compare total demand falling on i (in sector l) with the corresponding supply. In this way we can model the change in economic activity which will result from any disequilibrium between supply and demand in terms of a rate of desired expansion and a rate of price change:

$$\frac{\delta P_i^l}{\delta t} = 2 * P_i^L \left(\frac{D_i^l}{S_i^l} - 1 \right) \tag{11}$$

17.3.4 Population Movements

Now let us consider the second set of actors in our scheme—the households who migrate among regions according to perceptions of advantage. The way this is handled is for the model to calculate the number of new jobs in each region according to how well its sectors are generating profits per unit of output and consequently expanding (or reducing) production. The actual assumptions underlying this calculation are quite complicated and will be discussed in detail in future publications. At this stage however, we are able to generate an expansion or contraction of the vacancies (V) that could potentially be filled (or the number of people losing their jobs as a result of reductions in output). This equation relates changes in vacancies as a function of relative sectoral profitability, past trends in the labour market and the structure of the labour market with respect to specific sectors and zones.

If the sector is contracting, then V_i^l is less than 0 and:

$$\frac{\partial V_i^l}{\partial t} = \alpha * J_i^l \left(\frac{P_i^l}{C_i^l} - 1 \right) - \alpha * V_i^l \tag{12}$$

Otherwise the sector is expanding and:

$$\frac{\partial V_i^l}{\partial t} = \alpha * J_i^l \left(\frac{P_i^l}{C_i^l} - 1 \right) * \left[1 + \varepsilon * \sum_{l'} (J_i^{l'} - oldJ_i^l) \right] - \sigma * V_i^l * \left(\frac{0.75 * Pop_i}{\sum_{l'} oldJ_i^l} - 1 \right) \tag{13}$$

where $oldJ_i^l = $ jobs in the previous time period, $Pop_i = $ population in zone i, $\alpha = $ investment response parameter, $\varepsilon, \sigma = $ job history response parameters, and $C_i^l = $ costs, which are given by:

$$C_i^l = \sum_{l'} \sum_{j} Jcross^{l'l} * [P_j^{l'} + ts^{l'} * d_{ij}] * RA_{ji}^{l'} + \frac{wage_i^l + Rent_i * space_i^l}{Pry(t)_i^l} \tag{14}$$

where ts^l = transportation costs for a unit of l between i and j, d_{ij} = distance between i and j, RA_{ji}^l = relative attractivity of goods from i with respect to j, and $Pry(t)_i^l$ = labour productivity in zone i for sector l.

The model then calculates a term called 'residential attractivity' which reflects the relative attractiveness of each region to potential migrants from other regions. This is calculated as a function of the wages earned in a region relative to the wages earned in all other regions.

$$R_i = \exp\left(r_i * \frac{\sum_l (wage_i^l * J_i^l)}{\sum_l J_i^l}\right) \tag{15}$$

Demographic change for each region (in this example, region i) then becomes a computable function of births (b), deaths (m) and migration (Mig) in and out of the region moderated by a mobility parameter (mob).

$$\frac{\partial Pop_i}{\partial t} = (b - m) * Pop_i + Mig_i \tag{16}$$

where b = birth rate, m = death rate, Pop_i = population at i, and Mig_i = migration, which is given by:

$$Mig_i = mob * \sum_{j}\left(0.75 * Pop_j - \sum_l J_j^l\right) * \frac{R_i}{\sum_{i'} R_{i'}} \tag{17}$$

Wages then determine regional income as described above thus feeding back into the model and allowing it to continue to run. The end result is a (simple) economic model which generates continuous temporal data on economic output (by broad sector), jobs, land use and population.

17.3.5 Giving spatial dimensionality to economic activity

Once calibration has been completed and predictions checked with empirical data it is assumed that the model provides a working representation of the underlying dynamics of the Crete economy. However, since it is effectively a two-sector macro-model (i.e. abstracting from the rest of the world), further steps need to be taken to develop it into a useful policy tool. The way this is done is to start by characterizing the different landscapes of the island in terms of 'suitability classes' for each economic sector. Each suitability class is defined in terms of physiography, elevation, soil characteristics and proximity to amenities such as transportation facilities or airports. To calibrate the suitabilities which characterize the island in terms of its economic potential, a land-use map is generated based on the model results obtained for the 1971–81 decade using the suitabilities maps. This generated land-use map was then compared with current land-use patterns in Crete. The suitabilities are then used as a 'knowledge base' to predict where each economic activity will unfold in the future, based upon critical spatial requirements. Each economic sector is given spatial dimension in a real sense, linking changes in land-use patterns generated by the model, to specific locations and environmental attributes.

The desertification response units (DRUs) methodology is used to explore the possible response of the landscape to the changes in land use. The DRUs are of the order of 10–100 ha in area and are characterized in terms of their possible response to different land uses reflected in the structuring of soils in each unit. This is not a mechanical operation. To begin with apparently homogeneous regions are selected and investigated in terms of the response of soil, water redistribution and vegetation patterns (Imeson, Chapter 18). The results of these investigations are then analysed to determine the DRU boundaries with the help of geographic information systems. For each DRU socio-economic impact will affect the natural environment in a defined way, thus affecting economic productivity.

17.4 COMPLEX SYSTEMS MODELS AS DECISION-MAKING TOOLS

Considering the spatial dimension in a direct way is a necessity if we are to increase our understanding of the process of desertification. However, the spatial dynamics of living systems represents one of the biggest challenges that only geographers have dealt with until recently. Even though geographic information systems (GIS) tools have made significant progress in helping to describe the spatial dimension of environmental problems, they still fall short when it comes to addressing the conceptual and methodological aspects of processes which drive the spatial dynamics of systems (Pérez-Trejo, 1993). We have seen in Section 17.2 that the Crete model can generate simulated migration and employment results corresponding quite closely to real data. However, we should caution that the model is still relatively crude and still requires refinement and elaboration. The following discussion should enable readers to assess for themselves the strengths and weaknesses of the modelling approach and to stimulate thinking into how it could be improved.

17.4.1 Calibration

The first important distinction is that between substantive and intermediary variables. The former are variables for which actual data are available and which can be inserted directly into the model at various stages, particularly in defining the initial conditions. These include the rural and urban population of Crete (NOS, 1984), employment in each of the three economic sectors in the model (Chania Statistical Office, 1992; Blue Plan, 1992), available area in each region for each economic activity (Allbaugh, 1953; Kolodny, 1974), and the values for productivity (Chania Statistical Office, 1992).

A significant improvement might be to expand the number of sectors (and/or zones), provided data are available, to see if a richer modelling scheme provides more realistic intermediate quantities. In this way (and others) the model becomes a focus for more detailed research and exploration about the underlying realities of the Cretan economy and its development.

A second distinction is between parameters and variables. Where a system is well defined and structurally therefore relatively invariant, its parameters both define its structure and are capable of computation. Conversely its variables represent those aspects which change as the system functions. However, where the system in question is experiencing structural change (it is evolving) its parameters themselves will vary and hence it may be difficult in practice to determine empirically what is a parameter and what is a variable. In the Crete model parameters (r, α, etc.) are fixed initially with values which are consistent with a good calibration fit over the time period in question. The model presented here represents explicitly the spatial dynamics of changes in land-use patterns that result from structural change. We can describe the changes by counting the number of hotels, or the size of land holdings, or the number of people employed at different dates. However, these results should not be considered predictions of the actual number of job that might exist 20 years from now, but instead the model should be seen as a tool for exploring the many possible futures that could unfold. The spatial dynamics produced by the model are the result of processes such as technological changes, people's spatial perceptions of residential choice or investment, which are the key processes that link different spatial scales: local, national, regional, international. The resulting spatial patterns test the way we hypothesize that the processes might drive spatial dynamics in the future.

In general, trying to make the model re-create the historically observed path is an extremely thought-provoking experience, which in some ways is at the core of the method that we are presenting. Having supposed a 'reasonable' scheme of interacting processes, we must then discover whether realistic parameter values, and imputed exogenous changes really suffice to 'explain' the history of the system. Often, in this process we learn that 'something must have happened' in view of the figures, and later we find that this is indeed the case.

17.4.2 Assessing the effects of development scenarios and the link to desertification

Although we plan to increase the range of regions and economic sectors, the model at present consists of two zones (rural and urban) and three sectors (agriculture, industry and services). This might seem too coarse a resolution for assessing the effects of socio-economic dynamics on desertification. This is overcome by extending the structural changes resulting from the model (e.g. growth or decline in employment in any given sector in a particular region) over the areas where these activities are most likely to occur by means of the suitability maps described above. If, for example, the model shows that tourism is likely to increase over the next 20 years, then it will most likely occur in areas closest to the shore, with high road density, and with topographic characteristics not exceeding 40 m in elevation. This allows us to identify the areas most vulnerable to land degradation and desertification if this development scenario actually unfolds, by considering their possible response in terms of changes in the structure of the soils and water redistribution patterns of the DRUs (Imeson, Chapter 18).

Having described the Crete economy in terms of two regions and three sectors puts rather more strain on exogenous influences (i.e. the rest of the world) than is desirable at this stage although it is arguable that the open economic nature of Crete warrants it. The particular difficulty of course is that there is no real sense in which the external sector would respond to the evolutionary behaviour of the Crete economy so that exogenous influences have to be handled differently. For example tourism is not only beginning to play a growing role in overall economic activity but its impact on the environment is also going to be considerable in years to come. It is supposed that for Europe as a whole development of tourism in the Mediterranean has increased economic 'potential'. With mass air transport, a certain number of 'jobs' became possible in the tourist area. If there is a limited number of hotels or apartments for tourists to stay in, then the numbers that can come are limited. However, if some investment in tourist facilities is made, and the profits are found to be high, then more facilities will be developed.

In the model, we suppose an 'external demand' for tourist services, and that Crete captures a fraction of this depending on its competitiveness compared to that of rivals, and also depending on the amount of tourist space available. In fact, if required, the whole process can be modelled in some detail, with the growing 'attraction' of a destination, as its facilities develop, as well as its reputation among potential customers. Later, however, the activity may use up most of the prime tourist space, and hence begin to suffer from higher rent costs, bringing up prices. As this happens, demand itself may begin to decline, thus leading to a decrease in the relative competitiveness of the sector relative to other tourist destinations in the region.

Demand for hotel beds and accommodation will grow as long as prices are reasonable, and this will depend on the competitiveness of Crete in the market, compared to alternatives offering a similar experience. As long as tourist facilities grow this tends to translate into higher wages. Tourism will pull labour away from traditional activities which could provide a stimulus to agricultural mechanization, although this has not yet been explicitly modelled. The model can capture this whole process as additional tourist developments are begun, and are filled giving rise to employment, and to profits in the sector.

Water consumption of tourist developments is also of major importance, and often results in conflict with the needs of agriculture and industry. We will explore, in the future, how these water consumption dynamics evolve as development scenarios unfold. Other sources of disturbance may also affect the evolution of the tourism sector in Crete. As fashions change other types of holiday may become attractive, or the supply of holiday facilities may outstrip demand, and so lead to a decline in prices, and profitability. Recession may occur in northern Europe and lead to reduced demand. All these scenarios may be investigated using the model, and can be explored together with the resulting changes in land use, demand and supply of goods and services, and in the labour market.

17.4.3 A research methodology for spatially extended systems

Traditionally science has been about explaining a complex reality by means of experiment and validation rules. That is fine provided that reality does not change over the time period under investigation. The problem is, however, that human systems evolve fast and so our conclusion is that the best we can achieve is a limited engagement with rapid change of complex systems.

The model is simply a means of achieving greater understanding. Furthermore, because the people most concerned with the system are actually participating in its functioning, it is very important that they themselves should also be directly involved. This is so not only because they are participants *per se* but also because their particular experiences and knowledge need to be mobilized as part of the research process itself.

This approach makes modelling a more transparent process. What our computerized model does is to provide a means for all concerned agents to engage directly in a process of collective research about their own system's evolution. In a sense the model acts as an integrating node and a language of discourse for a range of people and organizations who are separated by a wide variety of interests and perceptions, including perceptions about the nature of reality itself and how it may be understood. It is this feature, we argue, that gives the approach we use a distinctiveness from older dynamical models of apparently similar kinds.

The approach of this chapter explicitly introduces behaviour through the use of 'attractivities' which change continuously as the system evolves. These attractivities are the mechanism by which small differences in productivity can be amplified through self-reinforcing processes (higher profits generating new job opportunities, attracting labour which improves productivity) and eventually lead to a spatial re-structuring of the system.

17.5 CONCLUSIONS

This chapter argues that the problem of understanding the evolutionary behaviour of desertification is only now being fully recognized, and this is probably because the reality of change, its pace and its potentially destructive consequences are beginning to impinge much more directly on many more people and organizations for the first time. In a very real sense the earth is becoming a 'global village' in which events in any part have irreversible consequences in all other parts. Under these circumstances it is no longer desirable, we believe, to attempt to understand its behaviour through the use of models derived from traditional approaches based upon single disciplines such as economics or geography.

In this chapter we have put forward a complex systems modelling framework which we believe may be a useful step towards integrating environmental considerations into development policy. The model has been applied to the island economy of Crete as part of an on-going research programme which has both methodological as well as substantive objectives. Although much remains to be done it is our argument that this (interdisciplinary) approach deserves further consideration in the search for more sustainable development options which can effectively reduce the threat of desertification.

17.6 REFERENCES

Allbaugh, L.G. (1953). *Crete: A Case Study of an Underdeveloped Area*. Princeton University Press, Princeton, NJ.

Allen, P.M. and Sanglier, M. (1979). Dynamic model of growth in a central place system. *Geographical Analysis*, **11**, 256–272.

Allen, P.M, Clark, N. and Perez-Trejo, F. (1992). Strategic planning of complex economic systems. *Review of Political Economy*, **4** (3), 275–290.

Blue Plan (1992). *Statistics on Population and Economic Activity for the Mediterranean*. Blue Plan Regional Activity Centre, Sophia Antipolis, Valbonne, France.

Chania Statistical Office (1992). *Demographic and Economic Statistics for the Nomos of Chania.* Greece.

Cincotta, R. and Perez-Trejo, F. (1990). A risk analysis methodology for assessing natural resources degradation. *Land Degradation and Rehabilitation,* **2,** 191–199.

IERC (International Ecotechnology Research Centre) (1991). *An Integrated Planning and Policy Framework for Senegal.* EEC Final Report, Article 8 946/89, IERC, October.

Kolodny, E.Y. (1974). *La population des îles de la Grèce,* Vol. 1. EDISUD, CNRS, France.

NOS (1984). *Results of the Census of the Population and Inhabitants, 5 April 1981. Volume 2: Social and Demographic Characteristics of the Population.* National Office of Statistics, Athens.

Pérez-Trejo, F. (1993). Landscape response units: process-based self-organising systems. In Haines-Young, R., Green, D.R. and Cousins, S.H. (eds) *Landscape Ecology and GIS.* Taylor and Francis, London, 288 pp.

Perez-Trejo, F., Clark, N.G. and Allen, P.M. (1993). An exploration of dynamical systems modelling as a decision tool for environmental policy. *Journal of Environmental Management,* **39,** 305–319.

18

The Response of Landscape Units to Desertification

A. C. IMESON

VFGB, Universiteit van Amsterdam, The Netherlands

F. PÉREZ-TREJO

UNITAR, Geneva, Switzerland

and

L. H. CAMMERAAT

VFGB, Universiteit van Amsterdam, The Netherlands

18.1 INTRODUCTION

This chapter describes the 'desertification response unit' (DRU) methodology that is being developed to explore how changes in climate, drought, land use and fire regime will generate changes in ecosystem processes and patterns that can influence the future progress of desertification at the MEDALUS field sites. The objective of the methodology is to be able to project how clearly defined land-units will respond dynamically to the changes induced by processes of desertification.

To project how the MEDALUS field sites will respond to desertification, three basic research problems must be confronted. The first is the need to extrapolate the data collected and modelled at the MEDALUS field locations to both larger geographical areas, and longer time-spans. This problem of scaling-up from point measurements to relatively large areas has to address issues of spatial variability, extreme events and highly variable process rates. The second problem is the need to integrate the impacts on the landscape, of both physically and socio-economically driven processes. For this reason, socio-economic models need to be used to project how future land-use changes will influence the operation of physical processes. The third problem concerns issues related to the stability, inertia and resilience of Mediterranean ecosystems. These ecosystem characteristics are being changed by desertification. It is these characteristics, and the

---

Mediterranean Desertification and Land Use. Edited by C. Jane Brandt and John B. Thornes.
© 1996 by John Wiley & Sons, Ltd.

changes in them, that need to be used to define and characterize 'desertification response units'.

The inertia (stability), and resilience (Westman, 1986) of desertification response units is defined in terms of two sets of variables. The first set refers to site-specific conditions that are constant during the relevant period of time but which influence water availability (e.g. rock fragments, slope position and aspect). The second refers to dynamic, site-specific characteristics that affect water availability (e.g. soil structure, porosity and vegetation pattern). These characteristics are affected by feedback processes related to both intrinsic and extrinsic conditions.

Water movement leads to dynamic interactions between the vegetation and soil that result in the evolution of heterogeneous, non-random structures and patterns. The presence and dynamics of the process-induced patterns and structures found on a response unit, therefore, not only express the degree of desertification or ecosystem degradation, they are also a sensitive measure of the inertia resisting change and the resilience should conditions be changed.

Essentially, the desertification response unit methodology is a bottom-up, process-oriented way of addressing the research problems stated above. It complements the other modelling approaches of MEDALUS (Kirkby et al., Chapter 14) that apply, mostly top-down, mathematical simulations of physical processes to forecast change.

The contents of this chapter are divided into four sections. Section 18.2 introduces the methodology; in Section 18.3, examples are given of the dynamic modelling of key process–response relationships. The third section (Section 18.4) describes the operational applications and the field data that can be used to establish the resilience of desertification response units. Finally, in Section 18.5, some conclusions are presented concerning the relative risk and susceptibility of different land units to desertification.

18.2 INTRODUCTION TO THE DESERTIFICATION RESPONSE UNIT CONCEPT

18.2.1 Definitions

A basic premise of the desertification response unit approach is that a landscape is composed of a mosaic of plant–soil associations, each of which is characterized by patterns and structures that have evolved in response to a hierarchical set of key processes. These will be explained with examples in Section 18.2.2. The patterns and structures that define these responses reflect both constant catchment attributes (e.g. lithology, relief, aspect, slope and drainage basin position) and dynamic processes (e.g. vegetation cover and soil structure evolution). Whereas traditional land evaluation focuses on the static attributes, the response-unit methodology focuses on those that are responding to key processes that generate change during the period of interest. In the context of desertification the key processes are those that regulate the dynamics of water availability.

The concept of desertification response units as developed in MEDALUS is derived from the idea of ecological and landscape response units described by Pérez-Trejo (1991) for the Llanos in Venezuela. The basic idea is that the plant–soil associations in an area have characteristic patterns that reflect the underlying properties of the soil, rocks and geomorphology, and land use in as much as these properties reflect water availability, radiant energy and nutrient availability. An important element of this concept is that it draws attention to the importance of spatial dependence of different locations; a particular pattern or soil–plant community has a spatial distribution characteristic of a particular location in the landscape that reflects inter-linkages with neighbouring locations.

If only the present is of interest, the distribution of ecological response units could be derived from classical air-photo interpretation, climatological data and from geomorphological and soil survey information. In the case of desertification, a type of site or land evaluation could be made that focuses on water storage potential, water availability and on actual evapotranspiration and soil temperature. The study presented in this volume by Boer (Chapter 19) is basically this kind of approach. Future projections of change ignore important feedback mechanisms and assume that the processes involved will remain unchanged.

The DRU approach presented here tries to be more realistic by incorporating a number of concepts from dynamic systems modelling and hierarchy theory (O'Neill et al., 1986). However, the work as done by Boer is essential as a starting point. Of importance is the idea that processes are hierarchically linked and that the effects of rapid processes are incorporated over a longer time-span into the effects (structures and patterns) of slower processes. Also taken from dynamic systems modelling is the idea that processes operating at a particular scale can lead to the appearance of emergent properties at a higher hierarchical level. Examples of these processes are considered in detail in the following section.

An objective of the DRU approach is to develop a scheme of classification based on the structures that have evolved through the dynamics of key processes responsible for change at different hierarchical levels. To achieve this goal, test the assumptions being made and to make quantitative spatial projections, modelling is required to describe how changes in process rates or process interactions impact on the evolutionary pathways of different structures and patterns. The modelling and field process studies described below have focused on two key processes, namely soil aggregation and water reallocation as a result of vegetation development.

18.2.2 The hierarchy of processes at different scales

The DRU approach requires the definition of the scales that are being studied. If water availability is taken as a key factor then the static and dynamic factors that influence this can be arranged with respect to the appropriate spatial and temporal scales (Dickinson, 1988). This has been done in Figure 18.1.

Figure 18.2 illustrates the three nested scales that are studied and indicates the key processes that affect water availability (Imeson et al., 1995). These scales are referred to as respectively the patch, plot (response unit), and slope (landscape response unit) scales. Figure 18.2 describes how these are related. At each scale, as shown in Figure 18.3, the position of any point or place is important in determining the water availability. At all spatial scales, site position affects the evaporative loss of water (amount of radiation and shadowing) and the degree to which the site is a net recipient or loser of moisture.

Key processes at the sub-patch and patch scales need to be considered at time scales of months, or on an event basis. Response-unit scale processes that link different patches have longer relevant time scales that vary from months to decades according to the frequency of events. The slope scale processes that transfer sediment and water at the

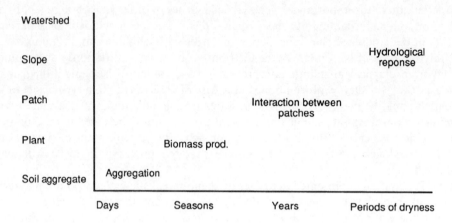

FIGURE 18.1   Hierarchy of spatial and temporal scales related to dominant processes regarding plant/soil hydrology interactions

hillslope and catchment scale have relevant time scales that may range from several years (farmland) to hundreds of years (natural vegetation).

In the following discussion the processes influencing water availability at these three scales are referred to as first (patch), second (response-unit/plot) and third-order (hillslope–landscape) processes. Similarly, the structures and patterns hypothesized as evolving in response to key processes at these three scales are termed first, second and third-order structures. Together with static site characteristics they provide what is described as first, second or third-order inertia and resilience.

18.2.3 Key first-order patch scale processes

At the patch scale the key first-order dynamic processes concern soil structure development in its widest sense (aggregation, disaggregation, biogenic crust development, etc.). Soil structure (including soil surface properties) is a key property because it regulates water acceptance, storage and movement. The degree to which climatologically driven biological processes have been able to alter the hydrological characteristics of the sediments and soils on a slope is reflected in various aspects of the soil structure. It has of course been demonstrated in many studies that soil aggregation evolves in time reducing erosion. This soil structure can equally well be lost. An early degree of desertification could be considered to occur when the patch scale aggregation processes locally fail to retain the same degree of aggregation as surrounding areas and a patchy vegetation pattern develops local bare areas that lead to water redistribution (as an emergent process) at the response-unit scale.

It is well established that organic matter is correlated with erosion. In a dynamic process-oriented approach in which the dynamics of structure are considered, it is production and degradation of soil stabilizing substances and fine roots that is important (Oades and Waters, 1991). These are reflected in the turnover rates of soil organic matter

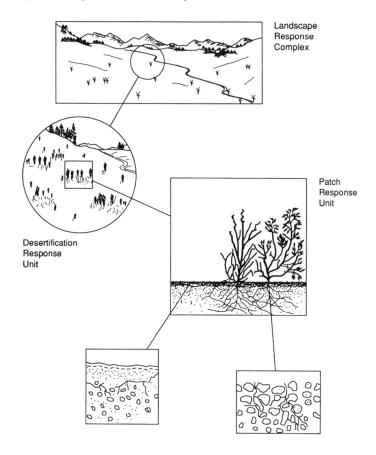

Landscape
Response
Complex

Patch
Response
Unit

Desertification
Response
Unit

FIGURE 18.2   Three nested scales as applied within the DRU concept

which are dependent on favourable temperature and moisture conditions for soil (micro) biological activity. The effects of relative rates of soil structure development S_d and soil structure breakdown S_b, which are related to the amount of subsoil biomass production, to the effects of micro-organisms and the mineralization of organic matter, can be studied by monitoring changes in soil structure. When the ratio between these two rates (aggregation ratio (A_r)) is < 1, the soil will, over several seasons, lose 'structure' and become more compact. In this degraded state it will have more unaggregated material, lower amounts of organic matter and a higher bulk density. At a shaded site, with an input of organic matter the soil is unlikely to experience structural degradation. If because of grazing, fire or drought bare open patches in the vegetation emerge, very local areas may develop with a value of $A_r < 1$. Figure 18.4 illustrates the positive feedback loop under plants, where $A_r > 1$. In between the plants open patches are shown with $A_r < 1$.

If this condition persists, structural degradation could lead to the development of areas of lower porosity and infiltration capacity. The key parameters chosen to characterize the importance and effect of these processes on soil structure generation and decay are: soil

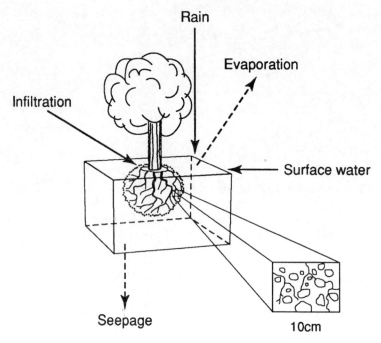

FIGURE 18.3   Water availability and moisture fluxes

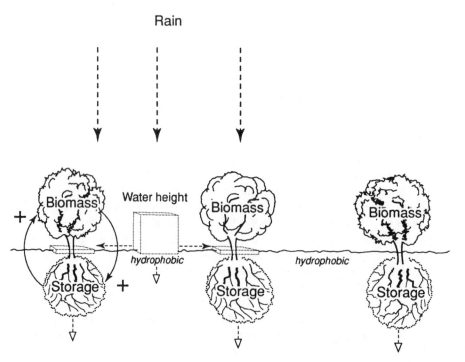

FIGURE 18.4   Arrangement of shrubs in relation to the reallocation of water

(micro-) aggregate size distributions, soil hydraulic conductivity and infiltration rates and the dispersion index. Because various soil properties such as texture and mineralogy also influence soil structure these have to be isolated when different sites are compared. For this reason it is sometimes necessary to work with ratios, comparing structure at different sites or as a proportion of a particular size fraction. Several static processes are also important in influencing water availability.

18.2.4 Key second-order response-unit processes

The response-unit areas have characteristic vegetation–soil patterns. The key parameters that determine the distribution of these patterns are (apart from land use) the radiation balance, and water storage and transmission characteristics. In other words patterns in vegetation–soil systems can be derived from information on geomorphology, soils, hydrology and topography. The response unit has geographic dependence so that it has to be seen in relation to other response units which may either drain to or from it. Some key variables (radiation balance and temperature) for example, can be derived from manipulations of a digital elevation model (DEM).

Because of space, this chapter restricts itself to key processes that can be related to water reallocation and radiation. In many cases, however, other processes may be equally or more important (erosion, soil surface hydrophobia, soil mesofauna activity, salt accumulation, grazing, fire and wood cutting) in explaining vegetation patterns. The vegetation–soil mosaic that characterizes the response unit and depends for its survival on local water reallocation may be thought of as second-order resilience. The loss of this structure leads to a third degree of desertification characterized by hillslope scale water transfers.

In the work described below, key static properties (during a period of tens of years) are the water storage potential at a site (derived from information on soil depth, stones and water retention characteristics), the rate of desiccation (derived from the radiation balance), and the position of major trees, relying on regular subsurface water supply (groundwater seepage, throughflow or exceptional storage possibilities).

The key dynamic properties are those that influence the amount of local water reallocation. These are due to the configuration and proportions of vegetated and bare surface. The properties of the surfaces themselves are dependent on the aggregation processes at the patch scale. The length of the boundaries between the various elements of the surface that make up the mosaic of rocks, biogenic crusts, different plants and bare areas is important. Key ratios, analogous to the aggregation ratio described above, include the ratio between the shaded and unshaded areas (S_r), the ratio between water storage potential and water supply (W_{sr}), and the ratio between bare patch and vegetated infiltration capacity (I_r). These are important because they describe the degree to which the patterns present at the patch-unit scale are able to benefit from and retain all of the water they receive. Degradation at this second stage of desertification occurs when the trapping efficiency of the patch scale processes is overwhelmed by extreme erosive events that destroy the structures. Also the structures at this scale are dependent on the supply of infiltrating water; should this supply be diminished then the ratio of unvegetated to vegetated areas can become so high that water will be able to move further downslope over longer distances. The response unit itself will experience soil structure degradation due to the effect of lower water and organic matter inputs and the higher temperature.

18.2.5 Third-order processes of erosion and runoff on catchment slopes

At the hillslope scale the effects of water movement (surface or subsurface) over long distances link together topographically contiguous desertification response units to create over very long time scales what may be called landscape response units. The more first and second-order inertia and resilience that the component response-unit elements on a slope have, the greater will be the extreme event required to initiate erosion and water movement that has a persistent impact on change. On farmland, tillage removes the first and second-order structures so that even under humid conditions low intensity events allow the long-distance water transfers (and erosion) to take place, and this can be important in terms of long-term slope development. In contrast, on matorral and forest slopes the high inertia and resilience of component response units means that over the time scales being considered landscape response scale change is negligible without major disturbances by very extreme events or by land-use change. The inertia and resilience are so great on matorral and abandoned land that measuring soil erosion as a means of estimating soil erosion rates is impractical because very extreme events with a return intervals of hundreds of years are needed to generate water transfer in the form of overland flow. Subsurface saturation overland flow may be on rare occasions a major process that is missed during research that focuses on runoff generation by Horton overland flow. The key processes at the hillslope scale are considered to be overland flow (all types) and soil erosion (including rill and gully erosion). The topographic control of these processes is important, in particular the degree to which at the hillslope scale some areas are net gainers or losers of water. The key parameters studied are those that can be related at the relevant scale to runoff generation and concentration. The first set of data is obtained from rainfall simulation experiments and the second from Geographic Information System (GIS) manipulations of Digital Terrain Models (DTMs) (Boer, 1996).

18.3 MODELLING THE EVOLUTION OF STRUCTURAL CHANGES IN MEDITERRANEAN SOILS

18.3.1 Introduction

The modelling component of the DRU research has concentrated on linking the key processes identified at the different scales. The aim of the research is not necessarily to make predictions about runoff and erosion but to explore the many evolutionary pathways which could unfold under different initial conditions and land-use history. The models also serve as a way of understanding how the processes at one scale can generate the spatial patterns of other processes at a higher hierarchical level, i.e. the water redistribution patterns of a particular DRU.

As illustrations of the approach, the first section below describe some of the processes that dominate soil structure dynamics at the patch scale. The next section goes on to describe the results of a slope-level model which explores the spatial interactions of the vegetation, water redistribution patterns and the spatial aggregation patterns such as those found on many matorral slopes. Differences in the aggregation of the surface layer of soils play a critical role in determining to a great extent their hydrological characteristics including water movement and retention capability. In Mediterranean environments, the patterns of aggregation of surface soil horizons under patchy vegetation present a high degree of spatial variability. This variability is non-random and strongly linked to the vegetation patterns. For example, in most cases larger and more stable aggregates are found in close proximity to plants and crusted less well-structured soils on bare surfaces. The dynamics of the aggregation processes that lead to these patterns in soil structure can provide important clues in better understanding the processes of desertification and loss of biological potential in Mediterranean soils. Several processes are being modelled, describing for example the effect of the meso-fauna (earthworms), shrinking and swelling and particle agglomeration. This last process is considered here because of its general importance.

The simulation of the dynamics of the soil/vegetation complex in a Mediterranean environment illustrates how the dynamics of aggregation, over a 20-year time horizon, are strongly determined by the vegetation dynamics which provide the primary source of organic carbon for soil micro-organisms in developing soil structure. In the case of the finer aggregate fraction, microbial activity in transforming organic material is one of the primary driving forces in the agglomeration process. The process of mineralization transforms organic material into nutrients that can then be utilized by plants for maintenance and growth: and in turn generate more organic matter for the micro-flora to live on. The causal-loop diagram shown in Figure 18.5 illustrates the major feedback relationships that drive soil/plant structuring dynamics over a medium-term horizon (20–30 years). A heuristic analysis of the feedback structure provides valuable insight into the possible behaviour patterns of the system.

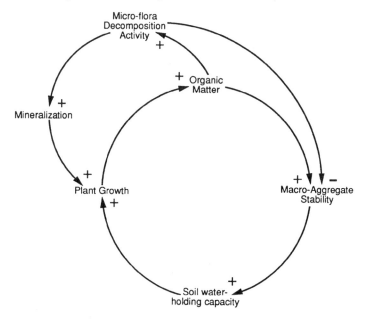

FIGURE 18.5   Causal-loop diagram illustrating the feedback relationships in aggregate growth from primary particles

There are two major feedback loops, which, depending on the initial conditions and the possible disturbances they might be subjected to, will dominate the trajectory of the dynamics over time. If the balance between the amount of organic carbon being produced by plants that leads to aggregation is positive in relation to that being consumed by micro-flora, then the soil will produce either more or larger agglomerations of particles, generating the positive feedbacks of increased water and nutrients for sustained plant growth. If, however, the negative feedbacks of decomposition of organic matter dominate, with a decrease in organic matter input, then the carbon input of the vegetation is not enough to sustain the aggregation. These aggregates then break down into either dispersed particles, or to finer micro-aggregates that produce surfaces with lower rates of water acceptance. This in turn weakens the self-reinforcing cycle of plants and aggregation, locally reducing the amount of water available for plant growth and, therefore, reducing the rate of carbon input which is the 'fuel' for the whole system dynamics.

18.3.2 The patch level model

Aggregate dynamics

The model consists of a representation of aggregation from small micro-aggregates to large micro-aggregates. Micro-aggregates are bound by very fine roots and fungal hyphae entanglement into large macro-aggregates.

The model explores the feedback mechanisms illustrated in Figure 18.5. The model equations are based on a representation of aggregate mass dynamics. The rate of change in aggregated mass is represented as a dynamic disequilibrium between organic carbon and the breakdown of organic cementing agents by the micro-flora (Equation 7). The breakdown of organic matter has a complicated effect: on the one hand it has a potentially destabilizing effect, consuming the organic network that holds larger aggregates together; on the other it mineralizes nutrients, making them available for plants to generate more organic matter to maintain aggregation.

The model simulates the dynamics of a spherical agglomeration by means of the sequence of equations presented below:

The computational sequence begins by calculating an initial aggregate volume based on the initial values of mass and density:

$$Volume_{Agg} = Mass_{Agg}/Density_{Agg} \tag{1}$$

Then the model calculates the radius for that volume:

$$R_{Agg} = (3/4*\pi))^{(1/3)} * Volume^{(1/3)} \tag{2}$$

Then the volume (V_{Agg}) to area (A_{Agg}) ratio is calculated:

$$V_{Agg}/A_{Agg} = R_{Agg}/3 \tag{3}$$

Then the effectiveness term l is calculated:

$$l = K_l/VdivA \tag{4}$$

where K_l is the system reaction rate coefficient (0.01 L^{-1}) and $VdivA$ is the volume divided by the area of the aggregate (L).

The term l expresses the limitation of the increasing aggregate size to the flow of water and substrate across a surface area which is only increasing by the square of the radius, to support a volume of aggregate micro-flora which is increasing by the cubed power of the radius.

The inflow of water and substrate into the aggregate is calculated as:

$$Agg\_Inflow = K_0 * l * Soil\_Water * (4\pi R^2) \tag{5}$$

where K_0 is the liquid-phase mass transfer coefficient (LT^{-1}) (1.0).

Depending on the vegetation dynamics (see equations below), the organic matter in the soil is changed by the amounts of carbon flowing in and the organic carbon consumed by micro-organisms in the soil. The new value of the change in soil organic matter (d_OM) is then used to calculate the new aggregate mass.

The aggregate mass equation is:

$$AggMass = AggMass + (dAg\_dt) * dt \tag{6}$$

$$dAgg\_dt = (Agg\_Inflow * l + d\_OM * 0.7) * D_{Agg} - ((4/3) * \pi * R^3) * s \tag{7}$$

where $dAgg\_dt$ is the change in the aggregate mass over a given time step, Agg_Inflow is the amount of substrate and water flowing into the aggregate during the time step dt (see Equation 5), d_OM is the change in organic matter over the time period (see Equation 13), D_{Agg} is the density of the aggregate (0.88), s is the maintenance coefficient of the micro-organisms associated with a given aggregate mass, assuming a fixed biomass concentration of organisms per cm^3 (0.02).

This new value of the aggregate mass is used to recalculate the volume and radius at each time step of the simulation.

Examples of model outputs are shown in Figures 18.6 and 18.7. These show the simulated change in aggregate dimension and the effect on this growth of a disturbance removing biomass, analogous to the effect of fire.

Modelling vegetation dynamics

The complicated dynamic processes which affect vegetation community dynamics and aggregation in the soil are important in the evolution of the system. For the purpose of this model, the interactions between the process of aggregation and the vegetation are embodied in the growth of a shrub biomass as the source of organic carbon to the aggregation process in the soil. The flow diagram in Figure 18.8 illustrates the conceptual formulation of the carbon flows in the system. The biomass dynamics were modelled using logistic growth curves (Equations 8 and 9) to simulate a growth curve that reaches a maximum at $500\ g\,m^{-2}$. The temporal dynamics of the biomass is described thus:

$$Biomass(t) = Biomass(t - dt) + (dBio\_dt) * dt \tag{8}$$

$$dBio\_dt = (Rg * Biomass * (1 - (Biomass/Bmax)) - Dead \tag{9}$$

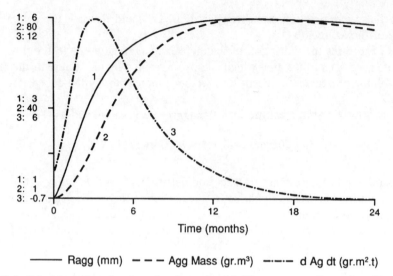

FIGURE 18.6   Model results showing the dynamics of (1) aggregate dimension (R_{agg}, mm), (2) aggregate mass (*Agg Mass*, $g\,m^{-3}$) and (3) the change in aggregate mass ($d\,Ag\_dt$, $g\,m^{-2}\,t^{-1}$)

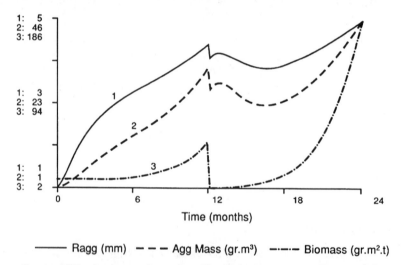

FIGURE 18.7   Aggregate dynamics affected by an event analogous to fire

where *Rg* is growth rate (1.0) ($g\,m^{-2}\,t^{-1}$), *Dead* is *Biomass* $*$ *Rd*, *Rd* is death rate (0.05), *Bmax* is maximum standing crop (500 $g\,m^{-2}$).

The dynamics of the litter compartment of the model is given by:

$$Litter(t) = Litter(t - dt) + (d\_Litter)*dt \qquad (10)$$
$$d\_Litter = Dead - Litter(t)*Decomp\_R \qquad (11)$$

where *Decomp_R* is the rate of decomposition (0.05).

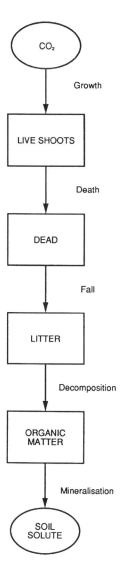

FIGURE 18.8   Flow diagram of the vegetation sub-model processes, illustrating the conceptual flows of carbon in the system

Organic matter change is simulated by the equations:

$$Organic\_matter(t) = Organic\_Matter(t - dt) + (d\_OM) * dt \qquad (12)$$
$$d\_OM = Decomp\_R * Litter - Organic\_Matter * MinR \qquad (13)$$

where *MinR* is the mineralization rate of organic matter (0.05).

The model can be used to explore the possible restructuring of DRUs to provide representation of what the soil structural properties might be like under different land-use scenarios.

The results of the model show how the growth of aggregates is strongly affected by their changing geometry, which determines their maximum growth rate. The linkage between the vegetation dynamics and the aggregation process was explored by simulating a decrease in biomass which reduces the organic carbon input into the soil as a single event (Figure 18.7).

Inter-patch transfer of water

Differences in aggregation under plants and on open patches leads to the transfer of water along the slope. The response of the vegetation to water transfer was also modelled. Figure 18.4 also illustrates the flows modelled when different infiltration rates resulting from the different aggregation dynamics are used to model the transfer of water. Linking different cells along a hypothetical 100 m slope with the initial model conditions described in Figure 18.9a, at $t = 1$, resulted after the emergence of the structured biomass pattern in Figure 18.9b at $t = 1500$.

18.4 FIELD DATA REQUIREMENTS AND MODEL VALIDATION

18.4.1 Introduction

The two key physical processes described above are soil aggregation dynamics and inter-patch water transfer. This section considers the problems that need to be addressed in translating the DRU concept from an abstract concept to an operational methodology.

The main problem concerns the data that are required for applying the approach. Two types of data are required; data needed for parametrizing the equations and data that can be used to define the nature and degree of structuring that expresses resilience at the different scales. The first data are used to establish process mechanisms and rates needed to relate processes to patterns and structures. The second data are used to describe existing patterns and structures and to monitor how these change over time and from place to place.

18.4.2 Data options at different scales

At the first-order scale, parameters of soil structure are required that describe water-regulating behaviour. Further the parameters need to be sensitive to change, so that the dynamics of the soil property can be followed, and quick to obtain, because spatial patterns need to be established from many measurements. Based on these criteria, it was decided to collect the following first-order data:

(a) naturally occurring particle size distributions,
(b) soil surface crust descriptions and infiltration behaviour and crust development,
(c) soil moisture retention characteristics,
(d) water-soluble salts and carbonate contents, and
(e) organic matter characteristics.

Ideally, these parameters should be measured seasonally or on an event basis.

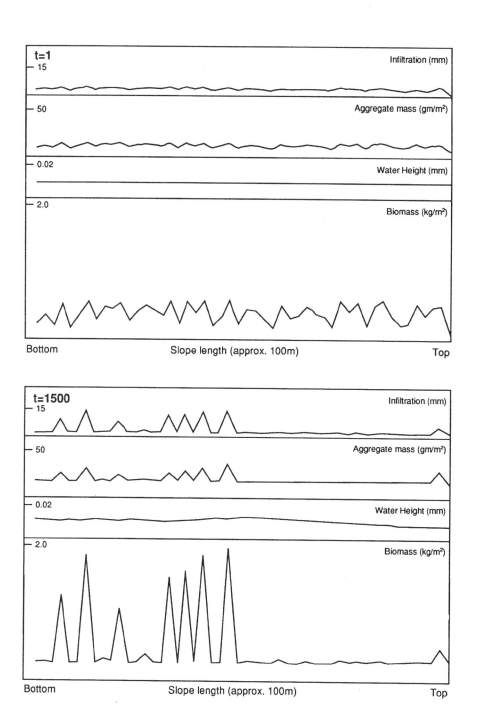

FIGURE 18.9   Emergence of the structured biomass pattern

The second-order data requirements relate to the transfer of water and sediment between the patches that make up the response units. Data are required for modelling inter-patch transfers and differences in process rates. The collected second-order data related to:

(a) the effects of shading by plants and stones,
(b) the nature of boundaries between patch elements,
(c) inter-patch water and sediment transfers, and
(d) soil aggregation and cover patterns.

The data required can be obtained from rainfall simulation experiments and field observations. Special attention had to be given to the nature of the changes that take place across vegetation–soil patch boundaries.

At the catchment or hillslope scale it is necessary to be able to define and map the response units that can be recognized on the slope. This can be done partly with second-order data. The longer-term changes however can be approached by infiltration–runoff experiments that focus on determining infiltration envelopes. The rate of sediment transport at this scale can only be inferred from studies of colluvium, sediment yields during extreme events and from soil erosion measurements on open plots. In the absence of this data simple hydrological transfer models of runoff can be used based on comparisons between the infiltration envelopes and the rainfall duration–frequency diagrams. To apply the above, data are also required that can be input into a geographic information (GIS), such as a DEM, maps of physiographic–soil–vegetation landscape units derived from air-photo interpretations and field surveys. Relationships between patterns and processes need to be underpinned and quantified by longer-term research.

It is stressed that on different types of parent material, land use or climate, different processes dominate water and sediment transfers according to the scale of consideration.

18.4.3 Examples of soil aggregation first-order data

As an example of the data collected, measurements are described that also illustrate the patterns in aggregation and water transfer projected by the modelling described in Section 18.3.

The procedure applied was as follows. At a patch boundary along a gradient or at a location, such as that shown in Figure 18.10, a detailed site description was made to establish surface characteristics and cover conditions, and samples were collected for laboratory analyses. At the same time infiltration experiments were carried out to study water infiltration patterns and rates. In the laboratory, attention was firstly given to establishing three simple parameters that could be rapidly determined for many sites. These were:

1.   The proportion of the soil consisting of naturally occurring aggregated particles.
2.   The size distribution of the aggregated particles.
3.   The water-stability of the aggregated material.

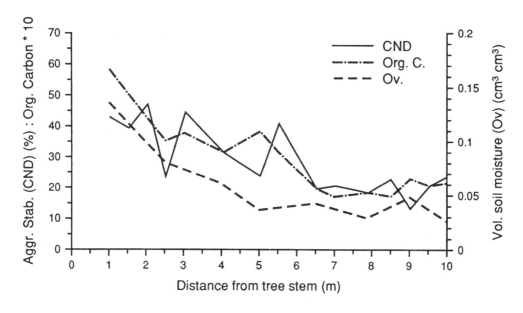

FIGURE 18.10   Transect under *Pinus halopensis* in El Ardal in June 1991

In addition various chemical and other analyses were made that were thought necessary to help understand the aggregation characteristics. These were organic matter content, soil water-soluble salt and carbonate content and the contents of stones and root fragments. The above three sets of aggregation parameters were complemented by a fourth set of analyses directed at the fine aggregate fraction (<106 or $250\,\mu m$).

The fourth set of aggregation parameters was obtained using an analytical procedure developed for the Quantrachrome Microscan Particle Size Analyser to analyse both the size distribution of fine primary particles and water-stable micro-aggregates. Size distributions of particles were measured either without (water-stable micro-aggregates) or with a dispersive and ultrasonic pre-treatment (primary particles) in one run. The mass distribution, total, mean, sorting and skewness of the micro-aggregate and primary particle fractions can be deduced from the instrument's results. Using this method approximately 15 samples can be handled in one day and various soil aggregation parameters calculated. The results of many hundreds of different analyses of the aggregation characteristics, examples of which are shown in Figures 18.10 and 18.11, have always shown the following.

(a) A close correspondence between different species of plant and both aggregate size distribution and stability.
(b) Sharp boundaries between finely or poorly aggregated soil on bare patches and strongly aggregated material under plants, indicating two contrasted sub-systems.
(c) A more rapid response of large aggregates to seasonal change than that of the smaller aggregates.
(d) The greater the proportion of stones the better the aggregation of the soil.

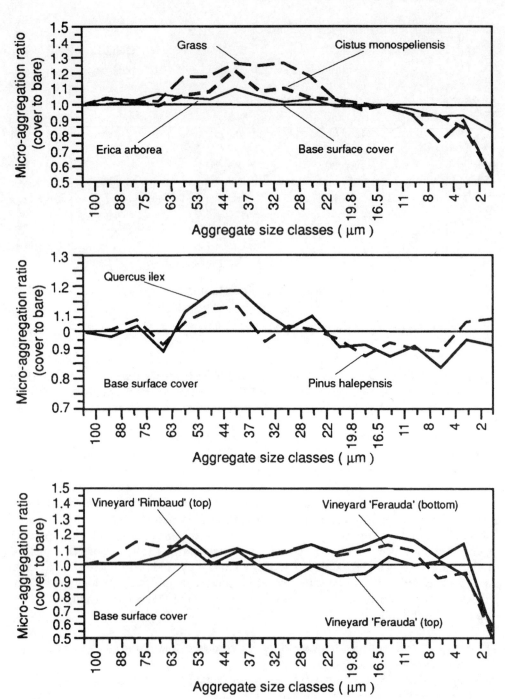

FIGURE 18.11 Differences in micro-aggregation under different types of land use. (a) Shrubs compared with bare soil; (b) trees compared with bare soil; (c) vineyards compared with bare soil

18.4.4 Examples of hydrological response

Rainfall simulation experiments illustrate data collection at the response-unit level. Two types of experiment were made. Firstly, key experiments were made whereby sites were instrumented in detail so that infiltration could be modelled from soil moisture and a whole range of other data that were collected. Secondly, routine experiments were made for establishing erosional and infiltration response. A major problem with the key experiments is that Mediterranean soils are often both very stony and compact. This makes the collection of undisturbed samples and the measurement of soil water retention and hydraulic conductivity very difficult.

For the key experiments, time domain reflectrometry (TDR) could be used for the *in-situ* determination and monitoring of the volumetric soil moisture content (Topp et al., 1980). Drungil et al. (1989) reported that TDR could be applied in stony soils and indeed this was found to be the case, provided measurements were corrected retrospectively for stone content. *In-situ* measurements of soil moisture tension were made parallel with the TDR data collection so that *in-situ* conductivity and water retention characteristics could be determined.

The DRU concept means that the scale of the rainfall simulation experiment is very important. The measurements are not only scale dependent in terms of the expected results; the results also reflect water transfer processes at higher hierarchical scales. Experiments were carried out at two scales. Large experiments on 10-m long slopes looked at inter-patch transfers and smaller experiments (1-m long plots) at in-patch infiltration. Results reported in Farenhorst (1991) and Imeson et al. (1992) point in the same direction.

Small-scale rainfall simulations at key sites were carried out along catenas at five core project sites (Almocreva, Vale Formoso, El Ardal, Canteranne and Réart; their locations are given in Cammeraat, Chapter 5) with a standard 0.5 m² variable intensity dripping plate simulator with raindrop randomizer. From the experiments, determinations of infiltration rates and envelopes and the runoff coefficients were made. In total approximately 130 simulations have been carried out, of which seven were in combination with continuous soil moisture registration with TDR equipment.

Large-scale rainfall simulations were carried out with a sprinkler type of simulator, covering an area of 20 m². The intensity was kept at 70 mm per hour and falling height of the drops was 3.5 m. From the results of these (Bergkamp et al., 1996) and other experiments (Abrahams and Parsons, 1991; Farenhorst, 1991; Cerdá-Bolinches, 1993), the following conclusions can be made that are relevant to characterizing DRU surfaces.

(a) There is always a close correspondence between the vegetation–soil cover and the infiltration rate.
(b) The depth of wetting and the irregularity of the wetting front increase with an increase in soil aggregation or stoniness.
(c) As the amount of structure increases the transfer of water across patch boundaries decreases.
(d) Well-aggregated zones of soil on shaded areas trap all excess runoff that moves as overland flow on slopes that have structures associated with long-abandoned land, forest and matorral.

18.4.5 Simple field procedures for characterizing DRUs

The above discussion refers to a methodology requiring substantial field and laboratory analysis. This analysis is necessary initially for identifying and analysing key processes and later for modelling the effects of change. Once processes have been linked to patterns DRU characterization can then be defined by the structures present.

In the examples given here, the patterns and structures being studied result from feedback processes resulting from discontinuities in hydrological processes. The DRU can, therefore, be characterized by simply quantifying the morphology and juxtaposition of the pattern components. The basic data required are maps of the patterns in:

1. Runoff sinks and sources within the DRU.
2. Shaded and unshaded areas.
3. The patterns of emergent or higher-order process.

In considering these patterns a distinction can be made between the static and dynamic DRU attributes. This is because of the need to distinguish between stability and resilience.

18.5 CONCLUSION: THE EVOLUTION OF SOIL STRUCTURE AND INFILTRATION RATES IN TIME, ECOSYSTEM RESILIENCE AND DESERTIFICATION

18.5.1 Evolution of soil structure

Both the modelling and field data are consistent in as much as the projected patterns of aggregation and infiltration correspond with the DRU methodology. However, at the moment although too little data has been collected to establish the dynamics of processes in time, some process rates can be deduced from comparisons of conditions during different successional stages.

The evolution of soil aggregation in time is suggested by relationships of aggregation parameters to the successional stage of the vegetation. This is the case, for example, by the study of Hin (1993) at Réart, illustrated in Figure 18.12. Similar observations were made in southeast Spain, on successively abandoned agricultural areas. From this and other studies it appears that soil aggregate stability responds very rapidly to change; within 2 years of shading or litter accumulation a stable aggregation is present. However, for the smaller building blocks of aggregation it seems that more time is needed. Thus Hin showed at Réart that at least 5 years were required for the first signs of changes in micro-aggregation to appear after land abandonment. This is in agreement with observations of Oades and Waters (1991).

Like aggregate stability, organic carbon content is always associated with the successional stage of the vegetation (see Figure 18.10).

It has already been mentioned that, as a successional stage advances, so the heterogeneity of the infiltration rates increase. This was found not only for all of the MEDALUS sites studied, but also by Cerdá-Bolinches (1995) in Valencia. He shows how a uniform wetting front on cultivated soils becomes more complex and crenellated over a 20-year period as the spatial structuring of the soil and the discontinuities in the runoff and infiltration processes evolve.

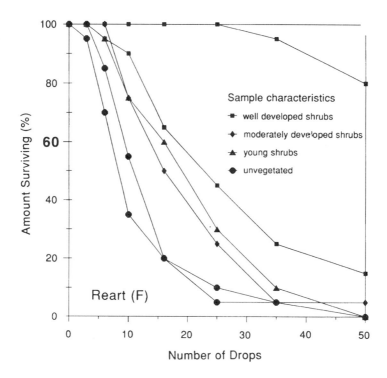

FIGURE 18.12 Aggregate stability frequency distributions for some samples from the Réart area (Roussillon). I = well-developed shrubs, II = moderately developed shrubs, III = juvenile shrubs and IV = bare area

In a review of the resilience of Mediterranean ecosystems, Westman (1986) discussed how ecosystem resilience could be quantified. As indicators of resilience, ecologists have used parameters of species composition and diversity (Fox and Fox, 1986). Developing and using species composition to establish the rate of ecosystem recovery has many practical difficulties related to the complexity of ecosystem dynamics and lack of knowledge about the species concerned.

An alternative more general method for quantifying the resilience of Mediterranean ecosystems would be through the quantification of the soil aggregation and infiltration patterns described above. It would seem that these patterns not only describe inertia (stability) but also the stage of ecosystem recovery following both land abandonment and fire, towards a final state that is characterized by the best possible utilization of water on a hillslope.

An important question with respect to desertification is the degree to which the loss of the structures described above, rather than their recovery, is in fact an expression of desertification. If this is assumed to be the case, then the first, second and third-order structures are indicative of the degree to which desertified conditions exist. This idea can be tested by examining vegetation conditions in more arid regions than found at the MEDALUS sites and by comparing the similarity between the patterns and structures found on desert slopes, with those on the desertified slopes being studied (Imeson et al., 1995).

18.5.2 The DRU concept and Mediterranean desertification

For the reasons outlined in the introduction, the DRU concept is believed to offer a way of overcoming several problems that make the assessment of desertification impact difficult. Although the procedures used to develop and validate the concepts relating key processes to patterns may seem complex and involved, the basic concepts are nevertheless quite straightforward. The DRU approach can be used to define simple indicators of desertification response at different scales, and in this way be easily applied to larger areas. Desertification response indicators derived from the DRU investigation are now being tested in the Guadalentín drainage basin and elsewhere. Because the indicators reflect the process-based concepts expressed in the DRU methodology, the concepts can be applied in all areas undergoing desertification to produce appropriate to the region. Together with the physical indicators described in this chapter, socio-economic ones are also being developed in order to investigate the implications for various policy scenarios on the physical impact of desertification.

18.6 REFERENCES

Abrahams, A.D. and Parsons, A.J. (1991) Relation between infiltration and stone cover on a semiarid hillslope, Southern Arizona. *Journal of Hydrology*, **122**, 49–59.

Bergkamp G., Cammeraat, L.H. and Martinez-Fernandez, J. (accepted; 1996) Water movement and vegetation patterns on shrubland and an abandoned field in two desertification threatened areas in Spain. *Earth Surf. Proc. and Landforms*.

Cerdà-Bolinches, A. (1995) Factores y variaciones espacio-temporales de la infiltración en los ecosystemas mediterráneos, Monografías Científicas 5, *Logroño: Geoforma Ediciones*, p. 1–151 (in Spanish).

Dickinson, R.E. (1988) Atmospheric systems and global change. In Roswall, T., Woodmansee, R.G. and Risser, P.G. (eds) *Scales and Global Change*. Wiley & Sons, Chichester, p. 57–80.

Drungil, C.E.C., Abt, K. and Gish, T.J. (1989) Soil moisture determination in gravelly soils with Time Domain Reflectometry. *Transactions ASEA*, **32**, 177–180.

Farenhorst, A. (1991) Spatial diversity of burned mediterranean forest eco-systems. MSc Thesis, University of Amsterdam, 151 pp. (in Dutch).

Fox, B.J. and Fox, M.D. (1986) Resilience of animal and plant communities to human disturbance. In Dell, B., Hopkins, A.J.M. and Lamont, B.B. (eds) *Resilience in Mediterranean-Type Ecosystems*. Junk Publishers, Dordrecht, 39–64.

Hin, J. (1993) Relationships between soil aggregation parameters and visually differentiated Mediterranean landscape units. MSc Thesis, University of Amsterdam, 65 pp (in Dutch, with English summary and figure captions).

Imeson, A.C., Verstraten, J.M., van Mulligen, E.J. and Sevink, J. (1992) The effects of fire and water repellency on the infiltration and runoff under Mediterranean type forest. *Catena*, **19**, 345–361.

Imeson, A.C., Cammeraat, L.H. and Pérez-Trejo, F. (1995) Desertification response units. In Fantechi, R., Peter D., Balabanis, P. and Rubio, J.L. (eds) *Desertification in a European Context: Physical and Socio-economic Aspects*. Luxembourg: Office for Official Publications of the European Communities, p. 263–277.

Klemmedson, J.O. (1989) Soil organic matter in arid and semiarid ecosystems: sources, accumulation, and distribution. *Arid Soil Res. and Rehabilitation*, **3**, 99–114.

Oades, J.M. and Waters, A.G. (1991) Aggregate hierarchy in soils. *Aust. J. Soil Res*, **29**, 815–828.

O'Neill, R.V., DeAngelis, D.L., Allen, T.F.H. and Waide, J.B. (1986) *A Hierarchal Concept of Ecosystems*. Monographs in Population Biology 23. Princeton University Press, Princeton, NJ.

Pérez-Trejo, F. (1991) Landscape response units: Process-based self-organizing systems. In Haines-Young, R., Green, D.R. and Cousins, S.H. (eds) *Landscape Ecology and Geographic Information Systems*. Taylor and Francis, London.

Topp, G.C., Davis, J.L. and Annan, A.P. (1980) Electromagnetic determination of soil water content: measurement in coaxial transmission lines. *Water Resour. Res.*, **16**, 574–582.

Westman, W.E. (1986) Resilience, concepts and measures. In Dell, B., Hopkins, A.J.M. and Lamont, B.B. (eds.) *Resilience in Mediterranean-Type Ecosystems*. Junk Publishers, Dordrecht, 5–20.

19

Geographical Information System-Based Application of the Desertification Response Unit Concept at the Hillslope Scale

M. M. BOER

Estación Experimental de Zonas Aridas, Almería, Spain (formerly at VFGB, Universiteit van Amsterdam, The Netherlands)

19.1 INTRODUCTION

In the previous chapter the concept of desertification response units (DRUs) has been described from a theoretical modelling point of view. Using examples from research at the MEDALUS field sites, three hierarchical levels were proposed and for each of these a number of key processes and parameters were identified as being principal determinants of the response to environmental disturbance (Imeson et al., 1996).

In this chapter the results of attempts to adapt and apply the DRU concept in a Geographical Information System (GIS)-based approach to desertification hazard zonation at the hillslope scale in a small test area near the MEDALUS field site at El Ardal (Rio Mula basin, southeast Spain) are discussed. Emphasis is given to the quantification of hillslope position, in terms of potential soil moisture availability and soil erosion, and to the evaluation of their sensitivity under a hypothetical scenario of increasing aridity.

The ultimate objective of this work, which is being continued in one of the MEDALUS II target areas, is to develop an operational method for identifying and mapping areas of a particular response to environmental disturbances, such as climate change (Houghton et al., 1990; Palutikof et al., 1992; Wigley, 1992) or land abandonment, in large drainage basins of poor data availability.

It is important to be aware that distributed environmental data at an adequate scale are generally sparsely available in the Mediterranean. For this reason research efforts are directed to developing sound landscape ecological methods to derive an optimal hazard assessment from existing sources of distributed data, such as (hard copy) maps of topography, geology or land use and remotely sensed images, rather than towards sophisticated physically-based modelling of the processes involved. Using digital terrain analysis, air-photo interpretation and field observations of soil hydrological behaviour as

Mediterranean Desertification and Land Use. Edited by C. Jane Brandt and John B. Thornes.
© 1996 by John Wiley & Sons, Ltd.

principal data sources this chapter reports on an attempt to assess the response to disturbance in terms of a possibility space for specific degradation processes. Within this approach DRUs are defined as land units of variable size whose combination of soil, vegetation, land use and terrain properties, determine possible changes in degradation process activity following a specified perturbation. So far, attention has focused on sensitivity to desiccation and soil erosion, two processes that have frequently been stated to dominate the degradation of Mediterranean landscapes (Fantechi and Margaris, 1986; Rubio and Sanroque, 1987; Albaladejo et al., 1990; López-Bermúdez, 1990; Imeson and Emmer, 1992). Future work aims at also including the sensitivity to other land degradation processes such as salinization.

19.2 DESCRIPTION OF THE RESEARCH AREA

The research area surrounds the MEDALUS field site near Mula in the upper Rio Mula basin (Murcia, southeast Spain) and includes an experimental first-order watershed at Campo Ardal operated by the University of Murcia since 1989 (Figure 19.1). Several experimental studies on plant–water–erosion relationships in this general area have been reported by López Bermúdez, Thornes and associates at Murcia and Bristol Universities (Francis, 1986; Romero Diaz et al., 1988; Francis and Thornes, 1990; López Bermúdez et al., 1991). López Bermúdez et al. (Chapter 8) report on the field observations and measurements that have been carried out as part of the core measurement programme of MEDALUS.

The upper basin of the Rio Mula is characterized by isolated hills of Eocene limestones reaching up to 600–1000 m a.s.l. with long straight footslopes and wide nearly flat valleys filled with Miocene marls and more recent colluvial material (IGME, 1972). The smooth topography of this part of the Mula basin differs significantly from the lower part, where gullies and rills are widespread and badlands are characteristic features of the landscape.

The climate is dry Mediterranean, with mean annual rainfall of about 300 mm, and a summer drought lasting generally from the end of May to the end of September. The temporal variability in rainfall is very high with annual precipitation ranging from less than 100 mm to more than 500 mm. The major part of the annual rainfall amount may fall in any of the four seasons and dry periods of more than 1 month may also occur at any time of the year.

The vegetation on the limestone hill, in the southeastern part of the study area, consists of scattered and discontinuous matorral (Tomaselli, 1981). On shallow stony soils (Leptosols), *Rosmarinus officinalis*, *Thymus vulgaris*, *Juniperus oxycedrus*, *Stipa tenacissima*, *Brachypodium retusum*, *Cistus clusii* and *Asphodelus* sp. are found as characteristic species. On the most favourable sites with a northerly aspect small pine stands (*Pinus halepensis*) occur on deeper (50–75 cm), more developed soils (Petric Calcisols). From the undulating footslope zones to the valley bottom agricultural land use predominates on soils developed in colluvial material (see Figure 19.2). In the footslope zone the shallow (*c.* 30–50 cm) soils have well-developed petrocalcic horizons. In this zone almonds, cereals and olives are grown in the traditional dry farming way (*secano*). Towards the valley bottom the colluvial deposits increase to more than 3 m in depth and are used for the cultivation of vines. The soils found in this lowest zone feature a calcic

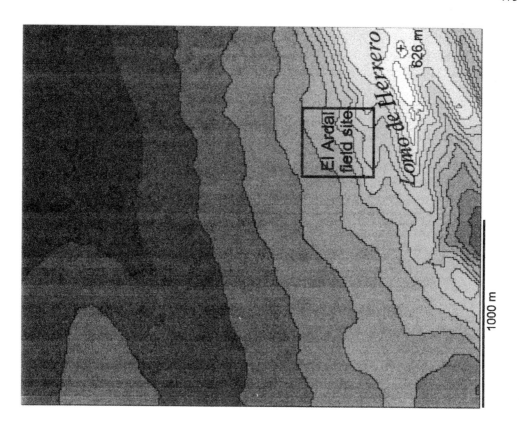

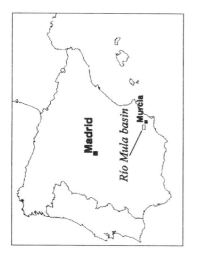

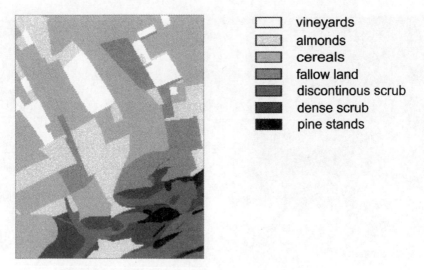

vineyards
almonds
cereals
fallow land
discontinous scrub
dense scrub
pine stands

FIGURE 19.2   Distribution of vegetation and land-use types

horizon over a rather well-developed palaeosol (Calcaric Cambisol). Most of the area, except the experimental basin itself, is extensively grazed by sheep and goats.

19.3 APPROACH

19.3.1 A conceptual model of ecological degradation

As explained in the previous chapter (Imeson et al.), a landscape undergoing degradation can be considered as a system loosing structure, first at the lower hierarchical levels (e.g. deterioration of soil structure) but progressively also at the higher hierarchical levels (e.g. soil–vegetation mosaic). Following the terminology of the previous chapter, structure refers to the degree of organization that enables an ecosystem to optimally use the available (water) resources.

This conceptual model of ecological degradation has been schematized in Figure 19.3. On a hillslope covered by a well-developed dense matorral, not affected by land degradation processes (Figure 19.3A), the vegetation cover is almost continuous and the soil is strongly aggregated, leading to high infiltration rates, no runoff, low evaporation losses and soil moisture availability being high on both north and south slopes. In an early stage of degradation (Figure 19.3B), for example caused by increasingly arid conditions, the vegetation cover has declined somewhat but still supports good soil structural and hydrological properties. Infiltration rates remain high throughout the hillslope but, due to evaporation losses, soil moisture availability has decreased in the areas of lower vegetation cover, especially on the south slope. As degradation proceeds the vegetation cover becomes discontinuous (Figure 19.3C) and gives way to the formation of patches of clumped shrubs with well-aggregated soils, high infiltration and high moisture availability, that contrast sharply with the bare areas in between having unfavourable soil structural and hydrological properties. Soil moisture patterns follow the distribution of these patches and the differences between north and south slopes become more pronounced. On the bare

(A)

Soil moisture availability

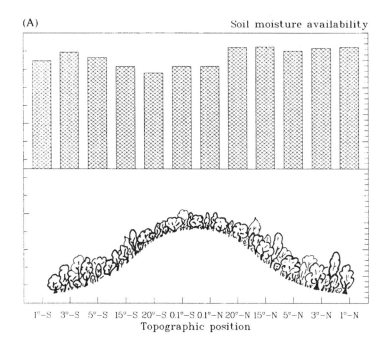

1°–S 3°–S 5°–S 15°–S 20°–S 0.1°–S 0.1°–N 20°–N 15°–N 5°–N 3°–N 1°–N

Topographic position

(B)

Soil moisture availability

⊡ stage 2   ⊡ stage 1

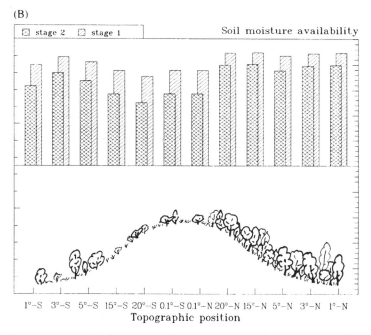

1°–S 3°–S 5°–S 15°–S 20°–S 0.1°–S 0.1°–N 20°–N 15°–N 5°–N 3°–N 1°–N

Topographic position

FIGURE 19.3   Conceptual model of ecological degradation on a Mediterranean hillslope undergoing aridification. The X axis shows the local slope angle and aspect of each of the hillslope positions, while the values on the Y axis indicate relative differences in soil moisture conditions. See text for further explanation

(C)

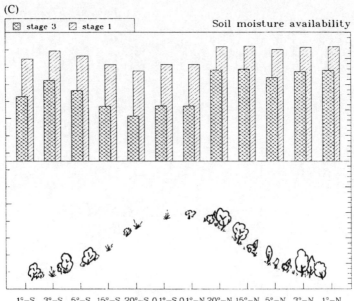

Topographic position

(D)

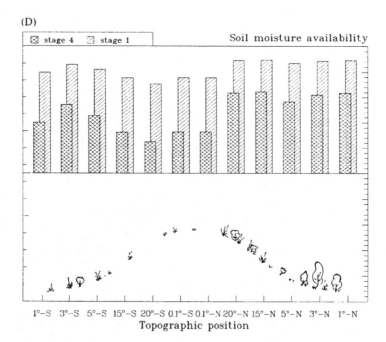

Topographic position

FIGURE 19.3   (*Continued*)

patches runoff is generated regularly and water, sediment, nutrients and seeds are transported over short distances from bare areas to vegetated patches (Bergkamp et al., 1994). The last figure (Figure 19.3D) corresponds to an advanced stage of degradation. The vegetation cover has declined severely and soil structural properties have been lost completely. With infiltration rates being uniformly low, water and sediments are transferred over the entire hillslope, and incision may occur at areas where runoff reaches critical transport capacities.

What we observe, as this landscape degrades and progressively loses first, second and third-level structure, is that the spatial patterns of important parameters, like soil moisture availability or potential sediment transport capacity, are controlled to an increasing extent by terrain attributes determining water movements at the hillslope scale, such as slope angle and slope aspect or specific catchment area, and to a decreasing extent by soil and vegetation properties, such as soil aggregation or canopy structure, controlling water and sediment fluxes at the scale of individual plants or shrubs. One could reason that in a hypothetical, final state of degradation, the hillslope patterns of soil moisture and soil erosion potential no longer depend on soil and vegetation properties, determining water and sediment movement at the lower hierarchical levels, but could be explained from the topography of the hillslope alone. Based on this hypothesis, one could use the topography-based pattern of soil moisture and soil erosion potential as a reference situation for assessing the potential magnitude of the changes that could occur due to environmental disturbances. The potential magnitude of change, or the degradation potential, could then be evaluated by comparing the reference pattern, that can be approximated by analysis of digital elevation models (DEMs) (e.g. Moore et al., 1991), with the 'actual' hillslope patterns, that incorporate properties controlling water movements at lower hierarchical levels (soil and vegetation properties) as well as hillslope topography.

Below, an attempt is made to operationalize this approach to DRU identification through the application of simple models of soil moisture distribution and soil erosion potential within a GIS. Firstly, the computation of the reference patterns from DEMs is discussed, while in the second part, the terrain-based patterns are combined with field observations on soil hydrological behaviour to approximate the actual patterns and evaluate the potential magnitude of change.

19.3.2 Simulating terrain-based patterns of soil moisture availability

The effect of topographic position on soil moisture distribution, runoff generation and soil erosion has been a subject of active research (e.g. Beven and Kirkby, 1979; Burt et al., 1986; O'Loughlin, 1986; Moore et al., 1988) and has recently been applied in a number of GIS-based studies (e.g. Moore et al., 1991, 1993; Ladson and Moore, 1992).

In sparsely vegetated areas of high relief, which correspond to many of the dry regions of the Mediterranean, topography is a major driving force of water movements through the landscape. Average soil moisture conditions reflect several aspects of topography, the most important being: (i) the effect of potential water concentration and drainage resulting from a particular position within the drainage network and the local surface slope angle, and (ii) the effect of local slope angle and aspect on solar radiation, as the major driving force for the evapotranspiration of soil water.

The potential concentration of subsurface water at a particular site in the landscape can be approximated by a simple soil wetness index $W_i = \ln (A_s/\tan\beta)$ introduced by Beven and Kirkby (1979), with A_s being the specific catchment area per unit contour length (m^2/m^{-1}) and β the local slope angle (degrees). This index has been widely applied in distributed parameter hydrologic models (e.g. Beven and Wood, 1983; Hornberger et al., 1985; Quinn et al., 1991; Durand et al., 1992; Robson et al., 1992) and several terrain-based studies of soil moisture distribution (Moore et al., 1988; Ladson and Moore, 1992) and catchment behaviour (Grayson et al., 1992a). To apply the wetness index one assumes that the soil surface properly represents the hydraulic gradient, that soil hydrologic properties and losses to the groundwater are spatially uniform and that there are no pipes causing the actual area draining to a site to be potentially very different from the specific catchment area determined from the surface topography (Jones, 1987).

The second aspect of topography, the effect on the potential short-wave radiation budget, can be quantified for a point i for any day of the year as:

$$R_i = \frac{24I}{\pi r^2} \cos\phi \cos\delta(\sin\eta - \eta\cos\eta) \tag{1}$$

where I is the solar constant, δ is the solar declination, r is the ratio of the earth-sun distance to its mean, and π and η are functions of the latitude, slope angle and slope aspect (Lee, 1978).

From these equations a simple topographic moisture index (TMI) can be derived that describes the topographic effect on average soil moisture conditions as a percentage of the index value of a horizontal surface:

$$TMI_i = \frac{W_i}{W_0} * \frac{R_0}{R_i} * 100 \tag{2}$$

where W_i and R_i are respectively the wetness index and the mean daily potential short-wave radiation budget of a point i and W_0 and R_0 are the same parameters for a horizontal surface.

In semi-arid Mediterranean environments the high temporal variability in rainfall poses specific problems to the application of static topographic indices. To adapt Equation 2 to the highly variable conditions of semi-arid Mediterranean landscapes it is proposed to weight both ratios according to the annual distribution in rainfall and temperature.

At the latitude of southeast Spain (c. 37–38°N) the spatial differences in short-wave radiation budgets are very small in summer, strongly increase from autumn to the winter season, reach their maximum in December and then decline again in spring. The extent to which spatial differences in radiation budgets are reflected in the spatial pattern of soil moisture conditions depends on soil moisture content itself: differences in insolation are, of course, only relevant when there is soil moisture to be evaporated. The typical Mediterranean distribution of annual rainfall implies that often the period of maximum spatial differences in solar radiation coincides with the major rainfall period, while the period of minimum spatial differences in insolation is normally characterized by summer drought. However, since there are many years with rainfall distributions that strongly deviate from the 'normal' pattern, spatial soil moisture patterns can vary significantly from one year to another, especially in areas of rugged topography.

To account for both the effect of the annual oscillation of potential short-wave radiation and the variability of the annual rainfall distribution, an alternative parameter was used, the so-called effective short-wave radiation (R'_i), which is calculated as a single annual value for each point i by:

$$R'_i = \sum_{j=1}^{12} P_j(a_i * R_{ij}) \tag{3}$$

where P_j is the percentage of annual rainfall falling in month j, R_{ij} is the mean daily potential short-wave radiation at point i in month j (calculated for the 21st of each month) and a_i is the ratio of bare soil and covered soil surfaces at point i.

The weight of the ratio W_i/W_0 also depends on the annual rainfall distribution and therefore its application in semi-arid regions is not straightforward. Application of the wetness index: $\ln(A_s/\tan\beta)$ not only assumes soil properties to be spatially uniform but also supposes that steady-state conditions of subsurface drainage equilibrium have been reached throughout the area (Jones, 1986; O'Loughlin, 1986). As steady-state conditions are rarely reached over an entire catchment, Barling et al. (1994) propose a quasi-dynamic wetness index that is based on the concept of time–area curves (Iida, 1984) and the calculation of an effective upslope contributing area per unit width of contour (A_e). Though physically a more realistic description of the topographic effect on soil moisture conditions, their quasi-dynamic wetness index requires hydraulic conductivity data that are usually not available at adequate scales. For the sake of simplicity, the former static expression was used in combination with a simple weighting factor to account for the fact that in dry Mediterranean areas processes of subsurface flow are, at least at the hillslope scale, restricted to very few, exceptionally wet periods when the soils reach field capacity.

Measurements of soil moisture content and soil moisture suction at the El Ardal field site during 1990 and 1991 revealed that even the soils under a protective vegetation cover rarely reach field capacity (López Bermúdez et al., 1992). These observations are supported by calculations of the potential soil moisture surplus at the nearby Embalse de la Cierva that have been carried out for five selected years using the standard formulae of Thornthwaite (see Table 19.1). The results show that monthly soil moisture surpluses are seldom large enough to bring soil profiles of 75 cm depth to a volumetric soil moisture content corresponding to field capacity (c. 35%).

Substituting the effective short-wave radiation R'_i for R_i, and including a criterion for the possible occurrence of subsurface water redistribution in Equation 2 produces a relative *TMI* that expresses the soil moisture conditions of a point in the landscape as the ratio of the wetness index and the mean effective solar radiation at point i relative to those of a horizontal surface:

$$TMI_i = \left(\frac{W_i}{W_0}\right) n \left(\frac{R'_0}{R'_i}\right) * 100 \tag{4}$$

TABLE 19.1   Potential moisture surplus or deficit (negative values) in mm, as a function of monthly rainfall and potential evapotranspiration (according to Thornthwaite). The selected years represent both situations of normal annual rainfall amounts (1960, 1962, 1964), as well as very dry (1981) and relatively humid conditions (1986)

	1960	1962	1964	1981	1986
January	18.5	−14.6	27.3	−16.1	2.2
February	31.6	−15.0	−12.3	−4.8	−19.3
March	−16.3	80.4	8.8	−23.7	6.1
April	−21.3	67.1	−24.0	−21.0	−37.3
May	−13.9	−64.0	−84.4	−79.9	−36.9
June	−35.5	−92.9	−77.0	−73.3	−117.9
July	−163.6	−163.6	−149.5	−163.6	−160.0
August	−150.2	−150.2	−127.9	−105.2	−147.7
September	−104.2	−76.9	−87.9	−96.6	−51.1
October	2.7	23.5	−54.7	−54.7	252.9
November	−18.4	18.8	−21.2	−30.3	228.0
December	18.2	−7.0	129.1	−15.2	−12.2
Annual rainfall (mm)	376.0	364.0	353.0	178.0	501.0

where n is a parameter for the proportion of the year with a potential moisture surplus exceeding the soil moisture volume at field capacity.

19.3.3   Simulating terrain-based patterns of potential soil erosion

That terrain form affects soil erosion is, of course, a well-established fact. Nevertheless, the physical and hydrological processes and feedback mechanisms involved are still incompletely understood and are, so far, only coarsely represented in even the most advanced, distributed parameter models (see for example discussions by Beven, 1989; Grayson et al., 1992b).

In land evaluation and soil erosion risk mapping, often dealing with large areas of poor data availability, application of physically based distributed parameter models (e.g. TOPMODEL, Beven and Kirkby, 1979; SHE, Abbott et al., 1986; ANSWERS, Beasely et al., 1989; MEDALUS model, Kirkby et al., this volume, Chapter 14) is nearly impossible due to the extensive data required to run these models. As a consequence it has been, and often still is, common practice in land evaluation to account for the topographic effects on soil erosion by calculating the LS factor of the Universal Soil Loss Equation (USLE) (e.g. Albaladejo Montoro et al., 1988). As Moore et al. (1992a) point out, application of the USLE (Wischmeier and Smith, 1978) or RUSLE (Renard et al., 1991) is restricted to specific conditions of uniform runoff generation through the infiltration excess mechanism. Though the Hortonian mechanism of runoff generation does indeed apply to many hillslopes in semi-arid Mediterranean environments, the condition of uniform runoff generation and the one-dimensional character of the equations seriously limit their utilization for identifying soil erosion hazard zones in real landscapes.

Addressing the apparent incompatibility of describing water and sediment movements across the landscape in a physically sensible way on the one hand and the need to cover large areas of low data availability on the other hand, Moore et al. (1992a,b) and Grayson et al. (1992b) advocate an index approach based on terrain attributes that can be derived from DEMs.

One of the indices proposed by these authors is a dimensionless expression of potential sediment transport capacity (T_c) derived from unit stream power theory (Moore and Burch, 1986). This index, which was also applied in this case study, describes potential soil erosion as a function of specific catchment area and local slope angle:

$$T_c = \left(\frac{A_s}{22.13}\right) m \left(\frac{\sin\beta}{0.0896}\right) n \tag{5}$$

where A_s is specific catchment area per unit contour length ($m^2 m^{-1}$); β is the local slope angle (°), and m and n are constants ($m=0.6$; $n=1.3$).

The use of this expression is restricted to situations of high sediment availability where soil loss is only limited by the transport capacity of the overland flow. On sparsely vegetated slopes in the dry regions of the Mediterranean where soil loss occurs during occasional events of high rainfall intensity this seems a safe assumption.

19.4 FIELD SAMPLING AND GIS CONSTRUCTION

Data collection for this case study at El Ardal consisted of:

— air-photo interpretation for the preliminary delimitation of soil types, vegetation cover and land-use types, and landforms;
— field surveys to check the air-photo interpretations, to describe and sample soil profiles, and to determine the hydrological behaviour of the soil using rainfall simulations;
— digitizing field maps and air-photo interpretations;
— laboratory analyses of soil samples;
— digital terrain analysis to compute topographic attributes.

In the field the preliminary maps, drawn from the air-photo interpretations, were checked, corrected and drawn on 1:10 000 topographic maps. Vegetation cover and land-use mapping focused on delimiting functional units, such as arable land, fruit trees, open, discontinuous or dense shrubland or forest, rather than on species composition or vegetation community. Checking of the air-photo interpretation map was done visually while surveying the area on foot. To check and improve the preliminary soil maps, soil profiles were described and sampled along representative catenas, following the FAO guidelines (FAO, 1977).

To determine the response of the soil upon wetting, rainfall simulations were carried out along catenas in different soil and vegetation/land-use units with the University of Amsterdam simulator described by Bowyer-Bower and Burt (1989). Rainfall simulations were performed on open plots of $0.5 m^2$ at three different intensities (approximately 2–3, 5–6, 7–8 cm h^{-1}) and were continued until 100% of the bare soil was ponded or runoff was observed. To avoid bias in the measured infiltration rate due to dispersive clays, distiled water was used for all experiments. At a number of sites along the catenas the saturated hydraulic conductivity was measured in an auger hole (Kessler and Oosterbaan, 1974), and undisturbed samples were taken for determination of water retention characteristics.

Primary data input to the GIS consisted of a digitized contour map, scale 1:10 000 with 5-m contour intervals, and digitized field maps of soils and vegetation cover/land use, also at scale 1:10 000. These vector maps were converted to raster maps with 10 m resolution using GRASS 4.0 (USA-CERL, 1991). The DEM was also constructed within GRASS using the r.surf contour program for interpolation of the digitized contours to the 10 m grid. All further GIS operations were carried out with PC-RASTER (Van Deursen, 1991), a PC package that was particularly developed for hydrological purposes.

Using both standard operations included in PC-RASTER and simple map algebra a geographic data-base was built that consisted of raster overlays derived from the digital elevation model on the one hand (slope angle, slope aspect, local drain direction, specific catchment area, wetness index, potential short-wave radiation, effective short-wave radiation, and topographic moisture index), and of the digitized field maps of soils and vegetation cover/land use on the other hand. For computing the topographic moisture index (*TMI.REF*), the climatic data of 1962 were used as an example of a year with a 'normal' rainfall distribution. Combination of the DEM with the data layers for soils, vegetation cover/land use and wetness index with the field and laboratory measurements and climate records produced the overlays *SOILDEPT* (soil depth), *STORAGE* (soil moisture storage) and *RUNOFPOT* (runoff potential).

The soil depth map (soil depth defined as the depth of the unconsolidated material) was computed by recoding the soil map, according to the minimum and maximum depth of each soil type. To model the pattern of soil depth as a continuous variable and to avoid unrealistic steps at the sharp boundaries of different soil units a more natural, smoothed map was computed by:

$$SOILDEPT = MINDEPTH + T(MAXDEPTH - MINDEPTH) \qquad (6)$$

where *MINDEPTH* and *MAXDEPTH* are maps of, respectively, the average maximum and minimum soil depth for each soil type, as observed in the field, and T is a map layer which approximates the hypothesized effect of topography on soil depth through the processes of erosion and deposition. T is computed as:

$$T = \frac{WETNESS}{WETNESS.AVG} \qquad (7)$$

with *WETNESS* being the map with the value of $\ln(A_s/\tan\beta)$ for each cell and *WETNESS.AVG* a map with the average value of $\ln(A_s/\tan\beta)$ for each soil type. *STORAGE*, a map of soil moisture storage capacity, was then computed by multiplying the simulated soil depth (*SOILDEPT*) with the volumetric water content at pF = 0 as determined for 100 cm^3 rings. *RUNOFPOT* is the map showing whether a cell does or does not have a runoff potential under current rainfall conditions. It was computed from a combination of two other maps: *SATFLOW* (saturated overland flow) and *HORTONFL* (infiltration-excess overland flow). The areas where saturated overland flow is probable under current climatic conditions were identified by selecting the cells with a soil moisture storage capacity of less than 5 cm. This value corresponds to the long-term rainfall distribution of the study area, which show that *c.* 90% of all the rain falls in daily amounts of less than 5 cm. Since the saturated soil moisture contents were found to average at least 35% ($cm^3\ cm^{-3}$), soil depths of more than *c.* 15 cm should be sufficient to store this rainfall volume (Boer M. M. Unpublished field data). Cells having less than 15 cm of soil were selected by reclassifying the soil depth map. The areas of regularly occurring Hortonian overland flow were identified by comparing the measured infiltration rates with the mean maximum rainfall intensity for the case study area, estimated at 4.5 cm h^{-1} for a 10-minute period (Elias Castillo and Ruiz Beltran, 1979).

19.5 ASSESSMENT OF THE IMPACT OF INCREASING ARIDITY

Assuming a bare hillslope with no vegetation cover and uniform runoff generation, application of Equations 4 and 5 produced the terrain-based, or so-called reference patterns, of average soil moisture availability (*TMI.REF*) and potential sediment transport capacity (*PSTC.REF*) (see Figures 19.4A and 19.5A).

A   B

FIGURE 19.4   Simulated patterns of potential soil moisture availability expressed as percentage of moisture availability for a horizontal surface. (A) Reference pattern based on hillslope topography alone. In most of the area potential soil moisture conditions resemble those of a horizontal surface (light grey). In the southeastern part of the area, where slope angles often reach more than 20°, potential soil moisture availability may range from less than half (very light grey) to more than twice (dark grey) of the amount available on a horizontal surface. (B) 'Actual' pattern based on hillslope topography and vegetation cover. Especially on south slopes the pattern differs from that in part A, as the vegetation cover very effectively increases potential soil moisture availability (dark grey tones). The 'banding' in the northwestern part of the area results from the interpolation algorithm used to compute the DEM that creates small, artificial steps in nearly flat areas

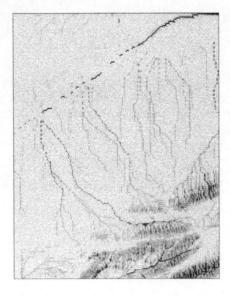 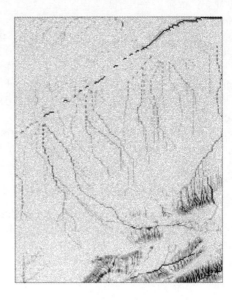

A   B

FIGURE 19.5 Simulated pattern of potential soil erosion. Light grey tones indicate a very low erosion potential, while the dark grey tones indicate a high erosion potential. (A) Reference pattern based on hillslope topography alone. (B) 'Actual' pattern based on hillslope topography and estimated effective contributing area. The most important differences between the two patterns occur in the southeastern part of the area, where dense vegetation actually creates favourable soil hydrological properties and only a small proportion of the specific catchment area effectively contributes to runoff discharge

Simulating the 'actual' patterns of soil moisture conditions or soil erosion potential from generally available data is, of course, much more complicated and is a subject of on-going research. So far, the effect of the vegetation cover on soil moisture distribution at the hillslope scale was modelled simply by assuming a linear relationship between vegetation cover, shading and soil moisture availability, and adapting the variable a_i of Equation 3 accordingly (Figure 19.4B). To identify areas of either a low or high potential for change in soil moisture conditions the difference between the values of the reference pattern and those of the actual pattern were calculated and stored as a separate raster overlay (*CHANGE.TMI*).

To calculate the actual pattern of potential sediment transport capacity (*PSTC.ACT*), the proportion of each cell's catchment area was assessed that may, according to the rainfall simulation results and long-term climate records, be expected to generate runoff under prevailing rainfall conditions. Equation 5 was then computed for each cell (Figure 19.5B). The areas where changes in soil hydrological behaviour at the patch scale can lead to either large or small modifications in the potential sediment transport capacity (*CHANGE.STC*) were identified again by calculating the difference between the values of the reference pattern (*PSTC.REF*) and those of the actual pattern (*PSTC.ACT*).

Evaluation of the potential magnitude of change in soil moisture availability or soil erosion potential still does not provide the information to identify hazard zones in terms of the areas where increasing aridity is most likely to cause desiccation or soil erosion. In order to do so, an evaluation of the potential rate of change is required as well, which is difficult to do quantitatively, as feedback mechanisms and aspects of ecosystem inertia and resilience come into play. Current knowledge of the ecosystems in question prevents the complexity of the ecological response, that determines the possible rate of change upon disturbance, to be characterized by a few parameters that can be readily derived from existing information or easily assessed in the field. For the moment therefore some simplifying assumptions are made. With respect to desiccation, the potential rate of change is assumed to be directly related to the actual soil moisture conditions: critical changes in soil moisture availability being assumed to be more likely reached in dry areas than in relatively humid areas. The areas most susceptible to desiccation (*DESICCA*) were then identified by the following step:

$$DESICCA = \left(1 - \frac{TMI.ACT}{TMI.MAX}\right)*(TMI.ACT - TMI.REF) \qquad (8)$$

where *TMI.MAX* refers to the maximum value of the *TMI.ACT* raster overlay (see Figure 19.6A).

Assessment of the potential rate of change in soil erosion potential was based on the simulated pattern of actual soil moisture availability and the actual runoff characteristics of each cell's specific catchment. Changes in soil erosion potential were assumed to result primarily from the impact of increasing aridity on the proportion of the specific catchment area that is likely to generate runoff on a regular basis. Hence, the effect of increasing aridity is most likely to be produced by a decline in soil hydrological behaviour at the sites within each cell's specific catchment that actually have high infiltration capacity and do not produce runoff under current climate conditions. The probability of a decline in soil hydrological properties at those 'good' sites was assumed to be related to the probability of changes in vegetation cover, that are, in turn, thought to be a function of actual soil moisture availability. The soil erosion hazard map was thus computed as:

$$EROSION = \left(1 - \frac{CATCHTMI.AVG}{CATCHTMI.MAX}\right)*(PSTC.REF - PSTC.ACT) \qquad (9)$$

where *PSTC.REF* and *PSTC.ACT* are, respectively, the maps of potential and actual sediment transport capacity, resulting from application of Equation 5. *CATCHTMI.AVG* is a map that shows the average value of *TMI.ACT* (actual soil moisture availability) for the part of each cell's specific catchment area that does not generate runoff under current conditions, and *CATCHTMI.MAX* refers to the maximum value of *CATCHTMI.AVG* (see Figure 19.6B).

19.6 DISCUSSION OF RESULTS

Figures 19.4 and 19.5 show the simulated patterns of soil moisture availability and soil erosion potential resulting from application of Equations 4 and 5. They show where specific combinations of soil hydrologic and topographic properties give rise to either

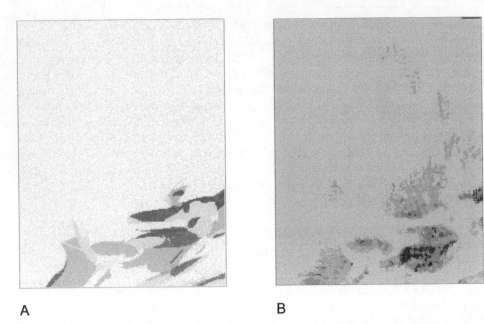

A   B

FIGURE 19.6 Simulated pattern of the areas most susceptible to desiccation in case of increasingly arid conditions (light grey = slight impact; dark grey = strong impact). (A) Effect on soil moisture conditions. (B) Effect on soil erosion potential. The most important impact is thought to occur in areas draining south slopes with potentially very unfavourable soil moisture conditions. In areas draining arable land increasing aridity would have only slight impact on potential soil erosion

favourable or unfavourable conditions for the occurrence of desiccation or soil erosion. The maps only provide a relative qualification of sensitivity to these processes since the computed patterns have not yet been compared with measured soil moisture or soil erosion patterns. A first visual comparison with field observations, however, suggests that the simulated soil moisture availability and soil erosion potential patterns agree rather well with the conditions in the field.

19.6.1 Soil erosion potential

According to the simulated pattern of soil erosion potential (Figure 19.5) the actual sediment transport capacity is rather low in most of the area except for sites along some particular flow lines where values can be very high. This is in accordance with evidence of actually occurring soil erosion being only observed in the gullies draining the Lomo de Herrero area (southeastern part: see Figure 19.1), which in turn is supported by the results of the rainfall simulation experiments, soil profile studies and analysis of rainfall records (M.M. Boer, unpublished field data). Areas where runoff is generated on a regular basis are to be found in the cultivated fields and in degraded open shrub on south slopes. In the former areas, very low slope angles and rather prominent micro-relief resulting from regular tillage prevent significant soil erosion during most of the events while, according to the incision of the colluvial deposits, soil erosion does occur in the latter areas that have higher slopes angles and lower surface roughness. The thick colluvial deposits and the buried palaeosol that were found at the valley bottom indicate that in the past soil erosion processes must have been very active, at least for some time, possibly coinciding with the beginning of agriculture in this area.

19.6.2 Soil moisture availability

The soil moisture availability pattern (Figure 19.4) is more difficult to validate, since monitoring of soil moisture conditions has only been carried out at the MEDALUS research plots (López Bermúdez et al., this volume, Chapter 8). The generated pattern describes the following, rather logical, distribution of soil moisture conditions during a year with average rainfall conditions:

— very dry sites on south slopes or nearly flat areas with no or little vegetation cover (arable land or scattered matorral);
— the most humid sites on north slopes with dense shrub or pine forest and in the valley bottom area;
— intermediate moisture conditions in the footslope zone under agricultural land use, being somewhat better along flow lines than on the hydrologically more isolated interfluve areas.

Soil moisture contents of samples taken in July 1991 at four depths at the rainfall simulation sites roughly support the computed moisture pattern. At 25 and 30 cm depth the highest volumetric soil moisture contents were found at the valley bottom location (vineyard), the lowest values at an interfluve position in the footslope zone (recently harvested cornfield) and at a clearing in discontinuous matorral, and intermediate moisture contents in two sites with favourable aspect in the footslope zone (fallow field and almond grove).

19.6.3 The response to increasing aridity

Figures 19.6A and B show which areas are most likely to respond to increasing aridity in terms of changes in soil moisture conditions or soil erosion potential. The shrubland and forest areas have the highest potential magnitude of change due to the strong effect of the vegetation cover on both soil moisture conditions and specific catchment runoff characteristics. Most of the arable land is classified as being rather inert to increasing aridity. In the shrubland and forest communities, the well-developed vegetation cover and the favourable soil structural and hydrological properties, have led to an improvement of the environmental conditions far above those described by the terrain-based 'reference pattern'. Due to structuring at the lower hierarchical levels, these areas take full advantage of the scarcely available water resources and could hence initially be rather resistant to increasing aridity. However, once the magnitude of the climatic changes go beyond the resistance of these systems, the potential for deterioration in local environmental conditions could be very significant and restoration of the former situation, being the result of long-term ecological capitalization, would be difficult. In terms of soil moisture availability and sediment transport conditions shrubland and pine forest thus have a low resilience (Westman, 1986).

In the arable land, on the other hand, current environmental conditions deviate very little from those described by the terrain-based 'reference pattern': the crops only slightly improved soil structural and hydrological properties, and provide little shade or protection against the sun and erosive forces. At the lower hierarchical levels agricultural land use has removed much of the structure (soil aggregation, vegetation–soil mosaic) and local environmental conditions are thus being determined to a large extent by topographic land properties. Consequently, the resistance to increasing aridity, at least in terms of soil moisture conditions, is very low and conditions will decrease proportionally to the changes in rainfall, temperature or other climate parameter that produces the increasing aridity. As climatological conditions improve soil moisture availability would quickly follow and hence the resilience of the system could be said to be high in this respect. The soil erosion potential, on the other hand, is thought to remain largely unchanged by increasingly arid conditions since only minor changes in the hydrological response of the contributing area are expected.

19.6.4 Desertification hazard

Finally, what can be concluded about the (distribution of the) desertification hazard under the hypothetical trend of increasing aridity? There is no single answer to this question, since it depends on the very definition of the term desertification (see, for example, Verstraete, 1986, for a discussion). If desertification is understood as the direct or indirect loss of potential bioproductivity, for instance by declining soil physical properties and subsequently decreasing soil moisture availability, then the desertification hazard could be said to be greater in the well-developed shrubland and forest areas on north slopes, where the potential magnitude of change is large, than in the areas of scattered, low, matorral or in the arable land. In the latter areas the loss of structure has already occurred to an important degree and the remaining system is rather inert to increasing aridity.

If we define desertification, however, as the decline of bioproductivity to a level of, so-called, desert-like conditions different conclusions could be drawn. Such an advanced stage of degradation is likely to occur first at sites with very shallow soils on steep south slopes where accelerated soil erosion will easily expose the parent rock. In the arable land of the footslope zone, where soils consist of less than 40 cm of unconsolidated material over rock-hard petrocalcic horizons, such marginal conditions could also occur rather easily once the upslope drainage area became effective. On the more densely vegetated north slopes this process would take longer or might not occur at all. With the thick, more clayey, colluvial deposits and, possibly increasing, lateral water input, the valley bottom area would obviously be the last part of the landscape to develop 'desert-like' conditions.

19.7 CONCLUSIONS

In this chapter the potential application of the DRU concept at the hillslope scale has been explored for a case study area undergoing a hypothetical change in climate conditions. The response to an increasing aridity was discussed in terms of the impact on hillslope patterns of soil moisture availability and soil erosion potential, that were derived from simple models describing the movement of surface and subsurface water across the landscape.

Based upon the idea that degradation causes landscape patterns of soil moisture and soil erosion potential to become increasingly determined by hillslope form and position and to a lesser extent by soil and vegetation properties, reference patterns were derived from a DEM. Subsequently, these topography-based reference patterns were used to assess the potential degradability of the land by comparing them with simulated maps of the actual distribution of soil moisture availability and soil erosion potential.

Validation of the simulated pattern of desertification response is difficult as it concerns a hypothetical future situation. The accuracy of the models underlying the DRU identification can be tested, however, by comparing the simulated patterns of soil moisture availability and soil erosion potential with measured values of soil moisture content and evidence of actually occurring soil erosion. To improve these models, current research therefore aims at mapping evidence of actual erosion activity and assessing mean annual soil moisture conditions, for land units corresponding to a 10-m grid cell, by periodically measuring soil moisture content along several transects across the case study area.

From the results obtained in this case study we conclude that by recognizing hierarchical structures in potential resource utilization and identifying corresponding stages of land degradation, the DRU concept may provide a framework to evaluate, at first theoretically, the response of different land units to different disturbances. Operationalization of the concept for application in land evaluation and hazard zonation in large areas of low data availability requires, however, further research at a range of nested scales. As data availability will continue to be an important limitation to the use of advanced distributed models in hazard zonation, future efforts should preferably take the line of establishing simple, but physically and ecologically sound, indices that determine the distribution of local environmental conditions in a semi-quantitative fashion and allow the relatively easy identification of the areas of most probable change.

19.8 REFERENCES

Abbott, M.B., Bathurst, J.C., Cunge, J.A., O'Connell, P.E. and Rasmussen, J. 1986. An introduction to the European Hydrological System–Systeme Hydrologique Europeen SHE, 2: Structure of a physically-based distributed modelling system. *J. Hydrol.*, **87**, 61–77.

Albaladejo Montoro, J., Ortiz Silla, R. and Martinez-Mena Garcia, M. 1988. Evaluation and mapping of erosion risks: an example from S.E. Spain. *Soil Technol.*, **1**, 77–87.

Albaladejo, J., Stocking, M. and Diaz, E. (eds) 1990. *Soil Degradation and Rehabilitation in Mediterranean Environmental Conditions*. CSIC-CEBAS, 235 pp.

Barling, R.D., Moore, I.D. and Grayson, R.B. 1994. A quasi-dynamic wetness index for characterizing the spatial distribution of zones of surface saturation and soil water content. *Water Resour. Res.*, **30**(4).

Bergkamp, G., Cammeraat, L.H., Martínez Fernández, J. 1996. Water movements and vegetation patterns in shrubland and an abandoned field in desertification threatened areas: results of large rainfall simulations in Spain. *Earth Surface Processes and Landforms* (in press).

Beasely, D.B., Huggins, L.F. and Monke, E.J. 1989. ANSWERS: a model for watershed planning. *Trans. American Society Agric. Engineers*, **23**(4), 938–944.

Beven, K. 1989. Changing ideas in hydrology—the case of physically-based models. *J. Hydrol.*, **105**, 157–172.

Beven, K. and Kirkby, M.J. 1979. A physically based variable contributing area model of basin hydrology. *Hydrol. Sci. Bull.*, **24**, 43–69.

Beven, K.J. and Wood, E.F. 1983. Catchment geomorphology and the dynamics of contributing areas. *J. Hydrol.*, **65**, 139–158.

Boer, M.M. and Puidefabregas, J. 1995. Assessing spatial patterns of precipitation and soil moisture in dry Mediterranean landscapes using digital terrain analysis. In: Ibañez, J.J. and Machado, C. (eds). *Análisis de la variabilidad espacio-temporal y procesos caóticos en ciencias medio ambientales. Geoforma Ediciones*, Logrofia, Spain, 259–276.

Bowyer-Bower, T.A.S. and Burt, T.P. 1989. Rainfall simulators for investigating soil response to rainfall. *Soil Technol.*, **2**, 1–16

Burt, T.P. and Butcher, D.P. 1986. Development of topographic indices for use in semi-distributed hillslope runoff models. *Z. Geomorph. N.F., Suppl. Bd*, **58**, 1–19.

Durand, P., Robson, A.M. and Neal, C. 1992. Modelling the hydrology of subMediterranean montane catchments (Mont-Lozère, France) using TOPMODEL: initial results. *J. Hydrol.*, **139**, 1–14.

Elias Castillo, F. and Ruiz Beltran, L. 1979. *Precipitaciones maximas en España*. ICONA, Monografía 21, Ministerio de Agricultura, Spain, 555 pp.

Fantechi, R. and Margaris, N.S. (eds) 1986. *Desertification in Europe*. Commission of the European Communities, Reidel, Dordrecht, 231 pp.

FAO 1977. *Guidelines for Soil Profile Description*. Soil Survey and Fertility Branch, Land and Water Development Division, FAO, Rome.

Francis, C. 1986. Soil erosion on fallow fields: an example from Murcia. *Papeles de Geografía Física*, **11**, 21–28. Universidad de Murcia.

Francis, C.F. and Thornes, J.B. 1990. Matorral: erosion and reclamation. In Albaladejo, J., Stocking, M.A. and Diaz, E. (eds) *Soil Degradation and Rehabilitation in Mediterranean Environmental Conditions*. CSIC, Murcia, 87–115.

Grayson, R.B., Moore, I.D. and McMahon, T.A. 1992a. Physically based hydrologic modelling, 1. A terrain-based model for investigative purposes. *Water Resour. Res.*, **28** (10), 2639–2658.

Grayson, R.B., Moore, I.D. and McMahon, T.A. 1992b. Physically based hydrologic modelling, 2. Is the concept realistic? *Water Resour. Res.*, **28** (10), 2659–2666.

Hornberger, G.M., Beven, K.J., Cosby, B.J. and Sappington, D.E. 1985. Shenandoah Watershed Study: calibration of a topography-based, variable contributing area hydrological model to a small forested catchment. *Water Resour. Res.*, **21** (12), 1841–1850.

Houghton, J.T., Seck, M. and Moura, A.D. (eds) 1990. *Scientific Assessment of Climatic Change*. Report I of IPCC Working Group I, 365 pp.

IGME 1972. *Mapa Geológico de España*, 1:50 000, sheets 911 and 912.

Iiada, T. 1984. A hydrological method of estimation of topographic effect on saturated throughflow. *Trans. Japanese Geomorphological Union*, **5** (1), 1–12.

Imeson, A.C. and Emmer, I. 1992. Implications of climatic change for land degradation in the Mediterranean. In Jeftic, L., Milliman, J.D. and Sestini, G. (eds) *Climatic Change in the Mediterranean*. Edward Arnold., 105–128.

Imeson, A.C., Perez Trejo, F. and Cammeraat, L.H. 1996. The response of landscape units to desertification. In this volume.

Jones, J.A.A. 1986. Some limitations of the a/s index for predicting basin-wide patterns of soil water drainage. *Z. Geomorph. N.F., Suppl. Bd*, **60**, 7–20.

Kessler, J. and Oosterbaan, R. 1974. Determining hydraulic conductivity of soils. In *Drainage Principles and Applications*. Int. Inst. for Land Reclamation and Improvement. Publication 16, III, 253–256.

Ladson, A.R. and Moore, I.D. 1991. Soil water prediction on the Konza Prairie by microwave remote sensing and topographic attributes. *J. Hydrology*, **138**, 385–407.

Lee, R. 1978. *Forest Microclimatology*. Columbia University Press, New York, 276 pp.

López Bermúdez, F. 1990. Soil erosion by water and the desertification of a semi-arid Mediterranean fluvial basin: the Segura basin, Spain. *Agric., Ecosyst. Environ.*, **33**, 129–145.

López Bermúdez, F., Romero Díaz, M.A. and Martínez Fernández, J. 1991. Soil erosion in a semi-arid Mediterranean environment: El Ardal experimental field (Murcia, Spain). In Sala, M.,

Rubio, J.L. and García Ruiz, J.M. (eds) *Soil Erosion Studies in Spain*. Geoforma Ediciones, Logroño, Spain, 137–152.

López Bermúdez, F. Martínez Fernández, J. and Martínez Martínez, J. 1992. *MEDALUS Interim report*. March 1992, University of Bristol.

Moore, I.D. and Burch, G.J. 1986. Physical basis of the length–slope factor in the Universal Soil Loss Equation. *Soil Sci. Soc. Am. J., 50*, 1294–1298.

Moore, I.D., Burch, G.J. and Mackenzie, D.H. 1988. Topographic effects on the distribution of surface soil water and the location of ephemeral gullies. *Trans American Soc. Agric. Engineers, 31* (4), 1098–1107.

Moore, I.D., Grayson, R.B. and Ladson, A.R. 1991. Digital terrain modelling: a review of hydrological, geomorphological and biological applications. *Hydr. Processes, 5*, 3–30.

Moore, I.D., Wilson, J.P. and Ciesiolka, C.A. 1992a. Soil erosion prediction and GIS: linking theory and practice. *Proc. Geographic Information System for Soil Erosion Management*, Taiyuan, Shanxi Province, China, 2–11 June 1992.

Moore, I.D., Turner,A.K., Wilson, J.P., Jenson, S.K. and Band, L.E. 1992b. GIS and land surface-subsurface process modelling. In Goodchild, M.F., Parks, B. and Steyert, L.T. (eds) *Geographic Information Systems and Environmental Modelling*. Oxford University Press.

Moore, I.D., Norton, T.W. and Williams, J.E. 1993. Modelling environmental heterogeneity in forested landscapes. *J. Hydrol., 150*, 717–747.

O'Loughlin, E.M. 1986. Prediction of surface saturation zones in natural catchments by topographic analysis. *Water Resources Research, 22* (5), 794–804

Palutikof, J.P., Guo, X., Wigley, T.M.L. and Gregory, J.M. 1992. *Regional Changes in Climate in the Mediterranean Basin due to Global Greenhouse Warming*. Mediterranean Action Plan Technical Reports Series, No. 66, UNEP, Athens, 172 pp.

Quinn, P., Beven, K.J., Chevallier, P. and Planchon, O. 1991. The prediction of hillslope flow paths for distributed hydrological modelling using digital terrain models. *Hydrol. Processes, 5*, 59–79.

Renard, K.G., Foster, G.R., Weesies, G.A. and Porter, J.P. 1991. RUSLE: Revised Universal Soil Loss Equation. *J. Soil and Water Conservation, 46*, 30–33.

Robson, A., Beven, K.J. and Neal, C. 1992. Towards identifying sources of subsurface flow: a comparison of components identified by a physically based runoff model and those determined by chemical mixing techniques. *Hydrol. Processes, 6*, 199–214.

Romero Diaz, M.A., López Bermúdez, F., Thornes, J.B., Francis, C. and Fisher, G.C. 1988. Variability of overland flow erosion rates in a semi-arid Mediterranean environment under matorral cover in Murcia, Spain. In Harvey, A.M. and Sala, M. (eds) *Geomorphic Processes in Environments with Strong Seasonal Contrasts*. Vol. II.: Geomorphic systems. Catena Supplement 13, 1–11.

Rubio, J.L. and Sanroque, P. 1987. *Soil Erosion and Desertification in Spain*. Proc. symposium on strategies to combat desertification in Mediterranean Europe. Commission of the European Communities, Consejo Superiór de Investigaciones Científicas, Generalitat de València.

Sestini, G., Jeftic, L. and Milliman, J.D. 1989. *Implications of Expected Climate Changes in the Mediterranean Region: An Overview*. Mediterranean Action Plan Technical Reports Series, No. 27, UNEP, Athens, 52 pp.

Tomaselli, R. 1981. Main physiognomic types and geographic distribution of shrub systems related to Mediterranean climates. In di Castri, F., Goodall, D.W. and Specht, R.L. (eds) *Mediterranean-Type Shrublands*. Elsevier Scientific Publishing Company, Amsterdam, 95–106.

USA-CERL 1991. *GRASS 4.0 User's Manual*. United States Army Corps of Engineers, Construction Engineering Laboratory, Champaign, IL.

Van Deursen, W.P.A. 1991. The PC-RASTER PACKAGE. Dept. of Physical Geography, University of Utrecht.

Verstraete, M. 1986. Defining desertification: a review. *Climate Change, 9* (1/2), 5–18.

Westman, W.E. 1986. Resilience, concepts and measures. In Dell, B., Hopkins, A.J.M. and Lamont, B.B. (eds) *Resilience in Mediterranean-Type Ecosystems*. Junk Publishers, Dordrecht, The Netherlands, 5–20.

Wigley, T.M.L. 1992. Future climate of the Mediterranean Basin with particular emphasis on
changes in precipitation. In Jeftic, L., Milliman, J.D. and Sestini, G. (eds) *Climate Change and the
Mediterranean*. Edward Arnold, London, 15–44.

Wischmeier, W.H. and Smith, D.D. 1978. *Predicting Rainfall Erosion Losses, A Guide to
Conservation Planning*. Agriculture Handbook No. 537, US Dept. of Agriculture, Washington,
DC.

20

Remote Sensing of Mediterranean Vegetation and Surface Lithology

A. R. HARRISON[1], J. MELIA[2], J. BASTIDA[2], S. GANDIA[2], M. A. GILABERT[2], S. J. HURCOM[1], A. LOPEZ BUENDIA[2], M. TABERNER[1] and M. T. YOUNIS[2]

[1]*Department of Geography, University of Bristol, UK*, [2]*Departamento de Termodinamica, Universitat de València, Spain*

20.1 INTRODUCTION

20.1.1 Background

Information on vegetation cover and surface lithology in semi-arid environments can provide two different inputs into desertification studies. First, as an input into models which aim to simulate and predict the effects of desertification. Second, by providing information on current and past changes in vegetation cover and terrain characteristics to enable areas which are threatened by desertification to be identified and monitored.

In terms of prediction, current attempts to model the interaction between vegetation and land degradation have been largely based at the spatial scale of the hillslope (Kirkby et al., 1990; Thornes, 1990) or catchment (Faulkner, 1990; Mitchell, 1990). A major requirement of such models is data on vegetation state (above-ground biomass, leaf area and cover) and surface state (composition, lithology, micro-topography). With attention shifting towards prediction of land degradation rates for planning and mitigation purposes, two major consequences ensue: first, the models must be extended to the regional scale, say 5000–10 000 km^2, and second, their run time must be increased to include periods of up to 100 years (Kirkby, 1993). To meet data needs at this new spatio-temporal scale, data collection methods which encompass both a large spatial extent and a rapid acquisition rate are required. Only remote sensing has this potential (Casmir et al., 1980; Hellden and Stern, 1980; Justice and Townshend, 1981; Pouget et al., 1984).

In terms of monitoring, remote sensing offers similar advantages in that it can monitor large areas, relatively frequently. For example, numerous studies have shown that remote sensing can be used to accurately monitor vegetation through time using the relationship between vegetation indices and green-leaf biomass (Wiegland et al., 1979; Holben et al., 1980).

Mediterranean Desertification and Land Use. Edited by C. Jane Brandt and John B. Thornes.
© 1996 by John Wiley & Sons, Ltd.

20.1.2 Remote sensing techniques employed in the MEDALUS project

This chapter describes the collection and analysis of ground spectroradiometric data, and the analysis of satellite sensor imagery, carried out by remote sensing groups at the University of Bristol (UK) and the University of Valencia (Spain). There are two main components to the work described here. First, the refinement of measurements of spectral signatures for different vegetation conditions and surface lithologies, based on *in-situ* radiometric observations, in order to understand and correct the spectral properties of satellite measurements from the Landsat Thematic Mapper (TM) sensor. Second, the use of multispectral classification procedures, based on Landsat TM imagery, to extend these site-based measurements of vegetation and surface lithology up to the regional scale. In addition, the Bristol group has investigated the use of vegetation indices and spectral mixture modelling to estimate vegetation amount and surface cover. Throughout, the primary objective was to provide basic data-sets on vegetation and surface characteristics to other groups within the MEDALUS project.

In most cases, techniques were based on fairly standard approaches which were then adapted to suit the specific problems presented by vegetation and surface conditions in semi-arid Mediterranean environments. A particular problem here is the complexity of the terrain where cover units are small and cover types are composed of complex mixtures of vegetation and bare surfaces. Moreover, many areas are not characterized by simple land cover types, nor frequently are the types separated by abrupt boundaries but merge one with another (Townshend and Justice, 1980; Harrison and Garg, 1991).

Three main groups of techniques were employed for extracting information on vegetation, land cover and surface characteristics from satellite sensor imagery:

1. Classification of multi-date Landsat TM imagery, using the maximum likelihood algorithm, providing information on the spatial distribution and extent of vegetation and land cover, and classification of single-date Landsat TM imagery, providing information on surface lithology.
2. Classification of sub-pixel fractional cover estimates using a spectral mixture modelling approach, providing information on percentage cover and composition within the vegetation and land cover distributions.
3. Production of normalized difference vegetation index (NDVI) images from calibrated multi-date Landsat TM imagery, providing information on vegetation amount (biomass) and inter-seasonal changes.

Each of these techniques depend upon the scaling-up of vegetation and surface characteristics, based on biophysical and mineralogical determinations, and *in-situ* spectral measurements, in relation to the spectral and statistical properties within calibrated Landsat TM image data. The approach developed here enables a coupling of these measurements—biophysical and mineralogical determinations, *in-situ* spectral reflectances and multispectral image data—to establish a quantitative means of relating *in-situ* measurements to digital values in Landsat TM imagery. Ideally, the coupling should be physically based. However, given the difficulties of absolute calibration of satellite sensor imagery, natural variation within vegetation communities and surface properties, and unexplained sources of measurement error, a statistically-based coupling often results.

A basic requirement for scaling-up is that a body of pre-processing techniques are available for converting raw radiation data to SI units to enable verification and inter-comparison between *in-situ* measurements and image data-sets. The provision of these techniques, along with a set of related analysis procedures, was a major task during Phase I of the MEDALUS project.

20.1.3 Organization of the chapter

The contents of this chapter are divided into two main sections. The first section (Section 20.2) presents the work of the University of Bristol which was focused on the development of procedures for deriving spatial distributions of vegetation and land cover, and estimates of surface cover and vegetation amount. The use of these procedures is demonstrated for a multi-date Landsat TM scene of the Rio Cobres basin (Portugal), containing the Almocreva and Vale Formoso field sites. The second section (Section 20.3) presents the work of the University of Valencia which concentrated on bare surfaces and, in particular, the spectral and mineralogical characterization of surface lithology as a means of mapping desertification potential. This is demonstrated for a single-date Landsat TM scene of the Rio Mula basin (southeast Spain), containing the El Ardal field site. Both studies were carried out largely independently of each other although there was some coordination of field spectroradiometry procedures to facilitate integration of *in-situ* spectral data-sets.

20.2 VEGETATION AND LAND COVER

20.2.1 Study area

The study area for vegetation and land cover is located in the Rio Cobres basin, south Portugal, and contains the MEDALUS field sites at Almocreva and Vale Formoso. Landsat TM image was acquired for the area on two dates: for the end of the growing season (13 April 1985) and the dry season (4 September 1985). A 2048×2048 pixel subscene covering the Almocreva and Vale Formoso field sites was extracted from the original data for analysis (Figure 20.1).

Ground data were collected during a series of field campaigns in Portugal (Rio Cobres basin), and Spain (Belmonte, Castilla La Mancha). The Belmonte site in central Spain is part of the EFEDA (European Field Site in Desertification Threatened Areas) test site and measurements taken at this site were used to supplement the Portuguese data-set. The final set of ground spectra and biophysical measurements is a composite of data from the April 1992 visit to the Castilla La Mancha field site, Spain, and the June 1992 visit to the field sites in the Rio Cobres basin, Portugal (Table 20.1).

20.2.2 Ground radiometry and field measurements

Data collection

Ground data consisted of two different types: ground radiometry and field measurements. Measurements of ground radiance were taken using the GER single field of view IRIS (SIRIS) spectroradiometer with the radiometer head placed vertically above the target at a distance of between 1 and 1.5 m. A single measurement consists of the recorded radiance from the sample along with the solar irradiance curve. This latter spectra was obtained by

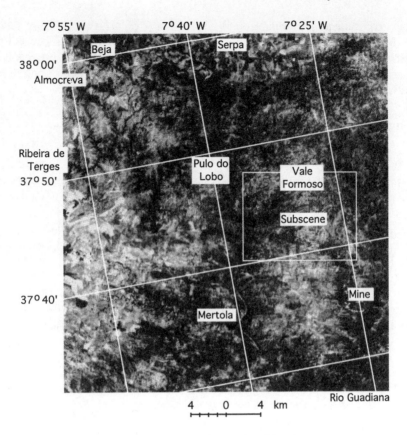

FIGURE 20.1   Extract from Landsat TM scene (path/row: 203/34, date: 14 April 1985), displayed as principal component (PC1) image, showing Portugese field sites at Almocreva and Vale Formoso. Vale Formoso subscene is located centre-right in image

TABLE 20.1   Summary of field spectroradiometric and biophysical measurements

Species	Sample	Date	SIRIS	Video	Leaf area index	Leaf chlor.	Plant height	Leaf moist
					Biophysical			
Quercus coccifera	16	Apr. 1992	✓	✓	✓	✓	✓	✓
Quercus rotundifolia	16	Apr. 1992	✓	✓	✓	✓	✓	✓
Cistus 1	12	June 1992	✓	✓	✓	✓	✓	X
Cistus 2	10	June 1992	✓	✓	✓	✓	✓	X
Rosmarinus	10	June 1992	✓	✓	✓	✓	✓	X
Eucalyptus	10	June 1992	✓	✓	✓	✓	✓	X

measuring the reflected upwelling solar irradiance from a Spectralon reference panel (Milton, 1987).

The SIRIS spectroradiometer covers the wavelengths from 290 to 3000 nm. Vegetation spectra, over this region, are usually considered to be subject to three main influences:

(i) 400–700 nm, the visible region—absorption by chlorophyll a and b and associated pigments is the dominating factor (Gausman et al., 1969).

(ii) 700–1200 nm, the near infra-red (NIR)—high reflectance has been shown to be associated with refraction caused by interfaces between cell walls and air (Gausman et al., 1969).

(iii) 1200–2500 nm, the short wave infra-red region (SWIR)—this is usually associated with the absorption properties of moisture.

Within these regions more localized absorption phenomena have been identified, associated with water absorption peaks (970, 1190, 1450, 1940 nm) (Curico and Petty, 1951; Hunt et al., 1987), lignin associations (1256, 1555, 1311 nm) and nitrogen (1240, 1522, 1559, 1587 nm) (Hergert, 1971; Wessman et al., 1987; Curran et al., 1992). A SIRIS plot of vegetation showing the distribution of the regions and absorption bands are given in Figure 20.2. This figure also shows the position and shape of the Landsat TM spectral band filters.

At each location three spectra were recorded in order to average out short-term atmospheric perturbations. After sampling at one location the projected SIRIS field-of-

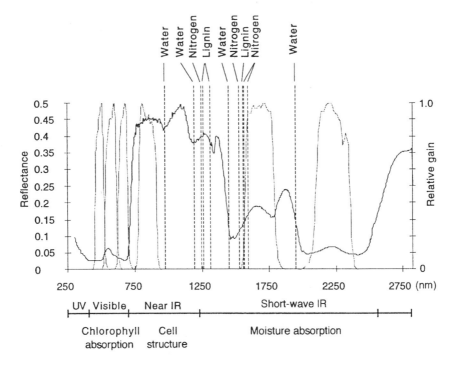

FIGURE 20.2   Significant spectral reflectance and absorbtion characteristics of green vegetation

view (FOV) was located and delimited by looking through the viewfinder. This area was recorded by video camera to allow the composition of the projected FOV to be measured. Having determined the FOV, an imaginary column was formed and projected through the vegetation canopy to the ground surface. All leaves which fell within this column were harvested and placed in sealed plastic bags ready for further analysis. Biophysical variables collected within this column include leaf area index (LAI) measurements, canopy chlorophyll data and canopy moisture content. Table 20.1 summarizes the data collected. The analysis of canopy chlorophyll and canopy moisture content data are presented in Harrison et al. (1993).

Pre-processing

To derive meaningful information from the raw data a certain amount of pre-processing must be undertaken to both the ground radiometric data and the field measurements.

In their raw form the spectroradiometric data consist of digital counts of both target and reference radiance. To convert these data to reflectance the following steps must be taken:

(i)   correct for changes in instrument gain during a scan;
(ii) correct for the reflectance properties of the reference panel;
(iii) convert from digital count to radiance ($W m^{-2} Sr^{-1} m^{-1}$) using, machine-specific calibration files; and
(iv) divide the target radiance by the reference to obtain bi-directional reflectance factors.

Since reflectance is a ratio term, when the input signal has a low signal-to-noise ratio (SNR) the resulting reflectance value often appears very noisy. To counteract this effect a differential Gaussian filter was applied to the reflectance curves when the target signal fell below a certain threshold. This threshold was determined from a graphical analysis of the noise contained within the target signal (Harrison et al., 1993). All spectral measurements have been documented and placed in a spectral library.

LAI measurements were made using the following method:

(i)   the leaves within the SIRIS FOV were harvested;
(ii) the leaves were laid on a white surface, so that none of them overlapped, and a video image recorded;
(iii) this video image was captured digitally using a video frame grabber and thresholded in order to classify the leaves; and
(iv) the leaf area was then calculated and ratioed by the area of the SIRIS FOV to give LAI.

Leaf biomass (dry weight of leaf matter per unit ground area) measurements were made by determining the dry weight of leaves within the column. After LAI determination, the leaves were dried in an oven for 24 hours at 100°C, and then weighed, giving the total dry weight of leaves within the column. This value was divided by the area of the column to yield biomass ($g cm^{-2}$).

Surface leaf area is defined as the fraction of directly irradiated green vegetation within the FOV of the spectroradiometer. Surface leaf area was calculated from a knowledge of the horizontal distribution of the elements of the FOV directly visible to the spectroradiometer. Elements included sunlit leaves, shaded leaves, twigs and understorey (shaded and unshaded). The FOV composition was determined by recording video images, through the SIRIS eye-piece, which were subsequently captured digitally in a similar way to the images used in the LAI calculations. From these images the surface leaf area within the FOV could be determined and expressed as a percentage via a simple pixel count.

Canopy chlorophyll content was determined spectrophotometrically (Arnon, 1949). Each sample consisted of 10 leaves which were harvested immediately before leaving the sampling site to minimize post-harvest breakdown of chlorophyll pigments. Canopy moisture content data were collected from the leaves within the column which were harvested for LAI and biomass determination.

Analysis

Two main techniques were used to analyse the ground data: (i) the application of filter functions to simulate the Landsat TM measurement space and (ii) factor analysis to aid analysis of data-sets with a high dimensionality.

Application of Landsat TM filter functions
To examine the ground-based SIRIS spectra in Landsat TM spectral space it is necessary that the SIRIS spectra are sampled in both wavelength and magnitude. This is achieved by applying the filter functions (Figure 20.2) to the SIRIS spectra. The filter functions scale the response of SIRIS spectra so that they mimic the Landsat TM spectra. However it should be noted that this procedure, while enabling the spectral response function of the Landsat TM sensor to be modelled, does not simulate the SNR of the Landsat TM sensor.

Factor analysis
Factor analysis is a technique which enables a multivariate data-set to be re-expressed in terms of factors which may be attributed to causing the variance (Malinowski and Howery, 1980). It has the ability to decouple spectral response curves into a set of simplified parameters that lend themselves to physical modelling. Here, the objective is to attempt to relate each factor to certain combinations of biophysical variables.

Factor analysis can express a spectral data matrix $\mathbf{D}$ in terms of both the factors causing the variance in the data-set $\mathbf{C}$ and the contribution of each factor to each sample in the original data matrix $\mathbf{R}$:

$$[\mathbf{D}] = [\mathbf{R}][\mathbf{C}] \tag{1}$$

where $\mathbf{D}$ is the experimental data matrix, $\mathbf{R}$ is the response or eigenspectra matrix of independent 'basis' spectral curves, and $\mathbf{C}$ is the eigenvector matrix consisting of the contributions or scalar multiples of each 'basis' curve to the experimental data.

Principal component analysis (PCA) (Richards, 1993) is initially used to decompose the data matrix $\mathbf{D}$ into an abstract eigenspectra matrix $\mathbf{R}_A$ and an abstract eigenvector matrix $\mathbf{C}_A$ such that $[\mathbf{D}] = [\mathbf{R}]_A[\mathbf{C}]_A$. Mathematically this is accomplished by solving the eigenvalue problem:

$$[Z]_o[C]_A = [\lambda][C]_A \tag{2}$$

where Z_o is a symmetric covariance matrix about the origin and λ is a diagonal matrix of eigenvalues. The abstract eigenspectra is then constructed according to:

$$[R]_A = [D][C]_A{}^T \tag{3}$$

where $C^{-1} = C^T$ for orthonormal matrices.

The eigenvalues, which are extracted in order of importance, are used to extract an intrinsic number of basis curves which account for all curve shape differences in the experimental data (Simonds, 1963). Knowing the dimensionality of the vegetation data-set also aids in determining the number of wavebands, their locations, and resolutions needed to fully characterize a vegetation species (Huete and Escadafal, 1991).

20.2.3 Image processing and analysis

Pre-processing

There were three stages of image pre-processing: image registration, image normalization, and image calibration. To carry out multitemporal image classification only image co-registration is necessary; however, for any other multitemporal study, the images need to be normalized to account for the varying effects of the atmosphere. Image normalization accounts for the relative magnitude of these effects within the same sensor band among scenes of different dates. Image correction, because it refers to an absolute standard, also takes into account the atmospheric effects among sensor bands for single-date imagery.

Image registration

Areas of interest were extracted from the two scenes to be co-registered and the images were shifted relative to each other until they were located in the x and y directions to within 0.5 pixels. Owing to the reasonably short time between the acquisition of the two scenes, it was unnecessary to warp the two images together. After co-registration it is possible to normalize the images.

Image normalization

Normalization is necessary to overcome the corruption caused by the atmosphere between dates. The overall effect of the atmosphere is two-fold: an additive term is introduced due to the path radiance and a multiplicative term causing range attenuation. Both terms vary with wavelength and their influence is greatest in the visible region of the electromagnetic spectrum.

The normalization procedure employed identifies a sample of the most consistent darkest and brightest pixels within each spectral band across the multitemporal Landsat TM data-set (Mather, 1987). It is assumed that the reflectance properties of these pixels are temporally invariant, and further, that any relative differences in reflectance are the result of atmospheric effects. In practice, these pixels define the intensity axis (i.e. the vector from the darkest to the brightest pixels) for each spectral band of the Landsat TM image. From the changes in orientation and intercept of these spectral intensity axes across the multitemporal data-set, the relative effects of the atmosphere for each spectral band are estimated (Harrison et al., 1993).

Image calibration
The calibration procedure makes use of ground spectroradiometric measurements to provide a 'standardized' set of reflectance measurements which can be compared with the Landsat TM images. SIRIS spectra were taken of selected bare soil fields which were sufficiently large to be located in the imagery. After applying filter functions to the spectra, to match the Landsat TM sensor bandwidths, the ground spectra were regressed with satellite spectra, extracted for the same area, and the gains and offsets required to convert the entire scene to radiance and reflectance units calculated. The method is often referred to as the empirical line method of atmospheric correction (Conel et al., 1987).

The performance of the normalization and calibration techniques was checked by comparing the results from each method and by comparing with other ground spectra.

Analysis

Three types of image processing algorithms have been employed in this research. Two of them, the maximum likelihood classifier and normalized difference vegetation index (NDVI) are used routinely. The third, the linear mixture model, is not yet in common use.

Maximum likelihood classification
The maximum likelihood algorithm (Richards, 1993) is one of the most common classification methods used with remote sensing image data. If it is assumed that the distribution of each class is multivariate normal within spectral space, then each class can be fully characterized by a mean vector and covariance matrix. A pixel x is assigned to each class w_i in accordance with the following rule:

$$x \in w_i \text{ if } p(w_i/x) > p(w_j/x) \text{ for all } j \text{ not } = 1 \qquad (4)$$

where $p(x/w_i)$ is the probability of finding a pixel from class w_i at position x (estimated from the multivariate normal distribution for class w_i). The term $p(x)$ represents the probability of finding a pixel from any class at x, and $p(w_i)$ is the probability that class w_i occurs in the image. In practice, these two terms are not required. In order to train the maximum likelihood classifier, areas in the image representing homogeneous classes must be located. This process was carried out by P. Casimiro of the Portuguese MEDALUS field group using a combination of local field knowledge and panchromatic aerial photography.

The output from a maximum likelihood algorithm is a thematic map with each pixel coded to represent the most likely class occurring within that pixel. When the majority of Landsat TM pixels contain more than one class, as is often the case given the spatial variability of the semi-arid Mediterranean landscape, this procedure is prone to high rates of misclassification. As a result, new methods have been devized with the explicit aim of estimating the proportion of each ground-cover class within each pixel. The most popular of these is the linear mixture model.

Linear mixture modelling
Linear mixture modelling is based upon a model of sub-pixel spectral mixing which, when certain constraints are met, can be unambiguously inverted to estimate sub-pixel class proportions (Horwitz et al., 1971). Spectral mixing is the process whereby reflectance spectra from different, spectrally distinct, surface components or 'endmembers' combine to produce the single radiance function measured by the detector. To model this process four assumptions are required:

  (i) spectral mixing is macroscopic, that is, each received photon interacts with just one endmember before reaching the sensor (Singer and McCord, 1979);

 (ii) changes in spectral reflectance are due to differences in endmember sub-pixel areal proportions only;

(iii) the number of endmembers cannot exceed the number of significant or intrinsic dimensions of the data-set (Kent and Mardia, 1986); and

(iv) endmembers are spectrally distinct.

For a sensor with one spectral band the spectral mixing process can be described by:

$$x = \mu_1 f_1 + \mu_2 f_2 + \ldots + \mu_c f_c + \epsilon \tag{5}$$

where x is the composite pixel signal, the μ_i are the mean vectors or endmembers characterising each class, the f_i are the proportions that each class occupies in the pixel, c is the number of such classes, and ϵ is the error term. When the sensor has more than one band (as is usually the case), this linear equation becomes a series of linear equations which can be expressed in matrix notation as:

$$\mathbf{X} = \mathbf{M}f + \epsilon \tag{6}$$

In this case, if n is the number of sensor spectral bands then $\mathbf{X}$ is the $(n * 1)$ observation vector for a pixel, $\mathbf{f}$ is the $(c * 1)$ vector of class proportions, $\mathbf{M}$ is the $(n * c)$ matrix of endmembers and ϵ is the error matrix. To obtain an estimate of $\mathbf{f}$ the sum of squares of e can be minimized by carrying out a standard least squares fit:

$$(\mathbf{X} - \mathbf{M}f)(\mathbf{X} - \mathbf{M}f) \tag{7}$$

This is the classical estimator (Pech et al., 1986).

In order to apply the mixture model the ground classes or spectral endmembers need to be determined. In this case the same classes used for the maximum likelihood classification were used for the linear mixture model. The output from a mixture model is a series of fraction images. Each image displays the sub-pixel proportions of a particular class (see Figure 20.16).

Normalized difference vegetation index
Vegetation indices are frequently used to provide an indication of vegetation amount based on some combination of the red and near infra-red (NIR) wavelengths (Curran, 1980). The red wavelengths are used as they are sensitive to chlorophyll absorption, and the NIR because it is reflected strongly by leaf structure. Although popular, the precise *meaning* of vegetation index numbers (VIN) is unclear in the absence of other information—differing canopy structures and variation in canopy geometry among

species change the response radically. Absolute calibration with plant biomass is difficult, and vegetation monitoring requiring multi-date imagery is ineffective without atmospheric corrections.

One of the most common vegetation indices is the NDVI which is defined by:

$$VIN_{NDVI} = (NIR - red)/(NIR + red) \tag{8}$$

As the NDVI is effectively a ratio about the sensor origin it never behaves in a totally predictable manner—path radiance always shifts the real origin in a wavelength-dependent manner. Furthermore, any atmospheric attenuation will change the real slope of the soil line so that the value of an index number becomes dependent to some extent on its brightness. If the images are normalized and account is taken of path radiance then the NDVI becomes more accurate and can be compared across dates.

20.2.4 Data and results

Field data

Reflectance spectra
The SIRIS reflectance spectra exhibited considerable within-species variation. For example, Figure 20.3 shows reflectance spectra based on 16 samples of *Quercus*

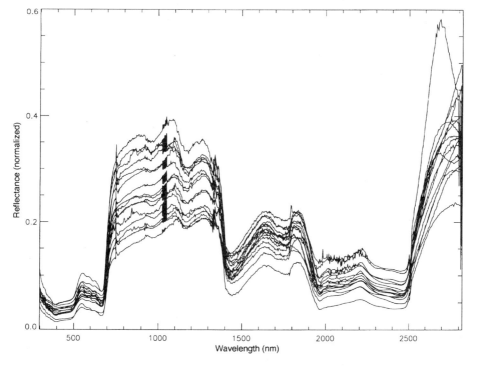

FIGURE 20.3 Ground reflectance spectra for *Quercus coccifera* (April 1992) showing natural variation based on 16 samples from Spanish field measurements

coccifera. The mean reflectance curves for the two *Quercus*, two *Cistus* and *Rosmarinus* species are given in Figure 20.4. These plots are of part-canopy reflectance and so are subject to canopy variables (canopy architecture, soil substrate effects, shadowing, etc) as well as leaf variables.

In the visible region, the reflectance is similar, and the range is very small, for all species. The mean reflectance peak at 550 nm is consistent among species. The slope of the line in the 400–500 nm region appears to be related to the maximum red absorption (680 nm). The edge at 700 nm is consistent among species, the only difference is the slope as determined by the red absorption at 680 nm and NIR reflectance at about 750 nm. In the NIR, the first *Cistus* species stands out distinctly from the other species by its much larger reflectance.

In the short-wave infra-red (SWIR) there is a sharp edge, at about 1400 nm, which complements the edge in the NIR (750 nm). Although within a species the location of the edge is extremely consistent there seem to be some differences in location among species. The edge is associated with water absorption at 1444 nm; however, with different species the bandwidth varies and shifts its location. The location of the edge is independent of the magnitude of the absorption.

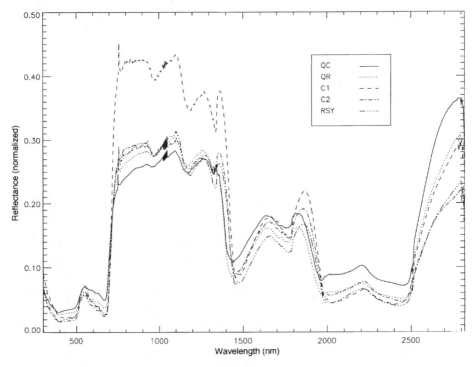

FIGURE 20.4   Mean reflectance spectra for each vegetation class (QC *Quercus coccifera*, QR *Quercus rotundifolia*, C1 *Cistus* 1, C2 *Cistus* 2, RSY Rosemary) derived from Portugese field measurements (June 1992)

Biophysical data

The leaf area indices (Figure 20.5) cover a range in all species. The minimum LAI is influenced by the desire, when taking SIRIS readings, to exclude background features. If the distribution of samples within the species is typical of the population then the data suggest that each species has a maximum LAI associated with it. *Quercus coccifera*, for example, has a low average LAI of 2.1 and a low range (0.9), whereas *Cistus* 2 had an average LAI of 3.5 and a much higher range (1.5). There would appear to be little difference between LAI and bush height (Figure 20.6), in the case of *Quercus rotundifolia* and *Quercus coccifera* at least.

Surface leaf area in the SIRIS field-of-view (Figure 20.7) is much less variable within a species than LAI yet there appear to be distinct species differences. This implies that, for a single species, the surface leaf area is more or less constant over a range of plant sizes even when the range in LAI is great.

Results

The SIRIS part-canopy spectra, transformed into TM space, are given in Figure 20.8. It can be seen in TM space that the response in the first three bands is low and any differences between species are also low. The major differences occur in the NIR (TM band 4) where the within- and between-species responses are much higher. At a standard deviation of 1.5 the species become increasingly statistically separable (Figure 20.9).

The correlation between LAI and NDVI was found to be low (Figure 20.10). However, as suggested by the TM band 3–4 feature space plot (Figure 20.9) there is a high correlation between NDVI and the area covered by directly irradiated green vegetation expressed as surface leaf area (Figure 20.11)—a relationship which is species dependent.

By examining the spectra using factor analysis the major axes of variation are defined. Factor one (vector 1 in Figure 20.12), when all the vegetation spectra are combined, resembles the mirror image of the mean spectra for each species (Figure 20.12). This is not surprising, since factor one never accounts for less than 98% of the original variance in the spectral data.

The major feature of factor two (Figure 20.13) is the sharp point of inflexion which occurs at the red region of the spectra at approximately 675 nm. After this point the factor steadily falls or rises (depending on the direction of the vector) until reaching a minima or maxima region extending between 1400 and 2500 nm. The shape of factor two, affected as it is by the red edge and influenced by the SWIR region suggests that it may be affected by some measure of vegetation amount and moisture content. The NDVI is largely associated with this vector (Figure 20.13). (There is no relationship between NDVI and the first vector.) It is likely, therefore, that this vector would be both associated with surface leaf area and provide an axis of discrimination between species.

A comparison of SIRIS, TM filtered, ground data with the TM subscene gives an idea of how successful the normalization and calibration has been. Figure 20.14 shows the excellent correspondence between wheat spectra obtained on the ground and that from the satellite sensor imagery. Figure 20.15 shows the correspondence of selected brushland ground spectra with those extracted from the imagery, with the fraction of cover as estimated from the mixture model.

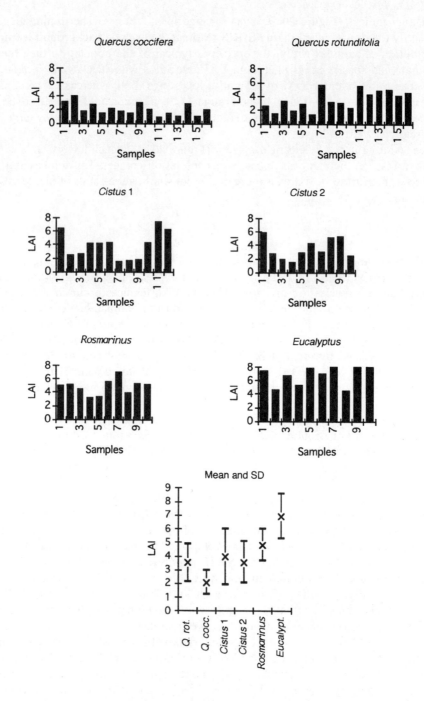

FIGURE 20.5 Leaf area index (LAI) distributions for *Quercus*, *Cistus*, Rosemary and *Eucalyptus* derived from Portugese field measurements (June 1992)

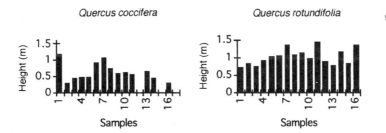

FIGURE 20.6   Plant height distributions for *Quercus coccifera* and *Quercus rotundifolia* derived from Portugese field measurements (June 1992)

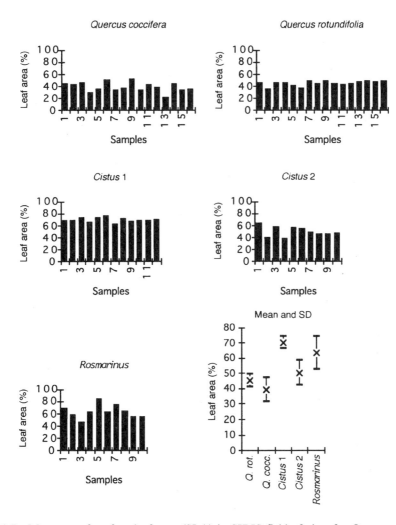

FIGURE 20.7   Measures of surface leaf area (SLA) in SIRIS field-of-view for Quercus, Cistus and Rosemary derived from Portugese field measurements (June 1992)

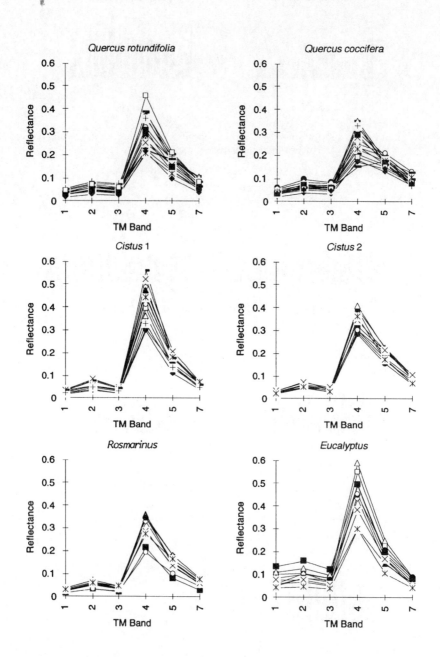

FIGURE 20.8   Ground reflectance spectra for *Quercus rotundifolia*, *Quercus coccifera*, *Cistus* 1, *Cistus* 2, Rosemary and *Eucalyptus* after application of Landsat TM filter functions

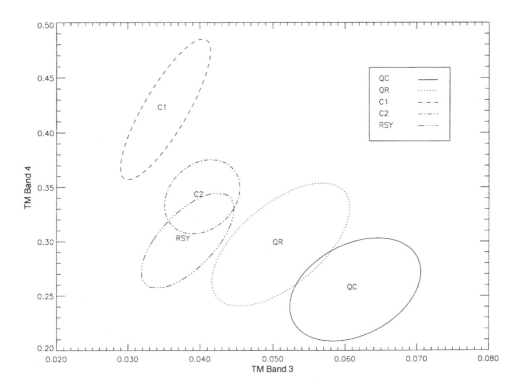

FIGURE 20.9   Ellipse plots for vegetation classes (QC *Quercus coccifera*, QR *Quercus rotundifolia*, C1 *Cistus* 1, C2 *Cistus* 2, RSY *Rosmarinus*). The major and minor axes of each ellipse are defined by the class variance ($s = 1.5$) within Landsat TM bands 3 and centred on the class mean vector

Maximum likelihood classification and mixture model images

Figure 20.16 shows the result of applying the maximum likelihood classifier to the Vale Formosa sub-scene. The major difference in land use occurs between the anthropogenic farming areas, such as cereal, bare soil fields and fallow, and the semi-natural land use such as brushland and scrubland. The former are distributed as regular field systems, the latter as a more spatially heterogeneous system. An accuracy assessment of the classified image was carried out by analysing the results from the classified training areas. The proportion of correctly to incorrectly classified pixels was determined and the results expressed as a percentage of correctly classified pixels. As the method uses the training areas to provide an accuracy assessment one would expect the results given in Table 20.2 to be an over-optimistic assessment. Based on this analysis the overall classification accuracy is in excess of 80%.

Given the spatial variability of the semi-natural cover types many different classes occur with a Landsat TM pixel. As discussed previously, the maximum likelihood classifier will not give a realistic measure of vegetation amount and distribution in such an environment, since it is assumed that only one cover class exists in each pixel. In contrast the mixture model provides a series of fraction images, each image showing the spatial distribution of

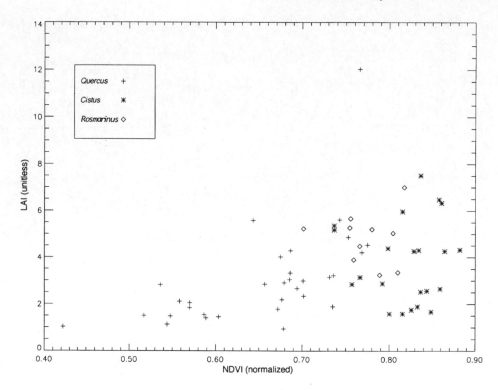

F<small>IGURE</small> 20.10   Scatterplot of normalized difference vegetation index (NDVI) versus leaf area index
(LAI) for *Quercus, Cistus* and *Rosmarinus*

the sub-pixel proportions of the given cover types. An example of such an image is given in
Figure 20.17 for the brushland class.

Compared with the results from the maximum likelihood algorithm, these images are
much more representative of the ground environment. For example, when comparing the
fraction image for brushland with the classified areas from Figure 20.16 it is immediately

T<small>ABLE</small> 20.2   Classification accuracy assessment, based
on training areas, for maximum likelihood classification

Class name	Classification accuracy (%)
Bare soil	90.56
Cereal	80.00
Rock	93.54
Scrubland	91.25
Fallow	95.45
Brushland	95.45
Water	100.00

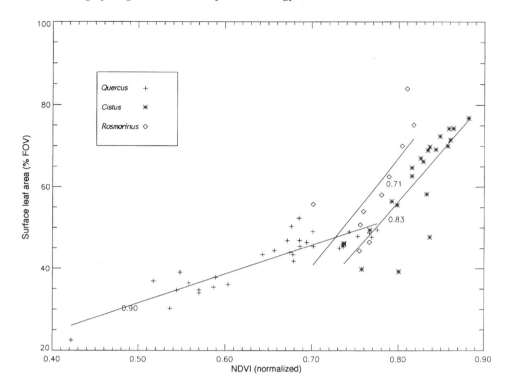

FIGURE 20.11   Scatterplot of normalized difference vegetation index (NDVI) versus surface leaf area (SLA), with regression lines, for *Quercus* ($r^2 = 0.90$), *Cistus* ($r^2 = 0.83$) and *Rosmarinus* ($r^2 = 0.71$)

clear how the spatial extent of this cover type is much greater than would be inferred from the classified image. In the case of the bare soil and cereal classes, there appears to be good correspondence between the fraction and classified images. Once again, however, the mixture model identifies areas of both soil and cereal which are lost in the classified image. Tables 20.3 to 20.5 demonstrate this point quantitatively. As can be seen, in the

TABLE 20.3   Fraction table derived from mixture model showing percentage cover within maximum likelihood fallow class

Pixel		Class						
x	*y*	Soil	Cereal	Rock	Scrub	Fallow	Brush	Water
45	97	0	0	64	59	111	0	20
123	47	17	0	76	37	118	0	8
509	111	0	0	61	53	123	0	16
345	56	0	0	11	0	208	0	34
444	335	0	0	0	0	205	9	39
247	195	4	0	0	18	229	0	2

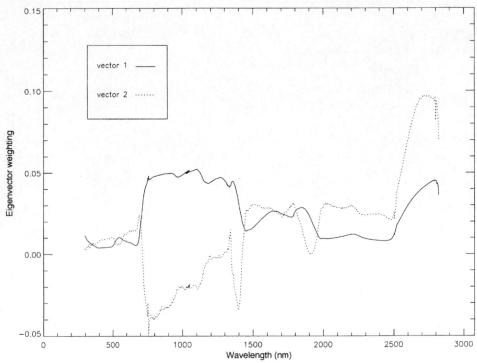

FIGURE 20.12   Plot of eigenvector one and eigenvector two computed for all vegetation spectra

TABLE 20.4   Fraction table derived from mixture model showing percentage cover within maximum
likelihood brushland class

Pixel		Class						
x	*y*	Soil	Cereal	Rock	Scrub	Fallow	Brush	Water
351	345	35	5	0	0	0	214	0
174	286	0	0	77	2	49	127	0
47	53	20	13	8	47	0	168	0
497	256	28	0	106	5	0	113	0
501	222	21	3	31	20	15	164	10

TABLE 20.5   Fraction table derived from mixture model showing percentage cover within maximum
likelihood wheat class

Pixel		Class						
x	*y*	Soil	Cereal	Rock	Scrub	Fallow	Brush	Water
125	87	0	105	59	88	0	0	3
501	367	0	151	50	8	47	0	0
278	458	0	101	102	17	0	0	34
448	150	5	121	0	2	55	72	0
289	101	0	132	59	3	0	62	0

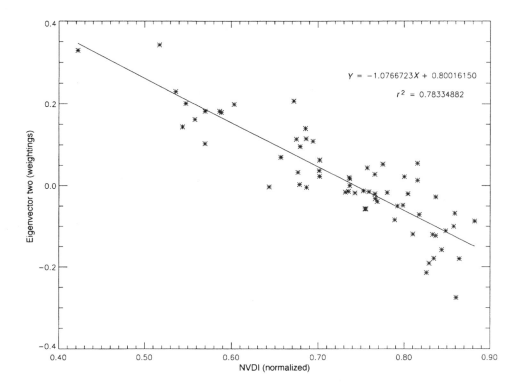

FIGURE 20.13 Scatterplot of normalized difference vegetation index (NDVI) versus eigenvector two for all vegetation spectra ($r^2 = 0.78$)

overwhelming majority of cases the class with the dominant proportion is classified as being that class by the maximum likelihood algorithm.

Normalized difference vegetation index images
The NDVI images subscenes for April and September are given in Figure 20.18. In April, the highest NDVI is associated with the cereal fields, while the semi-natural brushland has intermediate values. The vegetation in the numerous small valleys is particularly noticeable. In September there is no vegetation associated with the farmland—the only vegetation being in the semi-natural areas. As the images are normalized it is possible to examine the change between April and September by computing a difference image (Figure 20.19). This image highlights the change in the agricultural land, but also emphasizes the change occurring in the more highly stressed semi-natural vegetation.

20.2.5 Summary

The video data, of the SIRIS field-of-view, by providing accurate estimates of proportions, have enabled the relationship between surface leaf area and reflectance to be established. It has been shown that surface leaf area can be related to SIRIS NDVI, and this, in turn, associated with the second-most important factor in the factor analysis. LAI appeared not

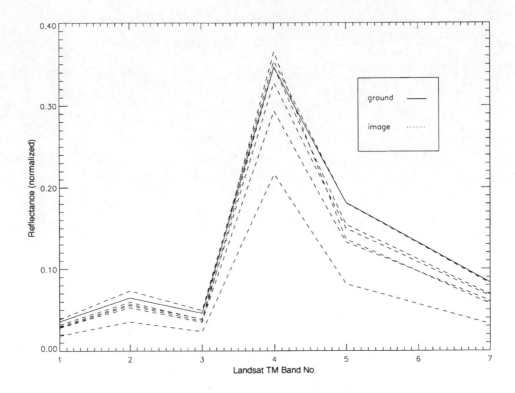

FIGURE 20.14 Comparison of Landsat TM and ground reflectance spectra for wheat showing spectral variation within April 1985 image

to be well correlated with NDVI. In the literature, the LAI relationship is generally restricted to LAIs less than about three which is lower than those determined for these plants.

From these data we can say that the NDVI responds to partial canopy surface leaf area. The response is species-specific—two plants from different species might well have different NDVI numbers. Variation in surface leaf area is also species-specific. LAI is not an important factor. More analysis needs to be carried out to examine the various interactions and why the response is species-specific. More fieldwork is necessary to determine whether surface leaf area can be used as a surrogate for plant biomass. At the moment, therefore, it would seem that satellite-derived NDVI numbers ought to be used in conjunction with the land classification although satellite classifications to the bush scale have not been achieved.

The maximum likelihood classification appeared to be good—despite the fact that semi-arid vegetation is reportedly difficult to classify (Townshend and Justice, 1980; Harrison and Garg, 1991). A combination of imagery from the growing and dry season maxima enables agricultural land to be distinguished easily from the semi-natural. Semi-natural vegetation can be classified based on its spectral position and the vegetation dynamics. Although good, there were some problems in the classification which it would be desirable to rectify: extensive eucalyptus stands were not classified as such because of the strong soil

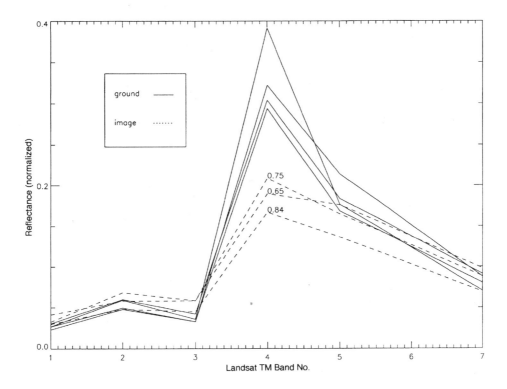

FIGURE 20.15 Comparison of Landsat TM and ground reflectance spectra for brushland showing spectral variation within April 1985 image. (Landsat TM spectral curves are labelled with percentage cover estimates from mixture model)

reflectance caused by lack of undergrowth, greater discrimination within the semi-natural vegetation would seem possible with more carefully chosen training sets, and cropped trees were not well distinguished from semi-natural ones.

The mixture model provides a much better representation of land cover but its performance needs to be examined in greater detail and, because accuracy depends on the separability of the classes, it is necessary to investigate how to express the sensitivity. Even in the worse case, though, the mixture model can perform no worse than the maximum likelihood classification. Ideally, training classes should be as spectrally distinct as possible and current research, using the spectroradiometric data, is examining the optimum selection of classes.

20.3 SURFACE LITHOLOGY

20.3.1 Study area

The study area for surface lithology is situated to the west of the city of Murcia, southeast Spain, and was subdivided into a training area and a test area corresponding to two map

FIGURE 20.16 Maximum likelihood land cover classification (eight classes) of Vale Formoso subscene (April 1985)

FIGURE 20.17 Fractional image from mixture model for brushland class in Vale Formoso subscene (April 1985). Image shows percentage cover from black (0%) to white (50%)

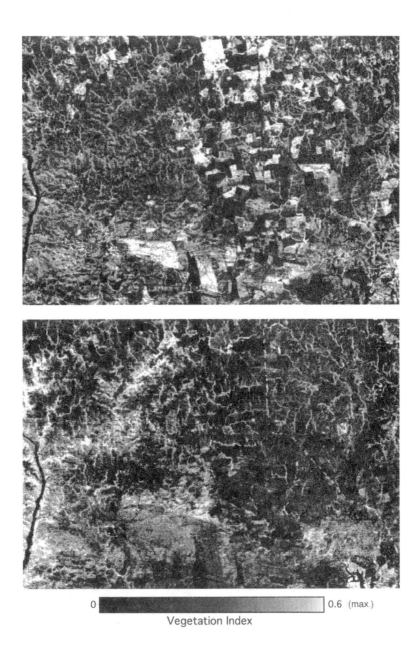

FIGURE 20.18   Top: April 1985 normalized difference vegetation index (NDVI) image for Vale Formoso subscene. Bottom: September 1985 normalized difference vegetation index (NDVI) image for Vale Formoso subscene

0 [_____] max.

Change in Amount of Vegetation

FIGURE 20.19.   Image showing change in vegetation amount between April 1985 and September 1985 derived from normalized difference vegetation index (NDVI) images for Vale Formoso subscene

sheets from the Spanish IGME 1:50 000 National Geological Map series. The training area (Alcantarilla, map sheet 933) lies within the geographical frame: latitude N 37°50′04″6–38°0′04″6 and longitude W 1°11′10″9–1°31′10″9. The test area (Mula, map sheet 912), situated to the north of the training area, lies within the geographical frame: latitude N 38°00′04″6–38°10′04″6 and longitude W 1°11′10″9–1°31′10″9. Figure 20.20 shows the general geological character of the study area.

Table 20.6 summarizes the lithological composition of the study area and identifies the different lithological units which are exposed at the surface. Special attention was paid to the Quaternary deposits in the Alcantarilla area (denoted by the letter Q in the published map series). These comprise travertins, conglomeratic red clays and silts with caliches of the Q13 unit, and undifferentiated deposits of terraces and river beds and alluvial fans (IGME, 1974a).

Spatially, the undifferentiated Quaternary deposits were found to be composed of different lithologies with some intercalation of one type to another. The lithological variability of the undifferentiated deposits ranges from uncemented grey silts, brownish and pale silts with some intercalation of caliche, grey to white calcic clayey silts and calcic pale brown silt deposits. These undifferentiated lithologies extend over a wide area, sufficient to be distinguished in Landsat TM imagery, and support a variety of vegetation cover types, from cultivated crops to low density 'matorral'. In order to better represent the spatial extent and variability of these deposits, they were further subdivided, based on field and lithological characteristics, into four units (Q11, Q12, Q14 and Q15), in addition

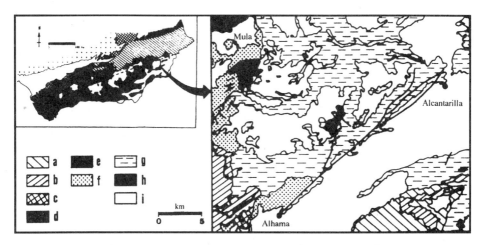

FIGURE 20.20 Geological map of the study area: (a) Alpujárride; (b) Maláguide; (c) Intermediate zone; (d) Jurassic sediments; (e) Nummulitic limestones (Eocene); (f) Calcarenite and marls (Middle Miocene–Lower Tortonian); (g) Calcarenite, marl and gypsum (Tortonian–Upper Pliocene); (h) Basalt, and (i) Quaternary deposits

to the differentiated caliche deposits which were already mapped separately in the published geology maps as unit Q13 (see Table 20.6).

The training and test areas have a number of important characteristics in common. Due to their location in the Mula basin, the lithological composition of both areas is very similar, based on limestones, marls, red clays and conglomerate, Quaternary and volcanic rocks. In addition, the pattern and spatial extent of the various lithological units exposed and outcropping at the surface is similar and capable of being resolved by the 30 m spatial resolution of the Landsat TM sensor. Both areas contain 'badland' landscapes, characterized by deeply incized gullies, occurring on marls. Finally, in terms of vegetation type and cover, there are many similarities between the areas. The carbonate units are partially covered by dense pine forests and low density 'matorral', the Quaternary deposits support areas of cultivated land, in addition to scattered low density 'matorral' and pine forests.

20.3.2 Ground radiometry and field measurements

Field measurements

Field spectral measurements were obtained to characterize the different *in-situ* spectral characteristics of the lithological classes exposed in the area. Measurements of surface radiance over bare rock surfaces were taken using a SIRIS spectroradiometer following the procedure already outlined for the collection of vegetation spectra in Section 20.2.2 above.

A series of exposures, representative of the different lithologies in the area, were selected for field measurement in order to characterize their spectral response and their mineralogical composition. These exposures were also selected on the basis of their

TABLE 20.6   Lithological description of exposed units within the study area (sampling code followed by lithology and lithological description)

Unit code	Lithological description of outcropping rocks
L1	White shiny gypsum, fibrous or massive, secondary in certain places with pale blue cavities of clays or marls and reddish-brown inclusions of iron oxides. It forms bare surfaces with very sparse vegetation.
Q1	Fibrous, secondary gypsum and marls. It occurs as disseminated fine crystals of 0.5–6 cm appearing at the surface of the soil. The appearance of this type of gypsum is local and distributed along the depth of the Quaternary or soil deposits reaching up to 80 cm of depth. It is covered by dense vine and citrus orchards.
P1	Pale red to yellowish marl and reddish silts. These sediments are not well cemented with high clay content and oxidized dotted inclusions in the marl surface. The vegetation cover is variable, consisting of low 'matorral' and some disperse pine forests.
A1	Marl and marly limestones, pale yellow in colour, friable with some interbedded dolomite. The vegetation cover is composed of citrus orchards and sparse dispersed pine forests.
B1	Pale blue, grey to deep grey marl with alternation of yellow and pale red–brownish marls, with the presence of calcite crystals of secondary origin. The vegetation cover is disperse matorral of about 50–60 cm of height.
Q11	Grey to white marls and clays (Lutites), friable with abundance of calcic garins (calcite or dolomite). Gypsum nodules and patches are seen in certain places of very restricted distribution. This unit was found to be of poor vegetation cover.
Q12	Red to dark brown soils and clays with a notable content of organic matter and carbonate pebbles. The sample colour is spatially changed and tends to be darker with increasing of the organic matter content. The vegetation cover is composed of citrus and fruit plots.
Q13	Caliches and clays sediments of red to brown colour relatively harder than that of the rest of the Quaternary units with notable amount of organic matter. The vegetation cover consists of almond plots.
Q14	Light to deep grey marls and soils with calcic grains and carbonate pebbles, friable and with some gypsum crystals in certain places. The vegetation cover is of olive plots.
Q15	Dark grey marls and soils with notable amount of fine friable carbonate pebbles. The vegetation cover is low seasonal grasses and very low matorral.
P2	Pale grey to reddish brown sandstone, well bedded with inclusions of iron oxide presenting brown colour, showing the presence of calcite matrix in the hand specimen. The vegetation cover is made of dispersed pine forests and some low matorral.
S1	Red to deep brown clays and clayey silt alternating with laminated or not outcrops of sand–siltstone with unbedded outcrops of clays and marls. The vegetation cover is of pine forests of sporadic distribution.
G1	Very pale brown calcarenite, composed of calcite and/or dolomite and a detrital fraction of quartz grains. Ghost structure and obliterated fossils are abundant and found in the thin sections. Recrystalization representing shiny calcite grains is also clear. Topographically, the unit forms high lands covered by sparse pine.
A2	Brecciated or aggregate of conglomeratic limestone, with different rock size and pale brown carbonate rocks. The vegetation cover is of dense pine forests in the north, less dense in the south with more citrus orchards.
P3	Nummulitic limestone, hard, poorly bedded, white to pale brown in colour, massive. Dolomite has also been observed to present. The vegetation cover is of pine forests and matorral.
D1	Black to deep green lamproite of a mineralogical composition of orthopyroxenes and orthoclase. Vesicles of various dimension are found filled by calcite or alteration products of the original rock. Phenocrystals of various size of olivine and plagioclase were also found. The vegetation cover consists of low matorral and dispersed seasonal grasses.
R1	Red to brown lamproite, massive with very little alteration at the surface. Some white phenocrysts are found. The vegetation cover is very sparse and consists of low plants, seasonal grasses and matorral.

spatial extent and vegetation cover so that spectral measurements from Landsat TM imagery could be obtained later. Areas covered by dense vegetation and forest were excluded as the vegetation effectively masks the spectral response of the underlying surface.

A series of experiments was conducted, as part of the field measurement exercise, in order to investigate the influence of surface weathering on spectral response. To do this, measurements of weathered and fresh surfaces were carried out both to identify differences in reflectance and absorption characteristics and to assess the sensitivity of satellite measurements to these features.

The following lithological units were selected for this purpose:

(i) the volcanic units including the D1 and R1 units,
(ii) the carbonate unit P3,
(iii) the gypsum unit L1, and
(iv) the terrigeneous units of P2.

The marl units (A1, B1 and P1), the gypsiferous unit (Q1) and the terrigenous unit (S1) were excluded from the experiments due to their friable texture. This made it very difficult to obtain satisfactory weathered and fresh surfaces in the field and to take undisturbed samples back to the laboratory. The results of these experiments are presented in Section 20.3.4.

Analysis

Mineralogical analysis
In order to study the relationship between spectral reflectance and mineralogical content, the mineralogical contents of the selected lithological units samples were analysed by x-ray diffraction and petrographical thin section analysis techniques. In practice, use of either the x-ray diffraction or petrographic thin analysis depends on the abundance of crystalline components in the sample. Hence, it was possible, given the predominant crystalline structure of the volcanic samples, units R1 and D1, to determine mineralogical content using both thin section analysis and x-ray diffraction.

The analysed samples were unorientated powder of the whole rock (grain size < 50 mm), prepared by the Niscanen procedure (Niscanen, 1964). For the shales, marls and clays, oriented aggregates were prepared for normal, glycolation and heated sample examination in order to identify the sheet silicates following Warshaw and Roy (1961).

After the identification of the crystalline phases of the rocks, several characteristic d_{hkl} spacings were selected in order to carry out a semi-quantitative estimation of the different mineral contents (Caballero and Martín-Vivaldi, 1975; Hooton and Giorgetta, 1977). The selected d_{hkl} spacings were: 7.5, 4.49, 4.26, 3.02, 2.88, 2.89 and 2.69 for the gypsum, sheet silicate, quartz, calcite, dolomite, pyroxene and hematite respectively. The relative error of the x-ray diffraction analysis was found to be of 8.9, 4.7, 6.3, 7.7 and 4.6% for sheet silicates, quartz, calcite, dolomite and hematite respectively.

Definition of lithological desertification response units

To incorporate the lithological factor formally in the desertification process, work was done to re-group the 18 lithological units, based on their susceptibility to erosion, into six lithological desertification response units or LDRUs. This grouping was performed on the basis of mineralogical composition, field characteristics, vegetation cover and susceptibility to erosion by water (Bastida et al., 1992). The resultant LDRUs are shown in Table 20.7, and consist of gypsum, marls, Quaternary deposits, terrigenous units, carbonates and volcanic units, arranged in ascending order of erodibility by water.

20.3.3 Image processing and analysis

Pre-processing

A single cloud-free Landsat-5 TM image (14 September 1987) centred in a window at 37°N Latitude, and containing the Alcantarilla (training) and Mula (test) areas, was selected for this study. The low level of vegetation cover present in the scene was an important criteria in the selection of the image. The spectral bands of the raw Landsat TM image were converted into reflectance by taking into account the calibration constants of the TM sensor and applying an atmospheric correction.

The atmospheric correction method employed here requires a series of commonly available inputs (date, hour, latitude, height, type of aerosol (continental or maritime)) and the presence of dark surfaces within the Landsat TM image to be corrected. Pixel values for these dark areas are extracted from Landsat bands TM1 and TM3 and enable the actual aerosol model to be estimated using the wavelength dependence characteristics of the aerosol path radiance. A horizontally homogeneous atmosphere is assumed, so that transmittance and path radiance are constant over the scene and their value can be

TABLE 20.7 Classification of the lithological units according to susceptibility of erosion by water (1 = high susceptibility, 6 = low susceptibility)

LDRU unit	Lithology	Vegetation cover	Consolidation	Relief
1-Gypsum	gypsum, gypsi-ferous sediments	very disperse	very soluble, friable	moderate
2-Marl	marls	disperse	friable, little consolidated	low
3-Quaternary	clayey conglomerate	matorral and cultivated lands (variable cover)	low consolidation	very low
4-Terrigenous	cemented conglomerate, sandstone	disperse pine forests	moderate consolidation	moderate
5-Carbonate	carbonates	natural vegetation of pine and matorral	well consolidated	high
6-Volcanic	volcanic rocks	low matorral, very disperse	very well consolidated	moderate

determined for each image. The correction procedure itself uses an inversion algorithm based on a simplified radiative transfer model in which the characteristics of atmospheric aerosols are estimated rather than assumed *a priori*; although as the Rayleigh component is known it is obtained from published data. Thus, on the basis of the supplied input data and extracted pixel values from the image, the retrieval of reflectances from Landsat TM images is possible. The method is applicable to Landsat TM data, assuming that dark pixel values are present, which is particularly advantageous in retrospective studies. Given care in the identification of dark pixel values, the relative error in retrieved reflectance values is usually quite low (10–20%), which can be considered acceptable for most practical applications.

Analysis

Two approaches to image analysis were employed. First, the use of a variety of image enhancement procedures to assist visual interpretation of the Landsat TM imagery. Second, the use of supervized multispectral classifiers to discriminate automatically between surface lithologies.

Preliminary analysis of the calibrated Landsat TM images was carried out visually in order to evaluate the use of false colour composites for identifying and discriminating between the different lithologies and LDRUs. These visual analyses were assisted and checked by comparison with the published geological maps, 1:30 000 aerial photography and fieldwork. False colour composites were produced using a variety of original and derived spectral bands based on:

(i)   atmospherically calibrated Landsat TM bands;
(ii) spectral band ratios derived from Landsat TM bands; and
(iii) images derived from principal component analysis (PCA) of Landsat TM bands.

Two classification approaches were tested and evaluated: the maximum likelihood classifier and the minimum distance classifier. Better results were obtained from the maximum likelihood classifier and this algorithm was selected for all of the lithological classifications carried out. The maximum likelihood classifier has already been described in Section 20.2.3.

In order to train the maximum likelihood classifier, areas were selected within the Alcantarilla training image representing homogeneous classes for each of the lithological desertification response units, LDRUs. Selection of training areas was carried out with the aid of aerial photographs to ensure each area was representative of the underlying lithology and as homogeneous as possible.

The resulting classified image of LDRUs was then geometrically rectified to the Universal Transverse Mercator projection and resampled to produce final hard copy products which coincided exactly with the published 1:50 000 geological maps. The geometric correction algorithm was based on the standard method of ground control points, applying a second degree polynomial equation and included a filtering process that selects the maximum value for each 3×3 window. The same process was applied to the Mula test image in order to evaluate the overall classification approach. Classification results were also tested by field sampling, particularly within those units which were absent from the geological map, for instance, areas of gypsum and gypsiferous marls.

20.3.4 Data and results

Mineralogical analysis of lithological units

Table 20.8 shows the results of mineralogical analysis of the lithological units. The gypsum and the fibrous secondary-gypsum units (L1 and Q1) have widely differing mineralogical composition. The L1 unit is predominantly gypsum (90.0%) with a considerable amount of dolomite (9.83%), whereas the Q1 unit is dominated by sheet silicate, calcite and dolomite and only a subsidiary amount of gypsum. This wide variation in the mineralogical content of the secondary gypsum unit suggests that it should more accurately be referred to as gypsiferous (13.03% gypsum) rather than straight gypsum.

The marl units (P1, A1 and B1) were found to be composed of sheet silicate and/or carbonate minerals (calcite and dolomite) and quartz, with smaller and varying amounts of feldspar and iron mineral (hematite) content. Calcite is directly proportional to the dolomite content and the two are inversely proportional to the sheet silicate and quartz content. The P1 unit has a higher sheet silicate and quartz content (70.0 and 11.5%), and a lower calcite and dolomite content (4.04 and 1.4%), while the B1 unit has a higher calcite and dolomite content (53.4 and 16.6%) and a lower sheet silicate and quartz content (26.8 and 2.17%).

The mineralogical compositions of the Quaternary deposits (Q11, Q12, Q13, Q14 and Q15) are comparable with the marl and terrigenous units (i.e. calcite, dolomite, sheet silicate and iron mineral content), but show a higher degree of variability in terms of calcite, dolomite and iron mineral content.

Analysis of the terrigenous sediments (P2 and S1) revealed that the two units are composed of similar amounts of sheet silicates (54.7 and 54.5%) and dolomite (5.8 and

TABLE 20.8   Semi-quantitative x-ray diffraction mineralogical analysis of the studied lithologies

LDRU Unit	Gypsum (%)	Sheet silicate(%)	Quartz (%)	Feldspar (%)	Calcite (%)	Dolomite (%)	Hematite (%)
L1	90.0	—	—	—	—	9.8	—
Q1	13.0	38.4	3.6	—	31.1	12.5	—
P1	—	70.0	11.5	13.1	4.0	1.4	—
A1	—	58.6	5.8	1.0	28.7	5.3	—
B1	—	26.8	2.2	1.0	53.4	16.6	—
Q11	—	40.3	3.4	—	52.6	3.5	—
Q12	—	31.8	3.3	—	58.8	7.5	1.6
Q13	—	36.3	3.3	—	48.6	10.0	0.9
Q14	—	40.3	3.4	—	46.7	8.8	0.8
Q15	—	43.5	1.7	—	43.7	9.7	0.3
P2	—	54.7	22.5	12.8	2.7	5.8	—
S1	—	54.5	3.5	2.3	33.8	4.9	1.1
G1	—	2.6	—	—	95.8	1.6	—
A2	—	7.7	1.5	—	70.3	20.5	—
P3	—	11.6	3.1	—	51.9	33.4	—
D1*	—	58.6	5.8	1.0	28.7	5.4	—
R1*	—	20.0	48.0	14.9	4.0	1.4	—

*Percentages of vitreous components are not considered.

4.9%) in P2 and S1 respectively. However, the units have quite different calcite and quartz contents (2.7, 33.8% calcite and 22.5, 3.4% quartz in P2 and S1 respectively). Feldspars were found to comprise 12.8% in P2 and 2.3% in S1. Normally, the extreme differences in the mineralogical composition of terrigenous units are explained by differences in the rock matrix and nature of cementing material. Both properties may have an important influence on the spectral reflectance of the P2 and S1 units.

The x-ray diffraction analysis revealed that the carbonate rocks (G1, A2 and P3) are almost entirely composed of calcite with small amounts of dolomite and sheet silicate minerals. The dolomite and sheet silicate content was found to increase across the G1 (lowest), A2 and P3 (highest) units, whereas the calcite content decreased across the G1 (highest) to A2 and P3 (lowest) units. This variation in mineralogical content is inversely proportional to the carbonate fraction (calcite plus dolomite) and the insoluble fraction (sheet silicate plus quartz).

The x-ray diffraction analysis of the volcanic rocks (D1 and R1) showed that the main crystalline components of these rocks are olivine, plagioclase and sheet silicate minerals. Thin section analysis showed that the dark lamproite is formed mainly by altered olivine crystals, phlogopite and plagioclase with a high content of glass material. The olivine crystals showed an alteration of the borders and transformation of the olivine to idingsite. The analysed red lamproite is composed of olivine, phlogopite and plagioclase; the olivine showing zoning structures. The predominant phenocrystals are olivine and phlogopite. The outstanding differences between the dark and red lamproite are: (i) a higher content of glass matrix in dark lamproite, and (ii) a higher content of phlogopite in the red lamproite. Another distinguishable difference is the higher alteration of olivine to idingsite in dark lamproite compared to that of the red lamproite.

Spectral characterization of fresh and weathered surfaces

Surface weathering will affect the spectral response of the lithological units as measured by satellite. In order to quantify this effect, spectral measurements of fresh and weathered surfaces were collected and comparisons made.

Overall, spectra of the fresh and weathered surfaces of P2, P3, D1 and R1 units show that the fresh surface spectra always have higher reflectance than the weathered ones. In the weathered surface spectra, the presence of iron absorption features, for example at 850 nm, was found to be more pronounced and intense than in the fresh surfaces as a result of iron oxidation through contact with the atmosphere (White et al., 1984). For the P3 unit, fresh and weathered surfaces show identical reflectance properties in the 400–900 nm region although there are some reflectance differences which increase towards the NIR region. Units P2, D1 and R1 have very similar reflectance characteristics between fresh and weathered surfaces.

The gypsum unit, which showed more marked differences in spectral response between fresh and weathered surfaces, will be discussed separately.

Gypsum unit (L1)

Figure 20.21 shows the weathered and fresh spectra for the gypsum unit L1. The two surface spectra show large differences (in reflectance) in the 400–1000 nm range. However,

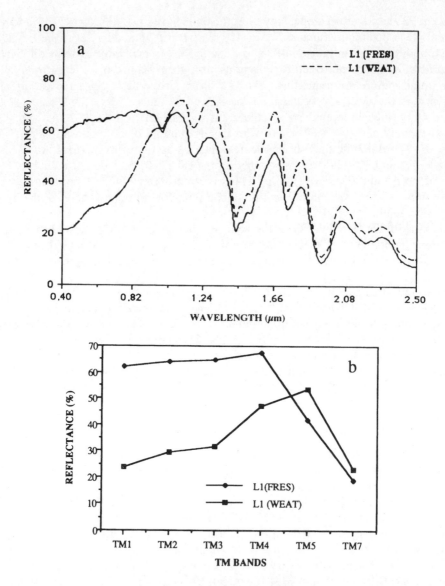

FIGURE 20.21   (a) Spectra of weathered and fresh gypsum surfaces; (b) corresponding Landsat TM
spectra for weathered and fresh gypsum surfaces

at longer wavelengths the spectra are identical. The main differences can be summarized
as:

(i) The 400–1000 nm region of the fresh surface reflectance spectrum is characterized by
    a high plateau (60%), which is almost constant and with very little change, with an
    absorption feature at 1000 nm, while the weathered surface spectrum is characterized

by a moderate–steep slope towards the ultra-violet (UV) region with pronounced broad absorption features around 450 and near 940 nm.

(ii) In the 1000–2500 nm region, while the two spectra are similar, the weathered surface spectra has a lower reflectance and the absorption feature in the weathered surface spectra is more pronounced.

The mineralogical composition of the weathered surface of the gypsum unit is characterized by the presence of clays, iron minerals (jarosite) and large quantities of fine quartz grains mixed with the remnants of the former fresh gypsum surface. While the weathered surface is composed of clays and fine quartz and iron minerals, the reflectance in the 400–900 nm region is completely controlled by the iron mineral content, in particular absorption features associated with iron oxides (Townsend, 1987). In contrast, the fresh surface is composed of shiny crystals of gypsum that have a strong lustre and present a constant reflectance in this region. The similarity of the fresh and weathered spectra in the 1000–2500 nm region is due to the weak effect of the iron content in this region and the strong absorption features of H_2O and OH^- around 1000, 1750, 2080 and 2200 nm.

A spectroscopic study of the weathered gypsum surface revealed that it is quite different to the other lithological units under investigation. The mineralogical composition of the weathered surface is characterized by the presence of jarosite and gypsum in addition to cryptocrystalline clays and fine-grained quartz.

Field spectral characterization of the lithological units

Field spectroradiometric measurements were collected for each of the exposed lithological units. Interpretation of these spectra was aided by field observation and the results of the mineralogical content of each lithological unit. Rather than consider each lithological unit separately in turn, they are grouped by LDRU, which enables similarities in spectral response, lithology and mineralogical properties to be considered together. Gypsum spectra are not included due to the marked differences in fresh and weathered gypsum surfaces noted above (see Figure 20.21). Spectra for the terrigenous unit are also not included as their spectral properties are very similar to the Quaternary units.

The main criteria for the spectral analysis is the relationship between the position of absorption features and mineralogical composition. This was carried out in two important spectral regions: 400–1100 nm (where the iron content effect is dominant) and 2080–2350 nm (where the carbonate and clay content effect is dominant).

Marl units (A1, B1, P1)
The marl unit spectra (A1, B1, P1) (Figure 20.22) are characterized by a steep fall in the visible region towards the UV and a nearly constant response in the NIR region, up to about 1600 nm, with a gentle fall beyond this region towards longer wavelengths. The spectra are also characterized by the presence of weak absorption features in the visible region which can be attributed to (i) the colour of the lithological formation (500 nm) and (ii) the presence of low iron content (below the detection limit of x-ray diffraction analysis which is 4–5% ferric iron minerals, in the form of oxide or hydroxides) either in the oxide/

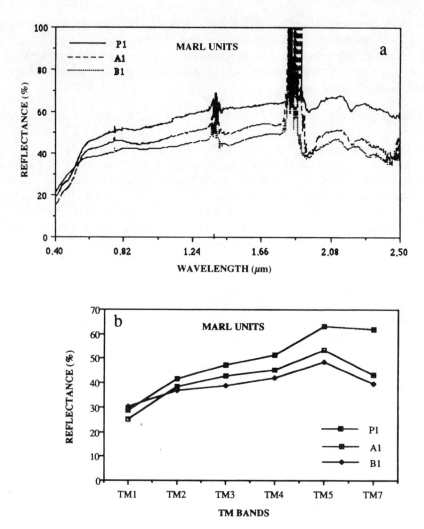

FIGURE 20.22   (a) Spectra of the marl units; (b) corresponding Landsat TM spectra for the marl
units

hydroxide form and/or as free iron incorporated in the lattice structure of the clays (Goetz
et al., 1983).

The absorption features attributed to the presence of iron minerals are discussed and
reported by many authors (Sherman et al., 1982; Morris et al., 1985; Townsend, 1987) and
are overtones of strong features in the UV region at 230, 290 and 350 nm. Although iron
comprises only a small percentage of the total marl composition, the shape of the spectra is
formed by the presence of small amounts of ferric iron minerals and a qualitative
determination is possible by analysis of the spectra. The position of the broad absorption
feature around 900 nm and the 'shelf-like' shape of the spectrum around 650 nm due to its

'trans-opaque' behaviour (associated with the crystal field effect and charge transfer of transition metals) indicates the presence of goethite (Goetz et al., 1985). In the NIR region the spectrum is characterized by more intense absorption features located at (1450, 1940, 2220, 2340 and 2450 nm). The most intense absorption features are located at 2220 and 2340 nm, which are the overtones of the hydroxyl group (OH^-) and calcite and/or dolomite respectively (Hunt et al., 1973; Lang et al., 1990).

As can be seen in Figure 20.22, the marl spectra are similar to each other in form and position of the absorption features but not in intensity and reflectance. This similarity is not surprising given that the marl units share the same lithology. The difference is found to be in the overall reflectance (albedo) which increases from B1 (lowest) through to A1 and P1 (highest). This could be due to the higher concentration of quartz in the P1 unit than the A1 and B1 units. Pure quartz is almost spectrally featureless, but its presence in certain cases can be noted from the increased lustre it can induce in the target (Hunt and Salisbury, 1970).

Quaternary sedimentary units (Q11, Q12, Q13, Q14, Q15)

The Quaternary deposits are composed predominantly of clay and carbonate minerals (calcite and dolomite), quartz, feldspars and iron oxides. The spectra (Figure 20.23) are similar in the presence and position of the absorption features. The predominant mineralogical composition is reflected in the presence of absorption features at 2180 nm and 2340 nm due to the presence of sheet silicate and carbonate minerals (see Table 20.9). Although, the spectra are generally similar in shape and form, certain differences exist between some of them (Figure 20.23). These differences can be summarized as follows:

(i) Q15 unit has the maximum albedo (56%) followed by the Q11, Q13, Q12 and Q14 units (53, 52, 48 and 43% respectively);

(ii) the presence of a 'shelf-like' absorption feature around 550 nm in the Q12 and Q13 units and the absence of the feature from the other units;

TABLE 20.9   Common mineral absorption features in the visible, near and middle infra-red region

Mineral	Absorption features (μm)	Reference
Calcite	1.88, 2.00, 2.16, 2.35, 2.50	Hunt and Salisbury, 1970
Dolomite	1.86, 1.99, 2.14, 2.33, 2.53	Hunt and Salisbury, 1970
Goethite	1.00, 1.45, 1.90, 2.30	Hunt et al., 1973
Limonite	0.55, 0.90, 1.40, 1.90	Hunt et al., 1973
Magnetite	Featureless in the visible and NIR region or very weak absorption features near 1.00 μm	Hunt et al., 1973
Gypsum	1.45, 1.75, 2.20–2.27	Hunt et al., 1973
Chlorite	0.70, 0.90, 1.40, 1.90, 2.20	Hunt and Salisbury, 1970
Montmorillonite	0.50, 0.97, 1.40, 1.90, 2.20	Hunt and Salisbury, 1970
Kaolinite	0.50, 0.97, 1.40, 1.90, 2.20	Hunt and Salisbury, 1970
Pyroxenes	0.77, 1.10, 1.40, 1.90, 2.20–2.50	Hunt et al., 1973
Smectite	1.10, 1.41, 1.46, 1.91, 2.22	Lang *et al.*, 1990
Quartz	Featureless in the visible and NIR region	Hunt and Salisbury, 1970

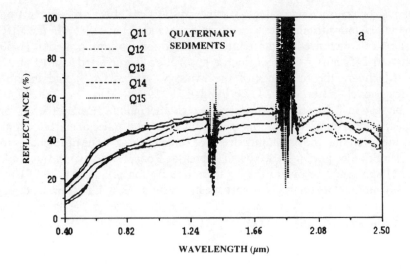

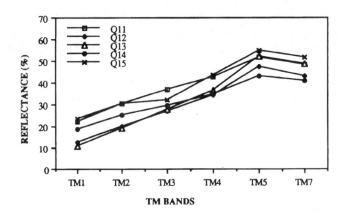

FIGURE 20.23   (a) Spectra of the Quaternary units; (b) corresponding Landsat TM spectra for the
Quaternary units

(iii) the presence of a very weak but wide absorption feature around 850 nm in the Q13
       unit; and
(iv)   absorption features around 2180 and 2340 nm are less pronounced in the Q14 unit
       than the rest of the units.

The interdigitating mineralogical composition of the Quaternary units in terms of sheet
silicates, carbonate and iron oxides is reflected in the slope of the reflectance spectra.

Carbonate units (G1, A2, P3)
The carbonate spectra of G1, P3 and A2 (Figure 20.24) are broadly similar in form. In
general, the spectra are characterized by a gentle slope on both sides towards the UV and

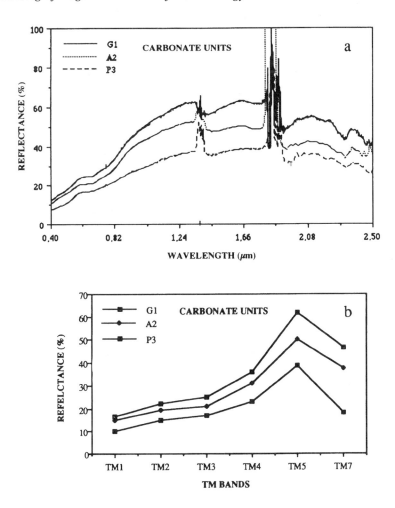

FIGURE 20.24   (a) Spectra of the carbonate units; (b) corresponding Landsat TM spectra for the carbonate units

IR regions. In the visible region, these spectra are characterized by absorption features at 630 and 650 nm which can be explained (see previous discussion) by the presence of iron oxides (Goetz, 1989). Although the spectra show strong similarities, closer examination reveals some subtle but important differences. These differences can be seen from the steeper slope of the A2 and G1 spectra, between 400 and 1000 nm, in the direction of the UV region, compared to the slope of the P3 spectra which is more gentle. These differences in slope can be attributed to the presence of the iron minerals. The other difference of the spectra in the visible region is the presence of a colour centre around 450–480 nm in the P3 spectra and the absence of this feature in the other two spectra. The colour of the P3 unit was recorded in the field as 'pale yellow to pale brown', which explains the above-mentioned absorption feature.

In the IR region, the carbonate spectra show two characteristic absorption features:

(i) a 2200 nm absorption feature, due to the presence of the OH^- group associated with the clay fraction of the carbonate rocks; and

(ii) a 2300 nm absorption feature, due to an overtone of a strong absorption feature associated with the molecular vibration of CO^2 carbonate radical; also a weaker carbonate combination overtone around 2000 nm is visible in the G1 and P3 spectra attributable to a weaker overtone of the same radical absorption at 2500 nm (Rowan and Kahle, 1982).

The albedo falls across the carbonate spectra, from G1 (high), to A2 and P3 (low). This may be due to the higher quartz content in the G1 and A2 units and the lower quartz content of the P3 unit, or to recrystallization present in the G1 lithological unit, which results in a shiny surface. In the NIR region, the three lithologies are characterized by the presence of an absorption feature at 2310–2340 nm (Rowan et al., 1977) which is attributed to the presence of calcite and/or dolomite.

Volcanic units (D1, R1)

Although these two volcanic units are of similar rock type, they show extreme spectral differences (Figure 20.25) which are particularly evident in the IR region. In the visible, from 400–700 nm, the D1 and R1 spectra are similar in shape and form although there is a difference in the form of a 'shelf-like' feature at around 660 nm in the R1 unit which is absent in the D1 unit. This results from the red colour of the R1 unit (probably hematite or geothite content, see Table 20.9) and the darker colour of the D1 unit.

The large differences between the two spectra start around 700 nm and increase towards the SWIR region (2500 nm). The spectrum of the R1 unit is characterized by increasing reflectance up to 1300 nm, after which it falls, while the D1 spectrum is characterized by a flat (nearly constant) reflectance up to 2500 nm. Absorption features at 2200 and 2300 nm in the R1 unit, due to Fe–Al and Mg–Al silicates respectively, are not present in the D1 unit spectra.

These large differences between the spectral characteristics of the R1 and D1 units, in spectral form, reflectance behaviour and the presence of absorption features at 2200 and 2300 nm, can be explained by the different mineralogical composition of the two units. These mineralogical differences were revealed by petrographic examination and x-ray diffraction analysis of the volcanic units and are summarized below.

(i) Higher dark glass and lower crystalline content in the D1 unit results in a lower reflectance component for the D1 unit in comparison with R1.

(ii) The alteration of the olivine borders to idingsite in the D1 unit, which is absent in the R1 unit, reduces and masks the reflectance response of other D1 components (plagioclase and phlogopite).

(iii) The presence of chlorite in the D1 unit, and its absence in the R1 unit, has a similar effect to idingsite, as noted above, in masking reflectance response. Reflectance masking by chlorite is mentioned by Hunt and Salisbury (1970) who have noted reflectance masking due to the presence of chlorite in certain basalt rocks.

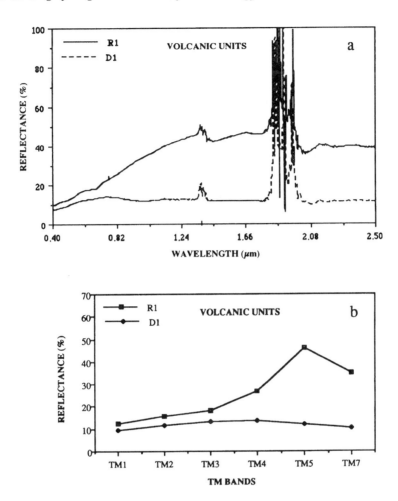

FIGURE 20.25   (a) Spectra of the volcanic units; (b) corresponding Landsat TM spectra for the volcanic units

As a summary to this discussion of the spectroradiometric data, Table 20.10 shows the main absorption features than can be detected by field radiometry measurements of the studied lithological units.

Image analysis

Selection of spectral bands for the false colour composition and spectral band ratio combinations was performed on the basis of the spectral and mineralogical characteristics of the LDRUs; in particular noting specific spectral shapes and absorption features. In specifying false colour composites here, the standard RGB ordering system is assumed for allocating a colour to each spectral band in the triplet.

TABLE 20.10   Absorption features and the mineralogical composition for each lithological unit obtained by x-ray diffraction

Unit	Absorption characteristic features (μm)	Mineralogical composition	d_{hkl} spacing code
L1	0.48, 0.99, 1.98, 1.35, 1.44, 1.54, 1.73, 2.21, 2.28, 2.45	gypsum, dolomite	7.65, 2.88
Q1	0.60, 1.00, 1.98, 1.44, 1.48, 1.54, 1.73, 2.21, 2.28, 2.45	clay*, quartz, calcite, gypsum	4.49, 4.26, 3.02, 7.65
P1	0.46, 0.89, 0.63, 1.93, 2.20, 2.31, 2.45	clay, quartz, calcite, dolomite	4.49, 4.26, 3.02, 2.88
A1	0.55, 0.65, 0.86, 1.95, 2.22, 2.34, 2.45	clay, calcite, dolomite, quartz	4.49, 3.02, 2.88, 4.26
B1	0.60, 1.45, 1.95, 2.22, 2.34, 2.47	clay, calcite, dolomite, quartz	4.49, 3.02, 2.88, 4.20
Q11	0.55, 0.65, 0.85, 1.95, 2.20, 2.34	clay, calcite, dolomite, quartz	4.49, 3.02, 2.88, 4.26
Q12	0.55, 0.65, 0.85, 1.95, 2.20, 2.34	clay, calcite, dolomite, quartz	4.49, 3.02, 2.88, 4.26
Q13	0.55, 0.65, 0.85, 1.95, 2.20, 2.34	clay, calcite, dolomite, quartz	4.49, 3.02, 2.88, 4.26
Q14	0.55, 0.65, 0.85, 1.95, 2.20, 2.34	clay, calcite, dolomite, quartz	4.49, 3.02, 2.88, 4.26
Q15	0.55, 0.65, 0.85, 1.95, 2.20, 2.34	clay, calcite, dolomite, quartz	4.49, 3.02, 2.88, 4.26
P2	048, 0.63, 1.44, 1.73, 2.08, 2.28, 2.31, 2.45	quartz, clay, calcite, hematite	4.26, 4.49, 3.02, 2.69
S1	0.58, 0.92, 1.95, 2.22, 2.34, 2.45	quartz, clay, calcite	4.26, 4.49, 3.02
G1	0.48, 0.60, 0.63, 1.41, 1.73, 2.14, 2.32, 2.45	calcite, clay	3.02, 4.49
A2	0.60, 0.63, 1.44, 1.73, 2.20, 2.34, 2.45	calcite, dolomite, quartz, clay	3.02, 2.88, 4.26, 4.49
P3	0.48, 0.63, 1.93, 2.02, 2.20, 2.31, 2.45	calcite, dolomite, clay, quartz	3.02, 2.88, 4.26
R1	0.49, 0.60, 0.63, 2.18, 2.27, 2.45	pyroxene, plagioclase	2.52, 4.03, 4.04
D1	0.75, 1.00, 1.94, 2.25	pyroxene	2.52

*Clays correspond to the assemblage of kaolinite, illite, chlorite, smectite ± mixed layers.

A false colour composite based on Landsat TM bands 3, 5, 7 covers the wavelength range of most of the absorption features and spectral properties of the LDRUs noted above. This band combination was able to identify and discriminate between the R1 (red lamproite), D1 (dark lamproite), and L1 (gypsum) units with a high degree of contrast, and the S1 (terrigenous sediments), B1 (gypsiferous marls), P3 (nummilitic limestone) and P1 (marls) units with a lower tonal contrast. Other false colour combinations were investigated and evaluated by visual interpretation. For example, composites based on Landsat TM bands: 2, 4, 7; 3, 4, 7 and 2, 4, 5 were found to provide equivalent levels of tonal contrast and discrimination for the recognition of the vegetation cover, various

Quaternary deposit types (L1, B1, Q1), D1 (dark lamproite) and in lower contrast the P1 and P2 units.

False colour composites based on band ratio combinations added further information, and additional tonal contrast, for the recognition of LDRUs. The particularly informative combinations were those which enhanced specific spectral absorption differences associated with the mineralogical composition of each lithological unit. A complex band ratio using Landsat TM bands 1/4, 3/5, 7/4 enabled discrimination between the B1, P1 and P2 units, an enhancement of the tonal contrast of the S1 unit and improved vegetation cover discrimination. The Quaternary deposits Q11, Q12, Q13, Q14 and Q15 were also enhanced, sufficient to be separately identified, using this complex band ratio.

PCA was used to combine information from spectrally related and adjacent Landsat TM bands to increase the potential amount of information submitted to the false colour composite. A composite based on the second principal component of the three visible bands (TM1, TM2 and TM3), TM4 and the second principal component of the SWIR bands (TM5 and TM7) produced an interpretable colour image which was capable of recognizing the majority of the LDRUs with higher sharp contrast, including discrimination of the S1 unit. The higher information content and contrast enhancement of this PCA-derived false colour composite is due to the informed selection of the bands input to the PCA. The first input, comprising Landsat TM bands 1, 2 and 3, covers most of the absorption features of the iron oxides and hydroxides (480, 530, 640 nm), while the third input, based on Landsat TM bands 5 and 7, covers the most important absorption features of the H_2O (gypsum) and OH^- (clay mineral) groups along with absorption features of the calcite and dolomite (2080, 2180, 2340, 2350 nm respectively).

The results of applying the maximum likelihood classifier to the Alcantarilla and Mula images are given in Table 20.11 which shows the classification results as percentage areas covered by each LDRU. The maximum likelihood approach is capable of discriminating between and mapping the LDRUs and the different vegetation cover types. Classification results were checked using four main procedures: (i) classification of the Mula test image, (ii) comparison with conventional geological maps, (iii) comparison with 1:30 000 aerial

TABLE 20.11 Extent of main lithological classes (LDRUs) in the Alcantarilla and Mula sub-images derived from Landsat TM maximum likelihood classification. The total area of each 1:50 000 sheet is about 538 km^2

Class	Alcantarilla (surface %)	Mula (surface %)
Gypsum	1.54	4.09
Marl	4.72	12.31
Quarternary	28.03	30.60
Terrigenous	28.88	18.49
Carbonate	0.99	0.86
Volcanic	0.63	0.01
Natural vegetation*	22.80	11.55
Cultivated area**	12.37	21.52

*Mainly pine forests.
**Mainly citrus orchards.

photographs, and (iv) ground verification using field sample points. All these confirm the high level of coincidence between the classified image and the ground truth. Figure 20.26 shows the classification result for the Alcantarilla image. As a final product, a classified image at the scale of 1:50 000 was produced as a hard copy map and distributed to users within the MEDALUS group.

Probably the most significant result from the maximum likelihood classification was the successful discrimination of the gypsum and gypsiferous marl outcrops. These exposures represent areas which are most prone to desertification and land degradation yet which are not mapped in the published IGME geological map series. Verification of this result was carried out by checking the location of gypsiferous exposures in the classified image with the distribution of gypsiferous exposures in the field. The classification approach was also found to be a very useful tool in the discrimination of vegetation cover in the study area (pine forests, matorral and citrus orchards).

20.3.5 Summary

The lithological units exposed at the surface within the study were grouped, according to their susceptibility to erosion processes, and taking into account their mineralogical composition, field situation and vegetation cover, into six lithological desertification response units or LDRUs. The resultant LDRUs identified were: gypsum, marls, Quaternary deposits, terrigenous, carbonates and volcanic units. Field spectral measurements were obtained to characterize the spectral properties of these lithological units. Mineralogical analysis of these units was carried out by x-ray diffraction and thin section analysis.

A series of experiments were conducted to investigate the influence of surface weathering on spectral response. For volcanic and terrigenous units, absorption features related to iron oxides are more pronounced in the weathered surface while the carbonate and clay absorptions are hardly affected. Therefore, field radiometry and satellite data could be considered as representative of the lithological composition of the two units. An exception was found in gypsum units, in which the fresh surface has a higher reflectance than the weathered one in the visible and NIR regions (to 1100 nm). However, for higher wavelengths the weathered surface presents similar reflectances to the fresh ones. For this reason, we recommend use of the SWIR region (TM5 and TM7) for the spectral characterization and differentiation of this lithological unit. On the other hand, the carbonate units show the higher dispersion in reflectance spectra of the two surfaces, which increases as wavelength increases, but preserving the main absorption features. This fact suggests the use of the visible region.

Spectral characterization and analysis of the lithological units was based on the form and shape of the spectra, position of absorption features and mineralogical composition. A higher spectral contrast between marl and carbonate units was found in the visible and NIR regions (TM1–TM4) where the reflectance was mainly determined by ferric iron content. Lithological units with considerable clay content (marl, terrigenous and Quaternary deposits) present spectral behaviour highly affected by the organic matter content, which results in lowering albedo. Also a decrease in albedo, for all the wavelengths, can be observed in volcanic units (D1) as a consequence of the chlorite, glass and iron oxide proceeding from the alteration of olivine.

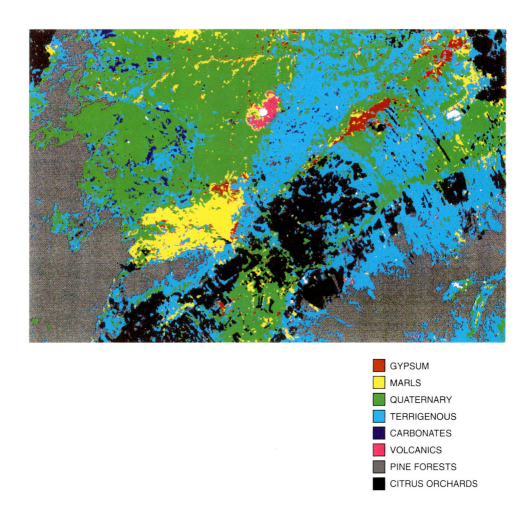

🟥	GYPSUM
🟨	MARLS
🟩	QUATERNARY
🟦	TERRIGENOUS
🟪	CARBONATES
🟥	VOLCANICS
⬜	PINE FORESTS
⬛	CITRUS ORCHARDS

FIGURE 20.26 Maximum likelihood classification of the Alcantarilla training area

Digital image processing of Landsat TM images was based on two main approaches: visual interpretation of false colour composites, based on atmospherically calibrated Landsat TM bands, band ratios and PCA, and supervized multispectral classification.

As a general result, the use of TM3, TM5 and TM7 bands is suggested for the global discrimination of the lithologies. Ratios of TM bands that control spectra slope in the visible and NIR regions (e.g. TM4/TM1; TM3/TM1) are also recommended for their sensitivity to ferric iron content. Other band combinations which are recommended include: (TM2–TM4–TM7); (TM4/TM1–TM5/TM3–TM7/TM4); (PC2(1,2,3)–TM4–PC2(5,7)) in order to take into account the effect of the vegetation cover of the different lithological units which, as expected, can mask some spectral properties (absorption features) when observed by satellite sensors. These combinations present pronounced tonal contrasts between the different lithologies.

With respect to the classification method used, it was proved that for our case and with a confidence level of 95%, the maximum likelihood method is more appropriate than the minimum distance method. This classification method was found to be very sensitive for mapping gypsiferous exposures not identified by the conventional geological map of the study area. This is a particularly significant result given that the gypsum units are the most sensitive and eroded units in the desertification process.

20.4 CONCLUSIONS

The work presented in this chapter has demonstrated the ability to relate quantitatively *in-situ* measurements of vegetation and surface lithology to the multispectral response measured by the Landsat TM sensor. This provides the potential to retrieve quantitative information pertaining to vegetation and surface lithology from satellite sensor imagery, and, in particular, to produce estimates of vegetation type, extent and amount over wide areas, and to extract information on the erodability and mineralogical composition of lithological units outcropping at the surface. Work is still required to verify and evaluate these results more fully, and to investigate the nature and magnitude of error and uncertainty associated with these measurements. Nonetheless, the results reported here are sufficiently promising to actively consider the inclusion of these data into physically-based erosion and land degradation models.

Finally, a number of more specific conclusions can be stated.

First, it has been shown that LAI is not related to the NDVI for the values of LAI determined in the field, and for the plant species considered here. Instead, it has been shown that NDVI is highly correlated with surface leaf area, and that surface leaf area is reasonably constant within a species.

Second, it has been demonstrated that high land cover classification accuracies (80–95%) can be obtained using a maximum likelihood classification of multi-date Landsat TM imagery. Further, a comparison of results from the maximum likelihood classifier with output from the linear mixture model have shown the ability of the mixture model to reveal realistic sub-pixel land cover proportions. While further verification of the mixture model results are required, this work has illustrated the potential of the mixture model for integrating *in-situ* measurements, essentially at a point, to pixel-based measurements, defined by the sensor FOV.

Third, image normalization and calibration procedures have been implemented which allow a very good correspondence between *in-situ* data and Landsat TM data to be achieved. This enables ground reflectance spectra to be accurately expressed as Landsat TM spectral signatures and opens up the possibility of more detailed work on scaling-up effects.

Fourth, the difficulty of relating NDVI, derived from Landsat TM imagery, to biomass has been illustrated. It has not been possible to use *in-situ* reflectance data to calibrate the NDVI images produced as these data relate to partial canopy reflectance. Further work is required to extend the spectroradiometric data-base to include full canopy reflectances.

Fifth, a series of Landsat TM false colour composites have been proposed for lithological discrimination. These band combinations are formulated on the basis of detailed field reflectance spectra, spectral absorption features and mineralogical analysis. It has been demonstrated that these false colour composites are capable of discriminating between different lithological units by enhancing specific spectral absorption differences associated with mineralogical composition.

Sixth, it has been demonstrated that it is possible to discriminate and map LDRUs, and associated land cover, using the maximum likelihood classifier with high a degree of mapping accuracy. The approach successfully discriminates between gypsum and gypsiferous marl outcrops, which are not mapped in existing Spanish IGME geological maps, and which correspond to areas which are particularly sensitive to desertification and land degradation processes.

Future research under the MEDALUS II programme will build upon the site-based ground measurements and analysis of satellite sensor imagery reported here. The basic goal is to operationalize the methods and techniques which have been developed, to meet the data requirements for modelling of land degradation at the regional scale (100–10 000 km^2), and for monitoring of vegetation and soils across the whole of the northern Mediterranean zone. This can be viewed essentially as a process of technology transfer designed to enable information products from satellite remote sensing to be used in support of inter-disciplinary scientific research, and ultimately, as a basis for policy formulation and evaluation.

20.5 ACKNOWLEDGEMENTS

The GER single field of view IRIS (SIRIS) spectroradiometer, used for field measurements of vegetation reflectance spectra by the Bristol group, was loaned from the NERC Equipment Pool for Field Spectroscopy (EPFS) based at the University of Southampton (UK) (Rollin and Milton, 1991). The Cecil Instruments spectrophotometer, used for field determination of chlorophyll concentrations, was loaned from the Department of Botany, University of Bristol (UK).

20.6 REFERENCES

Arnon, D.I. 1949. Copper enzymes in isolated chloroplasts. Polyphenoloxidase in *Beta vulgaris*. *Plant Physiology*, **24**, 123–132.
Bastida, J., Younis, M.T. and Buendia, A.M. 1992. Application of Landsat TM images for the study of the hydrographic network density and its relation to lithology in the area of Alcantarilla, SE Spain. *European Geophysical Society Annual Meeting*, Edinburgh, 24–26 April, 1992.
Caballero, M.A. and Martín-Vivaldi, J.L. 1975. Estudio mineralogico y genético de la fracción fina del Trias Español. *Memoria de IGME*, **87**, 277 pp.

Casmir, M.J., Winter, R.P. and Glastzer, B. 1980. Nomadism and remote sensing: animal husbandry and the Sagebrush Community in a nomad winter in western Afghanistan. *Journal of Arid Environments*, **3**, 231–254.

Conel, J.E., Green, R.O., Vane, G., Bruegge, C.J. and Alley, R.E. 1987. AIS-2 radiometry and a comparison of methods for recovery of ground reflectance. In Vane, G. (ed.) *Proceedings of the Third Airborne Imaging Spectrometer Data Analysis Workshop*, JPL, Pasedena, JPL Publication 87–30, 18–47.

Curico, J.A. and Petty, C.C. 1951. The near-infrared absorption spectrum of liquid water. *Journal of the Optical Society of America*, **41** (5), 302–304.

Curran, P.J. 1980. Multispectral remote sensing of vegetation amount. *Progress in Physical Geography*, **4**, 315–341.

Curran, P.J., Dungan, J.L., Macler, B.A., Plummer, S.E. and Peterson, D.L. 1992. Reflectance spectroscopy of fresh whole leaves for the estimation of chemical concentration. *Remote Sensing of Environment*, **39**, 153–166.

Faulkner, H. 1990. Vegetation cover density variations and infiltration patterns on piped alkali sodic soils: implications for the modelling of overland flow in semi-arid areas. In Thornes, J.B. (ed.) *Vegetation and Erosion: Processes and Environments*, Wiley, New York, 317–346.

Gausman, H.W., Allen, W.A. and Cardenas, R. 1969. Reflectance of cotton leaves and their structure. *Remote Sensing of Environment*, **1**, 19–22.

Goetz, A.F.H. 1989. Spectral remote sensing in geology. In Asrar, G. (ed.) *Theory and Application of Remote Sensing*, Wiley, New York, 491–525.

Goetz, A.F.H., Rock, B.N. and Rowan, L.C. 1983. Remote sensing for exploration. *Economic Geology*, **78**, 573–590.

Goetz, A.F.H., Vane, G. and Rock, B.N. 1985. Imaging spectrometry for earth remote sensing. *Science*, **228**, 950–969.

Harrison, A.R. and Garg, P.K. 1991. Multispectral classification for vegetation monitoring in semi-arid landscapes susceptible to soil erosion and desertification. In Barrett, E.C., Brown, K.A. and Micallef, J. (eds) *Remote Sensing for Marine and Coastal Hazard Monitoring and Disaster Assessment in the Mediterranean*. Gordon & Breach, New York, 109–138.

Harrison, A.R., Taberner, M. and Hurcom, S. 1993. Site-based remote sensing of vegetation and land cover. In Thornes, J.B. and Brandt, J. (eds) *MEDALUS I: Final Report to the European Commission (DG XII)*. Brussels, Research Contract EPOC-CT90-0014-(SMA), 225–263.

Hellden, U. and Stern, M. 1980. Evaluation of Landsat imagery and digital data for monitoring desertification indicators in Tunisia. *Proceedings of the Fourteenth International Symposium on Remote Sensing of the Environment*, **3**, 160–161.

Hergert, H.L. 1971. Infrared spectra. In Sarkanen, K.V. and Ludwig, C.H. (eds) *Lignin's Occurrence, Formation Structure and Reactions*, Wiley Interscience, New York.

Holben, B.N., Tucker, C.J. and Fan, C.J. 1980. Assessing leaf area and leaf biomass with spectral data. *Photogrammetric Engineering and Remote Sensing*, **46**, 651–656.

Hooton, D.H. and Giorgetta, N.E. 1977. Quantitative X-ray analysis by a direct calculation method. *X-ray Spectrometry*, **6**, 2–5.

Horwitz, H.M., Nalepka, R.F., Hyde, P.D. and Morganstern, J.P. 1971. Estimating the proportion of objects within a single resolution element of a multispectral scanner. NASA Contract NAS-9-9784, University of Michigan, Ann Arbor.

Huete, A. R. and Escadafal, R. 1991. Assessment of soil biophysical properties through spectral decomposition techniques. *Remote Sensing of Environment*, **17**, 37–53.

Hunt, G.R. and Salisbury, J.W. 1970. Visible and near-infrared spectra of minerals and rocks: I. Silicate minerals. *Modern Geology*, **1**, 283–300.

Hunt, G.R., Salisbury, J.W. and Lenhoff, C.J. 1973. Visible and near-infrared spectra of minerals and rocks: VI. Additional Silicates. *Modern Geology*, **4**, 85–106.

Hunt, G.R., Rock, B.N. and Nobel, P.S. 1987. Measurement of leaf relative water content by infrared reflectance. *Remote Sensing of Environment*, **22**, 429–435.

IGME 1974a. *Mapa geológico de España, E: 1/50000, Sheet No. 933 (Alcantarilla)*. Servicio de Publicaciones del Ministerio de Industria y Energía, Madrid.

IGME 1974b. *Mapa geológico de España', E: 1/50 000, Sheet No. 912 (Mula)*. Servicio de Publicaciones del Ministerio de Industria y Energía, Madrid.

Justice, C.O. and Townshend, J.R.G. 1981. The use of Landsat data for land cover inventories of Mediterranean lands. In Townshend, J.R.G. (ed.) *Terrain Analysis and Remote Sensing*, Allen & Unwin, London, 133–153.

Kent, J. and Mardia K.V. 1986. *Spatial Classification using Fuzzy Models*. Research Report, Dept. of Statistics, University of Leeds (UK).

Kirkby, M.J. 1993. Proposed large scale modelling scheme for MEDALUS II (MEDRUSH). Unpublished report, Department of Geography, University of Leeds, UK.

Kirkby, M.J., Atkinson, H. and Lockwood, J. 1990. Aspect, vegetation cover and erosion on semi-arid hillslopes. In Thornes, J.B. (ed.) *Vegetation and Erosion: Processes and Environments*, Wiley, New York, 25–40.

Lang, R.L., Bartholomew, M.A., Grove, C.I. and Paylor, E.D. 1990. Spectral reflectance characterization (0.4 to 2.5 and 8.0 to 12.0 nm) of Phanerozoic strata, Wind River basin and southern Bighorn areas, Wyoming. *Journal of Sedimentary Petrology*, **60**, 504–524.

Malinowski, E.R. and Howery, D.G. 1980. *Factor Analysis in Chemistry*. Wiley, New York.

Mather, P. 1987. *Computer Processing of Remotely Sensed Images*. Wiley, Chichester.

Milton, E.J. 1987. Principles of field spectroscopy. *International Journal of Remote Sensing*, **8**, 1807–1827.

Mitchell, D.J. 1990. The use of vegetation and land use parameters in modelling catchment sediment yields. In Thornes, J.B. (ed.) *Vegetation and Erosion: Processes and Environments*. Wiley, New York, 289–316.

Morris, R.V., Lauer, H.V., Lawson, C.A., Gibson, E. K., Ann Nac, G. G. and Stewart, J. 1985. Spectral and physiochemical properties of sub micron powders of hematite, maghematite, magnetite, geothite and lepidocrosite. *Journal of Geophysical Research*, **90**, 3126–3144.

Niscanen, E. 1964. Reduction of orientation effects in the quantitative X-ray diffraction analysis of kaolin minerals. *American Mineralogist*, **49**, 705–714.

Pech, R.P., Graetz, R.D. and Davis, A.W. 1986. Reflectance modelling and the derivation of vegetation indices for an Australian semi-arid shrubland. *International Journal of Remote Sensing*, **7**(3), 389–403.

Pouget, M., Lortic, B., Soussi, A. and Mtimet, A. 1984. Contribution of Landsat data to mapping of land resources in arid regions. *Proceedings of the Eighteenth International Symposium on Remote Sensing of Environment*, Paris, 1717–1725.

Richards, J.A. 1993. *Remote Sensing Digital Image Analysis: An Introduction*. Springer-Verlag, Berlin.

Rollin, E.M. and Milton, E.J. 1991. The UK Natural Environment Research Council Equipment Pool for Field Spectroscopy (NERC-EPFS). *Proceedings of the 5th International Colloquium—Physical Measurements in Remote Sensing, Courchevel, France*, ESA SP-319.

Rowan, L.C. and Kahle, A.B. 1982. Evaluation of 0.46 to 2.36 mm multispectral scanner images of the East Tintic Mining District, Utah, for Mapping Hydrothermally Altered Rocks. *Economic Geology*, **77**, 441–452.

Rowan, L.C., Goetz, A.F.H. and Ashely, R.P. 1977. Discrimination of altered and unaltered rocks in the visible and near infrared multispectral images. *Geophysics*, **42**, 522–535.

Sherman, D.M., Burns, R.G. and Burns, V.M. 1982. Spectral characteristics of the iron oxides with application to the Martian bright region mineralogy. *Journal of Geophysical Research*, **87**, 10169–10180.

Simonds, J.L. 1963. Application of characteristic vector analysis to photographic and optical response data. *Journal of the Optical Society America*, **53**, 968–974.

Singer, R.B. and McCord, T.B. 1979. Large scale mixing of light and dark surface materials and implications for the analysis of spectral reflectance. *Proceedings of the Lunar Planetry Science Conference*, **10**, 1835–1848.

Thornes, J.B. 1990. The interaction of erosional and vegetational dynamics in land degradation: spatial outcomes. In Thornes, J.B. (ed.) *Vegetation and Erosion: Processes and Environments*, Wiley, New York, 41–54.

Townsend, T.E. 1987. Discrimination of iron alteration minerals in the visible and near infrared reflectance data. *Journal of Geophysical Research*, **92**, 1441–1454.

Townshend, J.R.G. and Justice C.O. 1980. Unsupervized classification of Landsat MSS data for mapping vegetation in an area of complex terrain: principles and problems. *International Journal of Remote Sensing*, **1**, 105–120.

Warshaw, C. and Roy, R. 1961. Classification and scheme for the identification of layer silicates. *Geological Society of American Bulletin*, **72**, 1455–1492.

Wessman, C.A., Aber, D.L., Peterson, D.L. and Melillo, J.M. 1987. Foliar analysis using near infrared reflectance spectroscopy. *Canadian Journal of Forestry Resources*, **17**, 311–319.

White, I.D., Mottershead, D.N. and Harrison, S.J. (eds) 1984. *Environmental Systems*. Allen & Unwin, London, 495 pp.

Wiegland, C.L., Richardson, A.J. and Kanemasu, E.T. 1979. Leaf area index estimates for wheat from Landsat and their implications for evapotranspiration and crop modelling. *Agronomy Journal*, **71**, 336–342.

21

Summary and Prospects

J. B. THORNES

King's College, London

21.1 INITIAL RESULTS

In this volume we have presented an overview of the research carried out in the first 2 years (1991–92) of the MEDALUS project. In such a time there is a limit to what can be achieved. Nevertheless some themes are beginning to emerge of which the following appear most important.

Climate and climate change

i. The climate modelling reveals that potentially significant climate changes could occur in the Mediterranean region as a result of global warming in the next 60 years.

ii. There are clear indications of different reactions in the eastern and western Mediterranean in terms of globally induced changes, as a result of a putative 'Mediterranean oscillation'.

iii. The changes of the last century indicate variations of mean annual rainfall by as much as 50% between decades and large interannual variations. In both Italy and Spain intensive droughts in the 1980s and 1990s are the greatest in recorded history. There is also evidence of persistent runs of high and low rainfall which tend to obscure long-term effects.

iv. Extreme rainfall and temperature events are changing their incidence and are extremely important in the context of desertification.

v. Modelling plant cover variations seems to indicate that the impact of enhanced CO_2 is likely to prove relatively unimportant compared with the impact of natural or enhanced rainfall fluctuations in plant growth.

vi. Modelling flow and sediment yield from catchments appears to indicate that the effects of greenhouse gas warming and CO_2 doubling are likely to be small compared with natural fluctuations.

Field experiments

vii. Measurements of erosion at the experimental field sites, although for a very short period, support the view that the contrast between cultivated and uncultivated terrain is most important and that on non-cultivated land (both fallow and

---

Mediterranean Desertification and Land Use. Edited by C. Jane Brandt and John B. Thornes.

abandoned) vegetation cover is the overriding factor in determining erosion for rainfall of a given intensity, though the relationship of erosion to intensity is complex. Even at sites which have been instrumented for much longer, the rates are lower than those predicted by empirical models and only extremely rare, extremely large events are likely to provide catastrophic erosion rates.

viii. Rainfall exclusion experiments highlight the fundamental limiting character of water for cereal growth. Measurement of natural plant species in matorral reach the same conclusion.

 ix. In detailed field and laboratory studies, the impact of soil rock fragment content and the processes of aggregate formation on erosional and runoff processes are found to be very important.

  x. Field studies also reveal the importance of the complex spatial vegetation patterning at different scales, its relation to soil aggregation and development, and its evolution in relation to land degradation.

 xi. The impact of grazing and fire have been investigated. Grazing studies reveal the strong impact of grazing on growth rates and the emergence of critical bush size and spacing as a result of grazing activity. The fire studies support the view that there is a sharp reduction in sediment yield as vegetation cover commences.

Model development and implementation

xii. The MEDALUS hillslope model has been found capable of simulating reasonable plant growth and soil moisture conditions. Subsequent work has demonstrated that for runoff and sediment yield there is a systematic departure from observed field values for large events. These are thought to be due to the under-representation of high-intensity bursts at intervals lower that those modelled. The field site weather stations indicate that these short high-intensity bursts may last for only a few minutes. Field observations with simulated rainfall indicate that without proper replication of these bursts overland flow is very unlikely to occur in rainfall events on dry soils even after sustained rain.

xiii. SHETRAN-UK has been found to provide satisfactory replications of the runoff and sediment yields for the Cobres and Mula catchments despite the paucity of usable data at the event level.

xiv. Non-linear dynamical systems models have been used to predict the demands for water resources in Crete and the development of soil aggregates at the field scale.

xv. Groundwater models have been implemented at the small basin scale.

Remote sensing

xvi. Mixture models have been applied to separate different lithologies from each other under conditions of sparse plant cover; and

xvii. applications in Spain and Portugal have successfully associated NDVI with productivity and plant type on the ground.

Socio-economic factors in desertification

xviii. It is confirmed that there has been a long history of land degradation but it is argued that, contrary to generally held belief, many traditional agricultural practices, such as terracing, are in fact conservative measures and it may be that more attention must be paid to the breakdown of traditional agriculture.

xix. Under existing economic conditions the major crisis is seen as water supply rather than erosion and loss of soil productivity, reflecting the shift from extensive to intensive farming that has occurred in the last 40 years.

xx. The work of MEDALUS 1 nevertheless supports the view that there is a strong interaction between water supply and erosion–vegetation cover and that to appreciate and solve the water resources problem the more conventional desertification problem of land degradation has also to be solved.

21.2 LIMITATIONS OF THE FIRST PHASE OF MEDALUS

A number of clear signals emerge from the first phase of MEDALUS. The first is that despite a strong sense of purpose the team had not developed a truly interdisciplinary approach to the problem of desertification. This is never easy despite its obvious desirability. Focusing on the field sites successfully brought together those interested in observation and modelling, created a good interaction between the different physical and biological scientists involved and provided for intensive studies of small-scale processes. However, the scale of operation and the character of the work meant that there was less interaction between the field-scale research and the climate and remote sensing work. Second, it has become apparent that, notwithstanding the need to have a thorough appreciation of the local-scale processes, there is also a need to address the larger-scale dimensions of the problem, at both the catchment level and the regional scale. Third, it is clear that the socio-economic and political aspects of the problem had to be seen at the scale of administrative regions, and that an integrated approach to the whole problem would ultimately be required if positive recommendations for specific actions are to be developed. The complex interactions between physical and biological and social and economic forces cannot be understood in the context of local studies.

To accommodate these observations we have recognized the need to:

i. Maintain the excellent core observation programme and in-depth field studies of physical and biological processes.

ii. Provide erosion and runoff models based on a detailed knowledge of the key processes derived from the first phase, at the catchment scale (5 to 5000 km^2), compatible with available remote sensing, and other forms of data (such as DEMs and climatic records).

iii. Address more directly the development of mitigation plans at this scale, working with local authorities.

21.3 MEDALUS II

As this book is going to press the second phase of MEDALUS is just drawing to a close. This phase, lasting a further 2½ years, has involved four sections.

i. A continuation of the field research programme of the first phase, emphasizing the physical processes and their interactions and sustaining the core observations, with extension of the process studies to include more ecophysiology, greater emphasis on

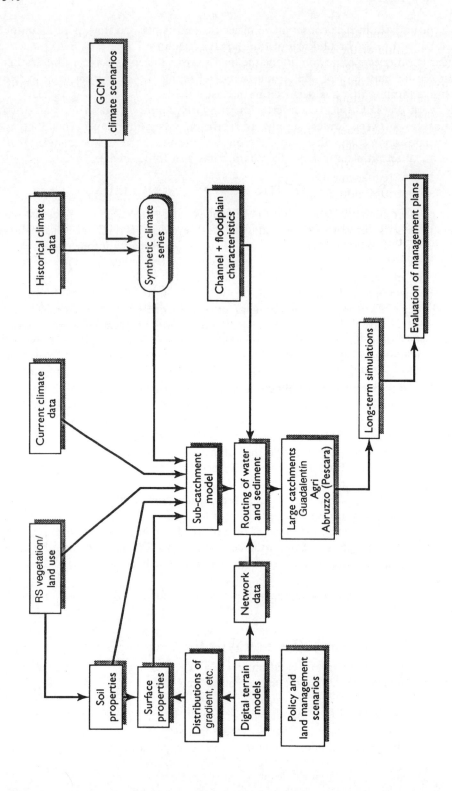

FIGURE 21.1 The modelling base for MEDALUS II has shifted from the hillslope to the basin scale and therefore requires a wider and more complex range of inputs, as depicted in this diagram

the recovery of land abandoned or in fallow, and specific investigations designed to provide validation of the MEDALUS hillslope model.

ii. An expanded modelling programme upscaling the hillslope model to the catchment scale (MEDRUSH) on a modified TOPMODEL-type of approach, linking it firmly with a channel routing model based on the SHETRAN-UK, together with accompanying field studies for calibration and validation (Figure 21.1). A further downscaling of the general circulation modelling results to accommodate regional-scale variability through statistical modelling and an emphasis on regional-scale modelling of weather types and their incidence. New studies of large area weather-GCM for the Mediterranean have also been incorporated.

iii. The development of more detailed socio-economic studies, especially into agricultural change, impacts of afforestation policy, the results of pollution of soils in desertified environments, the specific methods of mitigation and the role of economic change on the desertification problem.

iv. The development of three case studies designed to bring together socio-economic, physical and biological aspects of the desertification into a common regional framework. These regions have been called Target Areas, and include the Guadalentín valley in southeast Spain, the Sauro sub-catchment in the Agri valley in Basilicata, Italy, and the Piomba and Cigno sub-catchments in Pescara basin in the Abruzzo, eastern Italy. In each of these cases assessments of the desertification problem have been conducted through historical studies, through regionalization and classification of the physiographic units within the basins, through the application of modelling techniques and through expert assessments using local expertise.

The first results of the second phase of the programme are to be provided in the *Atlas of Mediterranean Desertification* to be published by Wiley.

21.4 ... AND MEDALUS III

Finally it is worth mentioning the third phase of the MEDALUS programme. Starting in early 1996, this will perform the final upscaling to regions of sub-national scale, with a heavy emphasis on regional-scale indicators of desertification based on GIS and remote sensing. The applications of the MEDALUS, MEDRUSH and SHETRAN models will be completed and the field sites will be reduced in number and intensified in scope. Some fresh work will be initiated to make up gaps in the earlier part of the programme, notably the investigation of gullies and ephemeral channels in a desertification context. The target areas have been changed to incorporate some integrated studies in the Alentejo, but the Agri and the Guadalentín will continue to form key components of the regional integrated modelling. Emphasis in the target areas and at the regional level will be on providing methodology for the recognition, representation and evaluation of Desertification Sensitive Areas (DeSAs) comparable to the Environmentally Sensitive Areas (ESAs) already recognized and used in European Union policy. This represents the end point of the goals of this project first spelled out in 1991, namely to provide a methodology for addressing the desertification problem in the southern states of the European Union.

Index